Encyclopedia of Alternative and Renewable Energy: Hydrogen Energy

Volume 31

Encyclopedia of Alternative and Renewable Energy: Hydrogen Energy Volume 31

Edited by **Rob Koslowski and David McCartney**

New York

Published by Callisto Reference,
106 Park Avenue, Suite 200,
New York, NY 10016, USA
www.callistoreference.com

Encyclopedia of Alternative and Renewable Energy: Hydrogen Energy
Volume 31
Edited by Rob Koslowski and David McCartney

International Standard Book Number: 978-1-63239-205-3 (Hardback)

Printed in the United States of America.

Contents

Preface

The concept of hydrogen energy is a comparatively novel and is an important part of developed nations' clean energy. Hydrogen economy symbolizes the potential future of humankind. Diminishing resources of our planet are urging us to look for renewable clean energy resources and hydrogen figures as distinguished energy carriers of a future sustainable energy system. There are important obstacles which need to be overcome to make hydrogen feasible in production, storage and power generations however, the safety of operation remains a crucial aspect which determines the success or failure of a proposed solution. Latest advancements in these areas have been reviewed in this book, together with current research in the field of hydrogen energy and its use.

After months of intensive research and writing, this book is the end result of all who devoted their time and efforts in the initiation and progress of this book. It will surely be a source of reference in enhancing the required knowledge of the new developments in the area. During the course of developing this book, certain measures such as accuracy, authenticity and research focused analytical studies were given preference in order to produce a comprehensive book in the area of study.

This book would not have been possible without the efforts of the authors and the publisher. I extend my sincere thanks to them. Secondly, I express my gratitude to my family and well-wishers. And most importantly, I thank my students for constantly expressing their willingness and curiosity in enhancing their knowledge in the field, which encourages me to take up further research projects for the advancement of the area.

Editor

Section 1

General Aspects of Hydrogen Energy

Chapter 1

Hydrogen Economy: Modern Concepts, Challenges and Perspectives

Vladimir A. Blagojević, Dejan G. Minić,
Jasmina Grbović Novaković and Dragica M. Minić

Additional information is available at the end of the chapter

1. Introduction

Identifying and building a sustainable energy system are two of the most critical issues for any modern society. Ideally, current energy system, based mostly on fossil fuels, which have limited supply and considerable negative environmental impact, would be replaced with a system based on a renewable fuel. Hydrogen, as an energy carrier primarily derived from water, can address the issues of sustainability, environmental emissions and energy security. If one assumes a full hydrogen economy the size of United States, the amount of hydrogen for just purposes of transportation would be about 150 million tons per year, which would amount to consuming, with current production efficiency, between 2 and 5 billion tons of water. As a comparison, current water consumption in United States for purpose of thermoelectric power generation in power plants is around 300 billion tons, with another 1.2 billion tons consumed for gasoline production. Therefore, rather than consume, a hydrogen economy would most likely significantly reduce water consumption for purposes of energy generation (Turner, 2004).

Hydrogen is the most abundant element in the universe, burns clean, producing only water and has the highest energy density per unit mass. This is why hydrogen is considered most suitable to replace fossil fuels as the primary energy material for the mobile industry (Šušić, 1997c). However, hydrogen is not an energy source, only an energy carrier, and it is not freely available in nature and needs to be produced, either from water or other compounds. If it is produced from water, it costs more energy to produce it than one would recover burning it. This is why, ideally, a hydrogen cycle would include hydrogen produced by splitting water using electrolysis with solar energy and stored reversibly in a solid. However, there are considerable difficulties associated with efficient hydrogen production, storage and use in fuel cells. Among them, hydrogen storage for mobile applications is currently the most difficult obstacle.

Gasoline has very high energy density of 31.6 MJ/L, compared to 4.4 MJ/L of compressed hydrogen and 8.8 MJ/L of liquid hydrogen. In addition, gasoline tank has extremely short filling time, is capable of providing energy at low temperatures and provides excellent control of energy discharge, allowing rapid acceleration, high sustained speed and considerable range. These are the challenges that a successful hydrogen tank has to meet, too. US Department of Energy (DOE) target requirements for a hydrogen tank require hydrogen gravimetric density of 7.5 mass% and volumetric density of 70g/L, operating temperature between 233 and 358K, minimum delivery pressure of 12bar (1.2MPa) and fueling time of 3min. In addition, the storage system should be safe, durable (1500 operational cycle life) and cost effective. None of the existing systems meet these requirements.

In order to achieve the hydrogen economy, there are some obstacles that need to be overcome to make hydrogen a viable energy carrier. They are characterized by four main aspects of hydrogen use and some of these will be addressed here:

1. Production – since hydrogen needs to be produced, ideally from water, it is necessary to develop production methods that would consume the least amount of energy and provide ability to produce hydrogen renewably on a large scale.
2. Storage – fuel needs to be easily stored for use and transport, where one of the main requirements is that it is readily available, which requires not just short charge/discharge times, but also excellent control of charge/discharge process coupled with sufficient energy and gravimetric/volumetric density.
3. Power generation – once hydrogen is ready to be consumed, it is necessary to do so in the most effective way: the power generation system that uses hydrogen needs to be both efficient and, for mobile application, lightweight.
4. Safety – hydrogen use and storage comes with some risks (flammability) which necessitate certain precautions and safety measures; another aspect related to this is environmental impact of the hydrogen cycle, which depends on the methods used to produce, store and use it.

Since hydrogen is thought to be a renewable fuel for the future, it is only appropriate that, when we consider all the challenges associated with its production, storage and use, we keep in mind that when we consider proposed systems, efficiency is only one of the factors that will determine the success of these systems. Other important aspects are production cost (both financial and in resource), durability, stability of operation and safety, and these can, more than efficiency, determine the success or failure of any of the proposed solutions for a part of the hydrogen cycle.

2. Hydrogen production

There are several potential sources of hydrogen on our planet, although these are exclusively hydrogen compounds, necessitating extraction of hydrogen at energy cost. The most abundant is water, while hydrogen can also be obtained from hydrocarbons, either fossil fuels or biomass. While production from water is clean and renewable, with no CO_2

emission, production from fossil fuels generates similar or even higher levels of CO_2 emissions as burning of coal and gasoline. Hydrogen production from biomass is carbon-neutral, since plants and organisms used during the process sequester approximately the same amount of CO_2 during their growth as it is emitted during the process of extraction of hydrogen from them. However, their negative environmental impact is considerable due to the fact that they require large land surfaces for growth.

From water. Although many technologies have been explored for production of hydrogen by splitting water into hydrogen and oxygen, these processes have yet to achieve the necessary efficiency and scalability for industrial application. The main advantage of hydrogen production from water is that it is clean, renewable and has little or no negative environmental impact, although the energy cost of its production is currently too high. There are several processes of interest, like electrolysis, catalytic thermolysis, photocatalytic water splitting and sulfur-iodine cycle.

Electrolysis of water is used today to produce around 5% of all industrial hydrogen. There are several types of cells in use: solid oxide electrolysis cell, alkaline electrolysis cell and polymer electrolyte membrane cell. These cells operate at elevated temperatures (350-570K) and contain high electrolyte concentrations and catalysts (typically yttrium-stabilized zirconium oxide mixed with nickel). Efficiency of typical electrolysis processes is usually between 50-80%, when inefficiencies of production of power used for electrolysis are not taken into account. With these taken into account, the energy efficiency would decrease to 30-45% for a typical nuclear or thermal power plant used as the power source, or even lower for a typical solar cell or array (Hauch et al., 2008).

Water thermolysis is thermal dissociation of water, which occurs spontaneously around 2800K. Although this temperature is too high for practical applications, significant effort has been invested into research of catalysts to reduce water thermolysis temperature and make it an industrially viable process. The goal is to use water thermolysis either in solar concentrators or in nuclear power plants to produce hydrogen directly using thermal energy. Solar concentrators can produce very high temperatures (over 1800K) by concentrating sunlight using a system of mirrors. Next generation nuclear power plants will be operating at lower temperatures (1000-1300K), but it is hoped that new catalysts will make it possible to use them for direct hydrogen generation using water thermolysis.

Photocatalytic water splitting is a process of directly producing hydrogen using solar energy. It relies on use of photocatalyst to capture the solar energy and use for water dissociation (Ni et al., 2007). There are two principal types of catalysts: photoelectrochemical and photobiological (discussed in biohydrogen production section). Photoelectrochemical (PEC) systems can be divided further into four groups:

- Type 1 – a single electrolyte-filled reactor containing a colloidal suspension of PEC nanoparticles producing a mixture of H_2 and O_2 gas;
- Type 2 – dual electrolyte-filled reactor beds containing a colloidal suspension of PEC nanoparticles, each carrying out half of the reaction process (one producing oxygen, the other hydrogen from H^+ produced in the first reactor bed);

- Type 3 – fixed PEC planar array using multi-junction photovoltaic/PEC cells immersed in electrolyte reservoir;
- Type 4 – PEC solar concentrator system, using reflectors to focus the solar flux at 10:1 intensity onto multi-junction photovoltaic/PEC cell receivers immersed in electrolyte reservoir and pressurized to 2MPa.

Estimated hydrogen production costs for these systems are: 1.63-10.36$/kg$H_2$ depending on the type of system. DOE target for 2015 price of hydrogen was originally set at 1.50$/kg$H_2$, but it was increased in 2005 to 2-3$ per kgH_2, which means that some of these systems already meet that modified requirement. The main issues in these systems remain separation of O_2 and H_2 from the product for Type 1, ionic conduction (through ionic bridge) between two reactor beds for Type 2, improvement of PEC cell structure to reduce cost for Type 3 and new composite concentrator structure and high pressure PEC operation for Type 4. Potential benefits are clean and renewable direct hydrogen production with relatively low cost, although efficiency, depending on the system, is in range of 5-25% for the entire system.

Sulfur-iodine cycle is a thermochemical process which produces hydrogen from hydrogen iodide at much higher efficiency than water splitting, and at lower temperature (700K). One of its advantages is that sulfur and iodine are recovered and reused during the process and not consumed. It is usually coupled with concentrating solar power systems to produce hydrogen using solar energy, providing clean and renewable source of energy.

From fossil fuels. Fossil fuels are the dominant source of industrial hydrogen today. Hydrogen can be produced from natural gas with efficiency of around 80% and from other hydrocarbon sources with a varying degree of efficiency.

Most widely used method of hydrogen production today is steam reforming of methane or natural gas. At high temperatures (1000-1300K), water vapor reacts with methane to yield syngas (mixture of hydrogen and carbon monoxide), which can be used to produce more hydrogen through reaction of water and carbon monoxide (also known as water gas shift reaction, performed around 400K). The drawback of this process is that it produces CO_2 waste (Lee & Lee, 2001).

Other methods of hydrogen production from fossil fuels are partial oxidation of hydrocarbons, which includes partial combustion of fuel-air mixture at high temperatures or in a presence of a catalyst, plasma reforming (Kvaerner process), which produces hydrogen and carbon black from hydrocarbons (no CO_2 waste), and coal gasification, where coal is converted to syngas and methane.

Biohydrogen production. Biological H_2 production represents an effort to harness biological processes to generate hydrogen on the industrial scale. Although they have found no industrial application, there are a number of processes for conversion of biomass and waste streams into biohydrogen. Some of them are the same as the ones described above for fossil fuels, except they use biomass in place of fossil fuel (biomass gasification, steam reforming), while others use biological conversion of solar energy (Tao et al., 2007). Biological conversion is process where biological organisms (usually plants) convert sunlight into hydrogen through their metabolic processes (Melis, 2002).

There are variety of pathways for biological conversion that include unicellular green algae, cyanobacteria, photosynthetic bacteria and dark fermentative bacteria. Photobiological production of hydrogen offers a perspective of operating with solar energy conversion efficiency to H_2 as high as 10%, if some of the barriers could be overcome, like slow H_2 production rate, or discontinuity of H_2 production due to co-generation of O_2 (Maness, 2002). Another challenge represents system engineering for cost-effective photobiological H_2 production. Because of an excellent variety of different bacteria, which absorb light in different spectral regions, it is hoped that, ultimately, a mixture of bacteria tailored to maximize sunlight absorption would be used to improve efficiency. This level of adaptability is one of the advantages of photobiological hydrogen production. There are no photobiological hydrogen production systems that are even close to being competitive with other methods of hydrogen generation, while relatively low overall efficiency would require large surface areas for harvesting and conversion of sunlight. On the other hand, biological production would probably have positive impact on environmental pollution and potentially serve as a source of high value bio-products, which could be useful in the food and synthetic chemistry industries. However, the main limitation of biological production is that it ultimately depends on availability of land for biomass production. This means that it cannot provide the amounts of hydrogen needed by an entire civilization, especially taking into account the fact that we live in a food-limited world with increasing population.

3. Hydrogen storage

Hydrogen storage is a problem that has been a focus of scientific research for decades (Minić & Šušić, 2002). During this time a variety of methods have been investigated, although, none of these have accomplished the required performance level so far. Current methods for hydrogen storage can be broadly separated into:

- mechanical storage: storage in a tank of compressed gas or liquid hydrogen;
- physisorbtion: storage in a solid material through physisorbtion; includes:
- graphene and other carbon structures
- metals and metallic nanocrystals and composites
- metal-organic frameworks, zeolites;
- hydrogen: storage in solid or liquid material of chemically bound hydrogen that is released on decomposition; includes:
- light metal hydrides (e.g. alkaline hydrides, alanates, alane)
- borohydrides
- amines and imides
- amino borane.

Each of these methods has its advantages and disadvantages, but all on-board storage technologies have to meet the following requirements:

- safety

- performance
- cost
- technical adaptation for the infrastructure
- scalability (application in both small and large vehicles).

3.1. Mechanical storage

Low-pressure gaseous form of hydrogen is preferable in terms of efficiency. However, since vehicles cannot store enough hydrogen in this form, compression or liquefaction of hydrogen represents a relatively straightforward way of increasing vehicle on-board capacity. Mechanical storage methods store hydrogen by confining its gas or liquid form in a mechanical tank, similar to how gasoline is stored. The advantages are relatively good charge/discharge time and durability, but the capacity of the tanks is limited by relatively low energy density of hydrogen gas and liquid. In addition, the weight of tank is considerably larger than the gasoline tank due to demands of safe hydrogen storage.

There are currently three broad methods of mechanical hydrogen storage:

- high-pressure tank systems
- hydrogen-absorbing alloy tank system
- liquid hydrogen tanks.

Most current vehicles using hydrogen are equipped with a composite high-pressure tank, due to its simple structure and ease of charge-discharge cycle. Standard high-pressure tanks store hydrogen at 35MPa (350bar) pressure, which provides vehicle autonomy of 300-350km. Tanks have been developed recently to store hydrogen at 70MPa (700bar) pressure.

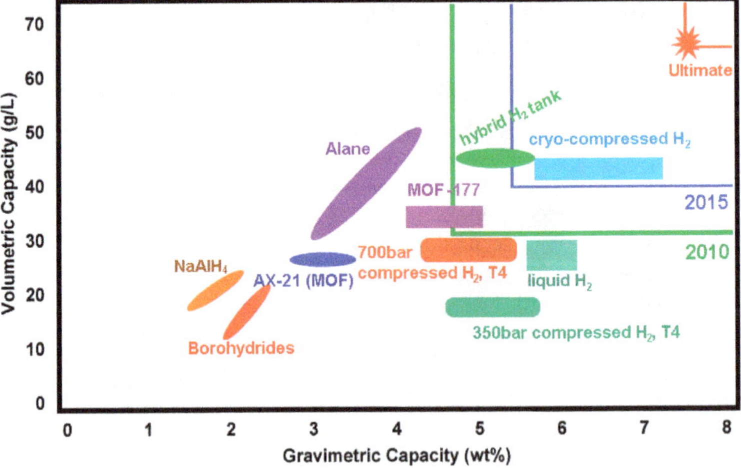

Figure 1. DOE set requirements (solid lines) and overview of existing developed hydrogen storage methods with respect to those requirements

However, pressure in this range makes the relationship between hydrogen pressure and amount non-linear. Therefore, doubling of pressure only leads to an increase of 40-60% more hydrogen, increasing vehicle autonomy to around 500km (Mori & Hirose, 2009).

Tank type	Type 1	Type 2	Type 3	Type 4
composition	all metal (carbon steel)	metal liner + GFRP layer	metal liner + CFRP layer	plastic liner + CFRP layer
thickness ratio (cylinder part)	1.0	0.7	0.32	0.28

Table 1. Different high-pressure (35MPa) tank types for hydrogen storage (GFRP – glass fiber reinforced polymer, CFRP – carbon fiber reinforced polymer)

In order to improve storage capacity of a high-pressure tank, hydrogen absorbing metal alloy is added to the tank to produce a hybrid high-pressure hydrogen absorbing tank system. The alloy increases capacity of the tank from around 3kg per 180L tank to 7-10kg per 180L tank, but hydrogen absorption on charging releases considerable amount of heat, requiring heat exchanger and radiator to be fitted with the tank, resulting in an increase in tank weight from around 100kg to over 400kg. High-pressure (35MPa) hydrogen environment helps metal alloy absorb hydrogen quickly (around 80% charge in 5min) and improves discharge speed, which is a common difficulty of classical metal hydrides, although high hydrogen pressure means that this system has the same safety issues as a regular high-pressure tank system. This system is also capable of operating at temperatures as low as 243K, which is not possible with low-pressure metal hydride systems. These hybrid systems offer a lot of promise offering good charge/discharge performance, but they still don't achieve the required hydrogen capacity and they retain the same safety issues associated with conventional high-pressure systems.

Another improvement of high-pressure system represents the use of cryo-compressed storage, where hybrid high-pressure tank is equipped with heat-transfer system, allowing it to maintain hydrogen gas at temperature around 50K and 0.4MPa pressure (around 6mass%). As the tank is emptied during use, temperature is gradually increased using the heat released during discharge to maintain a minimum 0.4MPa pressure. Figure 2 shows a comparison of gravimetric and volumetric densities of hydrogen for different high-pressure tank systems. Projected capacity of cryo-compressed hybrid system would meet DOE interim requirements for 2015, the first system so far to achieve this. However, this system is still well short of ultimate goals set by DOE.

Liquid hydrogen exists at atmospheric pressure at temperatures below 20K and its density is much higher than compressed gas. From the point of use in vehicles and infrastructure, liquid hydrogen is very attractive, because it offers an opportunity to easily transport and store large amounts of hydrogen. Some studies show that the extra energy consumption of liquefaction process can be compensated by the ease of delivery and storage. However, due to low temperatures involved, liquid hydrogen tanks require double wall construction for thermal insulation, minimizing heat conductivity with vacuum multi-layer insulation. This consists of a thin metal layer on the spacer material which prevents both radiation and thermal irradiation between individual layers and heat intrusion from irradiation and gaseous convection. The most advanced liquid hydrogen tanks have limited heat flow to a few watts per second, which results in hydrogen evaporation and, therefore, loss of a few percent per day. Further developments in thermal insulation, advanced liquefaction and charge/discharge strategy are necessary in order to make liquid hydrogen a viable commercial alternative to a gasoline tank.

3.2. Carbon materials

There have been a number of studies of use of different carbon materials for hydrogen storage. Although initial reports suggested high hydrogen storage capacity of carbon nanotubes and other related nanostructures, more recent results have shown that the maximum sorption capacity of carbon nanostructures is around 2mass% (Zuettel et al., 2004) and that it is dependent only on the surface area of individual carbon nanostructure. Of those nanostructures, graphene sheets have exhibited the highest surface area and represent the most promising carbon material for hydrogen storage. However, since its capacity in pure form is about 2mass%, recent work has been focused on improving both performance and capacity of graphene. One of the issues that need to be overcome is that binding energy of atomistic hydrogen to carbon is 0.8eV, much lower than the energy of H-H bond in H_2 molecule (2.3eV per hydrogen atom). That is why recent research has been focused on catalyst assisted hydrogen sorption on carbon through spillover effect. Hydrogen spillover refers to transport of an active species generated on one substance (activator) to another (receptor) that would not normally adsorb it. In this case, activator is commonly a metal or a metal oxide, while carbon acts as a receptor, overcoming the energy barrier associated with dissociation of hydrogen molecule. Enhancement of hydrogen adsorption via spillover effect is much more pronounced at lower hydrogen pressure (below 0.5MPa), while it saturates at high pressures. This suggests existence of two distinct mechanisms corresponding to different spillover behavior, which can be controlled by activation of catalyst (Tsao et al., 2010).

Recent first-principle calculations have suggested that use of graphene double-layer and multi-layer structures could increase capacity for hydrogen storage (3-4mass%) (Patchkovskii et al., 2005), while recent experiments of Pd-loaded single-wall carbon nanotubes show improved performance after Pd-loading and thermal treatment at temperatures around 700K (Kocabas et al., 2008). Additional theoretical calculations of spillover on graphene suggests that hydrogen atoms on graphene surface should create compact clusters so that the lowest-energy luster is composed of closed six-hydrogen rings, which would correspond, if entire surface area of graphene is used, to hydrogen storage capacity of about 7.7mass% (Lin et al., 2008). However, recent studies of kinetics of hydrogen adsorption/desorption kinetics in Pd/Pt/Ni/Ru doped ordered mesoporous carbon indicated that there is no difference in the kinetics of doped and pure carbon (Saha & Deng, 2009), suggesting that doping of carbon materials using transition metals might not be able to achieve significant increase in capacity and charge/discharge rates.

Although work on carbon materials is ongoing, it is unlikely that carbon materials will, in the foreseeable future, achieve the necessary performance levels required to replace fossil fuels. These types of materials have exhibited encouraging hydrogen capacity at low temperatures (20-80K), but their performance has regularly diminished at room temperature and this remains the biggest challenge in research of carbon materials for hydrogen storage.

3.3. Zeolites and metal-organic frameworks

Zeolites are crystalline microporous materials, usually alumosilicates or aluminophosphates. They consist of microporous framework, which, in principle, appears highly suitable for hydrogen adsorption, as the adsorption energies in the narrow pores are very low, allowing thermal cycling to be used for adsorption and desorption of hydrogen. Initial reports of their storage capacities (Weitkamp et al., 1992, 1995, 1997) indicated very low capacities at room temperature, but these substantially increased at 77K, to 1.5mass% (Jhung et al., 2007). However, these fall well short of technical requirements. Projected maximum capacities for zeolite systems (assuming hydrogen packing density equal to liquid hydrogen) are a maximal 2.5 mass%, indicating that the currently known systems would not be able to meet the requirements to serve as hydrogen storage materials in mobile applications, although new materials might offer better performance.

Metal-organic framework (MOF) materials are composed of metal ions as vertices connected by organic molecules, often polyvalent carboxylic acids, to create a porous material of exceedingly high surface area (Rowsell & Yaghi, 2005). Reported hydrogen storage capacities of these materials have been encouraging: material MOF-177 was reported to have saturation uptake of 7.5 mass% at 77K and atmospheric pressure (Wong-Foy et al., 2006), although this declines to 1.62 mass% at 0.1MPa pressure (Rowsell et al., 2004). These materials exhibit an interesting feature, which could be of great importance for hydrogen adsorption – the so-called gated adsorption (Kitaura et al., 2003; Zhao et al., 2004). This process relies on flexibility of the framework of some MOFs allowing the structure to expand upon adsorption of guest species and shrink back upon desorption. This leads to a pronounced hysteresis, which can be exploited to load the materials at high pressure and still be able to capture hydrogen at lower pressures or higher temperature. However, these materials have some disadvantages as the loading of the material has to be performed at low temperatures, which consumes additional energy, and the binding of hydrogen to MOF materials is stronger than in carbon materials like graphite and carbon nanotubes (Rosi et al., 2003). However, this is still a new class of materials and most of them have yet to be investigated as hydrogen storage materials. In addition, in the foreseeable future, there should be many more new materials of this type, therefore, this class of materials shows great promise when it comes to hydrogen storage capabilities and offers genuine prospect of a hydrogen storage system that could meet all the requirements necessary for mobile applications.

3.4. Metal hydrides

Many metals and alloys are capable of absorbing considerable amounts of hydrogen according to the reaction (1):

$$Me + \frac{x}{2} H_2 \Leftrightarrow MeH_x \tag{1}$$

Here Me is a metal, a solid solution, or an intermetallic compound and MeH_x is a hydride (x is the ratio of hydrogen to metal H/Me). In most cases this reaction is exothermic and reversible. Heating of the hydride causes hydrogen desorption. Charging can be done using molecular hydrogen gas or hydrogen ions from an electrolyte. If hydrogen is loaded from gas phase, several reaction stages of hydrogen with metal would occur, as shown in Fig.2. The metal is usually in form of powder, and can be amorphous or crystalline (Minić et al., 1996). Repeated thermal treatment during hydrogen absorption and desorption causes structural changes in amorphous metals and intermetallic compounds, leading to crystallization (Minić et al., 1995; Šušić et al., 1996).

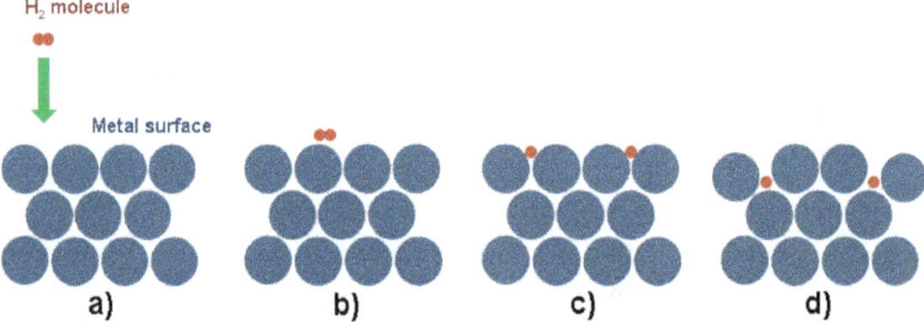

Figure 2. Reaction of H_2 molecule with metal surface: a) H_2 molecule move toward the metal surface. b) physisorbtion of H_2 molecule through Van der Waals interaction with metal surface c) chemisorption of hydrogen after dissociation d) occupation of subsurface sites and diffusion into bulk.

Most of the known metal hydrides exhibit unsatisfactory charge/discharge kinetics, which led to devotion of significant research effort on improving it using surface catalysts and taking advantage of spillover mechanism of hydrogen absorption (Fig. 3). This mechanism includes adsorption and subsequent dissociation of hydrogen molecule on surface catalyst, and migration and subsequent diffusion of adsorbed hydrogen atoms from surface catalyst into the metal (Minić et al., 1997). Since catalyst is a metal or intermetallic compound with superior hydrogen adsorption, it serves as a gateway for hydrogen absorption. This improves charging kinetics of the metal, while using relatively small amounts of catalyst (most commonly Pd) (Šušić, 1997a, 1997b).

The reaction of hydrogen gas with a metal can be described by one-dimensional Lennard-Jones potential curve, Figure 4 (Lennard-Jones, 1932). Far from the metal surface, the potential energy difference of a hydrogen molecule and that of 2 individual hydrogen atoms is equal to dissociation energy ($H_2 \rightarrow 2H$, E_D = 435.99 kJ/molH_2). The molecular hydrogen initially exhibits Van der Waals attractive interaction during approach to metal surface, leading to the physisorbed state ($E_{phys} \approx$ -5 kJ/molH) at a distance from the metal surface approximately equal to hydrogen molecule radius (≈0.2 nm).

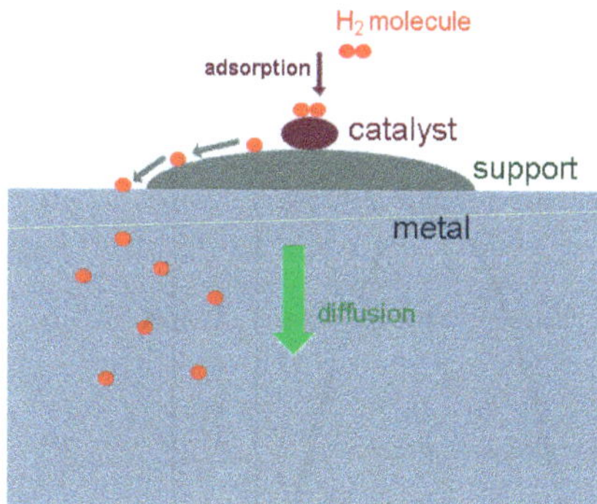

Figure 3. Illustration of spillover mechanism of absorption of hydrogen into a metal using a catalyst

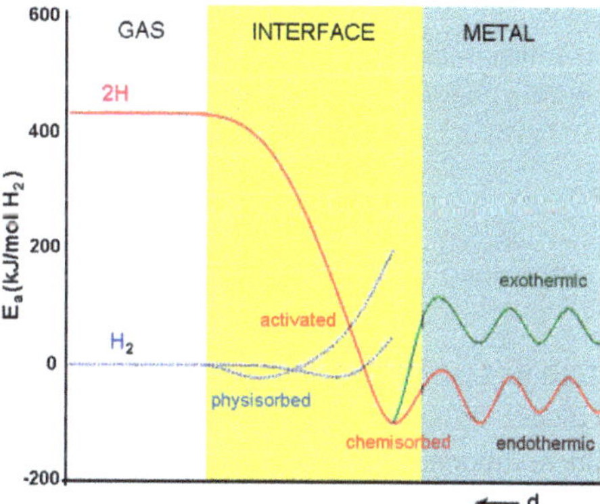

Figure 4. One-dimensional potential energy curve (one-dimensional Lennard-Jones potential for a hydrogen metal system.

Closer to the metal surface, hydrogen has to overcome an activation barrier for dissociation and formation of the hydrogen-metal bond (crossing point of dashed blue and solid red line). The height of the activation barrier depends on the chemical composition of the surface. When hydrogen atom becomes chemisorbed (E_{chem} ≈-50 kJ/molH_2) it shares its electron with the metal atoms at the surface. These hydrogen atoms have high surface mobility, interacting with each other and forming surface phases. In the next step chemisorbed hydrogen atom can migrate into the subsurface layer and, finally, diffuse into the interstitial sites through the host metal lattice, contributing their electrons to the band structure of the metal (Zuettel, 2003).

After dissociation on the metal surface, the H atoms generally diffuse rapidly through the bulk metal even at room temperature to form Me-H solid solution or α-phase. The thermodynamic aspects of hydride formation from gaseous hydrogen are described by pressure–composition isotherms (Fig. 5). After the maximum solubility of hydrogen in the

α-phase is reached, hydride phase (β-phase) will begin to form. Further increase in hydrogen pressure will result in substantial increase in the amount of absorbed hydrogen. This phenomenon can be explained using the Gibbs phase rule (2)

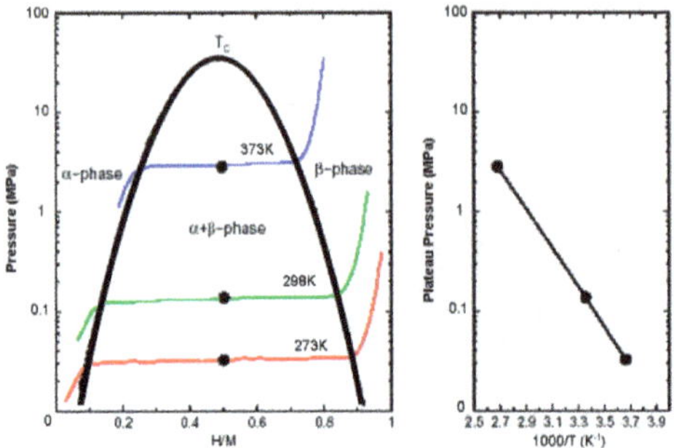

Figure 5. Left: Pressure-Composition-Isotherms (PCI) for a hypothetical metal hydride. Right: Van't Hoff plot for a hypothetical metal hydride derived from the measured pressures at plateau midpoints from the PCI's.

$$F = 2 - \pi + N \tag{2}$$

where F is the degree of freedom, π is the number of phases and N is the number of components.

Hydrogen pressure at which this transformation takes place is called the plateau pressure, where α- and β-phase co-exist. When the stoichiometric hydride is formed, completely depleting the α-phase, an additional degree of freedom is regained and the additional absorption of hydrogen would now require a substantial pressure increase, corresponding to the solid solution of hydrogen in β-phase.

The plateau pressure, described by Van't Hoff equation (3), gives us valuable information about reversible storage capacity. Width of the plateau and the position of the plateau at a given temperature give us a sense of the stability of the hydrides. Stable hydrides (enthalpy of formation, $H_f \ll 0$) would require higher temperatures than less stable hydrides ($H_f < 0$) to achieve a certain plateau pressure. Recording a series of PCI's at different temperatures makes it possible to construct a phase diagram from the end points of the plateaus in the individual PCI's.

$$\ln p = \frac{\Delta H}{RT} - \frac{\Delta S}{R} \tag{3}$$

where ΔH is enthalpy, ΔS is entropy, R is the gas constant and T is temperature.

Figure 6 represents a Van't Hoff diagram showing the dissociation pressures and temperatures of a number of hydrides. Light elements, such as Mg, have shown promising levels of stored hydrogen (about 7 wt% hydrogen), but require higher temperature for dehydrogenation. Conventional metal hydrides, which have been well characterized and their capacity for interstitial hydrogen storage is well-established, include type AB, AB_2, AB_5, A_2B (TiFe, $ZrMn_2$, $LaNi_5$, and Mg_2Ni) intermetallic compounds, and body-centered cubic metals. These materials typically store between 1.4 and 3.6 wt% hydrogen. (Table 2). However, the requirements for gravimetric capacity, fast kinetic and high storage capacity is barely satisfied, so development of new lightweight materials presents many scientific and technical challenges. Among metal hydrides, magnesium hydride appears to be the most promising, because of its high storage capacity and relatively low cost.

Type	Intemetallic compound	H/M	H_2 capacity (mass%)	Temperature (K) for 0.1MPa $P_{desorption}$
A_2B	Mg_2Ni	1.33	3.6	528
AB	TiFe	0.975	1.86	265
AB	ZrNi	1.4	1.85	565
AB_2	$ZrMn_2$	1.2	1.77	440
AB_5	$LaNi_5$	1.08	1.49	285
AB_2	$TiV_{0,62}Mn_{1,5}$	1.14	2.15	267

Table 2. Hydrogen storage properties of some intermetallic compounds

3.5. Magnesium hydride

High gravimetric (7.6 wt.%) (Hanada et al., 2004) and volumetric density (130 kg H_2/m^3) and relatively low price make magnesium hydride (MgH_2) an attractive hydrogen storage material. However the wide industrial application is still not feasible, due to its high enthalpy of formation ($\Delta H° = -75$ kJ/molH_2) and dehydrogenation temperature (720 K), as well as slow kinetics of hydrogenation reactions/dehydrogenation reaction. For example, there is no detectable hydrogen desorption at temperature of 573 K, while at 623 K it takes more than 3000s for complete decomposition of MgH_2 (Varin et al., 2009). Furthermore, in order to allow for the formation of MgH_2, it is necessary to perform an activation process of Mg metal by consecutive heating and cooling of metal in vacuum and in hydrogen atmosphere, which makes the material permeable to hydrogen. The efforts to overcome these deficiencies have made MgH_2 one of the most investigated materials in last two decades.

At moderate hydrogen pressure the only hydride phase existing in equilibrium with Mg is magnesium hydride, β-MgH_2. Pure Mg has a hexagonal crystal structure, while its hydride has a tetragonal lattice unit cell (rutile type) (Zeng et al., 1999; Yu & Lam, 1988). Figure 7 shows the crystal structure of β-MgH_2 ($P4_2/mnm$ space group), were each Mg atom is coordinated with six H atoms forming an irregular slightly distorted octahedron (Noritake et al., 2003). Each H atom is coordinated with three planar Mg atoms. Synchrotron X-ray diffraction gives the parameters of crystal a = 0.45180(6) nm and c = 0.30211(4) nm (Lide, 2006; Lide, 2007) while the powder diffraction file (JCPDS 12–0697) provides similar values for a = 0.4517 nm and c = 0.30205 nm. Density of MgH_2 is 1.45 g/cm^3 (Bastide, 1980).

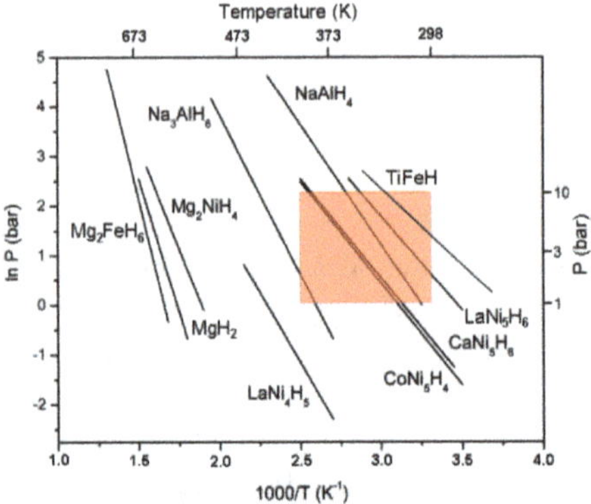

Figure 6. Van't Hoff plots of various metal hydrides, showing hydrogen dissociation pressures and temperatures (red shaded area represents desirable operating conditions).

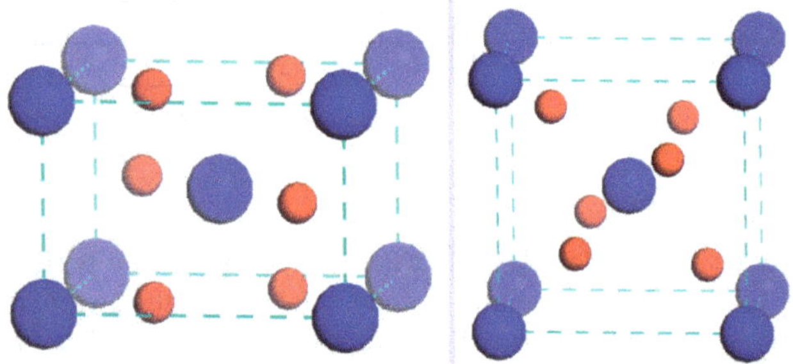

Figure 7. Two perspectives of crystal structure of β-MgH_2 showing the positions of Mg (blue) and H atoms (red)

Although there is considerable literature dealing with structure and properties (Novaković et al., 2009; Schimmel et al., 2004, 2005; Kelkar & Pal, 2009; Basseti et al., 2005) of MgH_2, there are still some uncertainties about H-desorption kinetics, since the Mg–H interaction strongly depends on method of synthesis and presence of additives. For instance, ball milling causes mechanical deformation, surface modification, and metastable phase formation which generally promote the solid–gas reaction. It also introduces defect zones which may accelerate the diffusion of hydrogen (Shan et al., 2004; Jensen et al., 2006; Bobet et al., 2001; Montone et al., 2006, 2007; Aguey-Zinsou et al., 2007). Addition of transition metals, metal oxides or ternary hydrides to mechanically milled MgH_2 decreases its thermal stability and decomposition temperature. It has also been established that nanosized powders provide a possible solution for the problem of hydrogen desorption kinetics (Varin et al., 2006). *Ab initio* DFT calculations (Kurko et al., 2011) have been used to determine possible ways of improving the performance of MgH_2. They have shown that destabilization temperature of MgH_2 reduces with decreasing cluster size, and that a crystallite size of 0.9 nm could result in a desorption temperature below 500 K (Wagemans et al., 2005). However ball milling can also cause agglomeration and cold welding of particles (Montone et al., 2005), therefore experimentally obtained decrease in hydrogen adsorption temperature in ball-milled samples is relatively low, keeping MgH_2 far from practical application (Milovanović et al., 2008). Based on *ab initio* calculations, Novakovic *et al.* (Novaković et al., 2009) examined the possible pathways for adsorption/desorption of hydrogen from MgH_2, showing that different H-filling patterns influence initial hcp-Mg structure in different ways, qualifying wurtzite MgH as a probable intermediate phase between the hcp-Mg and MgH_2 at 1:1 stoichiometry. Du et al. suggested that surface desorption is rate-determining since the activation barrier computed for H-vacancy diffusion from the surface into sublayer is less than 0.70 eV, much smaller than the activation energy for desorption of hydrogen on the MgH_2(110) surface (1.78-2.80 eV/H_2) (Du et al., 2007). Furthermore, the first principle calculations of MgH_2-*TM* (*TM*=Al, Ti, Fe, Ni, Cu, Nb, Co) were carried out to investigate their influence on the stability of the magnesium hydride (Novaković et al., 2010; Song et al., 2004). It was found that TM-H bonding is stronger than the Mg–H bond, but at the same time it weakens other bonds in the second and third coordination around TM atom, which leads to overall destabilization of the MgH_2 compound. Due to a higher number of d-electrons, this effect is more pronounced in late transition metals. In case of Co doping, spin polarization has an additional stabilizing influence on the compound structure.

3.6. Alanates

Alanates, or aluminohydrides, are a family of compounds containing aluminum and hydrogen, with $NaAlH_4$ being the most popular material in this family (Bogdanović & Schwickardi, 1997; Downs & Pulham, 1994; Fichter et al., 2003; Jain et al., 2010). One of the attractive features of alanates is their easy accessibility, since sodium and lithium alanate are commercially available and magnesium alanate can be easily synthesized. They have relatively high gravimetric hydrogen capacity (5-9 mass%), however, these materials usually undergo multi-step thermal decomposition (Eq. 4) to release hydrogen and these reactions require relatively high temperatures and considerable reaction time ($NaAlH_4$ releases 3.7 mass% H_2 in 3h at 483K).

$$NaAlH_4 \rightarrow NaH + Al + \frac{1}{2}H_2 \tag{4}$$

Additionally, some of the alanates are meta-stable, making their first decomposition (dehydrogenation) step exothermic and irreversible. This makes their direct rehydrogenation impossible, which would rule them out as candidates for on-board hydrogen storage applications.

In order to improve their charge/discharge performance and reduce decomposition temperatures, alanates have been doped with transition metals, with Ti- and Ni-doping exhibiting promising results thus far in improving kinetics and dehydrogenation rates, while preserving very high hydrogen capacity.

Material	H_2 capacity (mass%)	Dehydogenation temperature (K)	Dissociation enthalpy (kJ/mol H_2)
$NaAlH_4$	5.6	480-490 (I step) >525 (II step)	37 (I step, 3.7 mass% H_2) 42 (II step, 1.9 mass% H_2)
$LiAlH_4$	7.9	430-450 (I step) 450-490 (II step)	-10 (I step, 5.3 mass% H_2) 25 (II step, 2.6 mass% H_2)
$Mg(AlH_4)_2$	9.3	380-470 (I step) 510-650 (II step)	41 (I step, 7 mass% H_2) 76 (II step, 2.3 mass% H_2)
$KAlH_4$	5.7	570 (I step) 610 (II step) 650 (III step)	55 (I step, 2.9 mass% H_2) 70 (II step, 1.4 mass% H_2)
$Ca(AlH_4)_2$	5.9	400 (I step) 520 (II step)	-7 (I step, 2.9 mass% H_2) 28 (II step, 2.9 mass% H_2)

Table 3. Overview of characteristics of some of the alanates for hydrogen storage [18]

Alanates exhibit very high hydrogen storage capacity, although they cannot be charged and discharged easily and their working temperatures are too high. There is some hope that doping could alleviate some of these difficulties, but it is clear that if these materials become utilized for hydrogen storage, it will probably be in a hybrid high-pressure system, which could additionally improve their charge/discharge performance and be adapted to account for relatively slow discharge rates, while taking advantage of high hydrogen capacities.

3.7. Borohydrides

Borohydride, or tetrahydroborate, refers to a group of complex hydride compounds in which hydrogen in covalently bonded to the central atom in $[BH_4]^-$ complex anion (Eq. 5).

$$NaBH_4 + (2+x)H_2O \rightarrow 4H_2 + NaBO_2 \cdot xH_2O \tag{5}$$

They exhibit high gravimetric and volumetric hydrogen capacity, making the materials of interest for hydrogen storage research (Schlesinger et al., 1953). One of the issues that would have to be resolved is the fact that they exhibit exothermic desorption reactions (30-300 kJ/mol), making their rehydrogenation more difficult than that of other materials and making direct rehydrogenation thermodynamically impossible. In addition, they exhibit high dehydrogenation temperatures (520-670°C), which practically means that only part of their hydrogen capacity is accessible at normal working temperatures.

Material	H_2 capacity (mass%)	Dehydogenation temperature (K)	Dissociation enthalpy (kJ/mol H_2)
$NaBH_4$	10.8	670	-217 to -270
$LiBH_4$	13.4	650	-177
$Mg(BH_4)_2$	13.7	530-670	-39.3 to -50
$Ca(BH_4)_2$	9.6	620	32

Table 4. Overview of characteristics of some of the borohydrides for hydrogen storage [18]

Therefore, most of the work associated with borohydrates is focused on use of doping and catalysts to improve charge/discharge kinetics, lower dehydrogenation temperature and increase hydrogen discharge capacity and cycling ability of these materials (Soler et al., 2007, Pena-Alonso et al., 2007; Lee et al., 2008; Demirci et al., 2009).

3.8. Amides and imides

Amides and imides have attracted a lot of interest, due to their high hydrogen capacity and relatively low operating temperature (Chen et al., 2003; Hu et al., 2003). However, their poor absorption kinetics limits their current practical application (Eq. 6). Therefore there has been a lot of focus on overcoming this by doping with a catalyst, usually through mechanical ball milling (Liu et al., 2008).

$$LiNH_2 + LiH \rightarrow Li_2NH + H_2 \tag{6}$$

Another direction of research has been combining alkaline metal amides with other hydride materials using ball milling. Although $LiNH_2$ and $LiAlH_4$ mixture released approximately 4 atom equivalents of hydrogen, its rehydrogenation was unsuccessful at H_2 pressures up to 80bar (8MPa) (Xiong et al., 2006).

Metal ion of amide/imide system	H_2 capacity (mass%)	Dehydogenation temperature (K)	Dissociation enthalpy (kJ/mol H_2)
Li	5.4		148
	6.5	550	45
Mg	7.4	470	3.5
Ca	3.5	620	
	2.1	770	
Li-Mg	4.5	470	39
Li-Al	4	360	
	2	440	
	2.1	470	

Table 5. Overview of characteristics of some of the amida/imide systems for hydrogen storage [18]

3.9. Amino borane

Ammonium borane, NH_3BH_3, is considered a promising candidate for chemical storage of hydrogen, due to its low molecular weight and high hydrogen content of 19.6 mass%. Its thermal decomposition occurs in three stages (Eqs. 7 and 8), around 363, 420 and 970K, respectively, with one molar equivalent (6.5 mass%) of hydrogen released during each step (Gutowska et al. 2005; Baitalow et al., 2002).

$$NH_3BH_3 \rightarrow (BN)_n + 3nH_2 \tag{7}$$

$$NH_3BH_3 + H_2O \rightarrow NH_4^+ + BO_2^- + 3H_2 \tag{8}$$

Dehydrogenation rates are very slow due to long induction period (around 200min at 353K) (Chandra & Xu, 2006a, 2006b). The enthalpy is -21 kJ/mol, which is in the range that is suitable for on-board application. Therefore, most of the scientific effort has been focused on overcoming the slow reaction kinetics using additives like silica scaffolds, metal catalysts and ionic liquids (Umegaki et al., 2009).

3.10. Alane

Aluminum hydride (alane, AlH_3) has an average enthalpy of formation of -11.4 kJ/mol and freshly synthesized non-solvated alane is reported to desorb around 10 mass% of H_2 at temperatures below 373K (Graetz & Reilly, 2006). However, while the dehydrogenation reaction occurs readily (Eq. 9), the reverse process does not, requiring 2.5GPa pressure of H_2 to rehydride (Baranowski & Tkacz, 1983).

$$AlH_3 \rightarrow Al + \frac{3}{2}H_2 \tag{9}$$

In addition, dehydrogenation kinetics is too slow for practical applications, as it is limited by nucleation and growth of aluminum particles. Therefore, research has been focused on decreasing the dehydrogenation temperature and improving its kinetics using additives like alkali metal hydride (Sandrock et al., 2006) and particle size reduction using ball milling (Orimo et al., 2006). Best results have been achieved by doping AlH_3 with small amounts of LiH, NaH or KH.

4. Hydrogen power generation – Fuel cells

As early as 1839, William Grove discovered the basic operating principle of a fuel cell by reversing the electrolysis of water to generate electricity from hydrogen and oxygen. This principle remains unchanged today, that a fuel cell is an electrochemical device which continuously converts chemical energy into electric energy (and heat) for as long as fuel and oxidant are supplied. A fuel cell does not need recharging, operates quietly and efficiently and, with hydrogen as fuel, generates only power and water. This is why it is called a zero-emission engine. Unlike thermal engines, it is not limited by the Carnot efficiency; therefore, while thermal engines typically achieve efficiency of 24-32%, fuel cells achieve efficiencies of 35-60% (Minh & Takahashi, 1995).

One of the major obstacles in fuel cell application so far has been their insufficient lifetime: most fuel cells exhibit major performance decay after 1000 hours of operation, while DOE target for 2015 is 5000 hours of operation at 60% efficiency for mobile applications. Additionally, their price per kW is almost twice that of an internal combustion engine (Hoogers, 2003). There is a wide variety of fuel cell types, which can be distinguished by the electrolyte used, but they all function in the same basic way. At the anode a fuel (usually hydrogen) is oxidized into electrons and protons, and at the cathode, oxygen is reduced to

Fuel Cell Type	Electrolyte	Charge Carrier	Operating temperature (K)	Fuel	Electric Efficiency (system)	Power Range (kW)
Alkaline	KOH	OH^-	330-390	H_2	35-55%	<5kW
Proton Exchange Membrane	Solid polymer	H^+	320-373	H_2 (tolerates CO_2)	35-45%	5-250
Phosphoric acid	Phosphoric acid	H^+	~590	H_2 (tolerates CO_2)	40%	200
Molten carbonate	Li and K carbonate	CO_3^{2-}	~1020	H_2, CO, CH_4 (tolerates CO_2)	>50%	200-1000
Solid oxide	Solid oxide electrolyte (Y, Zr)	O^{2-}	~1270	H_2, CO, CH_4 (tolerates CO_2)	>50%	2-1000

Table 6. Overview of different types of fuel cells and their characteristics

oxide species. Depending on the electrolyte, either protons or oxide ions are transported through the electrolyte (ion-conducting, but electronically insulating) to combine with oxide species or protons, respectively, to generate water and electric power. Alkaline fuel cells do not tolerate presence of CO_2 in the system, which forms alkaline carbonates, meaning that they can only use carbon-free fuel.

Of the currently developed types of fuel cells, proton exchange membrane fuel cells (PEMFC) and alkaline fuel cells (AFC) satisfy the required range of operating temperature and provide sufficient current density for mobile applications, although the power range for AFC is sufficient only for specialty applications (like power generation of space crafts). PEMFC, also known as solid polymer fuel cell, takes its name from the special plastic membrane it uses for electrolyte, combining all the key parts (anode, cathode, electrolyte) in a very compact unit, not thicker than few hundreds microns. The membrane requires presence of liquid water, limiting its operating temperature to below 373K, which means that, to achieve a good performance, this fuel cell requires a good catalyst (Wang et al., 2011).

Two high-temperature types of fuel cells (molten carbonate and solid oxide) are mainly considered for large scale stationary power generation. They achieve higher efficiencies than low-temperature systems and high operating temperatures allow for direct internal processing of fuels like natural gas, reducing the system's complexity (Tucker, 2010). However, this also means that they cannot easily be turned off.

In addition to using pure hydrogen of hydrocarbons and carbon monoxide as fuel, in recent times, fuel cells have been coupled with biomass-derived fuel processors. This makes it possible to use biomass-derived fuels, such as ethanol, methanol, biodiesel or biogas, and feed them to a fuel processor as a raw fuel, which would, after reforming, be used by a fuel cell (Xuan et al., 2009).

Although all of types of fuel cells described above operate on gas fuel, recently, PEM fuel cells have been constructed to operate using liquid fuel – methanol (Ismail et al., 2011). This has advantages for mobile applications and fuel transportation, but these direct methanol fuel cells are not yet advanced enough to generate high enough current and power density to compete with gas fuel cells.

5. Conclusion

The limitations of our world's natural resources mean that our civilization has to find an alternative energy source and energy carrier to replace fossil fuels. Taking into account available fossil fuel reserves and that such a conversion of energy source/carrier historically takes 75-100 years, it is clear that this is crucial time for alternative energy research. Hydrogen is the obvious candidate for the renewable energy carrier for the future, due to its availability in form of water, high energy density and lack of negative environmental impact. However, current state of technology is such that significant advances are needed in the next decade in order to make hydrogen economy viable. There is still lack of an effective large-scale hydrogen production process from water, while hydrogen storage still falls well behind the requirements set by gasoline. When it comes to power generation, hydrogen fuel cells are still lagging behind internal combustion engines and conventional batteries in performance and power/weight ratio. On the other hand, significant strides have been made in recent times, which give hope that hydrogen will, in the foreseeable future, be able to challenge fossil fuels as the primary energy carrier, but only through a sustained and focused effort.

Author details

Vladimir A. Blagojević and Dragica M. Minić
University of Belgrade, Faculty for Physical Chemistry, Serbia

Dejan G. Minić
Kontrola LLC, Austin, TX, USA

Jasmina Grbović Novaković
Laboratory for Material Sciences, Institute for Nuclear Science Vinča, University of Belgrade, Belgrade, Serbia

Acknowledgement

The research has been supported by the by the Serbian Ministry of Education and Science under grants 172015 and III45012.

6. References

Aguey-Zinsou K-F., Ares Fernandez J.R., Klassen T. & Bormann R. (2007), *Effect of Nb_2O_5 on MgH_2 properties during mechanical milling*, International Journal of Hydrogen Energy, vol. 32, No. 13, pp. 2400-2407

Baitalow F., Baumann J., Wolf G., Jaenicke-Rößler K. & Leitner G., *Thermal decomposition of B–N–H compounds investigated by using combined thermoanalytical methods*, Thermochimica Acta, vol. 391, pp. 159–168

Baranowski B. & Tkacz M. (1983), *The Equilibrium Between Solid Aluminium Hydride and Gaseous Hydrogen*, Zeitschrift für Physikalische Chemie N.F., vol. 135, pp. 27-38

Bassetti A., Bonetti E., Pasquini L., Montone A., Grbovic J. & Vittori Antisari M. (2005), *Hydrogen desorption from ball milled MgH_2 catalyzed with Fe*, European Physics Journal B, vol. 43, No. 1, pp. 19-28

Bastide J.P., Bonnetot B., Letoffe J.M. & Claudy P. (1980), Materials Research Bulletin, vol. 15, pp. 1215

Bobet J-L., Akiba E. & Darriet B. (2001), *Study of Mg-M (M=Co, Ni and Fe) mixture elaborated by reactive mechanical alloying: hydrogen sorption properties*, International Journal of Hydrogen Energy, vol. 26, No. 5, pp. 493-501

Bogdanovic B. & Schwickardi M. (1997), *Ti-doped alkali metal aluminium hydrides as potential novel reversible hydrogen storage materials*, Journal of Alloys and Compounds, vol. 253–254, pp. 1–9.

Demirci U.B., Akdim O. & Miele P. (2009), *Aluminum chloride for accelerating hydrogen generation from sodium borohydride*, Journal of Power Sources, vol. 192, No. 2, pp. 310–315.

Chandra M. & Xu Q. (2006), *A high-performance hydrogen generation system: Transition metal-catalyzed dissociation and hydrolysis of ammonia–borane*, Journal of Power Sources, vol. 156, No. 2, pp. 190–194.

Chandra M. & Xu Q. (2006), *Dissociation and hydrolysis of ammonia-borane with solid acids and carbon dioxide: An efficient hydrogen generation system*, Journal of Power Sources, vol. 159, No. 2, pp. 855-860.

Chen P., Xiong Z., Luo J.Z., Lin J.Y.& Tan K.L. (2003), *Interaction between Lithium Amide and Lithium Hydride*, Journal of Physical Chemistry B, vol. 107, pp. 10967–10970.

Downs A.J. & Pulham C.R. (1994), *The hydrides of aluminium, gallium, indium, and thallium: a re-evaluation*, Chemical Society Review, vol. 23, No. 3, pp. 175–184.

Du A.J., Smith S.C. & Lu G.Q. (2007), *First-Principle Studies of the Formation and Diffusion of Hydrogen Vacancies in Magnesium Hydride*, Journal of Physical Chemistry C, vol. 111, No. 23, pp. 8360-8365

Fichtner M., Fuhr O. & Kircher O. (2003), *Magnesium alanate—a material for reversible hydrogen storage?*, Journal of Alloys and Compounds, vol. 356–357, pp. 418–422.

Gutowska A., Li L., Shin Y., Wang C.M, Li X.S., Linehan J.C., Smith R.S., Kay B.D., Schmid B., Shaw W., Gutowski M. & Autrey T. (2005), *Nanoscaffold Mediates Hydrogen Release and the Reactivity of Ammonia Borane*, Angewandte Chemie. International Edition, vol. 44, No. 23, pp. 3578-3582.

Graetz J., Reilly J.J. (2006), *Thermodynamics of the α, β and γ polymorphs of AlH_3*, Journal of Alloys and Compounds, vol. 424, pp. 262–265.

Hanada N., Ichikawa T., Orimo S-I. & Fujii H. (2004), *Correlation between hydrogen storage properties and structural characteristics in mechanically milled magnesium hydride MgH_2*, Journal of Alloys and Compounds, vol. 366, pp. 269–273

Hauch A., Ebbesen S. D., Jensen S.H. & Mogensen M. (2008), *Highly Efficient High Temperature Electrolysis,* Journal of Materials Chemistry, vol. 18, pp. 2331-2340

Hoogers G. (editor), *Fuel Cell Technology Handbook,* CRC Press, New York, U.S.A., 2003

Jain I.P., Jain P. & Jain A. (2010), *Novel hydrogen storage materials: A review of lightweight complex hydrides,* Journal of Alloys and Compounds, vol. 503, pp. 303–339

Hu Y.H. & Ruckenstein E. (2003), *Ultrafast Reaction between LiH and NH_3 during H_2 Storage in Li_3N,* Journal of Physical Chemistry A, vol. 107, No. 46, pp. 9737–9739.

Ismail A., Kamarudin S.K., Daud W.R.W., Masdar S. & Yosfiah M.R. (2011), *Mass and heat transport in direct methanol fuel cells,* Journal of Power Sources, vol. 196, pp. 9847– 9855

Jensen T.R., Andreasen A., Vegge T., Andreasen J.W., Ståhl K., Pedersen A.S., Nielsen M.M., Molenbroek A.M. & Besenbacher F. (2006), *Dehydrogenation kinetics of pure and nickel-doped magnesium hydride investigated by in situ time-resolved powder X-ray diffraction,* International Journal of Hydrogen Energy, vol. 31, No. 14, pp. 2052-2062

Jhung S.H., Yoon J.W., Lee J.S. & Chang J.S. (2007), *Low-Temperature Adsorption/Storage of Hydrogen on FAU, MFI, and MOR Zeolites with Various Si/Al Ratios: Effect of Electrostatic Fields and Pore Structures,* Chemistry A European Journal, vol. 13 No. 22, pp. 6502 – 6507

Kelkar T. & Pal S. (2009), *A computational study of electronic structure, thermodynamics and kinetics of hydrogen desorption from Al- and Si-doped α-, γ-, and β-MgH_2,* Journal of Materials Chemistry, vol. 19, No. 25, pp. 4348-4355

Kitaura R., Seki K., Akiyama G. & Kitagawa S. (2003), *Porous Coordination-Polymer Crystals with Gated Channels Specific for Supercritical Gases,* Angewandte Chemie International Edition, vol. 42, No. 4, pp. 428 – 431.

Kocabas S., Kopac T., Dogu G. & Dogu T. (2008), *Effect of thermal treatments and palladium loading on hydrogen sorption characteristics of single-walled carbon nanotubes,* International Journal of Hydrogen Energy vol. 33, pp. 1693 – 1699

Kurko S., Paskaš Mamula B., Matovic Lj., Grbovic Novakovic J., Novakovic N. (2011), The Influence of Boron Doping Concentration on MgH_2 Electronic Structure, Acta Physica Polonica A, vol. 120, No. 2, pp. 238-241

Lee W-J. & Lee Y-K. (2001), *Internal Gas Pressure Characteristics Generated during Coal Carbonization in a Coke Oven,* Energy & Fuels, vol. 15, No. 3, pp. 618-623

Lee Y., Kim Y., Jeong H. & Kang M. (2008), *Hydrogen production from the photocatalytic hydrolysis of sodium borohydride in the presence of In–, Sn–, and Sb–TiO_2s,* Journal of Industrial Engineering and Chemistry, vol. 14, No. 5, pp. 655–660.

Lennard-Jones J.E. (1932), *Processes of adsorption and diffusion on solid surfaces,* Transactions of Faraday Society, vol. 28, pp. 333

Lide D.R. (editor),*Physical constants of inorganic compounds,* CRC Handbook of Chemistry and Physics, Editor: 87th on-line edition (2006-2007), CRC Press, New York, USA

Lin Y., Ding F. & Yakobson B.I. (2008), *Hydrogen storage by spillover on graphene as a phase nucleation process,* Physical Review B vol. 78, pp. 041402(R)

Liu Y., Zhong K., Gao M., Wang J., Pan H. & Wang Q. (2008), Chemistry of Materials, vol. 20, pp. 3521–3527

Maness P-C., Smolinski S., Dillon A.C., Heben M.J. & Weaver P.F. (2002) *Characterization of the oxygen tolerance of a hydrogenase linked to a carbon monoxide oxidation pathway in Rubrivivax gelatinosus*, Applied Environmental Microbiology, vol. 68, pp. 2636-2636.03

Melis A. (2002), *Green alga hydrogen production: progress, challenges and prospects*. International Journal of Hydrogen Energy, vol. 27, pp. 1217-1228

Milovanović S., Matović Lj., Drvendžija M. & Grbović Novaković J. (2008), *Hydrogen storage properties of MgH_2 - diatomite composites obtained by high energy ball milling*, Journal of Microscopy, vol. 232, No. 3, pp. 522-525

Minh N. & Takahashi T. (1995), *Science and Technology of Ceramic Fuel Cells*, Elsevier, The Netherlands

Minić D. & Šušić M. (1995), *Thermal Behaviour of 82Ni-18P Amorphous Powder Alloy in Hydrogen atmosphere*, Materials Chemistry and Physics, vol. 40, pp. 281-284.

Minić D., Šušić M. & Maričić A. (1996), *Absorption of Hydrogen by Amorphous and Crystalline 89Fe-11P Powder. Deformation of the Powder under Pressure and Relaxation on Heating*, Materials Chemistry and Physics, vol. 45, pp. 280-283.

Minić D., Šušić M., Tešić Ž. & Dimitrijević R. (1997), *Investigation of the Thermal Behaviour of Ag-Pd Intermetallic Compounds in Hydrogen Atmosphere*, Studies in surface and catalysis, vol. 112, pp. 447-456.

Minić D., Šušić M.V. (2002), *Modern Concepts of Conversion and Storage of Energy by Disperse Material Absorption*, Science of Sintering, vol. 34, pp. 247-259

Montone A., Grbović J., Bassetti A., Mirenghi L., Rotolo P., Bonetti E., Pasquini L. & Vittori Antisari M. (2005), *Role of Organic Additives in Hydriding Properties of Mg-C Nanocomposites*, Materials Scence Forum, vol. 494, pp. 137-142

Montone A., Grbović J., Stamenković Lj., Fiorini A.L., Pasquini L., Bonetti E. & Vittori Antisari M. (2006), *Desorption Behaviour in Nanostructured MgH_2-Co*, Materials Science Forum, vol. 518, pp. 79-84

Montone A., GrbovićNovaković J., Vittori Antisari M., Bassetti A., Bonetti E., Fiorini A.L., Pasquini L., Mirenghi L. & Rotolo P. (2007), *Nano-micro MgH_2-Mg_2NiH_4 composites: Tayloring a multichannel system with selected hydrogen sorption properties*, International Journal of Hydrogen Energy, vol. 32, No. 14, pp. 2926-2934

Mori D. & Hirose K. (2009), *Recent challenges of hydrogen storage technologies for fuel cell vehicles*, International Journal of Hydrogen Energy vol. 34, pp. 4569 – 4574

Ni M., Leung M.K.H., Leung D.Y.C. & Sumathy K. (2007), *A review and recent developments in photocatalytic water-splitting using TiO_2 for hydrogen production*, Renewable and Sustainable Energy Reviews, vol. 11, pp. 401–425

Noritake T., Towata S., Aoki M., Seno Y., Hirose Y., Nishibori E., Takata M. & Sakata M. (2003), *Charge density measurement in MgH_2 by synchrotron X-ray diffraction*, Journal of Alloys and Compounds vol. 356–357, pp. 84-86

Novaković N., Matović Lj., Grbović Novaković J., Manasijević M. & Ivanović N.(2009), *Ab initio study of MgH_2 formation*, Materials Science and Engineering B, vol. 165, No. 3, pp. 235-238

Novaković N., Grbović Novaković J., Matović Lj., Manasijević M., Radisavljević I., Paskas Mamula B. & Ivanović N. (2010), *Ab initio calculations of MgH_2, MgH_2:Ti and MgH_2:Co compounds*, International Journal of Hydrogen Energy, vol. 35, No. 2, pp. 598-608

Orimo S., Nakamori Y., Kato T., Brown C. & Jensen C.M. (2006), *Intrinsic and mechanically modified thermal stabilities of α-, β- and γ-aluminum trihydrides AlH_3*, Applied Physics A, vol. 83, No. 1, pp. 5-8.

Patchkovskii S., Tse J.S., Yurchenko S.N., Zhechkov L., Heine T. & Seifert G. (2005), *Graphene nanostructures as tunable storage media for molecular hydrogen*, Proceedings of National Academy of Sciences, vol. 102 No. 30 pp. 10439-10444

Pena-Alonso R., Sicurelli A., Callone E., Carturan G. & Raj R. (2007), *A picoscale catalyst for hydrogen generation from $NaBH_4$ for fuel cells*, Journal of Power Sources, vol. 165, No. 1, pp. 315–323.

Rosi N.L., Eckert J., Eddaoudi M., Vodak D.T., Kim J., O'Keeffe M. & Yaghi O.M. (2003), *Hydrogen Storage in Microporous Metal-Organic Frameworks*, Science, vol. 300, No. 5622, pp. 1127-1129.

Rowsell J.L.C., Millward A.R., Park K.S. & Yaghi O.M. (2004), *Hydrogen Sorption in Functionalized Metal–Organic Frameworks*, Journal of the American Chemical Society vol. 126, No. 18, pp. 5666 – 5667

Rowsell J.L.C. & Yaghi O.M. (2005), *Strategies for Hydrogen Storage in Metal–Organic Frameworks*, Angewandte Chemie International Edition, vol. 44 No. 30, pp. 4670 – 4679

Saha D. & Deng S. (2009), *Hydrogen Adsorption on Ordered Mesoporous Carbons Doped with Pd, Pt, Ni, and Ru*, Langmuir vol. 25, No. 21, pp. 12550-12560

Sandrock G., Reilly J.J., Graetz J., Zhou W-M., Johnson J. & Wegrzyn J. (2006), *Alkali metal hydride doping of α-AlH_3 for enhanced H_2 desorption kinetics*, Journal of Alloys and Compounds, vol. 421, pp. 185–189.

Schimmel H.G., Johnson M.R., Kearley G.J., Ramirez-Cuesta A.J., Huot J. & Mulder F.M. (2004), *The vibrational spectrum of magnesium hydride from inelastic neutron scattering and density functional theory*, Materials Science and Engineering B, vol. 108, No. 1-2, pp. 38-41

Schimmel H.G., Johnson M.R., Kearley G.J., Ramirez-Cuesta A.J., Huot J. & Mulder F.M. (2005), *Structural information on ball milled magnesium hydride from vibrational spectroscopy and ab-initio calculations*, Journal of Alloys and Compounds, vol. 393, pp. 1-4

Schlesinger H.I., Brown C.H., Finholt A.E., Gilbreath J.R., Hoekstra H.R. & Hyde E.K. (1953), *Sodium Borohydride, Its Hydrolysis and its Use as a Reducing Agent and in the Generation of Hydrogen*, Journal of the American Chemical Society, vol. 75, No. 1, pp. 215-219

Shan C.X., Bououdina M., Song Y., Guo Z.X. (2004), *Mechanical alloying and electronic simulations of (MgH_2+M) systems (M=Al, Ti, Fe, Ni, Cu and Nb) for hydrogen storage*, International Journal of Hydrogen Energy, vol. 29, No. 1, pp. 73-80

Soler L., Macanás J., Munoz M. & Casado J. (2007), *Synergistic hydrogen generation from aluminum, aluminum alloys and sodium borohydride in aqueous solutions*, International Journal of Hydrogen Energy, vol. 32, No. 18, pp. 4702–4710.

Song Y., Guo Z.X. & Yang R. (2004), *Influence of selected alloying elements on the stability of magnesium dihydride for hydrogen storage applications: A first-principles investigation*, Physical Review B, vol. 69, No. 9, pp. 094205

Šušić M., Minić D., Maričić A., Jordović B. & Krsmanović D. (1996), *Structural Changes During Heating a Cold Sintered Amorphous Powder of 82Ni and 18P Alloy*, Science of Sintering, vol. 28, pp. 105-110.

Šušić M.V. (1997), *Hydriding and dehydriding of a graphite powder doped with palladium*, Journal of Serbian Chemical Society, vol. 62, No. 12, pp. 1183-1186.

Šušić M.V. (1997), *Hydriding and dehydriding of palladium-doped charcoal*, Journal of Serbian Chemical Society, vol. 62, No. 8, pp. 631-634.

Šušić M.V. (1997), *Kinetics of the process of isothermal hydriding and dehydriding of hydrogen absorbers*, International Journal of Hydrogen Energy, vol. 22, No. 6, pp. 585-589.

Tao Y., Chen Y., Wu Y., He, Y. & Zhou Z. (2007). *High hydrogen yield from a two-step process of dark- and photo-fermentation of sucrose*, International Journal of Hydrogen Energy, vol. 32, No. 2, pp. 200-206

Tsao C.S., Tzeng Y.R., Yu M.S., Wang C.Y., Tseng H.H., Chung T.Y., Wu H.C., Yamamoto T., Kaneko K. & Chen S.H. (2010), *Effect of Catalyst Size on Hydrogen Storage Capacity of Pt-Impregnated Active Carbon via Spillover*, Journal of Physical Chemistry Letters, vol. 1, pp. 1060-1063

Turner J. (2004), *Sustainable hydrogen production*, Science vol. 305, pp. 972-974

Tucker M.C. (2010), *Progress in metal supported solid oxide fuel cells: A review*, Journal of Power Sources, vol. 195, pp. 4570–4582

Umegaki T., Yan J.M., Zhang X.B., Shioyama H., Kuriyama N. & Xu Q. (2009), *Boron- and nitrogen-based chemical hydrogen storage materials*, International Journal of Hydrogen Energy, vol. 34, No. 5, pp. 2303–2311

Varin R.A., Czujko T. & Wronski Z. (2006), *Particle size, grain size and γ-MgH_2 effects on the desorption properties of nanocrystalline commercial magnesium hydride processed by controlled mechanical milling*, Nanotechnology, vol. 17, No. 15, pp. 3856-3865

Varin R.A., Czujko T. & Wronski Z.S. (2009), *Nanomaterials for Solid State Hydrogen Storage*, Springer Science+Business Media, LLC, ISBN 978-0-387-77711-5

Wagemans R.W.P., van Lenthe J.H., de Jongh P.E., van Dillen A.J. & de Jong K. P. (2005), *Hydrogen Storage in Magnesium Clusters: Quantum Chemical Study*, Journal of the American Chemical Society, vol. 127, No. 47, pp. 16675-16680

Wang Y., Chen K.S., Mishler J., Cho S.C. & Adroher X.C. (2011), *A review of polymer electrolyte membrane fuel cells: Technology, applications, and needs on fundamental research*, Applied Energy, vol. 88, pp. 981–1007

Weitkamp J., Fritz M. & Ernst S. (1992), *Zeolithe als Speichermaterialien für Wasserstoff*, Chemie Ingenieur Technik, vol. 64, No. 12, pp. 1106 – 1109.

Weitkamp J., Fritz M. & Ernst S. (1995), *Zeolites as media for hydrogen storage*, International Journal of Hydrogen Energy, vol. 20, No. 12, pp. 967 – 970.

Weitkamp J., Ernst S., Cubero F., Wester F. & Schnick W. (1999), *Nitrido-sodalite Zn6[P12N24] as a material for reversible hydrogen encapsulation*, Advanced Materials vol. 9, No. 3, pp. 247 – 248.

Wong-Foy A.G., Matzger A.J. & Yaghi O.M. (2006), *Exceptional H_2 Saturation Uptake in Microporous Metal–Organic Frameworks*, Journal of the American Chemical Society vol. 128, No. 11, pp. 3494 – 3495.Xiong Z.T., Wu G.T., Hu J.J. & Chen P. (2006), *Investigation* *on chemical reaction between $LiAlH_4$ and $LiNH_2$*, Journal of Power Sources, vol. 159, No. 1, pp. 167-170

Xuan J., Leung M.K.H., Leung D.Y.C. & Ni M. (2009), *A review of biomass-derived fuel processors for fuel cell systems*, Renewable and Sustainable Energy Reviews, vol. 13, pp. 1301–1313

Yu R. & Lam P (1988), *Electronic and structural properties of MgH_2*, Physical Revew B vol. 37, No. 15, pp. 8730-8737

Zhao X., Xiao B., Fletcher A.J., Thomas K.M., Bradshaw D. & Rosseinsky M.J. (2004), *Hysteretic Adsorption and Desorption of Hydrogen by Nanoporous Metal-Organic Frameworks*, Science vol. 306, No. 5698, pp. 1012 – 1015.

Zeng K., Klassen T., Oelerich W. & Bormann R. (1999), *Critical assessment and thermodynamic modeling of the Mg–H system*, International Journal of Hydrogen Energy, vol. 24, No. 10, pp. 989-1004

Zuettel A. (2003), *Materials for hydrogen storage*, Materials Today, vol. 6, No. 9, pp. 24-33

Zuettel A., Wenger P., Sudan P., Mauron P. & Orimo S. (2004), *Hydrogen density in nanostructured carbon, metals and complex materials*, Materials Science and Engineering B vol. 108, pp. 9–18

Section 2

Hydrogen Production

Chapter 2

Hydrogen Generation by Treatment of Aluminium Metal with Aqueous Solutions: Procedures and Uses

J.M. Olivares-Ramírez, Á. Marroquín de Jesús,
O. Jiménez-Sandoval and R.C. Pless

Additional information is available at the end of the chapter

1. Introduction

The use of energy on a large scale has been a determining aspect of the world economy in modern times. For the last several centuries, this energy has come mainly from the transformation of combustible fossils, in the form of peat, coal, petroleum, and natural gas. During the year 2003 the supply of primary energy ran to 10,579 Mtoe (443 EJ), and this number represented a 75% increase over 30 years [1]. Of this, 80% came from fossil fuels. In 2003, the production of electricity worldwide was still dominated by coal (40%), followed by natural gas (19%), and nuclear and hydro generation (15% each) [1].

In contrast to the overall energy requirements stated above, transportation depends almost exclusively (to 95%) on petroleum and its derivatives; the problems surrounding this resource will necessitate the investment of an estimated 16 trillion U.S. dollars to develop and update new ways to power vehicles [1]. Bauen [1] mentioned that, in order to reduce CO_2 emissions, in the future we could use a pre-combustion at the coal plants, producing 250 Mt of hydrogen per year, which is six times the present production. He held out a vision of use of fuel cells for the transformation of energy and its use in hybrid vehicles which would use as their fuel hydrogen produced from water. With respect to the energy invested during a process, Liu et al. [2] state that apart from a qualitative understanding of the energy, a quantitative understanding is essential, analyzing an industrial process in terms of the material flow and amount of energy, using as example an analysis of the statistical data from an aluminum refinery at Zhenzhou, China.

Research on renewable energy sources is, in large measure, driven by the expectation of future shortages in the supply of crude oil. In the U.S.A., maximal crude-oil production occurred in the 1990s [3]; this event was termed "Peak Oil" for the American case. Other important oil producers, such as Mexico, Indonesia, United Kingdom, Norway, Romania, are also clearly beyond their date of maximal production. As to the timing of Peak Oil on the worldwide scale, different workers have, at different times, made their predictions; some of these are summarized in Table 1. Most of these forecast that Peak Oil was less than a dozen years away from the date of their prediction, though some predictions are slightly more sanguine [15], [17]. But even if, by dint of new exploration, new crude-oil production technologies, more efficient use of this resource, and the development of alternative energy technologies the actual Peak Oil date has continually been deferred, this signal event will probably take place at some point in the relatively near future. This represents a substantial menace for the world economy, in light of the increasing energy demand of major, rapidly industrializing national economies, especially in Asia and South America.

Date of forecast	Peak Oil date	Reference
2000	2004-2014	Bartlett [4]
2001	Beyond 2020	Deffeyes [5]
2002	2011-2016	Smith [6]
2003	Around 2010	Campbell [7]
2004	2006-2007	Bakhtiari [8]
2005	After 2010	Koppelaar [9]
2006	After 2010	Skrebowski [10]
2007	2008-2018	Robelius [11]
2008	2035	CERA, [12]
2008	2010-2011	Hirsch [13]
2009	2015	de Almeida [14]
2010	2030	Aleklett [15]
2011	2015-2020	Murphy [16]
2011	2035	Winch [17]

Table 1. Peak Oil dates projected by various authors.

The wide use of crude oil, and of fossil fuels in general, has also given rise to a separate, but equally pressing concern: that the liberation of carbon dioxide to the atmosphere attendant on the use of these fuels is having a major, and potentially accelerating, deleterious climate effect, which may impinge in critical ways on the world economy and the well-being of mankind. The two considerations are independently important: even if easily accessible, vast new oilfields were to be found in the near future, relieving the supply concerns in this area, the continuous increase in atmospheric CO_2 over the past decades is deemed to be of sufficient concern to warrant worldwide, concerted efforts to reduce production of this gas, mainly by switching energy production on a large scale to new, alternative forms, which minimize or entirely avoid the formation of this by-product.

2. Overview

2.1. Hydrogen production

A promising part in the new developments of energy technologies is played by hydrogen. It is not a primary fuel, as no hydrogen can be mined on our planet; rather it has to be considered an energy vector, a material produced by an endergonic chemical process starting from hydrogen-containing compounds, and whose chemical potential can in turn be converted to other forms of energy, by combustion in air, through fuel cells, or by other means. Hydrogen is, thus, useful as a storage form of energy derived from intermittent sources such as wind power, to later be transformed to electricity when and where required. An important application would be as a portable energy source in automotive transport, where hydrogen fuel cells now operate with an efficiency of about 40%, and perhaps 50% in the near future, while gasoline- or diesel-operated internal-combustion engines have efficiencies of 25%-30% under real driving conditions [18]. Also, fuel-cell powered vehicles are mechanically simpler, facilitating new designs in automobile construction. Finally, and importantly, fuel cells with proton membranes only release water to the environment, an innocuous product.

On our planet, hydrogen occurs naturally in the form of compounds such as water or hydrocarbons. It can be produced in elementary form through partial combustion of fossil fuels, as in reforming of natural gas, in coal gasification, through high-temperature electrolysis, e.g. at operating temperatures of 800°C in nuclear reactors [19], or from renewable energy by processes such as water electrolysis, water photoelectrolysis, or biomass gasification. In 2007 about 50 million metric tonnes of hydrogen were used worldwide, mainly in the production of ammonia fertilizers, in chemical syntheses, and in refining processes [20]. 95% of the hydrogen production is in a captive mode (i.e. it is produced where it is used) starting from fossil fuels (50% from natural gas, 30% from crude oil, and 20% from coal). However, from an economic-environmental point of view, the most favourable method of hydrogen production is starting from renewable energies, mainly from hydro and wind energies through electrolysis; nonetheless, water electrolysis only accounts for 4% of total worldwide hydrogen production [21], even though as far back as 1977 there was already an international consensus about the need for a concerted initiative to develop water-electrolysis technologies such as aqueous water electrolysis, solid-polymer and solid-oxide water electrolysis, and electrocatalysis [22].

Another method of hydrogen production is the reduction of water using aluminium or its alloys. Because of its light weight and great strength, aluminium is widely used for structural purposes, but it also offers important advantages for its potentially wide employment as an energy carrier: its high energy density, of 29 MJ/kg, the fact that it is the most abundant metal in the earth's crust, and its highly negative standard redox potential (ε_0 = -1.66 V) which makes it, in principle, an excellent reducing agent, capable of producing hydrogen gas upon contact with water, in a corrosion process which does not entail production of CO_2. This type of reaction can be carried out with the help of alkali, or under neutral conditions, or at elevated temperatures, or via reaction of aluminium with alcohol [23]. As a practical example for the high energy density of aluminium, it can be calculated that an electrical automobile energized by fuel cells could run a distance of 400 km with about 4 kg of hydrogen which could be obtained from 36 kg of aluminium via an aluminium-water reaction [23]. This example points out the advantages inherent in the use of aluminium metal as a compact source of hydrogen; by contrast, storage of hydrogen at standard temperature and pressure requires volumes 3000-fold larger than for gasoline, and in liquid state it is necessary to take it to temperatures of -253°C through cooling systems, which entails additional important energy costs.

The aluminium-water reaction produces energy in different forms, all of which are potentially usable: heat, water vapour, and hydrogen gas. The water vapour and hydrogen gas formed can be used to power a turbine, and the hydrogen gas furthermore represents an energy reservoir to be used by high-temperature combustion or to feed fuel cells. Such systems have been quantitatively modeled [24], [25] and [26].

2.2. Aluminium production

The main prime material for aluminium production is alumina (Al_2O_3), found in a large number of natural minerals. 98% of metallurgical alumina is produced by the Bayer process [27] starting from bauxite, a mineral composed to 50%-80% of hydrous alumina (aluminium hydroxides and oxyhydroxides). A simplified rendition of the chemistry underlying the Bayer process is given in the equation:

$$Al_2O_3 \bullet nH_2O + 2NaOH \rightarrow 2NaAlO_2 + (n+1)H_2O \quad (1)$$

The production of 1 kg of aluminium takes about 2 kg of alumina, which would be derived from 4 kg of bauxite. The energy investment in 1 kg of aluminium is 29.6 MJ for the obtention of the prime material (i.e. bauxite mining, alumina refining, and production of the carbon anodes) and 56 MJ for the electric work of alumina reduction. As the heat content required by the carbon for each kg of aluminium is about 14 MJ, the total energy investment of the aluminium is about 100 MJ/kg [28].

As stated before, it is of advantage to produce the hydrogen at the point of use, therefore, portable systems are required, based on electrochemical oxidation of the aluminium [29], or on the aluminium-water reaction in alkaline conditions, or on the oxidation of mechanochemically or mechanically activated aluminium [29]. Theoretically the electrochemical oxidation has an energy density of 4300 Wh/kg and an electrical efficiency of 55%. Chemical-oxidation technologies with aluminium can reach energy densities of 1040 Wh/kg and electrical efficiencies of 25%. As a reference in the comparison of efficiencies Schwarz et al. [30] stated that power plants which use heat derived from fossil fuels have efficiencies of 53% in the case of natural gas, 48% for crude oil, and 43% for coal.

3. Performance of the aluminium-water reaction

Because of its low equivalent weight, aluminium is an excellent potential producer of hydrogen by weight, though it is outdone in this regard by sodium borohydride, which in hydrolysis can produce 2.4 litres of hydrogen gas per gramme of substance. Aluminium and magnesium have values of 1.245 L and 0.95 L per gramme respectively, while their corresponding hydrides, AlH_3 and MgH_2, produce 2.24 L g^{-1} and 1.88 L g^{-1}, respectively [31].

Due to its highly negative redox potential, aluminium should react easily with water, producing hydrogen gas and $Al(OH)_3$ according to the equation:

$$2Al + 3H_2O \rightarrow 3H_2 + Al_2O_3 \quad (2)$$

In practice however, aluminium is generally passivated by formation of a tightly adhering surface film. For example, when exposed to air, the surface of aluminium metal is rapidly oxidized to form a tight layer of aluminium oxide, which prevents further penetration of oxygen, thus protecting the underlying metal from further oxidation. This passivation of the aluminium also interferes with the aluminium-water reaction at the interface between metal and liquid. Surface oxidation of the aluminium also comes about as a result of the aluminium-water reaction itself, and may also constitute a limitation to the rate and yield of the reaction. Overcoming these impediments is a central problem in all practical applications of the aluminium-water reaction.

The passivating surface layer of Al_2O_3 is an amphoteric material, soluble in both strongly acidic and strongly basic aqueous solution. Therefore, both acidic and basic hydrolysis of aluminium can be employed for hydrogen production. However, the corrosive character of the solutions employed in these instances is of disadvantage in practical applications. Accordingly, a large part of the recent research in this area has centred on methods to promote the aluminium-water reaction in neutral or near-neutral conditions, by employing special aluminium alloys, by addition of activators, by mechanical pretreatment of the aluminium, or by irradiation.

For several decades, aluminium alloys with gallium and indium have been studied as highly reactive materials in the aluminium-water reaction, as the minor-metal admixtures cause embrittlement of the metal and destruction of the intergranular bonds and of the passivating aluminium oxide film. The activation of aluminium powders by grinding or co-milling with gallams of various compositions, Ga-In (70:30), Ga-In-Zn (70:25:3), Ga-In-Sn (62:25:13), Ga-In-Sn-Zn (60:25:10:5), leads to significant rates of hydrogen gas evolution on water contact [31] [32]. These are in the range of 14-20 mL g^{-1} min^{-1} with powders prepared by soft mechanical treatments, but can be two orders of magnitude higher, at 1000 - 2500 mL g^{-1} min^{-1}, in the case of the milled powders. The rate of the reaction rises with increasing admixture of the gallam, up to 10 wt%, relative to the aluminium, and with the reaction temperature. The cost of the gallams is, obviously, a consideration; still, a 2%-3% admixture of gallam and a temperature of 60°C bring about hydrogen evolution rates of 2000 mL g^{-1} min^{-1}. Materials with lower gallium content, however, perform more poorly, and an increase in the concentration of cheaper dopants (Sn, Zn, Pb) is ineffective [33].

Certain aluminum-based composites, generally containing aluminium at 80 wt%, doped with gallium, indium, zinc, or tin, obtained by co-melting of the metallic components, have been studied by powder X-ray diffraction, differential thermal analysis, and EDX [33]. The tests indicated the presence of a crystalline aluminium metal phase, an intermetallic compound, $InSn_4$, and a eutectic in the Ga-In-Sn-Zn system, melting at 6°C, distributed over the grain boundary space. Some of these alloys performed satisfactorily in the aluminium-water reaction, forming hydrogen and amorphous precipitates of the corresponding metal hydroxides; however, they lost effectiveness upon storage at room temperature, and were completely stable only in vacuum at liquid-nitrogen temperature.

Shaytura et al. [34] studied the aluminium-water reaction in the presence of an undisclosed chemical activator which interacts with the OH-groups of the $Al(OH)_3$ formed during the reaction, thereby affecting the pore size distribution in the newly made oxide layer, which facilitates permeation through this layer and thus the hydrogen-forming reaction. Similar effects of better pore-size distribution and faster hydrogen evolution were also obtained by applying an ultrasonic field to the reacting mixture.

Czech and Troczynski [35] found that the passivation of aluminium metal in water in the pH range from 5 to 9 is significantly suppressed when the metal has been milled with inorganic salts, such as KCl o NaCl. Corrosion of this type of aluminium in tap water, with production of hydrogen and precipitation of aluminium hydroxide, at normal pressure and a temperature of about 55°C, is rapid and substantial. By way of example, 92% of the aluminium in the Al-KCl system (milled for 1 h) is corroded in 1 h, in aqueous solution at neutral or near-neutral pH, with liberation of 1.5 mol of hydrogen for each mole of aluminium consumed in the reaction. Apart from gaseous hydrogen, only solid aluminium hydroxides are formed as by-products of the reaction; this is a promising aspect for direct recycling from this system.

Similarly, Alinejad and Mahmoodi [36] proposed a simple method for hydrogen generation, based on highly activated aluminium and water. Activation was achieved by milling aluminium powder with sodium chloride as a nano-miller. The mean rate of hydrogen release per gram of aluminium was 75 ml/min, for powder prepared with a salt/aluminium molar ratio of 1.5, and 100% yield was reached after 40 minutes.

Mahmoodi and Alinejad [37] also reported preparing highly active material for the aluminium-water reaction by ball-milling the metal with a large amount of sodium chloride as a nano-miller, in a dilute argon atmosphere. The powder obtained could be stored for a long time; once submerged in hot water, it was rapidly hydrolysed, and hydrogen was produced in a 100% yield. The rate of hydrogen generation was found to depend critically on the initial water temperature. The heat released in the exothermal aluminium-water reaction was employed to raise the water temperature during the reaction. The mean rate of hydrogen production was ~101 and ~210 ml/min per 1 g de Al, using initial water temperatures of 55°C y 70°C, respectively. However, distinctly higher rates of hydrogen evolution were achieved (713 mL g^{-1} min^{-1}) when the aluminium was activated by ball milling with 7 wt% bismuth.

Wang et al. [38] used nanocrystals of metal oxides such as TiO_2, Co_3O_4, Cr_2O_3, Fe_2O_3, Mn_2O_3, NiO, CuO, and ZnO as modifiers in aluminium metal powders to produce hydrogen by a room-temperature reaction with deionised water or tap water. They studied the effect of TiO_2 nanocrystals of different crystal size on hydrogen production in the reaction with tap water, and showed that the water quality and the metal-oxide nanocrystals such as TiO_2, Co_3O_4, and Cr_2O_3 increase hydrogen production by the reaction of the aluminium powder in neutral water at ambient temperature.

Mercury or zinc amalgam have been used as activators to promote the hydrolysis of aluminium [39]. The results showed that in the presence of mercury or zinc amalgam aluminium hydrolysis to generate hydrogen could take place at room temperature, in distilled water. The rate of hydrogen release depended on the reaction temperature, and the maximum rate of hydrogen generation was 43.5 $cm^3\ h^{-1}\ cm^{-2}$, obtained at 65°C with the zinc-amalgam technique. For this method, the apparent energy of activation for aluminium hydrolysis induced by zinc amalgam had a value 43.4 kJ mol^{-1}, while in the case of the mercury method it was 74.8 kJ mol^{-1}. The results of the X-ray diffraction analysis showed that the subproduct formed is bayerite.

Deng et al. [40] used three different modification agents, γ-Al_2O_3, α-Al_2O_3, and TiO_2, to surface-modify aluminium particles. The selected oxide was ball-milled with the aluminium metal in ethanol suspension, using highly pure Al_2O_3 spheres. They investigated the effect of different modification agents on hydrogen generation in the reaction of aluminium powder with water. The different modification agents were seen to have different effects on the dynamics of the aluminium-water reaction. In particular, induction times for the reaction increased in the series: γ-Al_2O_3-modified aluminium < α-Al_2O_3-modified < TiO_2-modified, interpreted as the series of increasing energy barriers for nucleation of hydrogen bubbles on the variously modified metal surfaces.

Fan et al. [41] prepared a series of aluminium-based materials by ball-milling and/or fusion; they used the techniques of XRD, SEM, and TG-DTA to characterise the samples. They evaluated the effects of different alloying metals, such as Zn, Ca, Ga, Bi, Mg, In, and Sn, on hydrogen generation through hydrolysis in pure water. They found mechanical milling to be preferable to melting as a method to produce aluminium alloys containing metals with low melting points (Ga, In). The addition of Sn, Ga, or In could reduce the hydrolysis rate of the Al-Bi alloy, but the addition of Zn accelerated the hydrolysis of this alloy. Best yields of hydrogen were obtained with quinternary alloys containing Al, Bi, Zn, Ga, and calcium hydride.

Parker et al. [42] developed a process to obtain hydrogen by the reaction of mixtures of finely divided magnesium and finely divided aluminium with seawater, at normal pressure and temperature. The procedure is appropriate for fixed applications where no electric net is available, it requires no or only a minimal supply of electricity. As a side product a mixture of magnesium hydroxide and aluminum hydroxide are obtained; these side products have a high market value as prime materials for the production of thermal and electrical insulation (i.e. heatproof sheathings or linings).

Soler et al. [43] experimented with aluminium strips in aqueous solutions of NaOH or KOH, at alkali concentrations of 1 to 5 M and temperatures between 290 K and 350 K, and found a maximum hydrogen production per gramme of aluminium of 3100 cm^3 min^{-1} with 5 M NaOH at 350 K, while the use of 5 M KOH at 350 K gave a rate of 2900 $cm^3 min^{-1}$ per gramme of aluminium.

Macanás et al. [44] examined the effect of the presence of various inorganic salts, at a concentration of 0.01 M, as corrosion promoters in the reaction of metallic aluminium in 0.1 M NaOH at 75°C and found the advantages of 100% yield of hydrogen production, self-initiation without heating, and significant accelerating effects (almost two-fold in the case of NaF, and 1.5-fold with $MgCl_2$, $Fe_2(SO_4)_3$, Na_2SO_4, and $FeCl_3$). High rates and good yields of hydrogen were also achieved in the absence of NaOH with the use of mixed solutions of sodium aluminate, sodium stannate, and sodium metaborate, each in concentrations between 0.1 M and 0.5 M. In almost all of these experiments, beginning and end values of pH were between 12 and 13.

The use of solutions containing only sodium aluminate was examined by Soler et al. [45] in comparison with NaOH solutions giving the same initial pH value; similar results both in terms of yields and of maximum hydrogen flow rates were obtained, leaving in doubt whether there is a specific chemical effect of the aluminate, especially as this salt was used in high concentration (0.5 M). For the sodium aluminate case, an Arrhenius energy of activation of 71 kJ mol^{-1} was determined, interpreted as indicating that the rate-controlling step in the process was a chemical event.

The aluminium-water reaction in the presence of sodium stannate at 0.075 M (Soler et al., [46]) gives interesting results in terms of acceleration. In this case, the reaction is affected from its earliest stages by formation of metallic tin as a by-product, noticeably reducing the hydrogen yield and complicating the product mixture. The energy of activation of the hydrogen formation was determined at 73 kJ mol^{-1}.

Uehara et al. [47] observed that when aluminium metal was cut under water effervescence occurred at the freshly cut surface, which however soon subsided. The method is obviously not applicable as a practical way to produce hydrogen gas; nonetheless, the observation serves to illustrate the problem of rapid passivation under water and the need to address this issue in a realistic manner to produce hydrogen in the aluminium-water reaction, e.g. through addition of appropriate activators.

Watanabe [48] studied the mechanism of the aluminium-water reaction in aluminium powders with particle sizes in the micron and sub-micron range, obtained by mechanical grinding, and came to the following conclusions. Micro-cracks formed at grain boundaries at the surface of the metal particles grow inward, due to corrosion by water. Inside these fissures, unsaturated aluminium atoms (with one free bond or two free bonds, i.e. (Al=) and (Al-)) may conform, with other such atoms, clusters which split the water molecules present in the crack, with initial formation of AlH_3, and hydroxylated aluminum species in the crevice wall. The formation of these species creates internal stresses in the metal, extending the micro-cracks, by a kind of micro-tribochemical effect. The AlH_3 further reacts with water, producing diatomic hydrogen.

In a simulation using the reactive force-field method (ReaxFF), Russo et al. [49] examined the dynamics of water dissociation at the surface of aluminium nanoclusters, over time periods of up to 70 ps, and with different numbers of total participating water molecules. Interestingly, an optimum water concentration (of about 30 water molecules per Al_{100} cluster) was found, indicative of two countervailing effects: the need for some additional solvating water molecules to assist the reaction of the aluminium-adsorbed water undergoing reaction, and surface saturation of the aluminium surface at high water concentration, sterically hindering the binding of hydrogen to the cluster.

Table 2 summarises further results reported with different mixtures as prime materials for the aluminium-water reaction. A few examples of the related magnesium-water reaction are also included.

Mixture	Process	Rate of hydrogen production	Source
Hydrogenated Mg_3La + water	Hydrolysis, during the first 20 minutes	43.8 ml min^{-1} g^{-1}	Ouyang et al. [50]
Hydrogenated La_2Mg + water		40.1 ml min^{-1} g^{-1}	
Milled Mg + sea water	Hydrolysis, during the first 10 min	90.6 ml min^{-1} g^{-1}	Zou et al. [51]
Milled Mg/Co (95:5) + sea water	Hydrolysis, during the first minute	575 ml min^{-1} g^{-1}	Zou et al. [51]
Milled Al/Bi/NaCl (80:15:5) + water	Hydrolysis, during the first 30 min	300 ml min^{-1} g^{-1}	Fan et al. [52]
Milled Al/Bi/CaH_2 (80:10:10) + water	Hydrolysis, during the first 3 min	340ml min^{-1} g^{-1}	
Milled Al/Bi (80:20) + water	Hydrolysis, during the first 30 min	24 ml min^{-1} g^{-1}	
Al-20%wt CaH_2	Hydrolysis, during the first 5 min	24 ml min^{-1} g^{-1}	
$Ni_{20}Al_{80}$ + 0.46 M NaOH	Hydrolysis, during the first 15 min	63 ml min^{-1} g^{-1}	Hu, et al. [53]
$Ni_{30}Al_{70}$ + 0.46 M NaOH	Hydrolysis, during the first 15 min	54 ml min^{-1} g^{-1}	
$Ni_{40}Al_{60}$ + 0.46 M NaOH	Hydrolysis, during the first 20 min	47 ml min^{-1} g^{-1}	
$Ni_{50}Al_{50}$ + 0.46 M NaOH	Hydrolysis, during the first 60 min	7.3 ml min^{-1} g^{-1}	

Table 2. Mixtures of water, aluminium, and other materials, for hydrogen production

4. Applications of the aluminium–water reaction

4.1. Thermal energy

A special type of application of the aluminium-water reaction, which emphasizes the thermal aspect, was described by Sabourin et al. [54], who carried out the combustion of nano-aluminium (38 nm diameter particle size) in mixtures of liquid water and hydrogen peroxide. They obtained, at 3.65 MPa argon pressure, mass burning rates per unit area between 6.93 g cm^{-2} s^{-1} (at 0% H_2O_2) and 37.04 g cm^{-2} s^{-1} (at 32% H_2O_2), corresponding to linear burning rates of 9.58 cm s^{-1} y 58.2 cm $s^{-1,}$ respectively. The difficulty lies in the high cost of the preparation of aluminium on the nanoscale for its combustion [55].

There are various techniques to prepare aluminium on the nanoscale, such as the electro-exploded wire method cited by Kotov [56] and by Kwon et al. [57], explosion in plasma [58], plasma electro-condensation process [59], sol-gel [60], heating evaporation [61], and evaporation [62], for which two routes can be used: induction heating evaporation (IHE) and laser-induction complex heating evaporation (LCHE). For the "IHE" method one uses a chamber which contains an induction coil; in the centre of the coil are located two crucibles, one made of graphite, the other of alumina. The alumina crucible is charged with aluminium of 99.6% purity, while the interior of the chamber holds a dilute argon atmosphere at 10 Pa pressure. The coil is energized at high frequency at an initial power of 5 kW; after several minutes the aluminium metal has melted. The coil is deenergized, and the molten liquid rapidly evaporates; these atoms of evaporated aluminium are collected through collisions with the argon gas, producing in this manner aluminium nanopowders. For the "LCHE" method the equipment is fitted with a continuous-wave 1.6-kW CO_2 laser; this laser is switched on when the aluminium has molten and vaporises it rapidly, in the subsequent condensation step aluminium nanopowder is again obtained. The thermal properties were determined by Chen et al. [62] using thermogravimetric (TGA) techniques and differential thermo-analysis (DTA), finding that the temperature peaks were at 560°C and 565°C, and the enthalpy increases were 1.18 kJ/g and 3.54 kJ/g for "IHE" y "LCHE", respectively.

Shafirovich et al. [63] proposed the use of $NaBH_4$/metal/H_2O, were aluminium metal in powder form can be used, thus reducing the extra cost entailed in the preparation of nanoparticulate aluminium. In their experiments they used different stoichiometric ratios, with the result that the mixtures containing powdered aluminium do not burn as easily as when nanoscale aluminium is used. To resolve this problem they added magnesium in powder form as an activator, thus achieving combustion. Here again, the main result is the thermal effect, with only a minor yield of hydrogen.

A different type of thermal application with Al/H_20/NaOH, where the aluminium used was the shavings from lathes and milling machines, i.e. aluminium pieces on the centimeter scale, was described by Olivares-Ramírez et al. [64]. In this system, in a first stage hydrolysers are used to obtain hydrogen through the chemical reaction between aluminium, sodium hydroxide, and water, the hydrogen obtained is then passed through water to trap the vapour generated in the original exothermal reaction; finally the hydrogen is burned in air. The second stage consists of a refrigerator based on the ammonia-water absorption principle, energized by the heat produced in the combustion of the hydrogen.

4.2. Fuel cells

Hydrogen is the fuel most frequently used in fuel cells. At present, microfuel cells are being developed for applications in portable electronic devices, to address the problem of hydrogen storage in a portable mode. The difficulty lies in that we do not yet have portable systems for hydrogen storage; for instance, even a small hydrogen container, to energize a laptop computer, would not be allowed onto an airplane. Small aluminium-water reactors, however, could supply this need [65]. The hydrogen could be produced from variegated sources, such as electrical wire, metal hydride, or aluminium foil.

Jung et al. [66] describe a small-scale hydrogen generator, in which two types of additives, NaOH or CaO, are used to control the flow of the hydrogen generated, with the idea that this could be connected to a small fuel cell. In their experiment they feed the reactor, through a micro-pump, with NaOH solution at a rate of 0.2×10^{-6} m^3 min^{-1}, at a pressure of 1500 kgf cm^{-2}, and NaOH at concentrations of 5%, 10%, 15% y 20% (w/w), achieving a maximum hydrogen production of 2.65×10^{-3} m^3 using 15% (w/w) NaOH. They also tested mixtures of aluminium with 5% (w/w) NaOH, obtaining the maximal production of hydrogen (3.22×10^{-3} m^3) with a CaO:Al mass ratio of 0.1; best performance was obtained at 2000 kgf cm^{-2}.

Shkolnikov et al. [67] presented a 2-W cell-phone consisting of a micro fuel cell and a micro generator for hydrogen based on the aluminium-water reaction. Each gramme of aluminium produces 1.2 -1.8 Wh of electrical energy. In their experimental work they examined three ranges of current density: low current density, i.e. <200 mA cm^{-2}; medium current density, 200–500 mA cm^{-2}, and high current density, >500 mA cm^{-2}. The micro fuel cell used was of the air–hydrogen polymer electrolyte membrane type, the air being supplied by a micro fan which was itself energised by the micro fuel cell.

Wanga et al. [68] used a mini-reactor containing an aqueous sodium hydroxide solution and strips of an alloy consisting to 99% (w/w) of aluminium, connected to a fuel cell, to constitute a portable device. In their work they tried NaOH concentrations of 9%, 17% y 25% (w/w); with the latter concentration they observed the maximum rate of hydrogen production, 247 ml min^{-1}, at an initial temperature of 20°C; once this was coupled to the cell, the latter used hydrogen at a rate of 40 ml min^{1}, producing power values of up to 0.15 W.

In an attempt to use lower-cost aluminium for hydrogen production, Silva et al. [69] used aluminium from empty soft-drink cans. To speed up the reaction they used 2 M NaOH, removing the paint cover of the cans with sulfuric acid, then cutting them into strips to introduce them into the reactor, which produced experimentally 0.049 moles of hydrogen per gramme of aluminium. This hydrogen was then used in a PEM-type fuel cell. The oxygen for the fuel-cell operation was obtained by water electrolysis energized by photovoltaic cells. The fuel cell worked with efficiencies of up 10.14%, and achieved a cell voltage of 150 mV.

A portable generator of 2050 mL capacity has been described by Fan et al. [70], charged with Al-Bi-NaCl, in which the aluminium and the bismuth were in powder form (13 microns) mixed with sodium chloride particles. The mixtures originally evaluated were Al with 10 wt% Bi and 1, 3, 5, or 10 wt% NaCl, where the composition Al-10wt.%Bi–5wt.% NaClgave the best hydrogen yield (1063 mL/g Al). Later experiments examined the addition of zinc, when the highest rate of hydrogen production (1026 mL/g Al) was obtained with the mixture Al-10 wt.%Bi-1 wt.%Zn-2 wt.%NaCl. The authors propose this generator as suitable for use with hydrogen fuel cells.

4.3. Power plants

Vlaskin et al. [71] designed a co-generation power plant in which they used aluminium powders with mean particle sizes of up to 70 μm as the main fuel and water as the main oxidizing agent. The plant can function autonomously (i.e. without connection to the electrical net), without ceasing production of hydrogen, electrical energy, and heat. One of the key components of the pilot plant is water-aluminium in a high pressure reactor projected for hydrogen production at a rate of 10 $nm^3\ h^{-1}$. The hydrogen formed flows through a condenser and a dehumidifier with a dew point of -25°C, and enters then a 16-kW hydrogen-air fuel cell. Using 1 kg of aluminium the experimental plant produces 1 kWh of electrical energy and 5 – 7 kWh of heat. Total efficiency of the power plant is 72%, and electrical efficiency is 12%. The electrical efficiency of power plants based on the aluminium-water reaction can be raised by developing devices which use vapour-hydrogen at high temperature to produce electrical energy. They reported that the cost of electrical-energy production in power plants fuelled by aluminium is comparable to the energy costs involved in power generation via the traditional liquid hydrocarbons, while energy generation based on aluminium is a more ecological option.

Rosenband and Gany [72] carried out a parametric study on the aluminium-water method, using a special type of aluminium powder activated by a thermo-chemical process involving a small proportion of a lithium-based activator [73]. The experiments showed a rapid, sustained self-starting reaction between aluminium and water, proceeding even at room temperature, with hydrogen yields of virtually 100% under appropriate conditions. The method demonstrates a safe and compact method of hydrogen storage (11 wt%, based on aluminium). They propose that possible applications would be in fuel cells, as well as in land and sea traffic. The rate of the reaction is independent of the type of water (distilled water, tap water, seawater). The apparent activation energy of the process is about 16.5 kcal/mol.

Silva et al. [74] studied the recycling of aluminium for green energy production, producing high-purity hydrogen gas by the reaction between aluminium and sodium hydroxide at different molar ratios. The results showed acceptable hydrogen yields of sufficient purity for use in PEM fuel cells for electricity generation. A test with 100 aluminium cans reacting with caustic soda showed that hydrogen production would be possible with a scale-up to 5 kWh in few hours. This work is environmentally friendly and shows that green energy can be produced from aluminium residues at low cost. The hydrogen was easily liberated through a spontaneous chemical reaction, and at relatively low cost, through contact of aluminium from discarded cans with aqueous sodium hydroxide solution. They obtained hydrogen of high purity, which they used in a commercial fuel cell to produce electricity. The hydrogen was produced from a recyclable material, without input of energy and without any additional release of contaminants to the air. They also successfully used the by-product obtained, $NaAl(OH)_4$, to produce an aluminium hydroxide gel to treat water contaminated with arsenic.

Zhuk et al. [75] investigated the generation of electricity using low-cost aluminium, and found that suppressing parasitic corrosion while at the same time maintaining the electrochemical activity of the metal anode is one of the main problems affecting the energy efficiency of aluminium-air batteries. The need to employ aluminium alloys or high-purity aluminium causes a significant increase in the cost of the anode, and therefore in the total cost of the energy produced in the aluminium-air battery; this limits possible applications for this type of power source. They stated that the process of parasitic corrosion is itself a possible method of hydrogen production. Hydrogen produced in this manner in an aluminium-air battery can be used in a fuel cell or burnt to produce heat. Different anode materials would be suitable, such as commercial aluminium, aluminium alloys, or secondary aluminium, which are much cheaper than special alloys for aluminium anodes or high-purity aluminium. Their work consisted mainly in a comparison of the cost of energy production with commercial aluminium alloys, high-purity aluminium, and a special Al-In anode alloy, as anode materials for an aluminium-air battery and for the combined production of electrical energy and hydrogen.

Davoodi et al. [76] tested a microscopic device, based on the principles of scanning electrochemical microscopy (SECM), for collecting electric energy off the surface of an aluminum alloy, AA3003, immersed in 0.01 M aqueous sodium chloride. Using a nanometer-scale atomic-force microscope (AFM), the probe would gather the electrochemical current on the aluminium surface, searching for the regions of highest current density. Their results are interesting from a mechanistic point of view, as they document a large variation in local current densities (from 0.45 pA μm^{-2} for matrix regions to 14.2 pA μm^{-2} for trenches at intermetallic particles); they probably do not hold out realistic prospects for using this method in an economic sense for energy production.

According to Namba et al. [77] the process of remelting of scrap aluminium produces a slag which contains not only metal oxide, but also other by-products. He reported that in Japan 0.35 million tons of this dross are produced yearly. It is remelted to recover aluminium metal with a concentration of between 10 and several dozen percent by weight. Hiraki and Akiyama [78] proposed a novel system to treat aluminium residues, such as this slag. They evaluated from the point of view of the life cycle the total exergy loss (EXL) in comparison to the EXLs attendant on the co-production of 1 kg of hydrogen at 30 MPa and 26 kg of aluminium hydroxide. The exergy flow diagram shows that the exergy of aluminium residues containing only 15% of the metal by weight is still large, while the exergy of pure aluminium hydroxide is relatively small. The exergy of the proposed system (150.9 MJ) is inferior by 55% compared to the conventional system (337.7 MJ), in which the gas compressor and the production of aluminium hydroxide consume significantly more exergy. The results also show that the exergy analysis should be applied to the life-cycle assessment (LCA) as a critical consideration for practical use, additional to the conventional LCA on the emission of carbon dioxide. When this concentration becomes smaller than 20% by weight due to the remelting, the slag is chemically treated as an inocuous substance by methods such as high-temperature fusion in the electric arc, at a cost of US$200-300 per tonne.

5. Patents on reactors for the aluminum-water reaction

Houser [79] developed an invention on heating systems, and more specifically a device in which the chemical oxidation-reduction reactions would proceed producing heat, but without flame formation, in contrast to most heating systems. The system has a housing whose lower part holds the liquid reagent, e.g. NaOH solution. The solid reactant, e.g. aluminium metal, is immersed in the solution to produce heat and gaseous product, e.g. hydrogen. The position of the container within the deposit is regulated to control the extent of immersion, and the solid reagent inside the container can react with the liquid, producing heat. The hydrogen gas produced rises to the top of the conical cover and exits via a topside conduit to be washed with water and then transferred to storage or used directly via combustion. The water bath insulates the interior of the system, thus precluding combustion of the hydrogen in the system. The sodium aluminate produced in the reaction forms a mud, which is diverted to the drain by an inclined conical wall in the lower part of the housing. To start the system functioning, the lower part of the housing is filled with the solution and the container is raised above the level of the solution. To activate the heating system, the solution is first ohmically heated by an electrical resistance to about 140°C for total efficiency.

Houser [80] describes another gas generator which utilizes spherical pieces of a solid reactive material in contact with a second, liquid reagent. The spheres are moved through the reaction chamber via inclined channels with holes, which could be made e.g. from screen mesh tubing. The liquid reagent is sprayed from dispersion nozzles onto the spheres which move upward via the channels. The solid residues are removed from the spheres through the rolling action of the spheres against the channel walls and washed away by gravity. The liquid reagent is again conducted to a sump, whence it is pumped through a filter to remove the waste product of the reaction; it then is returned to the nozzles in the reaction chamber. The concentration of the liquid reagent is regulated through a sensor in the sump, which controls the addition of concentrated make-up solution. The temperature of the liquid reagent is also controlled by a sensor in the sump or a thermostatic valve in the sediments filter, allowing the liquid to flow through an indirect heat exchanger. The gas generated is delivered from the reaction chamber by its own pressure. Not surprisingly, the preferred embodiment of the invention is the production of hydrogen from aluminium spheres and sodium hydroxide solution.

Andersen and Andersen [81] built a prototype device to produce hydrogen by making aluminium react with water in the presence of sodium hydroxide at between 0.26 M and 19 M. The reaction takes place with effervescence at the metal surface; at the same time, a precipitate collects at the bottom of the container. The zone of effervescence is kept separate from the zone where the precipitate collects. This reduces the possible hindrance of the hydrogen-generating reaction by the precipitate. Satisfactory results are obtained with sodium hydroxide concentrations between 1.2 and 19 M and at temperatures from 4°C to 170°C, with best performance seen at NaOH values between 5 and 10 M and solution temperatures around 75°C.

Andersen and Andersen [82] also constructed another prototype which produces heat and hydrogen gas, all at ambient temperature. Here, aqueous NaOH (18 wt%) half-fills an expandable container which adjusts to the momentary pressure and temperature of the reaction by expanding and contracting, thus controlling the level of immersion of a fuel cartridge containing aluminium filings or shavings or aluminium foil, to manage the intensity and duration of the reaction. The upper part of the container is built from flexible material and joined to the support of the fuel cartridge, so that, as the hydrogen is produced, the internal pressure in the container increases, moving the upper lid and thus lifting the aluminum out of the aqueous sodium hydroxide, thereby stopping the chemical reaction. As the hydrogen is consumed, e.g. by combustion, the internal pressure in the container drops, and the aluminum again gets in contact with the lye, and hydrogen production restarts. This cycle repeats itself until the aluminium is used up.

Troczynski [83] described a hydrogen-generating system based on hydrolysis of a composite aluminium material, at a pH close to neutral, to supply hydrogen to fuel cells or other devices. The process of use involves: (a) Supply of water to the composite material in the reactor vessel to produce the hydrogen (b) Passage of the hydrogen formed from the reactor to a buffer vessel, and (c) liberation of the hydrogen from the buffer vessel to the fuel cell, at a second, lower, pressure which is compatible with the fuel cell. Several reactor vessels and several buffer vessels are monitored by pressure sensors and connected by processor-controlled valves to assure a continuous supply of hydrogen to the fuel cell.

Fullerton [84] developed a recyclable hydrogen generator which uses aluminium, an alkali hydroxide, and water, connected to a user, which could be an internal-combustion engine, a turbine, or a fuel cell. The hydrogen generator safely supplies the hydrogen demand at atmospheric pressure, and it can be stored anywhere at low cost, as it is safe and inert.

6. Conclusions

Interest in hydrogen as an energy carrier is based on considerations of the finite nature of our hydrocarbon resources and of the worldwide deleterious effects of the carbon dioxide emitted during hydrocarbon-based energy production. In this context, the aluminium-water reaction shows interesting possibilities as an energy-providing small or mid-scale reaction. In contrast to gasoline-air based systems, which benefit from the ubiquitous presence of air, the aluminium-water reaction requires that the latter be specifically supplied or brought along; however, the ready availability of water at most places minimizes the importance of this consideration.

An attractive feature of the aluminium-water reaction as a way to produce hydrogen gas on demand is its essential simplicity. Obviously, in its most basic form, the reaction is fundamentally hampered by passivation of the metal, but this can be overcome by the use of strongly alkaline conditions. However, the corrosiveness of such conditions is a distinct drawback; for this reason, extensive research has centred in recent years on means of maintaining an active metal surface in aqueous conditions at neutral or near-neutral pH, e.g. by the use of aluminium alloys. Research in these aspects is likely to remain important in the immediate future.

Large-scale application of the aluminium-water reaction appears less promising and is unlikely to replace to any significant extent the more established modalities, e.g. energy storage in battery banks.

Author details

J.M. Olivares-Ramírez and Á. Marroquín de Jesús
Universidad Tecnológica de San Juan del Río, San Juan del Río, Querétaro, México

O. Jiménez-Sandoval
Centro de Investigación y de Estudios Avanzados del Instituto Politécnico Nacional, Unidad Querétaro, México

R.C. Pless
Centro de Investigación en Ciencia Aplicada y Tecnología Avanzada del IPN, Unidad Querétaro, México

7. References

[1] Bauen A. (2006) Future energy sources and systems-Acting on climate change and energy security. Journal of Power Sources. 157: 893-901.

[2] Liu L, Lu A, Lu Z, Zhang P. (2006) Effect of material flows on energy intensity in process industries. Energy. 31: 1870–1882.

[3] Verbruggen A, al Marchohi M. (2010) Views on peak oil and its relation to climate change policy. Energy Policy. 38: 5572-5581.

[4] Bartlett A. (2000) An analysis of US and world oil production patterns using Hubbert-style curves. Mathematical Geology. 32: 1–17.

[5] Deffeyes K.S. (2001) Hubbert's peak: The impending world oil shortage. Princeton.

[6] Smith M.R. (2002) Consultancy Report from The Energy Network/Energy files. Available from: "http://www.energyfiles.com/" http://www.energyfiles.com/ .
[7] Campbell C. (2003) Industry urged to watch for regular oil production peaks, depletion signals. Oil and Gas Journal. 101: 38–45.
[8] Bakhtiari A. (2004) World oil production capacity model suggests output peak by 2006–07. Oil and Gas Journal. 102: 18–19.
[9] Koppelaar R. (2005) World oil production and Peaking outlook "Peak Oil Netherlands Foundation/ASPO Netherlands. http://www.peakoil.nl."
[10] Skrebowski C. (2006) Prices holding steady, despite massive-planned capacity additions. Petroleum Review. 28–31.
[11] Robelius F. (2007) Giant Oil Fields-The High way to Oil: Giant Oil Fields and their Importance for Future Oil Production. Doctoral thesis, Uppsala University, Sweden.
[12] CERA A. (2008) Available from: "CERA/HIS Study, Cambridge Energy Research Associates, Inc., Cambridge, MA
http://www.cera.com/aspx/cda/public1/news/press
[13] Hirsch R.L. (2008) Mitigation of maximum world oil production: Shortage scenarios. Energy Policy. 36: 881–889.
[14] de Almeida P, Silva PD. (2009) The peak of oil production-Timings and market recognition. Energy Policy. 37. 1267–1276.
[15] Aleklett K, Höök M, Jakobsson K, Lardelli M, Snowden S, Söderbergh B. (2010) The Peak of the Oil Age–Analyzing the world oil production Reference Scenario in World Energy Outlook 2008. Energy Policy. 38: 1398-1414.
[16] Murphy DJ, Hall CAS. (2011) Energy return on investment, peak oil, and the end of economic growth. Ecological Economics Reviews. 1219: 52-72.
[17] Winch P, Stepnitz R. (2011) Peak Oil and Health in Low- and Middle-Income Countries: Impacts and Potential Responses. American Journal of Public Health. 101: 1607-1614.
[18] Ball M, Wietschel M. (2009) The future of hydrogen–opportunities and challenges. International Journal of Hydrogen Energy. 34: 615-627.
[19] Fujiwara S, Kasai S, Yamauchi H, Yamada K, Makino S, Matsunaga K, Yoshino M, Kameda T, Ogawa T, Momma S, Hoashi E. (2008) Hydrogen production by high temperature electrolysis with nuclear reactor. Progress in Nuclear Energy. 50: 422-426.
[20] Lattin WC, Utgikar VP. (2007) Transition to hydrogen economy in the United States: A 2006 status report. International Journal of Hydrogen Energy. 32: 3230-3237.
[21] Kothari R, Buddhi D, Sawhney RL. (2008) Comparison of environmental and economic aspects of various hydrogen production methods. Renewable and Sustainable Energy Reviews. 12: 553-563.
[22] Vanderryn J, Salzano FJ, Bowman MG. (1977) International cooperation on development of hydrogen technologies. International Journal of Hydrogen Energy. 1: 357-363.

[23] Wang HZ, Leung DYC., Leung MKH, Ni M. (2009) A review on hydrogen production using aluminum and aluminum alloys. Renewable and Sustainable Energy Reviews. 13: 845-853.

[24] Franzoni F, Milani M, Montorsi L, Golovitchev V. (2010) Combined hydrogen production and power generation from aluminum combustion with water: Analysis of the concept. International Journal of Hydrogen Energy. 35: 1548-1559.

[25] Franzoni F, Mercati S, Milani M, Montorsi L. (2011) Operating maps of a combined hydrogen production and power generation system based on aluminum combustion with water. International Journal of Hydrogen Energy. 36: 2083-2816.

[26] Milani M, Montorsi L, Golovitchev V. (2008) Combined Hydrogen, Heat, Steam, and Power Generation System. Proceedings of the ISTVS-2008.

[27] International Aluminium Institute. http://www.world-aluminium.org/?pg=85. (2012 [cited 2012 April 20)

[28] Shkolnikov EI, Zhuk AZ, Vlaskin MS. (2011) Aluminum as energy carrier: Feasibility analysis and current technologies overview. Renewable and Sustainable Energy Reviews. 15: 4611-4623.

[29] Wang XW, Gao LR, Hua-Ben. (2008) Resources conservation-The alternative scenarios for Chinese aluminum industry. Resources, Conservation and Recycling. 52: 1216–1220.

[30] Schwarz HG, Briem S, Zapp P. (2001) Future carbon dioxide emissions in the global material flow of primary aluminium. Energy. 26: 775-795.

[31] Ilyukhina AV, Kravchenko OV, Bulychev BM, Shkolnikov EI. (2010) Mechanochemical activation of aluminum with gallams for hydrogen evolution from water. International Journal of Hydrogen Energy. 35: 1905-1910.

[32] Parmuzina AV, Kravchenko OV. (2008) Activation of aluminum metal to evolve hydrogen from water. International Journal of Hydrogen Energy. 33: 3073-3076.

[33] Kravchenko OV, Semenenko KM, Bulychev BM, Kalmykov KB. (2005) Activation of aluminum metal and its reaction with water. Journal of Alloys and Compounds. 397: 58-62.

[34] Shaytura NS, Laritchev MN, Laritcheva OO, Shkolnikov EI. (2010) Study of texture of hydroxides formed by aluminum oxidation with liquid water at various activation techniques. Current Applied Physics. 10: S66-S68.

[35] Czech E, Troczynski T. (2010) Hydrogen generation through massive corrosion of deformed aluminum in water. International Journal of Hydrogen Energy. 35: 1029-1037.

[36] Alinejad B, Mahmoodi K. (2009) A novel method for generating hydrogen by hydrolysis of highly activated aluminum nanoparticles in pure water. International Journal of Hydrogen Energy. 34: 7934-7938.

[37] Mahmoodi K, Alinejad B. (2010) Enhancement of hydrogen generation rate in reaction of aluminum with water. International Journal of Hydrogen Energy. 35: 5227-5232.

[38] Wang HW, Chung HW, Teng HT, Cao G. (2011) Generation of hydrogen from aluminum and water-Effect of metal oxide nanocrystals and water quality. International Journal of Hydrogen Energy. 36: 15136-15144.

[39] Huang XN, Lv CJ, Huang YX, Liu S, Wang C, Chen D. (2011) Effects of amalgam on hydrogen generation by hydrolysis of aluminum with water. International Journal of Hydrogen Energy. 36: 15119-15124.

[40] Deng ZY, Tang YB, Zhu LL, Sakka Y, Ye J. (2010) Effect of different modification agents on hydrogen-generation by the reaction of Al with water. International Journal of Hydrogen Energy. 35: 9561-9568.

[41] Fan MQ, Xu F, Sun LX. (2007) Studies on hydrogen generation characteristics of hydrolysis of the ball milling Al-based materials in pure water. International Journal of Hydrogen Energy. 32: 2809-2815.

[42] Parker JJ, Baldi AL, inventors; (2009) Composition and process for the displacement of hydrogen from water under standard temperature and pressure conditions. United States Patent Application US 2009/0280054 A1.

[43] Soler L, Macanás J, Muñoz M, Casado J. (2007) Aluminum and aluminum alloys as sources of hydrogen for fuel cell applications. Journal of Power Sources. 169: 144-149.

[44] Macanás J, Soler L, Candela AM, Muñoz M, Casado J. (2011) Hydrogen generation by aluminum corrosion in aqueous alkaline solutions of inorganic promoters: The AlHidrox process. Energy. 36: 2493-2501.

[45] Soler L, Candela AM, Macanás J, Muñoz M, Casado J. (2009) In situ generation of hydrogen from water by aluminum corrosion in solutions of sodium aluminate. Journal of Power Sources. 192: 21-26.

[46] Soler L, Candela AM, Macanás J, Muñoz M, Casado J. (2010) Hydrogen generation from water and aluminum promoted by sodium stannate. International Journal of Hydrogen Energy. 35: 1038-1048.

[47] Uehara K, Takeshita H, Kotaka H. (2002) Hydrogen gas generation in the wet cutting of aluminum and its alloys. Journal of Materials Processing Technology. 127: 174-177.

[48] Watanabe M. (2010) Chemical reactions in cracks of aluminum crystals: Generation of hydrogen from water. Journal of Physics and Chemistry of Solids. 71: 1251-1258.

[49] Russo MF, Li R, Mench M, Van Duin ACT. (2011) Molecular dynamic simulation of aluminum-water reactions using the ReaxFF reactive force field. International Journal of Hydrogen Energy. 36: 5828-5835.

[50] Ouyang LZ, Xu YJ, Dong HW, Sun LX, Zhu M. (2009) Production of hydrogen via hydrolysis of hydrides in Mg-La System. International Journal of Hydrogen Energy. 34: 9671-9676.

[51] Zou MS, Yang RJ, Guo XY, Huang HT, He JY, Zhang P. (2011) The preparation of Mg-based hydro-reactive materials and their reactive properties in seawater. International Journal of Hydrogen Energy. 36: 6478-6483.

[52] Fan MQ, Xu F, Sun LX, Zhao JN, Jiang T, Li WX. (2008) Hydrolysis of ball milling Al-Bi-hydride and Al-Bi-salt mixture for hydrogen generation. Journal of Alloys and Compounds. 460: 125-129.

[53] Hu H, Qiao M, Pei Y, Fan K, Li H, Zong B, Zhang X. (2003) Kinetics of hydrogen evolution in alkali leaching of rapidly quenched Ni-Al alloy. Applied Catalysis. 252: 173-183.

[54] Sabourin JL, Risha GA, Yetter RA, Son SF, Tappan BC. (2008) Combustion characteristics of nanoaluminum, liquid water, and hydrogen peroxide mixtures. Combustion and Flame. 154: 587-600.

[55] Shafirovich E, Diakov V, Varma A. (2007) Combustion-assisted hydrolysis of sodium borohydride for hydrogen generation. International Journal of Hydrogen Energy. 32: 207-211.

[56] Kotov YA. (2003) Electric explosion of wires as a method for preparation of nanopowders. Journal of Nanoparticle Research. 5: 539-550.

[57] Kwon YS, Jung YH, Yavorovsky NA, Illyn AP, Soon Kim J. (2001) Ultra-fine powder by wire explosion method. Scripta Materialia. 44: 2247-2251.

[58] Ivanov GV, Tepper F. (1997) "Activated" Aluminum as a Stored Energy Source for Propellants: Changes in propellants and combustion-100 years after novel, Kuo KK, ed. Begell House, inc. New York and Wallingford (UK): 636-645.

[59] Pivkina A, Ivanov D, Frolov Y, Mudretsova S, Nickolskaya A, Schoonman J. (2006) Plasma synthesized nano-aluminum powders structure, thermal properties and combustion behavior. Journal of Thermal Analysis and Calorimetry. 3: 733-738.

[60] Tillotson TM, Gash AE, Simpson RL, Hrubesh LW, Satcher JH, Poco JF. (2001) Nanostructured energetic materials using sol-gel methodologies. Journal of Non-crystalline Solids. 285: 338-345.

[61] Panda S, Pratsinis SE. (1995) Modeling the synthesis of aluminum particles by evaporation-condensation in an aerosol flow reactor. NanoStructured Materials. 5: 755-767.

[62] Chen L, Song WL, Guo LG, Xie CS. (2009) Thermal property and microstructure of Al nanopowders produced by two evaporation routes. Transactions of the Nonferrous Metals Society of China. 19: 187-191.

[63] Shafirovich E, Diakov V, Varma A. (2007) Combustion-assisted hydrolysis of sodium borohydride for hydrogen generation. International Journal of Hydrogen Energy. 32: 207-211.

[64] Olivares-Ramírez JM, Castellanos RH, Marroquín de Jesús Á, Borja-Arco E, Pless RC. (2008) Design and development of a refrigeration system energized with hydrogen produced from scrap aluminum. International Journal of Hydrogen Energy. 33: 2620-2626.

[65] Kundu A, Jang JH, Gil JH, Jung CR, Lee HR, Kim SH, Ku B, Oh YS.(2007) Micro-fuel cells-Current development and applications. Journal of Power Sources. 170: 67–78.

[66] Jung CR, Kundu A, Ku B, Gil JH, Lee HR, Jang JH. (2008) Hydrogen from aluminium in a flow reactor for fuel cell applications. Journal of Power Sources. 175: 490-494.

[67] Shkolnikov E, Vlaskin M, Iljukhin A, Zhuk A, Sheindlin A. (2008) 2 W power source based on air-hydrogen polymer electrolyte membrane fuel cells and water-aluminum hydrogen micro-generator. Journal of Power Sources. 185: 967-972.

[68] Wang ED, Shi PF, Du CY, Wang XR. (2008) A mini-type hydrogen generator from aluminum for proton exchange membrane fuel cells. Journal of Power Sources. 181: 144-148.

[69] Silva Martínez S, Albañil Sánchez L, Álvarez Gallegos AA, Sebastian PJ. (2007) Coupling a PEM fuel cell and the hydrogen generation from aluminum waste cans. International Journal of Hydrogen Energy. 32: 3159-3162.

[70] Fan MQ, Sun LX, Xu F. (2010) Experiment assessment of hydrogen production from activated aluminum alloys in portable generator for fuel cell applications. Energy. 35: 2922-2926.

[71] Vlaskin MS, Shkolnikov EI, Bersh AV, Zhuk AZ, Lisicyn AV, Sorokovikov AI, Pankina YV. (2011) An experimental aluminum-fueled power plant. Journal of Power Sources. 196: 8828-8835:

[72] Rosenband V, Gany A. (2010) Application of activated aluminum powder for generation of hydrogen from water. International Journal of Hydrogen Energy. 35: 10898-10904.

[73] Rosenband V, Gany A. (2010) Composition and methods for hydrogen generation. U. S. Patent Application US2010/0173225A1.

[74] Silva Martínez S, López Benites W, Álvarez Gallegos AA, Sebastián PJ. (2005) Recycling of aluminum to produce green energy. Solar Energy Materials & Solar Cells. 88: 237-243.

[75] Zhuk AZ, Sheindlin AE, Kleymenov BV, Shkolnikov EI, Lopatin MY. (2006) Use of low-cost aluminum in electric energy production. Journal of Power Sources. 157: 921-926.

[76] Davoodi A, Pan J, Leygraf C, Parvizi R. (2011) Minuscule device for hydrogen generation/electrical energy collection system on aluminum alloy surface. International Journal of Hydrogen Energy. 36: 2855-2859.

[77] Namba M, Ohnishi. (2003) Appropriate disposal and recycling of aluminum dross. Jounal of Japan Institute of Light Metallurgy. 53: 182–193.

[78] Hiraki T, Akiyama T. (2009) Exergetic life cycle assessment of new waste aluminium treatment system with co-production of pressurized hydrogen and aluminium hydroxide. International Journal of Hydrogen Energy. 34: 153-161.

[79] Houser CF, inventor; (1982 Apr. 23) Heating system. United Stated Patent 4,325,355.

[80] Houser CF, inventor; (1985 Sep. 24) Hydrogen generator. United States Patent 4,543,246.

[81] Andersen ER, Andersen EJ, inventors; (2003 Oct. 28) Method for producing hydrogen. United States patent 6,638,493 B2.

[82] Andersen ER, Andersen EJ, inventors; (2004 Oct. 5) Apparatus for producing hydrogen. United States Patent 6,800,258 B2.

[83] Troczynski T, inventor; (2007 Sep. 6) Power systems utilizing hydrolytically generated hydrogen. United States Patent Application US 2007/0207085 A1.

[84] Fullerton LW, inventor; (2009 Oct. 8) Aluminum-alkali hydroxide recyclable hydrogen generator. United States Patent Application US 2009/0252671 A1.

Chapter 3

Photofermentative Hydrogen Production in Outdoor Conditions

Dominic Deo Androga, Ebru Özgür, Inci Eroglu, Ufuk Gündüz and Meral Yücel

Additional information is available at the end of the chapter

1. Introduction

Today, we are consuming the solar energy accumulated on earth in million years as fossil fuels at a rate which is much faster than it is stored by photosynthesis. Alternative energy sources such as solar, wind, wave, geothermal and nuclear are today's need of carbon neutral technologies either as replacement of some of the existing ones or as producing new sources such as; biofuel, biogas and biohydrogen, to increase the energy supplement of the world. Besides the source, nowadays engineers are very much concerned about how to utilize these energies in a more efficient way. There is a solution for the future, a new energy carrier system that is hydrogen. Hydrogen can be produced from primary energy sources; it can be stored and directly converted to electricity in fuel cells efficiently when needed.

Hydrogen energy system is bio-analog strategy for the sustainable future. Photosynthesis is the most efficient way to store solar energy. Plants, algae and photosynthetic microorganisms have developed their energy transduction centers and they know how to do this energy transformation and storage. Man exploits photobiological and photobiomimetic production of hydrogen. Biological hydrogen production processes, namely biophotolysis, dark fermentation and photofermentation, offer the prospect of producing hydrogen from renewable sources. *Rhodobacter* species are photosynthetic PNSB that can produce hydrogen from small-chain organic acids derived from biomass at the expense of light energy.

Since fuel consumption rate is very high, biohydrogen needs to be produced at a much faster rate with new strategies basing on energy bionics. These processes should provide a net energy gain, be economically competitive, and be producible in large quantities without reducing the food supplies. Further research is needed in all fields to be competitive with conventional technologies. Hydrogen- powered fuel cell electric vehicle option is a clue for 21st century's people how to change their consumption habits for a sustainable future.

In this chapter we give the fundamentals of photobiological hydrogen production by PNSB and review the research published on photofermentative hydrogen production at outdoor conditions. We discuss the most critical factors in PBR design and compare the technological development, availability and economics of other biohydrogen production techniques with photofermentation.

2. Photobiological hydrogen production

Photobiological hydrogen production is a microbial process that requires light as energy source, an electron donating substrate, and a biological catalyst that generate H_2 by combining protons and electrons. Basically, a biocatalyst is used to convert light energy into H_2. The process can be categorized as oxygenic and non-oxygenic photobiological hydrogen production, depending on the formation of oxygen during the process (Table 1). Oxygenic photobiological hydrogen production is carried out by microalgae and cyanobacteria, which produce hydrogen during photoautotrophic growth using water as electron donor, CO_2 as carbon source and light as energy source. In microalgae, biohydrogen production, also termed direct biophotolysis, is catalyzed by [FeFe]-hydrogenase, which accepts electrons from ferrodoxin. Water is the proton and electron donor, which is split by PSII reaction center using light energy. Electrons travel through the Z-scheme (PSII-PQ-cyt b_6/f, PC and PSI, sequentially). Ferrodoxin is the final electron acceptor and it delivers electrons to [FeFe]-hydrogenase. The [FeFe]-hydrogenase combines protons and electrons to form molecular hydrogen. The hydrogenase enzyme is extremely sensitive to the presence of O_2, which limits industrial application of biohydrogen production using microalgae [1]. In order to circumvent the O_2 inhibition, Melis and co-workers [2] introduced 2-stage photobiological hydrogen production by *Chlamydomonas reinhardtii* by utilizing sulfur deprivation, which resulted in PSII inactivation, hence O_2 evolution. By this way, photosynthetic oxygen evolution and carbon fixation were temporally separated from consumption of cellular metabolites and H_2 evolution.

In cyanobacteria, biohydrogen production is catalyzed by nitrogenase enzyme. In filamentous cyanobacteria, the reaction proceeds in two spatially separated vegetative and heterocyst cells, hence, the process is called indirect biophotolysis. In vegetative (photosynthetic) growth mode, CO_2 is fixed into carbohydrates through photosynthesis using water as electron donor and light as energy source. Heterocyst cells are specialized in nitrogen fixation and contain nitrogenase enzyme. In heterocysts, under anoxic and nitrogen limited conditions, stored carbohydrates from vegetative cells are oxidized to form ATP. Electrons generated during oxidation reactions are delivered to PSI reaction center where they are excited by light and travel through several electron transfer proteins. The final electron acceptor, heterocyst-type Ferrodoxin, delivers the electrons to the nitrogenase enzyme that catalyzes H_2 formation from protons and electrons using ATP. In indirect biophotolysis, O_2 and H_2 production reactions are spatially separated, therefore eliminating O_2-induced nitrogenase repression [1]. Cyanobacteria also harbor bidirectional and uptake hydrogenases which function to maintain redox balance within the cell.

Non-oxygenic photobiological hydrogen production is carried out by PNSB, which produce hydrogen during photoheterotrophic growth on organic carbon sources utilizing energy from sunlight. Electrons generated during oxidation of substrates are converted to H_2 by nitrogenase. During photofermentative hydrogen production, O_2 is not produced and O_2-induced nitrogenase repression is not a concern. Photoheterotrophic H_2 evolution by PNSB is well characterized. Similar to cyanobacteria, PNSB also have uptake and bidirectional hydrogenases, which regulate H_2 cycling within the cell. The details of photofermentative hydrogen production by PNSB will be given in the following sections.

	Process	Organism	Enzyme	Reactions
Oxygenic	Direct Biophotolysis	Microalgae and Cyanobacteria	[FeFe] Hydrogenase	$2H_2O + Light \rightarrow 2H_2 + O_2$ *Hydrogenase reaction:* $2H^+ + 2e^- \rightarrow H_2$
	Indirect Biophotolysis	Filamentous Cyanobacteria	Nitrogenase	*In vegetative cells:* $6CO_2 + 6H_2O + Light \rightarrow C_6H_{12}O_6 + O_2$ *In heterocyst:* $C_6H_{12}O_6 + 6H_2O + Light \rightarrow 6CO_2 + 12H_2$ *Nitrogenase reaction:* $N_2 + 8H^+ + 8e^- + 16ATP \rightarrow 2NH_3 + H_2 + 16ADP + 16P_i$
Non-oxygenic	Photofermentation	Purple non-sulfur bacteria	Nitrogenase	$N_2 + 8H^+ + 8e^- + 16ATP \rightarrow 2NH_3 + H_2 + 16ADP + 16P_i$ *In the absence of N_2:* $2H^+ + 2e^- + 4ATP \rightarrow H_2 + 4ADP + 4P_i$

Table 1. Photobiological hydrogen production routes.

3. Photofermentative hydrogen production by purple non-sulfur bacteria (PNSB)

Photofermentative hydrogen production is a microbial process in which electrons and protons generated through oxidation of organic compounds are used to produce molecular hydrogen under anaerobic, nitrogen-limited conditions, utilizing light as energy source (Figure 1).

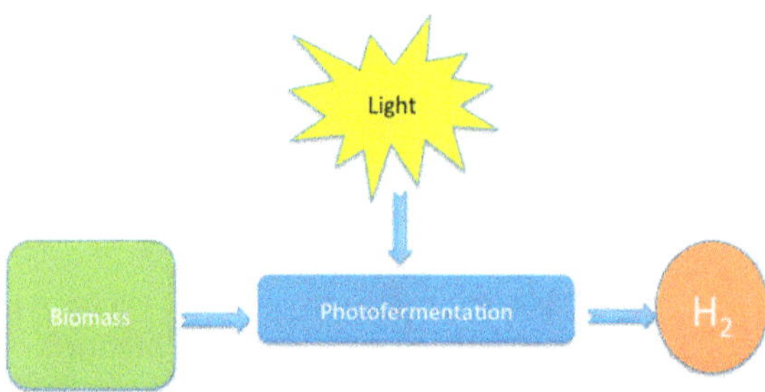

Figure 1. Photofermentative hydrogen production

The process is mainly mediated by nitrogenase enzyme, which catalyzes the reduction of N_2 to NH_3. Hydrogen production is an inherent activity of the nitrogenase enzyme, which forms 1 mole of H_2 per mole of N_2 fixed.

$$N_2 + 8H^+ + 8e^- 16ATP \rightarrow NH_3 + H_2 + 16ADP + 16P_i \quad (1)$$

However, under limited nitrogen source, the enzyme functions as hydrogenase and catalyzes the reduction of protons to form molecular hydrogen with the expense of 4 moles of ATP.

$$2H^+ + 2e^- + 4ATP \rightarrow H_2 + 4ADP + 4P_i \quad (2)$$

Hence, with the same energy requirement, 4 times more hydrogen can be produced under nitrogen-limiting conditions. There is also membrane-bound H_2-uptake [NiFe]-hydrogenase, which mainly catalyzes the oxidation of H_2 to protons and electrons by the following reversible reaction:

$$H_2 \rightarrow 2H^+ + 2e^- \quad (3)$$

A wide range of photosynthetic bacteria was reported to produce hydrogen. Among them, PNSB is the most widely studied and well characterized.

Purple non-sulfur bacteria (PNSB) are facultative anoxygenic phototrophs belonging to the class of *Alphaproteobacteria* and include several genera within order of *Rhodobacterales, Rhodospiralles and Rhizobiales* [3].They are a diverse group of photosynthetic microorganisms that are capable of photobiological hydrogen production under anaerobic, nitrogen limiting conditions. Various species of PNSB were utilized in hydrogen production studies, *Rhodobacter capsulatus, Rhodobacter sphaeroides, Rhodoseudomonas palustris* and *Rhodospirillum rubrum* being the most famous strains. They prefer photoheterotrophic growth in the presence of an organic carbon source, preferentially, small organic acids. Photoheterotrophic growth is the only growth mode that results in hydrogen production, however, PNSB are capable of growth under photoautotrophic, respiratory, fermentative or chemotrophic conditions, depending on the presence of light, type of carbon source and availability of O_2 (Table 2).

Growth Mode	C-source	Energy source	Notes
Photoheterotrophy	Organic carbon	Light	Only mode that results in H_2 production
Photoautotrophy	CO_2	Light	CO_2 fixation occurs. H_2 is used as electron donor
Aerobic respiration	Organic carbon	Organic carbon	O_2 is the terminal electron acceptor
Anaerobic respiration/chemoheterotrophy	Organic carbon	Organic carbon	Requires a terminal electron acceptor other than O_2 (N_2, H_2S or H_2)
Fermentation/anaerobic, dark	Organic carbon	Organic carbon	

Table 2. Various growth modes of PNSB [4].

This versatility of growth modes attracted research interest for many years, and made PNSB a model organism to study metabolic regulations of carbon, nitrogen and energy metabolism. There are three important external factors that determine the metabolic route: the carbon source, light and O_2 availability. PNSB are capable of growth on a variety of organic carbon sources including sugars (glucose, sucrose), short chain organic acids (acetate, malate, succinate, fumarate, formate, butyrate, propionate, lactate), amino acids, alcohols and even polyphenols. They also grow on inorganic carbon (CO_2) under photoautotrophy and chemoautotrophy. Under photoheterotrophic hydrogen production conditions, these bacteria preferentially use short chain organic acids as electron donors to obtain ATP for their metabolic processes. Short chain organic acids are assimilated through tricarboxylic acid cycle, which yields CO_2, protons and electrons, which are shuttled through electron transport chain that uses NAD/NADH and ferrodoxin.

Photosynthetic apparatus in PNSB is located in the intracytoplasmic membranes, the invaginations of cytoplasmic membrane forming a parallel lamella underlying the cytoplasmic membrane. It is composed of a photosystem, a series of electron transport proteins (cytoplasmic cytochrome *c*, lipid soluble quinones (Q/QH), cytochrome b/C_1 complex, and) and a transmembrane ATP synthase protein. The photosystem contains light harvesting complex 1 (LH1) and 2 (LH2) and a reaction center [5]. The LH complexes trap light in the visible (450-590 nm) and near infrared (800-875 nm) wavelength and transfer the excitation energy to the reaction center, and starts cyclic electron transfer. LH1, LH2 and reaction center are protein-pigment complexes that contain different types of carotenoids and bacteriochlorophyll *a*. Biosynthesis of photosynthetic apparatus is primarily controlled by the presence of O_2 and light [6,7,8]. During aerobic growth, the synthesis of bacteriochlorophyll is repressed. Once the O_2 tension is removed, the synthesis resumes. Light intensity and quality also controls the synthesis of photosynthetic apparatus. Under low light intensity, photosystem biosynthesis increases to gather more light energy, and at high light intensity, less photosystem is biosynthesized.

The photosystem of PNSB is not powerful enough to split water; hence, no O_2 evolves, which makes it very suitable for biohydrogen production. Electrons that are liberated through oxidation of organic carbon are funneled through a series of electron carriers, during which protons are pumped through the membrane. This leads to a development of a proton gradient across the membrane, which drives ATP production by ATP synthase. The electrons are either used for replenishment of quinone pool or donated to Ferrodoxin, which delivers electrons to nitrogenase enzyme to reduce molecular nitrogen to ammonia. When molecular nitrogen is not available, nitrogenase functions as hydrogenase and catalyzes the proton reduction with the electrons derived from ferrodoxin (Figure 2). By this way, electrons from organic compounds are stored in the form of H_2, by using light energy.

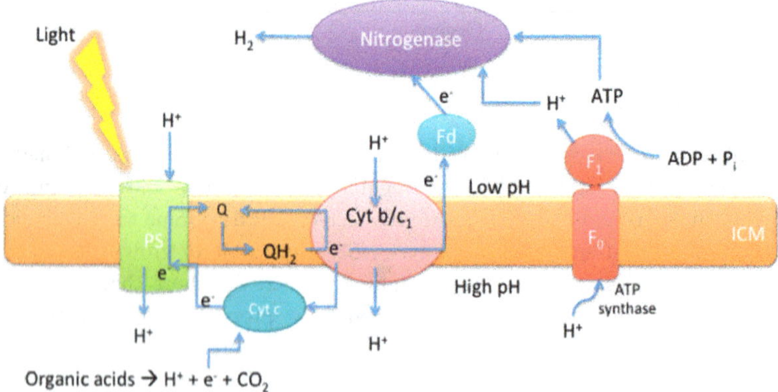

Figure 2. Photofermentative hydrogen production in PNSB. Oxidation of organic acids generates electrons, which are delivered to cytochrome *c* and travels through number electron transport proteins and delivered to ferredoxin. During this process, protons are pumped through the membranes forming a proton gradient. This proton motive force derives ATP production by ATP synthase. Ferredoxin delivers electrons to nitrogenase, which catalyzes the reduction of protons to molecular hydrogen using ATP.

The synthesized ATP is primarily used for biomass production. In order to produce H_2, ATP flux to the cell should surpass the amount of ATP necessary for growth. Bacteria produce hydrogen when there is an excess of reducing powers to maintain cellular redox balance. There are mainly three metabolic pathways that compete for electrons: CO_2 fixation, N_2 fixation/H_2 production and polyhydroxybutyrate (PHB) biosynthesis. PNSB use CO_2 as electron sink under photoheterotrophic conditions to get rid of excess reducing equivalents and maintain redox homeostasis. It uses Calvin-Benson-Bassham (CBB) pathway to fix CO_2 at the expense of ATP and NADH. The primary function of CBB pathway is to provide carbon for the cell under photoautotrophic growth on CO_2. However, under photoheterotrophic growth on organic carbon, it mainly functions for redox balancing [3,9,10]. The regulatory enzyme of the CO_2 fixation is ribulose-1,5-bisphosphate carboxylase/oxygenase (RuBisCO) which catalyzes the conversion of RuBP (ribulose-1,5-bisphosphate) into glyceraldehyde-3-phosphate. Genes involved in CO_2 fixation is located in *cbb* operon that is transcriptionally regulated by CbbR [11].

Another electron sink is the molecular nitrogen (N_2), which is fixed to NH_3 by nitrogenase enzyme at the expense of 16 moles of ATP. The primary function of nitrogenase is to fix N_2 to ammonia when the cells are grown on ammonia-free environment. However, nitrogenase also functions in redox balancing. Hydrogen production is an inherent activity of nitrogenase enzyme, which produces 1 mole of H_2 for 1 mole of N_2 fixed. When N_2 is not present, the enzyme functions as hydrogenase and catalyzes the proton reduction to form molecular hydrogen (H_2), with 4 times higher efficiency (4 ATP is utilized per mole of H_2 produced in the absence of N_2). This energy intensive process is under tight metabolic control that is regulated mainly by cellular nitrogen status. As discussed below, nitrogenase activity is tightly regulated by several environmental factors, including ammonia, O_2, and light. For this reason, hydrogen production studies are carried out under anaerobic, nitrogen-limited conditions in the presence of light.

Polyhydroxyalkanoates (PHAs) are the polymers of hydroxyalkanoates, which are accumulated as an energy storage material usually under the condition of limiting nutritional elements such as N, P, S, O, or Mg in the presence of excess carbon source [12]. Polyhydroxybutyrate (PHB) is the best-known representative. PHB synthesis and expenditure are closely connected with the energy requirements of the cell. Batch cultures of *R. palustris* growing photoheterotrophically on acetate with varying nitrogen sources and regimens of nitrogen supplementation, demonstrated that some competition for reducing equivalents exists between nitrogenase activity and PHB biosynthetic pathway [13]. Acetyl-CoA is the substrate for PHB biosynthesis in PNSB, hence, in cultures grown on acetate higher PHB accumulation was reported [14]. In *R. rubrum*, it is reported that S-deprivation caused inhibition of nitrogenase activity, including N_2 fixation and H_2 production, and a concomitant enhancement in PHB accumulation [15].

3.1. Enzymes involved

Nitrogenase: Nitrogenase is a metalloprotein complex that catalyzes the reaction of biological nitrogen fixation. At least three genetically distinct nitrogenase systems have been confirmed in PNSB, namely Nif, Vnf, and Anf, in which the active-site central metals are Mo, V, and Fe, respectively [16]. In general, the nitrogenase enzyme is composed of two oxygen labile and separable metalloproteins, dinitrogenase (component I; MoFe protein, VFe protein, FeFe protein) and dinitrogenase reductase (component II; Fe protein). Component I contain the active site for N_2 reduction, with a molecular weight of approximately 240 kDa and is composed of two heterodimers. Component II is a 60–70 kDa homodimer coupling ATP hydrolysis to inter-protein electron transfer. Mo-nitrogenase, coded by *nifHDK* genes, is the most widely distributed nitrogenase in PNSB, but many PNSB also contain alternative forms of nitrogenases. In *R. sphaeroides*, only Mo-nitrogenase is found, but in *R. capsulatus* and *R. rubrum* there is also Fe-only nitrogenase [17; 18]. *R. palustris* is known to contain all three forms of nitrogenases [19]. Alternative nitrogenases have been proposed to serve as a route for nitrogen fixation in situations where molybdenum is limited in the environment.

Due to highly endothermic nature of nitrogen fixation, bacteria developed a tight control of nitrogenase at both transcriptional and posttranslational level [20]. Availability of ammonium and cellular nitrogen status is the primary regulator of the nitrogenase synthesis and activity. Presence of high concentrations of ammonia inhibits nitrogenase activity and represses the expression of nitrogenase structural at transcriptional level in *R. sphaeroides* [21]. In addition, N_2 fixation is controlled by environmental factors, molybdenum, light, and oxygen.

Three levels of regulation in response to ammonium availability are proposed in *R. capsulatus* (i) transcriptional activation of the regulatory genes nifA1, nifA2 and anfA, (ii) posttranslational regulation of NifA and AnfA activity, and (iii) post-translational control of nitrogenase activity by reversible modification of NifH and AnfA. A diversity of regulatory proteins are involved in control of nitrogen fixation. Among these are two-component regulatory systems (NtrB/NtrC, RegB/RegA), two signal transduction proteins (GlnB, GlnK), three specific transcriptional activator proteins (NifA1, NifA2, AnfA), two molybdate-dependent repressor proteins (MopA, MopB), an ADPribosyl- transferase/glycohydrolase system (DraT, DraG), two (methyl)-ammonium transporter proteins (AmtB, AmtY), and a histone-like protein (HvrA) [20]. Nitrogen-fixing bacteria have been shown to regulate nitrogenase in the short term by post-translational covalent modification via reversible ADP-ribosylation of the Fe protein in response to different environmental stimuli: ammonium addition, darkness, and the absence of oxygen. This process is catalyzed by two non-*nif*-specific enzymes: dinitrogenase reductase ADP-ribosyltransferase (DRAT) and dinitrogenase reductase-activating glycohydrolase (DRAG). Another post-translational regulation that does not involve ADP-ribosylation was also proposed in *R. capsulatus* [22,23]. Due to the modulation of nitrogenase activity according to cellular nitrogen status, hydrogen production studies on PNSB are carried out on media with limited nitrogen source. Regulation of nitrogenase activity by posttranslational ADP-ribosylation has been shown to occur in response to light and O_2 status, as well. Rapid inhibition of nitrogenase activity by O_2 through ADP-ribosylation was reported in *R. capsulatus* [24].

Hydrogen production is an inherent activity of nitrogenase enzyme, producing one mole of H_2 per mole of N_2 fixed, at the expense of 16 moles of ATP. In the absence of nitrogenase, the enzyme acts as ATP-dependent hydrogenase and catalyzes the proton reduction for molecular hydrogen at the expense of 4 ATP. Nitrogenase mediated hydrogen production in PNSB also plays a role in maintaining cellular redox status.

Hydrogenase: Hydrogenases, the key enzymes of hydrogen metabolism, are metalloenzymes that catalyze either the oxidation of H_2 to form protons and reducing equivalents or, the reduction of protons to form molecular hydrogen. There are three distinct types of hydrogenases classified according the type of metal cofactor present in the active site: [FeFe]-hydrogenases, [NiFe] hydrogenases, and [Fe]-hydrogenase, the first two are divided into a variety of sub-types depending on the structures and functions in the cell [25]. [NiFe]-hydrogenases tend to be involved in H_2 consumption, while [FeFe]-hydrogenases are usually involved in H_2 production [1].

[FeFe]-hydrogenases are generally monomeric and consist of a catalytic subunit of ca. 45-48 kDa, which bidirectionally catalyze H_2 production [25]. The direction of the reaction is determined by the redox status of the cell. They are found in green algae and anaerobic prokaryotes, and characterized by high catalytic activity (turnover rates: 6000-9000 s^{-1}) [1]. However, they are particularly sensitive to O_2, which causes irreversible inactivation of the enzyme. Hence, anaerobiosis is a prerequisite for algal hydrogen production. PNSB bacteria do not contain [FeFe]-hydrogenase except *R. palustris*, which was reported to possess a [FeFe]-hydrogenase [26].

[NiFe]-hydrogenases are heterodimeric enzymes consisting of a large subunit (α-subunit) of ca. 60 kDa hosting the bimetallic active site and the small subunit (β-subunit) of ca. 30 kDa hosting the Fe-S clusters. [NiFe]-hydrogenases are present in cyanobacteria and PNSB. Membrane-bound uptake hydrogenases (i.e., HupSL and hynSL), hydrogen sensors (HupUV), NADP-reducing (HydDA), bidirectional NADP/NAD-reducing (hoxYH) and energy-converting membrane-associated H_2-evolving hydrogenases are the subgroups of [NiFe]-hydrogenases. Cyanobacterial H_2-uptake hydrogenases are cytoplasmic that are induced under N_2-fixing conditions. Cyanobacteria also contain bidirectional NAD(P)-linked [NiFe]-hydrogenase, which is responsible for catalyzing H_2 photoproduction in the absence of a functional nitrogenase. The bidirectional enzyme probably plays a role in fermentation and/or acts as an electron valve during photosynthesis [1,25,27].

Most PNSB harbor membrane-bound [NiFe]-hydrogenase, also called H_2 uptake hydrogenase (Hup) that catalyzes the oxidation of H_2 to protons and electrons. They are connected to the quinone pool of the respiratory chain in the membrane by a third subunit, which anchors the hydrogenase dimer to the membrane. Unlike those in cyanobacteria, H_2 uptake hydrogenases in PNSB are characterized by the presence of a long signal peptide at the N terminus of their small subunit. It serves as signal recognition to target the fully folded heterodimer to the membrane and the periplasm. In some PNSB (*R. capsulatus* and *R. palustris*) a cytoplasmic [NiFe]-hydrogenase is also present, which functions as H_2 sensors of the cell and trigger a cascade of cellular reactions controlling the synthesis of hydrogenases. In *R. rubrum*, there is also a CO-induced [NiFe]-hydrogenase (CooLH), which, together with CO-dehydrogenase, oxidizes CO to CO_2 with concomitant production of H_2. This allows *R. rubrum* to grow in the dark with CO as the sole energy source.

4. Photofermentation in outdoor conditions

4.1. Photobioreactors

Photobioreactors (PBRs) are systems designed to grow photosynthetic microorganisms under a given environmental condition [28]. They can be classified as open (raceway ponds, lagoons and lakes) or closed (flat plate, tubular) systems. Open systems are mostly suited to biomass production since they cannot provide the anaerobic conditions required for hydrogen production. Also, control of parameters like temperature, nutrients and pH is poor in such systems. On the other hand, closed systems allow better control of these parameters and result in higher biomass production and biohydrogen production [29].

Different types of PBRs are used in photofermentative hydrogen production studies. They are generally classified according to their: (i) Design - flat or tubular, horizontal, inclined, vertical or spiral and manifold or serpentine [30] (ii) Mode of operation batch, fed-batch and continuous [29]. In order to achieve sustainable photofermentative hydrogen production in outdoor conditions, the development of an optimized photobioreactor system that has the following properties is targeted: (i) a simply designed enclosed system that is impermeable to hydrogen (ii) a transparent system that allows maximum light penetration, preferably at high visible light or near red-infrared transmissions (iii) a system with high surface-to-volume ratio for better/wide distribution of light (iv) a system made from an unreactive material that is durable, easy to clean and sterilize [29,31-33]. Flat plate and tubular types of PBRs are commonly used in photofermentative hydrogen production (Figure 3). This is probably due to their high efficiencies brought about by their high illumination areas.

Flat-plate (panel) PBRs consist of frames placed in between two transparent rectangular plates (PMMA or glass plates). They generally have a depth of 1-5 cm and vary in height and width (smaller than 1 m in practice) [31]. These conventional panel PBRs are considered as the first generation plate-type bioreactors. The second generation is a flat panel airlift PBR made up of two deep-drawn plates glued together while the third generation comprises deep-drawn plates fused together under pressure and heat [29]. Panel PBRs have a short light path and facilitate the measurement of irradiance at the culture surface [28,3,34]. They can be placed vertically or tilted at optimal angles for maximum exposure to direct sunlight [35-37] and can be arranged in stacks close to each other, therefore providing a large illuminated area in a small ground area [38]. However, a drawback of this system is the lack of mixing. Mixing by aeration [31] or agitation via rocking motion [39] is suggested, but an impediment to these techniques is their high power consumption for pumping gas and shaking the photobioreactor. Also, mixing via aeration would lead to dilution of the gas produced and incur extra costs for gas separation. Recirculation of the evolved gas has been proposed [34], however, the PNSB have hydrogenase enzyme that can breakdown the produced hydrogen to protons and electrons, thus reducing the amount of gas produced. Moreover, reduction of hydrogen partial pressure by decreasing total gas pressure in the PBR headspace was shown to improve hydrogen production [40].

Tubular PBRs are made of long transparent tubes through which liquid culture is circulated using mechanical or gas-lift pumps. The tubes have diameters ranging between 3 and 6 cm and length between 10 and 100 m [31]. The PBRs fall under different categories: simple airlift or agitated bubble column (vertical type) [41-44], horizontal or nearly horizontal tubular PBRs [28,38,45] and helical type PBRs [46-48]. Tubular PBRs can be scaled-up by connecting a number of tubes to manifolds, but the length of the tubes is limited by the accumulation of gas [31]. A disadvantage of these PBRs is that they require large ground area. In comparing the hydrogen production performance of the panel and tubular PBRs, Gebicki et al. [39] reported that the ratio of the illuminated reactor surface to the installed ground area was 8:1 in the panel PBR, while that in the tubular PBR was 1:1.

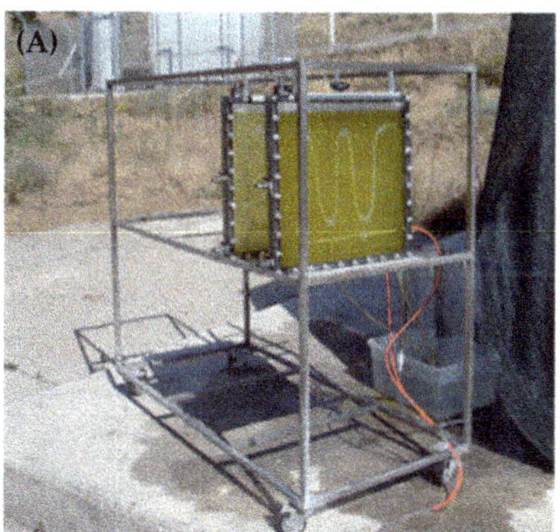

Figure 3. (A) Flat plate and (B) tubular photobioreactors.

4.2. Parameters affecting bacterial growth and hydrogen production in outdoor conditions

The ultimate goal of photobiological hydrogen production is to carry out the process in large scale PBRs operated at outdoor conditions, under natural sunlight. Solar light energy is warranted because it is a free resource that is abundant in nature. The earth receives about 5.7×10^{24} J of solar light energy per annum [49]. Its usage for photofermentative hydrogen production not only saves on operating costs but also supports the concept of sustainable hydrogen production from renewable resources. It promotes waste reduction and recycling [50].

Photofermentative hydrogen production in outdoors is affected by several conditions. The major parameters being physical variations like solar light energy and temperature, which are uncontrolled. These parameters regulate photosynthetic bacterial activity, therefore their daily (day/night cycle), seasonal and geographical variation greatly influence the amount of hydrogen produced [38,45,51-55]. In addition, other parameters such as PBR type, mode of operation, nutrients, carbon to nitrogen ratio, type and age of the microorganism are critical. Shown in Figure 4 are the general parameters influencing photofermentative hydrogen production.

4.2.1. Effects of the variation in solar light intensity and temperature on biomass and hydrogen production

The changing intensities of solar light energy and temperatures experienced in outdoor conditions greatly influence PNSB growth and hydrogen production. These enzymatic processes rely on chemical bond energy (ATP) [25] generated by the conversion of absorbed light energy to ATP as discussed in Section 3. This energy is utilized in different basic cellular metabolic activities such as biomass formation, biomass maintenance, hydrogen production and the excess is dissipated as heat [4,56]. Indoor studies have demonstrated that biomass increased with increasing light intensity [57] and temperature [58,59]. He et al. [58] investigated the growth and hydrogen production of two mutants of *Rhodobacter capsulatus* (JP91 and IR3) at different temperatures (26, 30 and 34 °C). They reported good cell growth and high substrate conversion efficiencies of 52.7% and 68.2% at 30°C for JP91 and IR3 strains, respectively. Likewise, in batch experiments using different light intensities (1.5-5 klux) and temperatures (20°C, 30°C and 38°C), Sevinç et al. [59] reported optimum light intensity for growth and hydrogen production to be 5 klux. The cell growth of the PNSB was found to fit the logistic model [4,58-60].

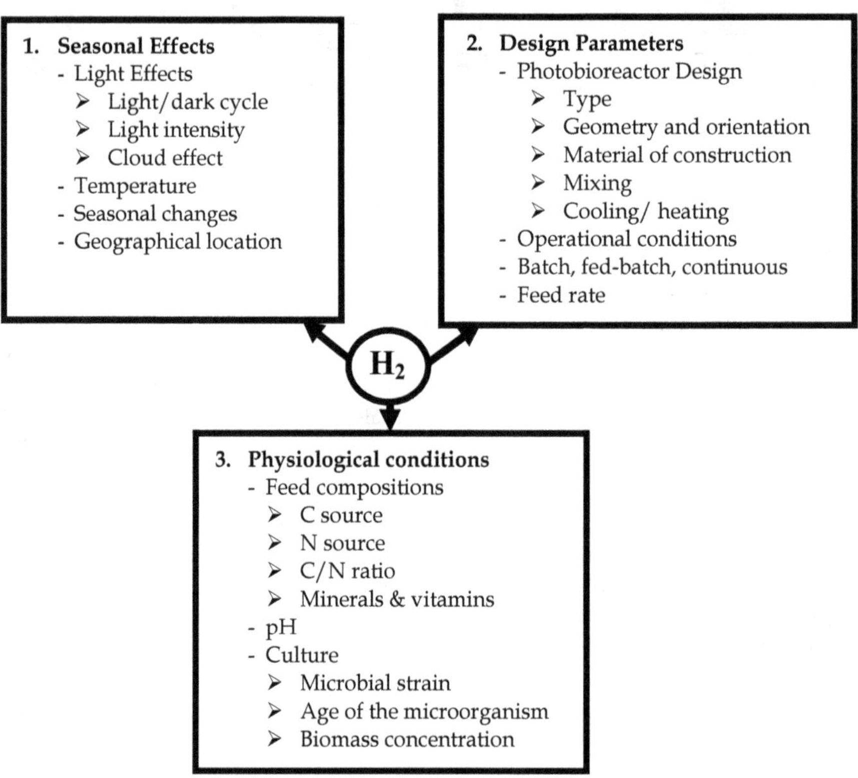

Figure 4. Parameters affecting photofermentative hydrogen production.

In outdoor operations, cell growth and hydrogen production were indicated to be dependent on the solar light intensity received [45,49,51-54,61]. Temperature and the intensity of solar radiation vary daily, seasonally and geographically. Northern Europe experiences lower temperatures and light intensities compared to Southern Europe; solar light intensities of up to 850-950 W/m^2 are reported to be common for most parts of Europe (40°-55°N) during summer [62]. The Mediterranean region, for example Turkey which lies between 36°-42°N latitudes receives abundant solar radiation (circa 3.6 kWh/m^2 day) [63].

Daily and seasonal variations in light intensity affect growth of PNSB. Eroglu et al. [51] observed that *R. sphaeroides* cell concentrations increased during the day but either remained the same or decreased at night. While investigating continuous hydrogen production using a fed-batch 90 L pilot tubular photobioreactor in outdoor conditions, Boran et al. [64] found that the specific growth rate of *R. capsulatus* increased exponentially with the total light intensity. Moreover, the growth rate of *R.capsulatus* YO3 (Hup^-) was found to be lower (0.0042 h^{-1}) during winter (low light intensity and temperatures) compared to that during summer (high light intensity and high temperature) (0.035 h^{-1}) [53].

Varying light intensities also influences the amount of hydrogen produced by the PNSB. Wakayama et al. [65] observed that the levels of hydrogen produced by batch cultures of *R. sphaeroides* depended on the irradiation intensity of sunlight. During the experiments, the light intensity and total irradiation (ranging between 6 to 7 kWh/m^2) fluctuated on a regular basis (up to 60%). Maximum hydrogen production rate of 4.0 L/m^2/h and light conversion efficiency of 2.2% was obtained. In another continuous hydrogen production study using an 8 L flat plate PBR, Androga et al. [53] observed that the daily amount of hydrogen produced decreased with the decreasing daily total global solar radiation (from 4000 Wh/m^2 to 2350 Wh/m^2). Avcioglu et al. [54] reported similar results using continuous cultures of *R. capsulatus*, fed with molasses dark fermenter effluent. Hydrogen yield factor was reported to linearly increase with increasing solar radiation [53,66].

4.2.2. *Effects of light/dark cycle*

Diurnal light/dark cycle adversely affects photofermentative hydrogen production [52,62,67]. During winter, shorter daylight periods (circa. 9 h) and shorter night periods are experienced while during summer, longer daylight periods (circa. 14 h) and shorter night periods are experienced [53]. Excessive light energy leads to photo-inhibition, which in turn reduces hydrogen production efficiency [62,67,68]. Sunlight intensity of over 1.0 kW/m^2 was found to be deterrent to hydrogen production [69]. Studies on light/dark cycle have demonstrated that little or no hydrogen was produced during the dark periods but bacteria survived and hydrogen production recovered once illumination resumed [4,40].

Wakayama et al. [67] observed that short intermittent light/dark cycles increased hydrogen production efficiency while longer intermittent periods reduced it. In experiments carried out under excessive light energy (1.2 kW/m^2), they found that a 30-min light/dark cycle improved efficiency to 150%, while a 12-h cycle reduced it to 73%, in comparison to the reference of continuous illumination. A 12-h light/12-h dark diurnal cycle yielded the same amount of biomass and volume of hydrogen as continuously illuminated bioreactors in batch studies using olive mill waste water, however, it resulted in a longer lag in biomass and hydrogen accumulation [70]. Under continuous illumination, maximum hydrogen production rate and substrate conversion efficiency of 92.41 ml H_2/L/h and 90.54% were obtained respectively. These values decreased to 89.96 ml H_2/L/h and 85.35% under 12-h light/12-h dark cycle and 86.91 ml H_2/L/h and 80.97% under 12-h dark/12-h light cycle, respectively [40]. Similar results were reported by Uyar et al. [62], who found that the average hydrogen production rate and the total hydrogen produced by *R. sphaeroides* cells exposed to light/dark cycles were lower compared to the continuously illuminated cultures. This could be attributed to lack of light energy to produce ATP (that is needed for hydrogen production) during the dark period [4]. Also, the metabolism of the bacterial cells may change to adapt to non-light conditions (fermentation) as observed by Eroglu et al. [51]. They reported that under limited sunlight intensity (<10 klux), no hydrogen production occurred, instead *R. sphaeroides* cells fed with malate performed fermentation, producing formate as the end product.

Fluctuating day and night temperatures have also been reported to significantly affect hydrogen production. In investigating the effect of temperature cycles and temperature plus light/dark cycle conditions on hydrogen production using batch cultures of *R. capsulatus* YO3 (Hup^-) grown outdoors, Özgür et al. [52] observed significantly lower substrate conversion efficiencies, yields and hydrogen productivities. The maximum hydrogen productivity and yield was obtained at reactor temperatures of 33°C. Light/dark cycle was reported to cause a further 50% decrease in hydrogen productivity.

4.3. Process technology for photobioreactors operation in outdoor conditions

4.3.1. Mode of operation: Batch, continuous and fed-batch systems

Photobioreactors used in biological hydrogen production can be classified depending on their mode of operation as batch, continuous and fed-batch. Batch PBRs have no flow of material in or out of the bioreactor and the reactions are time dependent. Continuous PBRs have both inflow and outflows (at the same time during operation) and operate at steady state (time independent). Fed-batch PBRs have either an inflow or outflow of material and the reactions are time dependent [33,71]. The mode of operation influences the growth of the microorganisms and hydrogen production rate and yield.

Batch PBRs are the most widely used bioreactors in photofermentative hydrogen production studies given in literature. The bacterial cells are left to grow, consuming the initially fed media, generating and accumulating products in a given time period. These systems are easy to operate, flexible and can be adapted to investigate various parameters. Most studies using batch reactors are carried out in small scale in the laboratory. However, they are usually liable to substrate and product inhibitions, , therefore resulting in low hydrogen production rates and yields [71]. There are a few outdoor batch studies reported in literature. Eroglu et al. [51] used a 6.5 L (working volume) temperature controlled flat plate solar photobioreactor to cultivate *R. sphaeroides* O.U 001 cells in batch mode using malate, lactate, acetate and olive mill waste water. They obtained the highest hydrogen production rate of 10 ml H_2/L/h using malate. Özgür et al. [53] investigated the effects of temperature fluctuations and day/night cycles on hydrogen production using *R. capsulatus* (a wild type and YO3 (Hup-) strain) grown in 550 ml glass bottle PBRs operated in batch mode under outdoor conditions. They found that temperature oscillations and day/night cycles greatly reduced hydrogen production in both strains and the YO3 (Hup-) strain performed better than the wild type.

PNSB have been shown to be able to produce hydrogen in the absence of growth. This avails the possibility of developing continuous photobioreactor systems that can produce hydrogen in long-term. Continuous systems operating under steady state conditions can generate products at constant rate, yield and quality, therefore are more advantageous than batch systems [71]. They may require less maintenance but long-term operation may cause contamination in bioreactors [33,72] and maintaining a stable cell concentration is challenging as the bacteria's cell growth and hydrogen production capability is highly susceptible to environmental (especially in outdoor conditions) and medium composition changes [32,53,73]. The hydraulic retention time (HRT) which is the average amount of time that a soluble compound spends in the reactor, is an important factor influencing the biomass, hydrogen production rate, hydrogen yield, and light conversion efficiency. Continuous PBRs can be operated as suspended cell processes or immobilized cell reactors. Suspended systems are prone to washouts and product inhibition, which could be overcome by cell recycle. Immobilized systems offer the advantages of cell longevity and have been shown to produce hydrogen at higher rates and yields [74,75]. Studies using different HRT values indicate that longer HRT is more suitable for photofermentation as the PNSB utilized the fed organic acids slowly [34,76-78]). It is postulated, especially for continuous systems, that HRT should be long enough to curtail cell growth so as to direct metabolic activity towards hydrogen production [33]. Chen et al. [99] carried out continuous fermentation at 96 h HRT employing *Rhodopseudomonas palustris* WP3-5, fed with dark fermenter effluent. The continuous culture ran stably for 10 days and produced an average yield of 10.21 mol H_2/mol sucrose. There are limited numbers of outdoor photofermentative hydrogen production studies using continuous systems reported in literature [80,81]. Most of the ones described were carried out in indoor conditions [34,46,56,82,83].

Fed-batch systems are one of the most promising processes for cell growth and metabolite production. They offer the prospects of having high cell densities and the feed flow rates and composition can be adjusted to control the reaction rates [84]. The substrates are added at controlled levels that adequately support cell growth, therefore alleviating substrate or product inhibition [71]. Repeated fed-batch operations, whereby products and parts of the settled bacteria are removed and replaced with fresh media (semi-continuous operation) to adjust the cell age and concentration of microorganisms, have been shown to be viable for hydrogen production [85]. Comparisons based on the effects of operation modes on biomass, substrate and product concentrations reveal fed-batch operation to be more promising for attaining high hydrogen rates and yields [71,79,85]. Controlled feeding and sequential product removal in fed-batch systems facilitate continuous photofermentative hydrogen production and allow feed media optimization. There are several outdoor fed-batch studies reported in literature [38,45,53-55,64,66,86].

4.3.2. *Photobioreactor positioning*

The performance of PBRs is highly dependent on the amount of sunlight received, therefore the location and orientation of the PBRs in outdoors is critical. An East-West positioning of

PBRs has been indicated to be the most suitable for receiving maximum sunlight energy and utilization [35,38,49,51,87]. PBRs facing East-West position were shown to receive higher amount of irradiance than South-North facing PBRs [87] and better utilization of long wavelength (red and infrared) that are prevalent in the mornings and evenings [72]. Orientation of the PBRs with inclinations of about 90° or less were demonstrated to be more suitable for cell growth and hydrogen production [35,38,49] than their counterparts without inclination. With inclination of flat plate PBRs towards the sun, major sunlight was received on the inclined surface of the PBR and the backside surface of the PBR was illuminated by diffuse and reflected light that may be good for photosynthesis. Vertical flat PBRs were suggested to be placed in a 30° and 60° inclinations for summer and winter operation [35]. In investigating methods of illumination to simulate the daily sunlight irradiation pattern, a Roux flask PBR with an irradiation area of 159 cm^2, working volume of 700 cm^3 and a light path of 4.5 cm was set at an angle of 30° from the horizontal. A solar tracking device was used to reposition the PBR to receive maximum sunlight every 30 minutes and maximum hydrogen productivity and maximum light conversion efficiency of 2.8 $L/m^2/h$ and 1.5% were obtained, respectively. Hydrogen production experiments using batch cultures of *R.sphaeroides* fed with different carbon sources (malate, lactate, and acetate and olive mill waste water) were carried out in flat plate PBR (8 L) inclined at 30°. Inclinations of 92°–93° were observed to provide the highest internal circulation and gas separation with a vertically plate photobioreactor [38]. In tubular PBRs, slight inclination (circa 10°) and pumping were applied to assist in gas separation [28,38,45]. Stacking of the flat pate PBRs next to each other provides a large illuminated area under small ground area. The packed arrangement caused a lamination effect where solar irradiance at the surface of the PBR was diluted therefore greatly improving the efficiency of conversion of solar radiance to biomass [88,89]. Temperature distribution experiments demonstrated that a 15 cm gap between the flat panel PBRs was optimal for hydrogen production. With the 15 cm spacing, the maximum temperature of the culture remained close to the physiological limit (31°C) at the occurrence of the highest daytime temperature [38].

4.3.3. Light distribution

Activity of the photosynthetic bacteria is dependent on the light distribution within the photobioreactor. Parameters such as the light path length [90,91], biomass concentration [92] and type and composition of the feed media [54,64] affect light distribution in the PBRs. Light intensity has been described to decay exponentially with the culture depth, following the Lambert-Beer law [38,93,94]. Nakada et al. [90] investigated the light penetration into a photobioreactor and its effect on hydrogen production using an A-type four-compartment bioreactor. They observed that 69% of the incident light energy was absorbed in the first compartment (0-5 mm) and 21% was absorbed in the second compartment (5-10 mm). Similarly, cells in the deeper parts of the bioreactor were demonstrated to be poorly illuminated as much of the incident light energy was absorbed as 69%, 21%, 7%, 2% in the first, second, third and fourth compartments of a 20 mm bioreactor [49]. The reduction of a flat panel bioreactor depth from 4 cm to 2 cm led to an increase in the overall light conversion efficiency from 0.13% to 0.53% [95]. Evaluations on the effect of bioreactor light path using a flat panel photobioreactor revealed that the highest hydrogen productivity was obtained at 20 mm depth with a mean biomass content of 0.7 g/L and illumination provided from both sides of the photobioreactor [38]. In comparing the percent penetration of light intensity in a photobioreactor, Boran et al. [96] observed that for each 1 cm of depth there was a decrease of 89 % for artificial medium, 70 % for thick juice DFE and 51 % for molasses dark fermenter effluent. Having a large tube diameter (6 cm) was concluded to have led to the decrease in light penetration in the tubular photobioreactor, thus decreasing hydrogen productivity. Decreasing the tube diameter to below 6 cm and increasing the wall thickness to prevent hydrogen loss was suggested to improve photofermentative hydrogen production in the pilot scale 90 L tubular photobioreactor [64].

Utilization of light energy by the bacterial cells is evaluated by the light conversion efficiency, which is the ratio of the total energy value of the hydrogen produced to the total energy input to the photobioreactor by light radiation as shown in Equation 4 [62].

$$\eta\ (\%) = (\text{Amount of } H_2 \text{ produced} \times H_2 \text{ energy content})/\text{Light energy input} \times 100$$

$$= (33.61 \times \rho_{H2} \times V_{H2})/\ (I \times A \times t) \times 100 \quad (4)$$

where 33.61 is the energy density of hydrogen gas in W.h/g, ρ_{H2} is the density of the produced hydrogen gas in g/L, V_{H2} is the volume of hydrogen gas produced in L, I is the light intensity in W/m^2, A is the irradiated area in m^2, and t is the duration of hydrogen production in hours.

Low light conversion efficiencies of 0.5 to 6% for solar and tungsten lamps, are considered to be the bottleneck in the scale-up of photofermentative hydrogen production systems. Generally, higher light conversion efficiencies were obtained at low light intensities [56]. Light conversion efficiency reduced from 1.11% to 0.25% as the light intensity increased from 88 to 405 W/m^2 [62]. In outdoor experiments using cultures of *R.sphaeroides* 8703, the solar light conversion efficiency was observed to decrease from 7% at low sunlight intensities (100 W/m^2) to 2% at high light intensities (1000 W/m^2). Average solar light conversion efficiencies of about 1% were obtained in photofermentative experiments carried out under natural sunlight [45,49].

Self-shading effect brought about by high biomass concentration is another important parameter affecting light distribution within the photobioreactor. Bacterial cells close to the illuminated surface prevent light from penetrating the PBR, therefore blocking the inner cells from receiving enough light energy. This reduces the PBR performance as it hinders cell growth and hydrogen production [92]. High sunlight intensities necessitated quick bacterial growth and higher biomass concentrations compared to low light intensities, which exhibited slow growth rate and lower biomass concentrations [53,54].

The use of supplementary light sources like LED or tungsten lamps at night to necessitate bacterial growth [45], use of optic fibers and solar tracking device to provide internal illumination at night and on cloudy days [49,79,97-99] and the use of covering material or optical fibers like Rhodamin B and $CuSO_4$ solution to reduce light intensity or block specific ranges of light wavelength [62,100,101] have been applied to improve light distribution in PBRs operated in outdoor conditions. Uyar et al. [62] suggested the use of artificial illumination in outdoor PBRs to necessitate growth during the night. Boran et al. [96] provided illumination using two 500 W halogen lamps during night to enable growth of bacterial cells and decrease lag time. Chen et al. [99] developed a solar-energy-excited optical fiber (SEEOF) PBR and observed improved hydrogen production using *Rhodopseudomonas palustris* WP3-5 cultures fed with acetate as sole carbon source. They showed that the provision of radiation using a combination of an optical fiber (excited by solar energy) and tungsten lamp improved hydrogen production and yield by 138% and 136%, respectively. Also combination of the optical fiber system with tungsten lamp and a light dependent resistor that monitored sunlight online and controlled irradiation intensity on the PBR, resulted in 27% increase in hydrogen production and yield. An experiment carried out using *R. sphaeroides* DSM 9483 cultures fed with lactate and grown in a column shaped 1.4 L bioreactor that was operated in outdoor conditions in the Sahara desert demonstrated that the use of fluorescent (laser) dye filters enhanced hydrogen production. Hydrogen production rate of 85 mL H_2/L/day was achieved. It was supposed that the laser dyes prevented the PNSB from being damaged by excessive sunlight and may have transformed the absorbed wavelengths into longer ones, which were more effective for photosynthesis [100]. In investigating the effect of wavelength on hydrogen production by *R.sphaeroides* O.U001, Rhodamin B and $CuSO_4$ solutions were used as optical filters. It was found that the blockage of infrared light negatively affected bacterial growth and reduced photofermentative hydrogen production by around 40% [62]. PNSB absorb light at near infrared (750-950 nm) for photofermentative hydrogen production [49]. Shading light bands were successfully employed to spatially disperse excessive light intensity and improve light conversion efficiencies of photosynthetic bacteria [101]. Experiments that were carried out in indoor and outdoor conditions using *R sphaeroides* RV strain resulted in an increase of 1.4 times (2.1%) and 1.3 times (1.4%) of light conversion efficiencies in indoor and outdoor conditions, respectively.

4.3.4. Temperature control

Most PNSB produce hydrogen optimally between 30 and 35°C [58,59,102]. They cannot grow or produce hydrogen above 38°C [59], except for a few such as *R. centenum*, which was able to grow optimally at 40-42°C but could not utilize substrate at high concentration, therefore limiting its usage in hydrogen production studies [103].

Fluctuating day and night temperatures significantly affect bacterial growth and hydrogen production in outdoors. During summer, the daytime outdoor temperatures fluctuate between 20 and 40°C, but during winter, it remains below 10°C [52,53]. During winter, electric heaters were used at night to maintain temperatures above freezing point. Slow bacterial growth rate attributed to low temperatures and low light intensities were reported in the experiments carried out a glasshouse during winter [45,53,54].

During the summer, due to the high temperatures experienced in outdoors (circa. 40°C), cooling of the PBRs is necessary to maintain temperatures at optimal PNSB growth and hydrogen production conditions. The major strategies that have been devised to control temperatures within the PBRs include: (i) external cooling by water spraying and shading [52] and submersion of the photobioreactor in a water bath/basin [86,104,105] (ii) internal cooling using internal coils [51-55,64,66].

Partial shading (60%) and cooling by water spraying were employed in hydrogen production studies carried out using batch cultures of *R. capsulatus* YO3. The cells were grown in a glass bottle photobioreactor operated in outdoor conditions. The PBR temperature was maintained at 33°C and a maximum hydrogen productivity and hydrogen yield of 0.63 $mgH_2/L/h$ and 0.045 $gH_2/g_{substrate}$ were obtained, respectively [52]. The temperature of a 0.8 L floating-type bioreactor was successfully controlled by submersion in seawater. When the atmospheric temperature rose to 36°C, the temperature of the seawater was 25°C and that of the reactor remained at 28°C [78]. Carlozzi and Sacchi [104] operated a temperature controlled underwater tubular PBR for 6 months in outdoor conditions to produce *Rhodopseudomonas palustris* strain 42OL biomass. They obtained an average productivity of 0.7 gram biomass dry weight per gram acetic acid. Adessi et al. [86] investigated hydrogen production using a 50 L tubular PBR submerged in a thermostated stainless steel water basin containing demineralized water set at 28±0.5°C. Maximum hydrogen production rate of 27.2 mL $H_2/L/h$ and substrate conversion efficiency of 49.7% were attained in the outdoor operated PBR. A glass tube internal coil measuring 1.70 m in length and 0.01m in internal diameter was used to cool a 6.5 L (working volume) flat panel PBR [51]. Flexible polyvinylchloride (PVC) cooling coils were integrated into flat panel PBRs and chilled water (5°C) was passed through to maintain the culture temperatures below 35°C [53,54]. PVC cooling coils were also utilized in a pilot-scale 90 L tubular PBR [64,66].

4.4. Comparison of photofermentative hydrogen productivities and yields obtained in outdoor conditions

The performances of outdoor operated PBRs are compared in Table 3. The parameters evaluated are hydrogen productivities, hydrogen yields (substrate conversion efficiencies) and light conversion efficiencies. Over the years, as advances in hydrogen production studies are made, more pilot studies are being carried out using PNSB and microalgae [38,45,64,66,86,105]. One of the highest hydrogen production rates was reported as 1.21 $mol/m^3/h$ by Adessi et al. [86] using a water basin cooled 50 L tubular PBR. The use of dark fermenter effluents of agricultural wastes such as molasses and sugar beet juice as feed media has also been shown to be viable in photofermentative hydrogen production [52,54,55,106]. In some cases it even led to better hydrogen production than the artificial feed media. This could be attributed to its multi-component nature in which the presence of extra nutrients enhanced hydrogen production and yield [54,55]. Different hydrogen productivities and hydrogen yields are reported (Table 3). Light conversion efficiencies not exceeding 1% are generally observed in outdoor studies. The differences in performances of the PBRs could be ascribed to the differences in: geometry and volume of the PBRs, type of microorganisms, mode of operation, nature of feed and composition, season and geographical location of the PBR operation.

Reactor type and volume	Mode of operation	Microorganism	C and N sources	H_2 Productivity	H_2 Yield	Substrate Conversion Efficiency (%)	Light Conversion Efficiency (%)	Ref.
Flat plate (33L)	Batch	*Rhodopseudomonas sphaeroides* B5/A	53mM lactate, 5 mM glutamate	0.13 mol $H_2/m^3/h$	-	69.1	-	[107]
	Batch	*Rhodopseudomonas sphaeroides* B5/B		0.18 mol $H_2/m^3/h$		78.1		
Flat plate (6L)	Semi-continuous	*Rhodopseudomonas sphaeroides* B6	60 mM lactate, 6 mM glutamate or and 150 mM lactate, 30 mM glutamate	0.84 mol $H_2/m^3/h$	-	62.5	-	[108]
Flat plate (4.4 L)	-	*Rhodobacter sphaeroides* RV	Lactate	0.56 mol $H_2/m^3/h$				[109]
Column (1.4L)	Batch	*Rhodobacter sphaeroides* DSM 9483	Lactate	0.16 mol $H_2/m^3/h$	-	-	-	[100]
Roux flask (0.7 L)	Batch	*Rhodobacter sphaeroides* RV	50 mM sodium lactate,10mM sodium glutamate	0.13 mol $H_2/m^2/h$	-	-	1.1	[49]
Helical tubular (4.35 L)	Batch	*Anabaena variabilis* PK84	2% Carbondioxide, air	0.82 mol $H_2/m^3/h$	-		0.14	[80]
Helical tubular (4.35 L)[a]	Chemostat			0.36 mol $H_2/m^3/h$	-		-	
Tubular (4.35 L)	Batch	*Anabaena variabilis* PK84	2% Carbondioxide, 98% air	0.31 mol $H_2/m^3/h$	-	-	0.38	[81]
	Continuous			0.50 mol $H_2/m^3/h$			0.57	
Flat plate (6.5 L)	Batch	*Rhodobacter sphaeroides* O.U.001(DSM 5864)	15 mM acetate, 2 mM glutamate	0.45 mol $H_2/m^3/h$	4.6 mol $H_2/mol_{substrate}$	-	-	[51]
			30 mM malate, 2 mM glutamate	0.01 mol $H_2/m^3/h$	0.6mol $H_2/mol_{substrate}$	-	-	
			30 mM acetate, 2 mM glutamate	0.36 mol $H_2/m^3/h$	1.2mol $H_2/mol_{substrate}$	-	-	
			20 mM lactate, 2 mM glutamate	0.09 mol $H_2/m^3/h$	0.8 mol $H_2/mol_{substrate}$	-	-	
			Olive mill waste water	0.13 mol $H_2/m^3/h$	-	-	-	
Glass bottle (0.55 L)	Batch	*Rhodobacter capsulatus* DSM 1710	30mM of acetate, 7.5mM of lactate,2 mM glutamate	0.14 mol $H_2/m^3/h$	0.027 $gH_2/g_{substrate}$	19	0.79	[52]
		Rhodobacter capsulatus YO3 (Hup-)		0.32 mol $H_2/m^3/h$	0.045 $gH_2/g_{substrate}$	33	2.41	

Tubular nearly horizontal[a,b,i] (80 L)	Fed-batch	*Rhodobacter capsulatus* DSM 1710	40 mM acetate, 2 mM glutamate	0.74 mol $H_2/m^3/h$	0.60 mol H_2/mol acetate	16	1	[45]
Flat plate (4×25 L)	Fed-batch	*Rhodobacter capsulatus* DSM 155	Acetate, lactate, glutamate	0.94 mol $H_2/m^3/h$	-	-	-	[110]
Tubular nearly horizontal[a] (65 L)	Fed-batch	*Rhodobacter capsulatus* DSM 155	Acetate, lactate, glutamate	0.74 mol $H_2/m^3/h$	-	-	-	
Flat plate[b] (8L)	Fed-batch	*Rhodobacter capsulatus* YO3 (Hup-)	40 mM acetate, (2-10) mM glutamate	0.30 mol $H_2/m^3/h$	-	44	-	[53]
Flat plate[c] (4L)	Fed-batch	*Rhodobacter capsulatus* YO3 (Hup-)	40 mM acetate, 4 mM glutamate	0.51 mol $H_2/m^3/h$	-	53	-	[53]
Flat plate (4 L)	Fed-batch	*Rhodobacter capsulatus* DSM 1710	molasses dark fermenter effluent	0.50 mol $H_2/m^3/h$	-	50		[54]
	Fed-batch	*Rhodobacter capsulatus* YO3 (Hup-)		0.67 mol $H_2/m^3/h$	-	78		
Tubular nearly horizontal (50 L)	-	*Rhodopseudomonas palustris* strain 42OL	malate, glutamate	1.21 mol $H_2/m^3/h$	-	49.7	0.92	[86]
Tubular nearly horizontal (90 L)	Fed-batch	*Rhodobacter capsulatus* YO3 (Hup-)	20 mM acetate, 2 mM glutamate	0.40 mol $H_2/m^3/h$	0.35 mol H_2/mol acetate	12	0.2	[64]
Tubular nearly horizontal (90 L)	Fed-batch	*Rhodobacter capsulatus* DSM 1710	Thick juice dark fermenter effluent	0.27 mol $H_2/m^3/h$	0.40 mol H_2/mol acetate fed	-	-	[66]
Flat plate (4 L)	Fed-batch	*Rhodobacter capsulatus* YO3 (Hup-)	Thick juice dark fermenter effluent	1.12 mol $H_2/m^3/h$	-	77	-	[55]
Tubular nearly horizontal (50 L)	-	*Chlamydomonas reinhardtii*	Tris–Acetate–Phoshpate medium (TAP)	0.03 mol $H_2/m^3/h$	-	-	-	[105]

[a] Continuous circulation,[b]operation during winter, [c]operation during summer, [i]llumination provided at night using artificial light.

Table 3. Comparison of photofermentative hydrogen production performance in photobioreactors operated at outdoor conditions.

4.5. Scale up

The eventual goal of photofermentative hydrogen production research is to produce hydrogen in large scale in outdoor conditions. For this to be realized, scale up of the hydrogen production systems, which have so far been used in small scale studies in the laboratory, need to be done. There are a few large-scale systems (pilot scale) studies reported in literature as shown in Table 3.

Scale up of the PBR systems faces several challenges. The first is related with the geometry of PBR. With scale up, the depth of the PBRs increases, this in turn increases the distance that light and evolved gas travels. Hence the possibility of hydrogen being used up by the microorganisms or diffusing through the PBR surface rises. Increasing the heights of the PBR may lead to build up of pressure at the bottom of the PBR, which may be detrimental to the bacterial cell growth. Scale up of flat plate PBR to 1 m in both height and width was reported suitable to reduce light deflection by the plates and allows gas-tightness of the enclosed volume without excessive pressure build [110,111]. Flat plate PBRs can be scaled up by cascading in stacks [38,53,54,88,89] while the size of tubular PBRs can be increased by connecting more tubes on the manifolds [28,45,64,66,86,105].

Another problem of scale up is that self-shading of cells becomes more prominent. The effect increases with increasing reactor size and cell concentration, negatively affecting cell growth and hydrogen production [92]. Due to the lack of sufficient light the bacterial growth rate is hampered and hydrogen production reduces as the organisms switch to alternative modes of growth [4,51]. The problem of self-shading can be alleviated by use of PBRs with larger illumination areas, use better light distribution methods such as integration of optic fibers within the PBR or genetically tailoring the photosynthetic apparatus of the photosynthetic bacteria [98,112]. These solutions may lead to the requirement of larger ground area, increasing the system costs and bring about ethical issues of using genetically modified microorganisms at industrial scale.

Light distribution in the scale up system could also be improved by providing artificial illumination at night or during cloudy days [62], however, extra expenses incurred will have to be considered.

Another difficulty of scale up is mixing. Mechanical mixing in large scale-systems is difficult because of the large surface to volume ratio in the reactors. Sparging of inert gas [31] or recycling of the produced gas through the culture [34] is preferred; however, the former leads to dilution of the total gas while the latter incurs extra operating costs due to pumping.

Sterilization of the PBRs and feed media is another concern of scale up. For large-scale application, it is suggested to operate the systems in non-sterile conditions. Constant hydrogen production was obtained for a period of almost two months using a semi-continuously operated PBR that was fed with non-sterile media [72]. Nonetheless, it is possible to operate large scale PBRs using sterilized PBR and media under optimal conditions, therefore reducing contamination risks [111]. However, further studies on sterilizing to avoid contamination in the scaled up PBR systems remain to be done.

5. Improvement of photofermentative hydrogen production

5.1. Photobioreactor design

5.1.1. Material of construction

PBRs can be constructed from a wide variety of materials such as; glass, low-density polyethylene film (LDPE), rigid acrylic or polymethyl methyl acrylate (PMMA), polycarbonate and transparent polyvinylchloride (PVC). Glass is considered a very good construction material because it is transparent, has low hydrogen permeability and a long lifespan (circa. 20 years). However, it is brittle, rigid, heavy, not easily workable and expensive, therefore not suitable for large scale systems [113]. PBRs constructed from glass have been reported in several studies [41,42].

LDPE is a flexible thin material that is mostly used in greenhouse covering. It has been applied in constructing tubular photobioreactors [38,45,64,66]. Its desirable properties include transmission of high visible and near infrared that is required by the PNSB for growth and hydrogen production, low UV transmission and low cost [114]. A major disadvantage is its thin wall thickness, which may lead to high hydrogen permeability [64]. Also, it has a short lifespan of about 3 years (maximum) [113].

PMMA is another suitable material for PBR construction. It is a highly transparent thermoplastic that is lighter, softer and easier to work with compared to glass. It transmits 92% of visible light (3 mm thickness) and filters ultraviolet (UV) light wavelengths below 300 nm while allowing infrared light of up to 2800 nm wavelength to pass. Also, it is weather resistant and can withstand outdoor conditions better compared to other plastics such as polycarbonate. However, it is brittle, inflexible and has higher hydrogen permeability compared to glass. A wall thickness of 4 mm minimum is needed to avoid leakage and cracking due to mechanical stress. It is stated to have a minimum lifespan of 10 years in outdoor operations [113]. PMMA has been used to construct flat plate [38,53-55] and tubular [36,115] PBRs.

Other materials such as polyethylene (PE), polypropylene (PP), polycarbonate (PC), polyvinyl chloride (PVC) have been used, especially in constructing tubular PBRs. However a major drawback in using them in outdoor conditions is that they can lose their transparency due to exposure to sunlight, thus are less durable [30].

Criteria in selecting the material of construction of PBRs for outdoor operations are such that they must: be transparent, be chemically unreactive and metal free, nontoxic, have low hydrogen and oxygen permeability, have high mechanical strength, have high durability and resistivity to sterilizing chemicals (i.e. hydrogen peroxide), be easy to clean and be available at low cost [29]. These factors affect the choice of material of construction and the overall system construction and sustainability costs.

5.1.2. Geometry, mixing and mode of operation

The geometry of the PBR impacts the illuminated surface area. Flat plate and tubular PBRs have been widely used in photofermentative hydrogen production studies because of their high illuminated surface area [33]. However, new PBR concepts with larger surface areas can be developed to improve hydrogen production. Some of the PBRs designed to enhance photofermentative hydrogen production studies include: helical tubular PBR [100], hollow channel plate PBR [72], a PBR with an integrated active gas separating membrane system [116], a multi-layered PBR [117] and a PBR with integrated solar excited optic fibers [79]. Moreover, in the design of the PBR geometry, PBR with a short light path is targeted to prevent exponential decay of light passing through the culture. A light path length of about 20 mm or less has been shown to be sufficient for the design of flat plate PBRs [38,49].

Mixing is another crucial parameter that needs to be considered in improving photofermentative hydrogen production. Good mixing facilitates the separation of evolved gas and assists in the homogenous distribution of cells, substrates and light within the PBR [118]. Stirring of the culture media using a magnetic stirrer was found to enhance hydrogen production and the amount of total gas produced. The experiment was carried out in a 400 ml water jacketed-glass column PBR using a combined system of *H. salinarum* packed cells with *R. sphaeroides* O.U.001 cells [42]. Shaking during the stationary phase of cell growth was shown to enhance hydrogen production more rather than mixing during growth (exponential) phase [40].However, vigorous mixing at high circulation rates may lead to cell damage and incur higher running costs. The use of mechanical agitators was depicted to be not suitable for mixing cultures in flat plate PBRs as their light path is short (narrow width). Sparging of inert gas [31], re-circulation of the evolved gas [34], and agitation through rocking motion [39] have been suggested. For tubular PBRs, mixing through continuous circulation of cell culture using a pump was stated to improve mass transfer between cells and necessitate easier hydrogen gas separation compared to intermittent circulation [64].

The choice of the mode of operation to be used in running outdoor PBRs for biohydrogen production is dependent on the production capacity, which in turn affects the capital investment and operating costs [33]. For outdoor operations, development of large scale continuous hydrogen production systems is targeted. Continuous and fed-batch systems are suitable for these processes as they enable long-term operations through regular feeding of the bacteria at certain dilution rates. However, fed-batch operation was described to be the most favorable mode of operation. It operates under high cell densities and allows the control of the reaction rates by adjustment of feed flow rates and compositions, thus preventing substrate and product inhibitions [71,84].

5.2. Immobilization studies

Hydrogen production can be improved by immobilizing cells within the PBR. Immobilized cultures have been shown to be more attractive for hydrogen production studies compared to suspended systems. The systems are easier to operate, the bioreactor effluent is cell-free and higher cell concentrations can be used, therefore enabling hydrogen production at higher rate and yields. Drawbacks of this system include substrates and products (H_2) diffusion limitation brought about by the high cell concentrations and difficulty in controlling parameters such as pH and hydrogen content. Continuous operation may be applied to curb this problem. Heterogeneity in parts of the PBR may exist, but this can also be overcome by growing the cells in matrix spaces [119].

Several techniques have been applied to immobilize the growing bacterial cells on the matrix spaces. They include adsorption, covalent bonding, cross-linking, entrapment and encapsulation [120]. Likewise, different materials have been used to immobilize the photosynthetic cells. Agar, alginate, carrageenan, cellulose and its derivatives, collagen, gelatin, epoxy resin, photo cross-linkable resins, polyacrylamide, polyester, polystyrene, polyurethane and porous glass are the most commonly used materials [74,82,120-122]. Gel entrapment was stated to be the best means of immobilizing bacterial cells [121,123].

Improvement of hydrogen production by immobilized systems has been reported in literature. Around 2-10 fold increase of hydrogen production rate was reported by immobilization [124,125]. A 4-fold increase in hydrogen production was observed in immobilized cells compared to suspended cells [126]. Entrapment of whole cells of *R. sphaeroides* inside reverse micelles led to 25-35 fold increase in hydrogen production [127]. *Rhodobacter sphaeroides* GL-1 was immobilized in polyurethane foam in continuously operated PBR. Hydrogen productivity and substrate conversion efficiency of 0.21 ml/h/ml foam-matrix and 86% were obtained, respectively [128]. Hydrogen production by immobilized *R.capsulatus* DSM1710 and *R.capsulatus* YO3 (Hup-) were investigated in agar immobilized systems. The optimization studies showed that artificial feed media containing 60 mM acetate (carbon source) and 4 mM glutamate (nitrogen source) produced the highest hydrogen yield of around 90-95%. Long-term hydrogen production of 67-82 days and 69-72 were also achieved in the *R. capsulatus* DSM 1710 and *R.capsulatus* YO3 (Hup-), respectively [120].

Moreover, immobilization may provide protection of bacteria cells against inhibitory effects of compounds like ammonium. In examining the suitability of using tofu wastewater for hydrogen production, it was found that agar protected the immobilized *R. sphaeroides* cells from the effect of ammonium [129]. Similar results were obtained in experiments carried out using *R. capsulatus* immobilized in agar gels and fed with artificial media containing different concentrations of ammonium chloride (2.5, 5 and 7.5 mM) [120].

5.3. Feed media: C/N and minor nutrients

PNSB are able to utilize a wide range of organic compounds as substrates for growth and hydrogen production [130]. They can use different C sources, preferably, volatile fatty acids (VFA) such as acetic, butyric, lactic and malic acid for hydrogen production and nitrogen sources such as glutamate and ammonium for growth [131]. Moreover, they can also use sugar containing wastes derived from various industries such as the tofu industry wastewater [129], olive mill wastewater [132], sugar refinery wastewater [133], dairy wastewater [134], palm oil mill wastewater [135] and ground wheat starch [136]. Effluents from dark fermentation have also been applied in several studies [54,55,66,106,137-139]. Studies using mixtures of VFAs demonstrated that the PNSB consume them at different sequences. *Rhodobacter sphaeroides* O.U.001 was found to initially deplete acetate, then propionate and butyrate [138]. *R.capsulatus* was found to initially consume lactic acid then acetic acid in a study using a mixture of the organic acids [59]. Ammonium ions have been reported to inhibit hydrogen production [21,140,141] and glutamate has been widely used for growth and hydrogen production by the PNSB [4,73,131].

The C/N ratio in the feed media is another crucial parameter affecting the growth, hydrogen productivity and yield. A low C/N ratio was reported to enhance microbial growth but decrease hydrogen production [41,46,73,79]. The rise in biomass concentration reduces light penetration into the PBR, thus decreasing hydrogen productivity. A high C/N ratio also increases biomass concentration and reduces hydrogen production [73,79,142]. Stable biomass concentration of 0.40 g dry cell weight per liter culture (gDCW/L) and maximum hydrogen productivity of 0.66 mmol hydrogen per liter culture per hour (mmol/L_c/h) were obtained with media containing 40 mM acetate and 4 mM glutamate (C/N = 25) for a period of over 20 days using fed-batch cultures of *R.capsulatus* YO3 (Hup-) [73]. The addition of minor nutrients, such as Fe and Mo which are co-factors to the nitrogenase enzyme, have been reported to enhance hydrogen production [137,138].

5.4. pH

The pH of the culture regulates growth and hydrogen production by the photosynthetic bacteria. PNSB such as *R. capsulatus* are reported to optimally grow between pH 6-9 [102]. Good cell growth and biomass yield is obtained between pH 6.0-7.0, but it decreases with further increase in pH. In investigating the effects of initial pH on photofermentation using cultures of *R. sphaeroides* O.U. 001, Nath and Das [143] obtained high biomass yield and maximum cumulative hydrogen production at pH 7.0. Biomass decreased as pH increased from 7.0 to 8.0. A drop in pH caused by the accumulation of VFAs during photofermentation of glucose by *R.capsulatus* JP91 was observed to reduce hydrogen and cell growth [144]. During exponential growth phase, high hydrogen production rates and yields were observed between the pH range 6.8-7.5 [73] and no hydrogen production was reported above pH 9.0 [141]. It is postulated that high pH values could have prevented the PNSB cells from maintaining their membrane potential, therefore affecting cellular metabolism and eventually hindering cell growth and hydrogen production [139].

5.5. Genetic modifications

Genetic engineering is a promising tool to increase the yield and productivity of photofermentative hydrogen production. Considering the H_2 metabolism of PNSB, genetic modifications can be done to (i) inhibit H_2 utilization, (ii) optimize the flow of reducing equivalents to nitrogenase by inhibition of PHB and CO_2 fixation, (iii) eliminate/decrease the effect environmental factors (NH_3, O_2, light, temperature), (iv) reduce the size of antenna pigments to increase light utilization efficiency.

Deletion of membrane-bound H_2-uptake hydrogenase (Hup-) gene of PNSB has been the major strategy to increase hydrogen production. In Rhodobacter sphaeroides KD131 deletion of Hup and PHB synthase genes resulted in an increase in H_2 production from 1.32

to 3.34 ml H_2/mg-dry cell weight, compared to the wild type strain [145]. Hup deletion in *R. rubrum* resulted in higher hydrogen production, however the rate of hydrogen production was not affected indicating that the capacity to recycle H_2 was not completely lost [146]. In *R. sphaeroides* O.U.001, HupSL deletion resulted in an increase in hydrogen production from 1.97 L H_2/L_c to 2.42 L H_2/L_c on malate containing culture, while on acetate both wild type and mutant strain produced hydrogen poorly [147]. In *R. capsulatus* MT1131, Hup deletion resulted in around 30% increase in H_2 production [148] in artificially illuminated batch cultures in indoor conditions, on malate as carbon source. The Hup- *R. capsulatus* MT1131 also showed enhanced photobiological H_2 production rates and yields in continuously operated outdoor photobioreactors on acetate as carbon source [53,64,66]. The strain also resulted in very promising H_2 production activities in studies carried out on real dark fermentation effluents in indoor batch studies [137,139] and in outdoor continuous PBR operations [54,55].

To optimize the flow of reducing equivalents to nitrogenase, genetic modifications were carried out targeting CO_2 fixation and PHB synthesis pathways, which compete for reducing equivalent. Spontaneous variants of *R. capsulatus* strains deficient in the CBB pathway have been shown to express nitrogenase structural genes to dissipate excess reducing equivalents, even in the presence of high concentrations of ammonia that is sufficient to repress nitrogenase expression in wild type [149]. In wild type and Hup- strains of *R. capsulatus* MT1131, inactivation of CO2 fixation pathway resulted in improvements in the yield and productivity of hydrogen production [150]. In *R. sphaeroides* KD131, inactivation of PHB synthase resulted in 2 fold increase in hydrogen production on acetate and butyrate, in spite of depressed cellular growth and lower substrate utilization.

Removal of NH_3 inhibition on nitrogenase activity is important especially for integrated dark and photofermentation studies, in which the dark fermenter effluent is usually rich in NH_3. To develop mutants with ammonia insensitive-nitrogenase activity, Pekgöz et al [140], deleted genes expressing two regulatory proteins of ammonia-dependent nitrogenase regulation, GlnB and GlnK in *R. capsulatus* DSM1710. However, glnB mutants showed lower hydrogen production, while glnK mutants were unviable. This observation suggests that GlnB/GlnK two component regulatory system most probably have role in other metabolic pathways, as well.

Increasing the light utilization efficiency of PNSB by reducing the size and quantity of photosynthetic pigments were addressed in *R. sphaeroides* RV. The mutant, which had lower LH2 content produced 50% more hydrogen compared to the wild type in plate-type photobioreactor [112].

Another strategy could be the development of recombinant PNSB, which express hydrogen-evolving hydrogenase. Kim et al. [151] developed a recombinant *R. sphaeroides* KCTC 12085 strain that harbor, with all the accessory genes necessary, Formate hydrogen lyase and Fe-only hydrogenase from *R. rubrum*, to enable dark fermentative hydrogen production from *R. sphaeroides*. The strain produced hydrogen during dark fermentative growth, and photofermentative hydrogen production increased by 2 fold.

6. Techno-economics of photofermentative hydrogen production

Biological hydrogen production through photofermentation is a sustainable way of producing hydrogen since it utilizes renewable resources (sunlight, water and biomass) and occurs under ambient conditions. Recently, a 6th framework EU integrated project, Hyvolution, entitled " Non-thermal production of pure hydrogen from biomass" was completed where the aim was to produce hydrogen from biomass with integration of dark and photofermentation. Effluents obtained from dark fermentation were used to produce H_2 by photofermentation [152]. These effluents contained both acetic acid and lactic acid as carbon sources that could be utilized in a consecutive photofermentation step. *Rhodobacter capsulatus* was the selected PNSB in the photofermentation stage since it produced hydrogen most effectively by breaking down organic acids such as acetic acid and lactic acid under anaerobic conditions and illumination [139].

The HYVOLUTION plant was assumed to be in operation 8000 hours per year producing 60 kg H_2/h (2MW thermal power). The PBR was assumedly in full operation during 10 hours per day, resulting in roughly 3330 hours of operation annually. The economic analysis included the capital cost for all four process steps, i.e. pretreatment, thermophilic fermentation, photofermentation and gas up-grading [153]. Four feedstocks (thick juice, molasses, potato steam peels and barley straw) were considered in the HYVOLUTION Plant. Aspen Plus was used to calculate mass and energy balances taking into account the integration of the processes. A net energy production, in form of hydrogen, showed that the production of hydrogen as an energy carrier was technically feasible with all the considered feedstocks. Thick juice had the lowest energy demand, but the other options required 20% more heat demand. This demonstrated that second generation biomass could compete with food biomass for the hydrogen production. Further investigations towards scale up and improved mass and energy balances (heat integration studies) for the various feedstocks would enable the betterment and selection of routes for the HYVOLUTION process [154].

From a costing perspective, the photofermenter was found to be bottleneck in the HYVOLUTION project. The final cost of photofermentative hydrogen production was estimated to be around 55-60 €/kg using a tubular photobioreactor and 385-390 €/kg in the case of a panel photobioreactor. These high production costs are mainly caused by the materials of construction of the PBRs; plastic and PMAA, at the current state-of-the-art. The total capital cost of the photofermenter was large, around €90 million and €320 million for the tubular and flat panel reactor, respectively. The cost of land was not considered; the ground area demand of the tubular and flat panel reactor was about 2.0 and 1.3 million square meters, respectively. Although photofermentation was the most expensive part of the HYVOLUTION process, other process steps have to be improved. To meet the proposed hydrogen cost of €10 /GJ, which corresponds to €1.21 /kg H_2 (based on the lower heating value, LHV) a maximum allowed capital investment of €5.3 million is necessary, neglecting all other costs (feedstock, labor, etc.). Presently the capital cost, excluding the photofermenter, is €24.6 million for the thick juice case and the capital costs of the tubular and flat panel photofermenter are €91 million and 332 million respectively. Clearly, vast improvements are required starting with the utilization of co-products. Forecasts for the improvement of the performance of the photo-fermentation promise a reduction of the cost to 20 €/kg hydrogen.

HYVOLUTION has also been compared, in terms of €/GJ, with the costs for a bioethanol production plant equipped with a biogas installation for the utilization of the residues and pentose sugars. With barley straw as feedstock, energy from state-of-the-art HYVOLUTION is 452 €/GJ and 18 €/GJ for ethanol with biogas. The estimation for HYVOLUTION after 6 years and at small scale is a decrease to 153 €/GJ. Even though these costs are still significantly higher than for ethanol, hydrogen production may be supplementary to bioethanol production when feedstocks with high moisture content are considered. In this case, downstream processing of ethanol will not be economical due to prohibitive costs for distillation leaving room for hydrogen to add to the future biofuel mixture. However, the studies carried out in a two-step process, dark fermentation followed by photofermentation (FP6 HYVOLUTION Project [152]), revealed that photofermentative hydrogen production cannot be competitive with current productivities and yields. The current estimated costs for hydrogen from HYVOLUTION are higher than anticipated at the start of the project, mainly but not solely, to the costs of the PBR.

Currently, the long-term scenario (2030) predicts a cost of 6 €/kg H_2 provided that an overall yield of 85% and productivity of 53 and 3.3 mmol H_2/L/h for thermophilic and photofermentation, respectively [106,152]. In order to obtain long-term operation, the reliability and the durability of the tubes should be increased by increasing the wall thickness of the tubes [64]. The tube diameter and wall thickness should be optimized for better light exposure of the cells. Decreasing the tube diameter can increase hydrogen productivity. However, feasibility of this approach should be investigated in terms of circulation energy [113] and land area requirements. Circulation should be continuous in order to increase the mass transfer between the cell (solid), liquid and the gas phases. This will also reduce the gas diffusion from the LDPE tubes.

7. Future prospects: Integration with other hydrogen production methods or alternative energy sources

To estimate the productivity target for photofermentation, one may consider data from photovoltaic and biogas as benchmark technologies. The output of the primary product (electricity, biomass) is in both cases directly linked to the available ground area. In using photovoltaic as a benchmark, one has to consider that electricity is obtained as the primary product. According to various calculations in the literature, the obtained electrical power per m^2 depends strongly on the location. The yearly global irradiation on optimally oriented PV-modules is ranging from ca. 1000 kWh/m^2.a in e.g. Germany and more than 2000 kWh/m^2.a in certain areas of Greece, Turkey, Italy and Spain. Calculations for Germany (which can be extrapolated to other regions like The Netherlands) results in an energy harvest of between 75 and 125 kWh/m^2.a (assuming a system efficiency of 75%) which corresponds to ca. 9-16 W/m^2 average (over the year) for Germany. For Nordrhein-Westfalen an energy harvest of ca. 80 kWh/m^2.a can be found in the literature and is used for any further calculations [155]. The calculation for a more southern country in Europe (ca. 150 kWh/m^2.a) results in a produced electrical power of 18 W/m^2. Converting finally the PV - electricity to hydrogen by electrolysis, one may obtain for the PV-system ca. 6.0 W H_2/m^2 (Nordrhein-Westfalen) and 11.3 W H_2/m^2 (Greece) assuming an efficiency of the electrolysis of 60% which corresponds to an hydrogen productivity between 180 mg/m^2.h (Nordrhein-Westfalen) and 338 mg/m^2.h (Greece) (HyLog – Project, E. Wahlmüller, Fronius Ltd.). Taking as another benchmark system the production of biogas from energy plants like maize, a production of 2.5 kW electricity out of 1 ha farm land (using a CHP with an electrical efficiency of ca. 39%) is typical. If the same amount of biogas is not converted to electricity but to hydrogen using methane steam reforming a final power of 0.54 W/m^2 can be obtained (assuming an efficiency of the reformer of 85% and neglecting the power demand of the reforming plant) which corresponds to a hydrogen productivity of 16.4 mg/m^2.h.

The technical and economic feasibility of dark fermentation followed by methane production via anaerobic digestion step have been investigated employing three base cases reflecting the different strategies that can be used when performing dark fermentation: high productivity, high yield, and low productivity-low yield. The production of pure methane was included as a reference case to investigate how the production of hydrogen affected the production cost. The cost estimates ranged from 50 to 340 €/GJ for the three base cases and the reference case for the process alternatives investigated. The capital costs and the nutrients used in the two biological steps were the main contributors to the cost in all base cases and the reference case. Utilization of the mixed biofuel (methane and hydrogen) may increase efficiency and lower environmental impact in terms of lower emissions [156].

Different hydrogen production technologies were evaluated based on renewable raw materials and/or renewable energy such as; alkaline electrolysis, steam reforming of both biogas and gasification gas, the coupled dark and photo fermentation as well as the coupled dark and biogas fermentation [157]. Each technology was investigated with different plant layouts and/or different raw materials. All examined technologies were designed to produce hydrogen in a quality suitable for the use in mobile fuel cells. The reforming of biogas gave good results regarding both hydrogen production and energy efficiency provided that the proper raw material was chosen. The reforming of gasification gas showed good production efficiencies but in contrast the energy efficiencies were low compared to the reforming of biogas. The production efficiency results of the coupled dark and photo fermentation were comparable to those of the reforming of biogas but their energy efficiencies were lower. However, since this technology is in an early stage of development it still has potential for development and might be a real alternative to reforming of biogas. Finally, the best results for the coupled dark and biogas fermentation regarding both hydrogen production and energy efficiency were obtained for the layout with on-site steam reforming of the produced biogas, showing efficiencies comparable to the dark and photo fermentation. The choice of the proper technology will have to be based on the availability of raw materials, since the kind of raw material had a strong influence on the performance of the technology, on the competition between food and energy production and on the development of raw material prices.

8. Conclusions

Scientific, project and market strategy is essential to lead in developing biological hydrogen production processes. Plans for future research should be made based on the knowledge and experience gained and techniques developed until now. Although considerable progress has been made, still many basic research questions remain unanswered and new ones were created. Therefore, more fundamental research is necessary besides the need to test several developed techniques at lab- and pilot scale, both as standalone and combined.

Outdoor production of hydrogen with photosynthetic bacteria is strongly affected by fluctuations in temperature and light intensity due to the day-night cycle and due to seasonal, geographic and climatic conditions. In order to forecast the hydrogen productivity at different places throughout the world and based on that to estimate the cost-effectiveness for a certain location, a model describing the dependency of hydrogen production from the natural parameters is necessary.

Operation of photobioreactor in outdoor is an energy requiring process due to the need for temperature control and recirculation. Exploitation of other renewable energy sources (sunlight, wind, geothermal energy, etc.) to supply energy to the PBR to be used for recirculation or for temperature control can be explored and implemented in the design of a biohydrogen plant. Heat economizing is necessary for the plant with the integration of cooling and heating streams in a heat exchange network. The light and substrate conversion efficiencies are low due to problems experienced in the bioreactor. Since the reactor design and materials used are still in research state, it would not be so useful to comment on a factor improvement for economical application of photofermentative hydrogen production by photobioreactors.

Nomenclature

Symbols

- A Irradiated photobioreactor surface area (m^2)
- I Light intensity (W/m^2)
- L Liter
- t Time (h)
- V_{H2} Volume of hydrogen gas in liters

Greek Letters

- η Light conversion efficiency (%)
- ρ_{H2} Density of hydrogen gas (g/L)

Acronyms

- ATP Adenosine tri phosphate
- CBB Calvin-Benson-Bassham
- Hup- Membrane bound uplake hydrogenase deficient (mutant)

- LDPE Low density polyethylene film
- LH Light harvesting
- NADH Reduced nicotinamide adenine dinucleotide
- PBR Photobioreactor
- PMMA Polymethyl methacrylate
- PNSB Purple non sulfur bacteria
- PS Photosystem,
- PU Polyurethane
- PVC Polyvinylchloride

Author details

Dominic Deo Androga
Department of Biotechnology, Middle East Technical University, Ankara, Turkey

Ebru Özgür
Micro-Electro-Mechanical Systems Research and Application Center,
Middle East Technical University, Ankara, Turkey

Inci Eroglu
Department of Chemical Engineering, Middle East Technical University, Ankara, Turkey

Ufuk Gündüz and Meral Yücel
Department of Biological Sciences, Middle East Technical University, Ankara, Turkey

Acknowledgement

Dominic Deo Androga acknowledges the Scientific and Technological Research Council of Turkey (TUBITAK-BIDEB) for providing financial support through the PhD Fellowships for Foreign Citizens (Code 2215) program.

9. References

[1] Ghirardi LM, Dubini A, Yu J, Pin-Ching Maness. Photobiological Hydrogen-Producing Systems. Chemical Society Reviews 2009;38(1)52-61.

[2] Melis A, Zhang L, Forestier M, Ghirardi LM, Seibert M. Sustained Photobiological Hydrogen Gas Production Upon Reversible Inactivation of Oxygen Evolution in the Green Alga *Chlamydomonas reinhardtii*. Plant Physiology 2000;122(1)127-135.

[3] Dubbs JM, Tabita FR. Regulators of Non Sulfur Purple Phototrophic Bacteria and the Interactive Control of CO_2 Assimilation, Nitrogen Fixation, Hydrogen Metabolism and Energy Generation. FEMS Microbiology Reviews 2004; 28(3) 353-376.

[4] Koku H, Eroglu I, Gündüz U, Yücel M, Türker L. Aspects of the Metabolism of Hydrogen Production by *Rhodobacter sphaeroides*. International Journal of Hydrogen Energy 27 2002;27(11-12)1315-1329.

[5] Verméglio A, Joliot P. The Photosynthetic Apparatus of *Rhodobacter sphaeroides*. Trends in Microbiology 1999;7(11)435-440.

[6] Zhu YS, Hearst JE. Regulation of Expression of Genes for Light-Harvesting Antenna Proteins LH-I and LH-II; Reaction Center Polypeptides RC-L, RC-M, and RC-H; and Enzymes of Bacteriochlorophyll and Carotenoid Biosynthesis in *Rhodobacter capsulatus* by Light and Oxygen. Proceedings of NationaI Academy of Science, USA 1986;83(20) 7613-7617.

[7] Firsow NN, Drews G. Differentiation of the Intracytoplasmic Membrane of *Rhodopseudomonas palustris* Induced by Variations of Oxygen Partial Pressure or Light Intensity. Archives of Microbiology 1977;115(3) 299-306.

[8] Pemberton JM, Horne IM, McEwan AG. Regulation of Photosynthetic Gene Expression in Purple Bacteria. Microbiology 1998;144(2) 267–278.

[9] Tichi MA, Tabita FR. Interactive Control of *Rhodobacter capsulatus* Redox-Balancing Systems during Phototrophic Metabolism. Journal of Bacteriology 2001;183(21) 6344-6354.

[10] Romagnoli S, Tabita FR (2005). Regulation of CO_2 Fixation in Non-Sulfur Purple Photosynthetic Bacteria. In: Omasa K, Nouchi I, De Kok LJ (eds). Plant Responses to Air Pollution and Global Change. Tokyo: Springer-Verlag. pp165-169.

[11] McKinlay JB, Harwood CS. Carbon dioxide Fixation as a Central Redox Cofactor Recycling Mechanism in Bacteria. Proceedings of the National Academy of Sciences 2010;107(26) 11669-11675.

[12] Lee SY, Choi J, Wong HH. Recent advances in Polyhydroxyalkanoate Production by Bacterial Fermentation: Mini-Review. International Journal of Biological Macromolecules 1999; 25(1-3) 31-36.

[13] De Philippis R, Ena A, Guastiini M, Sili C, Vincenzini M. Factors Affecting Poly-β-Hydroxybutyrate Accumulation in Cyanobacteria and in Purple Non-Sulfur Bacteria. FEMS Microbiology Letters 1992; 103(2-4) 187-194.

[14] Khatipov E, Miyake M, Miyake J, Asada Y. Accumulation of Poly-L-Hydroxybutyrate by *Rhodobacter sphaeroides* on Various Carbon and Nitrogen Substrates. FEMS Microbiology Letters 1998;162(1) 39-45.

[15] Melnicki MR, Eroglu E, Melis A. Changes in Hydrogen Production and Polymer Accumulation Upon Sulfur-Deprivation in Purple Photosynthetic Bacteria. International Journal of Hydrogen Energy 2009;34(15) 6157-6170.

[16] Zhao Y, Bian SM, Zhou HN, Huang JF. Diversity of Nitrogenase Systems in Diazotrophs. Journal of Integrative Plant Biology2006;48(7) 745-755.

[17] Davis R, Lehman L, Petrovich R, Shah VK, Roberts GP, Ludden PW. Purification and Characterization of the Alternative Nitrogenase from the Photosynthetic Bacterium *Rhodospirillum rubrum*. Journal of Bacteriology 1996;178(5) 1445-1450.

[18] Schnieder K, Gollan U, Drottboom M, Selsemeier-Voigt A, Muller A. Comparative Biochemical Characterization of the Iron-Only Nitrogenase and the Molybdenum

Nitrogenase from *Rhodobacter capsulatus*. European Journal of Biochemistry 1997;244(3)789-800.

[19] Oda Y, Samanta SK, Rey FE, Wu L, Liu X, Yan T, Zhou J, Harwood CS. Functional Genomic Analysis of Three Nitrogenase Isozymes in the Photosynthetic Bacterium *Rhodopseudomonas palustris*. Journal of Bacteriology 2005;187(22) 7784-7794.

[20] Masepohl B, Drepper T, Paschen A, Groß S, Pawlowski A, Raabe K, Riedel KU, Klipp W. Regulation of Nitrogen Fixation in the Phototrophic Purple Bacterium Rhodobacter capsulatus. Journal of Molecular Microbiology and Biotechnology 2002;4(3) 243-248.

[21] Akköse S, Gündüz U, Yücel M, Eroglu I. Effects of Ammonium Ion, Acetate and Aerobic Conditions on Hydrogen Production and Expression Levels of Nitrogenase Genes in *Rhodobacter sphaeroides* O.U.001. International Journal of Hydrogen Energy 2009;34(21) 8818-8827.

[22] Pierrard J, Ludden PW, Roberts GP. Posttranslational Regulation of Nitrogenase in *Rhodobacter capsulatus*: Existence of Two Independent Regulatory Effects of Ammonium. Journal of Bacteriology 1993;175(5) 1358-1366.

[23] Yakunin AF, Hallenbeck PC. Short-Term Regulation of Nitrogenase Activity by NH_4^+ in *Rhodobacter capsulatus*: Multiple In Vivo Nitrogenase Responses to NH_4^+ Addition. Journal of Bacteriology 1998;180(23) 6392-6395.

[24] Yakunin AF, Hallenbeck PC. Regulation of Nitrogenase Activity in *Rhodobacter capsulatus* Under Dark Microoxic Conditions. Archives of Microbiology 2000;173(5-6) 366-372.

[25] Vignais PM, Billoud B. Occurrence, Classification and Biological Function of Hydrogenases: An Overview. Chemical Reviews 2007;107(10) 4206-4272.

[26] Vincenzini M, Materassi R, Sili C, Balloni W. Evidence for an Hydrogenase-Dependent Hydrogen-Producing Activity in *Rhodopseudomonas palustris*. Annali di Microbiologia Ed Enzimologia 1985;35155-164.

[27] Kim DH, Kim MS. Hydrogenases for Biological Hydrogen Production. Bioresource Technology 2011;102(18) 8423-8431.

[28] Tredici MR (2004). Mass Production of Microalgae: Photobioreactors. In: Richmond A (ed). Handbook of Microalgal Culture, Biotechnology and Applied Phycology. Oxford: Blackwell Scientific. pp 178-214.

[29] Dasgupta CN, Gilbert JJ, Lindblad P, Heidorn T, Borgvang SA, Skjanes K, Das D. Recent Trends on the Development of Photobiological Processes and Photobioreactors for the Improvement of Hydrogen Production. International Journal of Hydrogen Energy 2010;35(19)10218-10238.

[30] Tredici MR (1999). Bioreactors. In: Flickinger MC, Drew SW (eds). Encyclopedia of Bioprocess Technology: Fermentation, Biocatalysis and Bioseparation, vol. 1. New York: Wiley. pp 395-419.

[31] Akkerman I, Janssen M, Rocha J, Wijffels RH. Photobiological Hydrogen Production: Photochemical Efficiency and Bioreactor Design. International Journal of Hydrogen Energy 2002;27(11-12)1195-1208.

[32] Chen CY, Liu CH, Lo YC, Chang JS. Perspectives on Cultivation Strategies and Photobioreactor Designs for Photo-fermentative Hydrogen Production. Bioresource Technology 2011;102(18) 8484-8492.

[33] Uyar B, Kars G, Yücel M, Gündüz U and Eroglu I (2011). Hydrogen Production via Photofermentations. In: Levin D, Azbar N (eds). State of the Art and Progress in Production of Biohydrogen: Danvers MA. Bentham Science Publishers. pp 54-77.

[34] Hoekema S, Bijmans M, Janssen M, Tramper J, Wijffels RH. A Pneumatically Agitated Flat-Panel Photobioreactor with Gas Re-circulation: Anaerobic Photoheterotrophic Cultivation of a Purple Non-sulfur Bacterium. International Journal of Hydrogen Energy 2002; 27(11-12)1331-1338.

[35] Hu Q, Guterman H, Richmond A. A Flat Inclined Modular Photobioreactor for Outdoor Mass Cultivation of Photoautotrophs. Biotechnology and Bioengineering 1996;51(1) 51-60.

[36] Tredici MR, Zitelli GC. Efficiency of Sunlight Utilization: Tubular vs Flat Photobioreactors. Biotechnololgy and Bioengineering 1998;57(2) 187-197.

[37] Richmond A, Zou N, 1999. Effect of Light-path Length in Outdoor Flat Plate Reactors on Output Rate of cell mass and of EPA in *Nannochloropsis sp.* Journal of Biotechnology 1999;70(1-3) 351-356.

[38] Gebicki J, Modigell M, Schumacher M, van der Burg J, Roebroeck E. Comparison of Two Reactor Concepts for Anoxygenic H_2 Production by *Rhodobacter capsulatus.* Journal of Cleaner Production 2010;18(1) S36-S42.

[39] Gilberta JJ, Ray S, Das D. Hydrogen Production Using *Rhodobacter sphaeroides* (O.U. 001) in a Flat Panel Rocking Photobioreactor. International Journal of Hydrogen Energy 2011;36(5) 3434-3441.

[40] Li X, Wang YH, Zhang SL, Chu J, Zhang M, Huang MZ, et al. Effects of Light/Dark Cycle, Mixing Pattern and Partial Pressure of H_2on Biohydrogen production by *Rhodobactersphaeroides* ZX-5. Bioresource Technology 2011;102(2)1142-1148.

[41] Eroglu I, Aslan K, Gündüz U, Yücel M, Türker L. Substrate Consumption Rates for Hydrogen Production by *Rhodobactereroides* in a Column Photobioreactor. Journal of Biotechnololgy 1999;70(1-3) 103-113.

[42] Zabut B, Kahlout KE, Yücel M. Hydrogen Gas Production by Combined Systems of *Rhodobacter sphaeroides* O.U. 001 and *Halobacterium salinarum* in a Photobioreactor. International Journal of Hydrogen Energy 2006;31(11) 1553-1562.

[43] Carlozzi P, Lambardi M. Fed-batch operation for Bio-H_2 production by *Rhodopseudomonas palustris* (strain 42OL). Renewable Energy 2009;34(12) 2577-2584.

[44] Lee CM, Hung GJ, Yang CF. Hydrogen Production by *Rhodopseudomonas palustris* WP 3-5 in a Serial Photobioreactor Fed with Hydrogen Fermentation Effluent. Bioresource Technology 2011;102(18) 8350-8356.

[45] Boran E, Özgür E, van der Burg J, Yücel M, Gündüz U, Eroglu I. Biological Hydrogen Production by *Rhodobacter capsulatus* in Solar Tubular Photo Bioreactor. Journal of Cleaner Production 2010;18(1) S29-S35.

[46] Tsygankov AA, Hall DO, Liu J, Rao KK. An Automated Helical Photobioreactor Incorporating Cyanobacteria for Continuous Hydrogen Production. In: Zaborsky OR. (ed.) Biohydrogen. London: Plenum Press; 1998. p431-440.

[47] Hai T, Ahlers H, Gorenflo V, Steinbüchel A. Axenic Cultivation of Anoxygenic Phototrophic Bacteria, Cyanobacteria and Microalgae in a New Closed Tubular Glass Photobioreactor. Applied Microbiology and Biotechnology 2000;53(4) 383-389.

[48] Watanabe Y. Design and Experimental Practice of Photobioreactor Incorporating Microalgae for Efficient Photosynthetic CO_2 Fixation Performance. Studies in Surface Science and Catalysis 2004; 153 445-452.

[49] Miyake J,Wakayama T, Schnackenberg J, Arai T. Simulation of the Daily Sunlight Illumination Pattern for Bacterial Photohydrogen Production. Journal of Bioscience and Bioengineering 1999;88(6) 659-63.

[50] Redwood MD, Beedle MP, Macaskie LE. Integrating Dark and Light Bio-Hydrogen Production Strategies: Towards the Hydrogen Economy. Reviews in Environmental Science and Biotechnology 2009;8(2) 149-185.

[51] Eroglu I, Tabanoglu A, Gündüz U, Eroglu E, Yücel M. Hydrogen Production by *Rhodobacter sphaeroides* O.U.001 in a Flat Plate Solar Bioreactor. International Journal of Hydrogen Energy 2008;33(2) 531-541.

[52] Özgür E, Uyar B, Öztürk Y, Yucel M, Gündüz U, Eroglu I. Biohydrogen Production by *Rhodobacter capsulatus* on Acetate at Fluctuating Temperatures. Resources, Conservation and Recycling 2010;54(5) 310-314.

[53] Androga DD, Özgür E, Gündüz U, Yücel M, Eroglu I. Factors Affecting the Long-Term Stability of Biomass and Hydrogen Productivity in Outdoor Photofermentation. International Journal of Hydrogen Energy 2011;36(17) 11369-11378.

[54] Avcioglu SG, Özgür E, Eroglu I, Yücel M, Gündüz U. Biohydrogen Production in an Outdoor Panel Photobioreactor on Dark Fermentation Effluent of Molasses. International Journal of Hydrogen Energy 2011;36(17) 11360-11368

[55] Özkan E, Uyar B, Özgür E, Yücel M, Eroglu I, Gündüz U. Photofermentative Hydrogen Production Using Dark Fermentation Effluent of Sugar Beet Thick Juice in Outdoor Conditions. International Journal of Hydrogen Energy 2012;37(2) 2044-2049.

[56] Hoekema S, Douma RD, Janssen M, Tramper J, Wijffels RH. Controlling Light-Use by *Rhodobacter capsulatus* Continuous Cultures in a Flat-Panel Photobioreactor. Biotechnology and Bioengineering 2006;95(4) 613-626.

[57] Obeid J, Magnin JP, Flaus JM, Adrot O, Willison JC, Zlatev R. Modelling of Hydrogen Production in Batch Cultures of the Photosynthetic Bacterium *Rhodobacter capsulatus*. International Journal of Hydrogen Energy 2009;34(1) 180-185.

[58] He D, Bultel Y, Magnin JP, Willison JC. Kinetic analysis of Photosynthetic Growth and Photohydrogen Production of

[59] Strains of *Rhodobacter capsulatus*. Enzyme Microbial Technology 2006;38(1-2) 253-259.

[60] Sevinç P, Gündüz U, Eroglu I, Yücel M. Kinetic analysis of photosynthetic growth, hydrogen production and dual substrate utilization by *Rhodobacter capsulatus*. International Journal of Hydrogen Energy 2012;doi:10.1016/j.ijhydene.2012.02.176.

[61] Nath K, Muthukumar M, Kumar A, Das D. Kinetics of Two-Stage Fermentation Process for the Production of Hydrogen. International Journal of Hydrogen Energy 2008;33(4) 1195-1203.

[62] Ogbonna JC, Toshihiko S, Tanaka H. An Integrated Solar and Artificial Light System for Internal Illumination of Photobioreactors. Journal of Biotechnology 1999;70(1-3) 289-297.

[63] Uyar B, Eroglu I, Yücel M, Gündüz U, Türker L. Effect of Light Intensity, Wavelength and Illumination Protocol on Hydrogen Production in Photobioreactors. International Journal of Hydrogen Energy 2007;32(18) 4670-4677.

[64] Sözen A, Arcaklioğlu E, Özalp M, Kanit EG. Solar Energy Potential in Turkey. Applied Energy 2005; 80(4) 367-381.

[65] Boran, Özgür, Yücel M, Gündüz U, Eroglu I. Biohydrogen Production by *Rhodobacter capsulatus* Hup- Mutant in Pilot Solar Tubular Photobioreactor 2012; doi:10.1016/j.ijhydene.2012.02.171.

[66] Wakayama T, Toriyama A, Kawasugi T, Arai T, Asada Y, Miyake J. Photohydrogen Production Using Photosynthetic Bacterium *Rhodobacter sphaeroides* RV: Simulation of the Light Cycle of Natural Sunlight Using an Artificial Source. In: Zaborsky OR. (ed.) Biohydrogen. London: Plenum Press; 1998. p375-381.

[67] Boran, Özgür, Yücel M, Gündüz U, Eroglu I. Biohydrogen Production by *Rhodobacter capsulatus* in Solar Tubular Photobioreactor on Thick Juice Dark Fermenter Effluent 2012b; doi: 10.1016/j.jclepro.2012.03.020.

[68] Wakayama T, Nakada E, Asada Y, Miyake J. Effect of Light/Dark Cycle on Bacterial Hydrogen Production by *Rhodobacter sphaeroides* RV. Applied Biochemistry and Biotechnology 2000;84–86(1-9) 431-440.

[69] Sasikala K, Ramana CV, Rao PR. Environmental Regulation for Optimal Biomass Yield and Photoproduction of Hydrogen by *Rhodobacter sphaeroides* O.U.001. International Journal of Hydrogen Energy 1991;16(9) 597-601.

[70] Miyake J, Mao X, Kawamura S. Photoproduction of Hydrogen from Glucose by a Co-culture of a Photosynthetic Bacteria and *Clostridium Butyricum*. Journal of Fermentation Technology 1984;62(6) 531-535.

[71] Eroglu E, Gündüz U, Yücel M, Eroglu I. Photosynthetic Bacterial Growth and Productivity Under continuous Illumination or Diurnal Cycles with Olive Mill Wastewater as Feedstock. International Journal of Hydrogen Energy 2010;35(11) 5293-5300.

[72] Argun H, Kargi F. Bio-Hydrogen Production by Different Operational Modes of Dark and Photofermentation: An Overview. International Journal of Hydrogen Energy 2011;36(13) 7443-7459.

[73] Modigell M, Holle N. Reactor Development for a Biosolar Hydrogen Production Process. Renewable Energy 1998;14(1-4) 421-426.

[74] Androga, Özgür E, Eroglu I, Gündüz U, Yücel M. Significance of Carbon to Nitrogen Ratio on the Long-Term Stability of Continuous Photofermentative Hydrogen Production. International Journal of Hydrogen Energy 2011;36(24)15583-15594.

[75] Tsygankov AA, Hirata Y, Miyake M, Asada Y, Miyake J. Photobioreactor with Photosynthetic Bacteria Immobilized on Porous Glass for Hydrogen Photoproduction. Journal of Fermentation and Bioengineering 1994;77(5) 575-578.

[76] Palazzi E, Perego P, Fabiano B. Mathematical Modelling and Optimization of Hydrogen Continuous Production in a Fixed-Bed Bioreactor. Chemical Engineering Science 2002;57(18) 3819-3830.

[77] Fascetti E, D'addario E, Todini O, Robertiello A. Photosynthetic Hydrogen Evolution with Volatile Organic Acids Derived from the Fermentation of Source Selected Municipal Solid Wastes. International Journal of Hydrogen Energy 1998;23(9) 753-760.

[78] Shi XY, Yu HQ. Continuous Production of Hydrogen from Mixed Volatile Fatty Acids with *Rhodopseudomonas capsulata*. International Journal of Hydrogen Energy 2006; 31(12) 1641-1647.

[79] Otsuki T, Uchiyama S, Fujiki K, Fukunaga S. Hydrogen Production by a Floating-type Photobioreactor. In: Zaborsky OR. (ed.) Biohydrogen. London: Plenum Press; 1998. p369-374.

[80] Chen CY, Lee CM, Chang JS. Feasibility Study on Bioreactor Strategies for Enhanced Photohydrogen Production from *Rhodopseudomonas palustris*WP3-5 using Optical-Fiber-Assisted Illumination Systems. International Journal of Hydrogen Energy 2006;31(15)2345-2355.

[81] Fedorov AS,Tsygankov AA, Rao KK, Hall DO. Production of Hydrogen by an *Anabaena variabilis* Mutant in a Photobioreactor Under Aerobic Outdoor Conditions. In: Miyake J, Matsunaga T, San Pietro A. (eds.) Biohydrogen II. New York: Pergamon; 2001. p223-228.

[82] Tsygankov AA, Fedorov AS, Kosourov SN, Rao KK. Hydrogen Production by Cyanobacteria in an Automated Outdoor Photobioreactor Under Aerobic conditions. Biotechnology and Bioengineering 2002;80(7) 777–783.

[83] Zürrer H, Bachofen R. Hydrogen Production by the Photosynthetic Bacterium *Rhodospirillum rubrum*. Applied and Environmental Microbiology 1979;37(5)789-793.

[84] Fascetti E, Todini O. *Rhodobacter sphaeroides* RV Cultivation and Hydrogen Production in a One-and Two Stage Chemostat. Applied Microbiology and Biotechnology 1995;44(3-4) 300-305.

[85] Soletto D, Binaghi L, Ferrari L, Lodi A, Carvalho JCM, Zilli M. Effects of Carbon dioxide Feeding Rate and Light Intensity on the Fed-Batch Pulse-Feeding Cultivation of *Spirulina platensis* in Helical Photo-Bioreactor. Biochemical Engineering 2008;39(2) 369-375.

[86] Ren NQ, Liu BF, Zheng GX, Xing DF, Zhao X, Guo WQ. Strategy for Enhancing Photo-Hydrogen Production Yield by Repeated Fed-Batch Cultures. International Journal of Hydrogen Energy 2009;34(18) 7579-7584.

[87] Adessi A, Torzillo G, Baccetti E, De Philippis R. Sustained Outdoor H_2 Production with *Rhodopseudomonas palustris* Cultures in a 50 L Tubular Photobioreactor. International Journal of Hydrogen Energy 2012; doi:10.1016/j.ijhydene.2012.01.081.

[88] Zhang K, Kurano N, Miyachi S. Outdoor Culture of a Cyanobacterium with a Vertical Flat-Plate Photobioreactor: Effects on Productivity of the Reactor Orientation, Distance Setting Between the Plates and Culture Temperature. Applied Microbiology and Biotechnology 1999;52(6) 781-786.

[89] Carlozzi P. Hydrodynamic Aspects and *Arthrospira* Growth in Two Outdoor Tubular Undulating Row Photobioreactors. Applied Mirobiology and Biotechnology 2000; 54(1) 14-22.

[90] Richmond A, Cheng-Wu Z. Optimization of a Flat Plate Glass Reactor for Mass Production of *Nannochloropsis sp*. Outdoors. Journal of Biotechnology 2001;85(3) 259-269.

[91] Nakada E, Nishikata S, Asada Y, Miyake J. Light Penetration and Wavelength Effect on Photosynthetic Bacteria Culture for Hydrogen Production. In: Zaborsky OR. (ed.) Biohydrogen. London: Plenum Press; 1998. p345-352.

[92] Kitajima Y, El-Shishtawy RMA, Ueno Y, Otsuka S, Miyake J, Morimoto M. Analysis of Compensation Point of Light Using Plane-Type Photosynthetic Bioreactor. In: Zaborsky OR. (ed.) Biohydrogen. London: Plenum Press; 1998. p359-367.

[93] Barbosa MJ, Rocha JMS, Tramper J, Wijffels RH. Acetate as a Carbon Source for Hydrogen Production by Photosynthetic Bacteria. Journal of Biotechnology 2001;85(1) 25-33.

[94] Katsuda T, Arimoto T, Igarashi K, Azuma M, Kato J, Takakuwa S, Ooshima H. Light Intensity Distribution in the Externally Illuminated Cylindrical Photo-Bioreactor and its Application to Hydrogen Production by *Rhodobacter capsulatus*. Biochemical Engineering Journal 2000;5(2) 157-164.

[95] Ogbonna JC, Tanaka H. Photobioreactor Design for Photobiological Production of Hydrogen. In: Miyake J, Matsunaga T, San Pietro A. (eds.) Biohydrogen II. New York:Pergamon; 2001. p245-261.

[96] Yoon JH, Shin JH, Kim MS, Sim SJ, Park TH. Evaluation of Conversion Efficiency of Light to Hydrogen Energy by *Anabaena variabilis*. International Journal of Hydrogen Energy 2006; 31(6) 721-7.

[97] Boran E. Process Development for Continuous Photofermentative Hydrogen Production. MSc thesis. Middle East Technical University; 2011.

[98] Mignot L, Junter GA, Labbé M. A New Type of Immobilized Cell Photoreactor with Internal Illumination by Optical Fiber. Biotechnology Techniques 1989;3(5) 98-107.

[99] Chen CY, Saratale GD, Lee CM, Chen PC, Chang JS. Phototrophic Hydrogen Production in Photobioreactors Coupled with Solar-Energy-Excited Optical Fibers. International Journal of Hydrogen Energy 2008, 33(23) 6886-6895.

[100] Chen CY, Yeh KL, Lo YC, Wang HM, Chang JS. Engineering Strategies for the Enhanced Photo-H_2 Production Using Effluents of Dark Fermentation Processes as Substrate. International Journal of Hydrogen Energy 2010;35(24) 13356-13364.

[101] Rechenberg I. Artificial Bacterial Algal Symbiosis (Project ArBAS): Sahara experiments. In: Zaborsky OR. (ed.) Biohydrogen. London: Plenum Press; 1998. p281-294.

[102] Wakayama T, Miyake J. Light Shade Bands for the Improvement of Solar Hydrogen Production Efficiency by *Rhodobacter sphaeroides* RV. International Journal of Hydrogen Energy 2002; 27(11-12) 1495-1500.

[103] Sasikala K, Ramana CV, Raghuveer RP, Kovacs KL. Anoxygenic Phototrophic Bacteria: Physiology and Advances in Hydrogen Production Technology. Advances in Applied Microbiology 1993;38 211-295.

[104] Favinger J, Stadtwald R, Gest H. *Rhodospirillum centenum*, sp. nov., a Thermotolerant Cyst-Forming Anoxygenic Photosynthetic Bacterium. Antonie Van Leeuwenhoek 1989; 55(3) 291-296.

[105] Carlozzi P, Sacchi A. Biomass Production and Studies on *Rhodopseudomonas palustris* grown in an Outdoor, Temperature Controlled, Underwater Tubular Photobioreactor. Journal of Biotechnology 2001;88(3) 239-249.

[106] Scoma A, Giannelli L, Faraloni C, Torzillo G. Outdoor H_2 Production in a 50-L Tubular Photobioreactor by Means of a Sulfur-Deprived Culture of the Microalga *Chlamydomonas reinhardtii*. Journal of Biotechnology 2012;157(4) 620-627.

[107] Claassen PAM, de Vrije T, Koukios EG, van Niel EWJ, Özgür E, Eroglu I, Nowik I, Modigell M, Wukovits W, Friedl A, Ochs D, Ahrer W. Non-thermal production of pure hydrogen from biomass: HYVOLUTION. In: Stolten D. (ed.) Hydrogen Energy. Weinheim:Wiley-Vch; 2010. p169-188.

[108] Kim JS, Ito K, Takahashi H. Production of Molecular Hydrogen in Outdoor Batch Culture of *Rhodopsedomonas sphaeroides*. Agricultural Biology and Chemistry 1982;46(4) 937-941.

[109] Kim JS, Ito K, Izaki K, Takahashi H. Production of Molecular Hydrogen by a Semi-Continuous Outdoor Culture of *Rhodopseudomonas sphaeroides*. Agricultural and Biological Chemistry 1987;51(4) 1173-1174.

[110] Arai T, Wakayama T, Okana S, Kitamura H. Open Air hydrogen Production by Photosynthetic Bacteria Used Solar Energy During Winter Seasons in Central Japan 1998: conference proceedings, June 23-26, 1997, Kona, Hawaii, USA. Tumer JA; ICBHP;1998.

[111] Gebicki J, Modigell M, Schumacher M, van der Burg J, Roebroeck E. Development of Photobioreactors for Anoxygenic Production of Hydrogen by Purple Bacteria. Chemical Engineering Transactions 2009;18 363-366.

[112] Janssen M, Tramper J, Mur LR, Wijffels RH. Enclosed Outdoor Photobioreactors: Light Regime, Photosynthetic Efficiency, Scale-up, and Future Prospects. Biotechnology and Bioengineering 2003;81(2):193–210.

[113] Kondo T, Arakawa M, Hirai T, Wakayama TG, Hara M, Miyake J. Enhancement of Hydrogen Production by a Photosynthetic Bacterium Mutant with Reduced Pigments. Journal of Bioscience and Bioengineering 2002;93(2) 145-150.

[114] Burgess G, Fernández-Velasco JG. Materials, Operational Energy Inputs and Net Energy Ratio for Photobiological Hydrogen Production. International Journal of Hydrogen Energy 2007;32(9) 1225-1234.

[115] Hussain I, Hamid H. Plastics in Agriculture. In: Andrady AL.(ed.) Plastics and the Environment. New York: Wiley; 2004. p185-209.

[116] Molina E, Fernández J, Acién FG, Chisti Y. Tubular Photobioreactor Design for Algal Cultures. Journal of Biotechnology 2001;92(2) 113-131.

[117] Teplyakov VV, Gassanova LG, Sostina EG, Slepova EV, Modigell M, Netrusov AI. Lab-Scale Bioreactor Integrated with Active Membrane System for Hydrogen Production: Experience and Prospects. International Journal of Hydrogen Energy 2002;27(11-12) 1149-1155.

[118] Kondo T, Wakayama T, Miyake J. Efficient Hydrogen Production Using a Multi-Layered Photobioreactor and a Photosynthetic Bacterium Mutant with Reduced Pigment. International Journal of Hydrogen Energy 2006;31(11) 1522-1526.

[119] Zhang K , Kurano N, Miyachi S. Optimized Aeration by Carbon dioxide gas for Microagal Production and Mass Transfer Characterization in a Vertical Flat-Plate Photobioreactor. Bioprocess and Biosystems Engineering 2002; 25(2) 97-101.

[120] Tsygankov AA. *Hydrogen Photoproduction by Purple bacteria* Immobilized *vs Suspension Culture.* In: Miyake J, Matsunaga T, San Pietro A. (eds.) Biohydrogen II. New York:Pergamon; 2001. p229-243.

[121] Kamal EME. Phototrophic Hydrogen Production by Agar-Immobilized *Rhodobacter capsulatus*. PhD thesis. Middle East Technical University; 2011.

[122] Sasikala K, Ramana Ch. V, Rao PR, Venkataraman LV. Hydrogen by Bio-Routs: A Perspective 1996: conference proceedings of the National Academy of Science of INDIA, Vol. LXVI, Section-B, Part I, 1-20.

[123] Tian X, Liao Q, Liu W, Wang YZ, Zhu X, Li J, Wang H. Photo-Hydrogen Production Rate of a PVA-Boric Acid Gel Granule Containing Immobilized Photosynthetic Bacteria Cells. International Journal of Hydrogen Energy 2009;34(11) 4708-4717.

[124] Bucke C, Brown DE. Immobilized Cells [and Discussion]. Philosophical Transactions of the Royal Society of London. Series B: Biological Sciences 1983; 300 (1100) 396-389.

[125] Vincenzini M, Florenzano G, Materassi R, Tredici MR. Hydrogen Production by Immobilizated Cells; H_2Photoevolution Waste Water-Treatment by Agar Entrapped Cells of *Rhodopseudomonas palustris* and *Rhodospirillum molischianum*. International Journal of Hydrogen Energy 1982;7(9) 725-728.

[126] Singh SP, Sirvastava SC, Pandey KD. Photoproduction of Hydrogen by a Non-Sulphur Bacterium Isolated from Root Zones Water Fern *Azolla Pinnata.* International Journal of Hydrogen Energy 1990;15(1) 403-406.

[127] Sasikala K, Ramana CV, Rao PR, Subrahmanyam M. Photoproduction of Hydrogen, Nitrogenase and Hydrogenase Activities of Free and Immobilized Whole Cells of *Rhodobacter sphaeroides* O.U.001. FEMS Microbiology Letters 1990;72(1-2) 23-28.

[128] Pandey A, Pandey A. Reverse Micelles as Suitable Microreactor for Increased Biohydrogen Production. International Journal of Hydrogen Energy 2008;33(1) 273-278.

[129] Fedorov AS, Tsygankov, AA, Rao KK, Hall DO. Hydrogen Photoproduction by *Rhodobacter sphaeroides* Immobilized on Polyurethane Foam. Biotechnology Letters 1998;20(11) 1007-1009.

[130] Zhu H, Suzuki T, Tsygankov AA, Asada Y, Miyake J. Hydrogen production from tofu wastewater by *Rhodobacter sphaeroides* Immobilized in Agar gels. International Journal of Hydrogen Energy 1999; 24(4) 305-310.

[131] Das D, Veziroğlu TN. Hydrogen Production by Biological Processes: A Survey of Literature. International Journal of Hydrogen Energy 2001. 26(1) 13-28.

[132] Li RY, Fang HHP. Heterotrophic Photofermentative Hydrogen Production. Critical Reviews in Environmental Science and Technology 2009;39(12):1081-1108.

[133] Eroglu E, Eroglu I, Gündüz U, Türker L, Yücel M. Biological Hydrogen Production from Olive Mill Wastewater with Two-Stage Processes. International Journal of Hydrogen Energy 2006;31(11) 1527-1535.

[134] Yetiş M, Gündüz U, Eroglu I, Yücel M, Türker L. Photoproduction of Hydrogen from Sugar Refinery Wastewater by *Rhodobacter sphaeroides* O.U.001. International Journal of Hydrogen Energy 2000; 25(11) 1035-1041.

[135] Seifert K, Waligorska M, Laniecki, M. Hydrogen Generation in Photobiological Process from Dairy Wastewater. International Journal of Hydrogen Energy 2010; 35(18) 9624-9629.

[136] Jamil Z, Mohamad Annuar MS, Ibrahim S, Vikineswary S. Optimization of Phototrophic Hydrogen Production by *Rhodopseudomonas palustris* PBUM001 via Statistical Experiment Design. International Journal of Hydrogen Energy 2009;34(17):7502-7512.

[137] Kapdan IK, Kargi F, Oztekin R, Argun H. Bio-hydrogen Production from Acid Hydrolyzed Wheat Starch by Photo-fermentation using Different *Rhodobacter sp.* International Journal of Hydrogen Energy 2009;34(5):2201-2207.

[138] Afşar N, Özgür E, Gürgan M, Akköse S, Yücel M, Gündüz U, Eroglu I. Hydrogen Productivity of Photosynthetic Bacteria on Dark fermenter Effluent of Potato Steam Peels Hydrolysate. International Journal of Hydrogen Energy 2011;36(1) 432-438.

[139] Uyar B, Eroglu I, Yücel M, Gündüz U. Photofermentative Hydrogen Production from Volatile Fatty Acids Present in Dark Fermentation Effluents. International Journal of Hydrogen Energy 2009;34(10) 4517-4523.

[140] Özgür E, Mars AE, Peksel B, Louwerse A, Yücel M, Gündüz U, Claassen PAM, Eroglu I. Biohydrogen Production from Beet molasses by Sequential Dark and Photofermentation. International Journal of Hydrogen Energy 2010;35(2):511-517.

[141] Pekgöz G, Gündüz U, Eroglu I, Yücel M, Kovacs K, Rakhely G. Effect of Inactivation of Genes Involved in Ammonium Regulation on the Biohydrogen Production of *Rhodobacter capsulatus*. International Journal of Hydrogen Energy 2011;36(21):13536-13546.

[142] Androga, Özgür E, Eroglu I, Gündüz U, Yücel M. Amelioration of Photofermentative Hydrogen Production from Molasses Dark Fermenter Effluent by Zeolite-Based Removal of Ammonium Ion. International Journal of Hydrogen Energy 2012;doi:10.1016/j.ijhydene.2012.02.177.

[143] Asada Y, Ohsawa M, Nagai Y, Ishimi K, Fukatsu M, Hideno A, Wakayama T, Miyake J. Re-evaluation of Hydrogen Productivity from Acetate by some Photosynthetic Bacteria. International Journal of Hydrogen 2008;33(19):5147-5170.

[144] Nath K, Debabrata Das D. Effect of Light Intensity and Initial pH During Hydrogen Production by an Integrated Dark and Photofermentation Process. International Journal of Hydrogen 2009;34(17)7497-7501.

[145] Keskin T, Hallenbeck PC. Hydrogen Production from Sugar Industry Wastes Using Single-Stage Photofermentation. Bioresource Technology 2012; 112:131-136.

[146] Kim MS,Baek JS, Lee JK. Comparison of H_2 Accumulation by *Rhodobacter sphaeroides* KD131 and its Uptake Hydrogenase and PHB synthase Deficient Mutant. International Journal of Hydrogen Energy 2006; 31(1) 121-127.

[147] Kern M, Klipp W, Klemme JH Increased Nitrogenase-Dependent H_2 Photoproduction by *Hup* Mutants of *Rhodospirillum rubrum*. Applied and Environmental Microbiology 1994;60(6) 1768-1774.

[148] Kars G, Gündüz U, Yücel M, Rakhley G, Kovacs K, Eroglu I. Evaluation of Hydrogen Production by *Rhodobacter sphaeroides* O.U.001 and its HupSL Deficient Mutant Using Acetate and Malate as Carbon Sources. International Journal of Hydrogen Energy 2009; 34(5) 2184-2190.

[149] Öztürk Y, Yücel M, Daldal F, Mandaci S, Gündüz U , Türker L, Eroglu I. Hydrogen Production by Using *Rhodobacter capsulatus* Mutants with Genetically Modified Electron Transfer Chains. International Journal of Hydrogen Energy 2006; 31(11) 1545-1552.

[150] Tichi MA, Tabita FR. Maintenance and Control of Redox Poise in *Rhodobacter capsulatus* Strains Deficient in the Calvin-Benson-Bassham pathway. Archives of Microbiology 2000; 174(5) 322-333.

[151] Öztürk Y, Gökçe A, Peksel B, Gürgan M, Özgür E, Gündüz P, Eroglu I, Yücel M. Hydrogen Production Properties of *Rhodobacter capsulatus* with Genetically Modified Redox Balancing Pathways. International Journal of Hydrogen Energy 2012; 37(2) 2014-2020.

[152] Kim EJ, Kim MS, Lee JK. Hydrogen Evolution Under Photoheterotrophic and dark Fermentative Conditions by Recombinant *Rhodobacter sphaeroides* Containing the Genes for Fermentative Pyruvate Metabolism of *Rhodospirillum rubrum*. International Journal of Hydrogen Energy 2008; 33(19) 5131-5136.

[153] Claassen PAM, de Vrije T, Koukios E, van Niel E, Eroglu I, Modigell M, Friedl A, Wukovits W, Ahrer W. Non-thermal Production of Pure Hydrogen from Biomass; HYVOLUTION Journal of Cleaner Production 2010;18(1) S4-S8.

[154] http://www.biohydrogen.nl/hyvolution. (accessed on 02/05/2012).

[155] Foglia D, Wukovits W, Friedl A, Ljunggren M, Zacchi G, Urbaniec K, Markowski M. Effects of Feedstocks on the Process Integrationof Biohydrogen Production. Clean Technologies and Environmental Policy 2011;13(4) 547-558. http://re.jrc.ec.europa.eu/pvgis/countries/europe/EU-Glob_opta_presentation.png. (accessed on 02/05/2012).

[156] Ljunggren M, Zacchi G. Techno-Economic Analysis of a Two-Step Biological Process Producing Hydrogen and Methane. Bioresource Technology 2010;101(20) 7780-7788.

[157] Miltner A, Wukovits W, Pröll T, Friedl A. Renewable Hydrogen Production: a Technical Evaluation Based on Process Simulation Journal of Cleaner Production 2010;18(1) S51-S62.

Chapter 4

Small Scale Hydrogen Production from Metal-Metal Oxide Redox Cycles

Doki Yamaguchi, Liangguang Tang, Nick Burke, Ken Chiang, Lucas Rye, Trevor Hadley and Seng Lim

Additional information is available at the end of the chapter

1. Introduction

The industrial production of hydrogen by reforming natural gas is well established. However, this process is energy intensive and process economics are adversely affected as scale is decreased. There are many situations where a smaller supply of hydrogen, sometimes in remote locations, is required. To this end, the steam-iron process, an originally coal-based process, has been re-considered as an alternative. Many recent investigations have shown that hydrogen (H_2) can be produced when methane (CH_4) is used as the feedstock under carefully controlled process conditions. The chemistry driving this chemical looping (CL) process involves the reduction of metal oxides by methane and the oxidation of lower oxidation state metal oxides with steam. This process utilises oxygen from oxide materials that are able to transfer oxygen and eliminates the need of purified oxygen for combustion. Such a system has the potential advantage of being less energy intensive than reforming processes and of being flexible enough for decentralised hydrogen production from stranded reserves of natural gas. This chapter first reviews the existing hydrogen production technologies then highlights the recent progress made on hydrogen production from small scale CL processes. The development of oxygen carrier materials will also be discussed. Finally, a preliminary economic appraisal of the CL process will be presented.

2. A brief overview on hydrogen production

Hydrogen can be produced from the reaction of feedstock including fossil fuels and biomass with water. Today, 96 % of hydrogen is derived from fossil fuels of which 48 %, 30 % and 18

% originates from natural gas, higher hydrocarbons and coal, respectively and the remaining 4 % comes from electrolysis. Fossil fuel based hydrogen production processes are mature technologies and are currently the most economic routes for large scale hydrogen production. Because coal, natural gas and biomass all contain carbon, carbon dioxide is inevitably produced as a by-product of the energy released. A pictorial overview of the available hydrogen production processes is given in Figure 1. The basics of two commercialised processes, namely steam methane reforming and partial oxidation, are considered in this section. A brief discussion on emerging hydrogen production technology will also be presented.

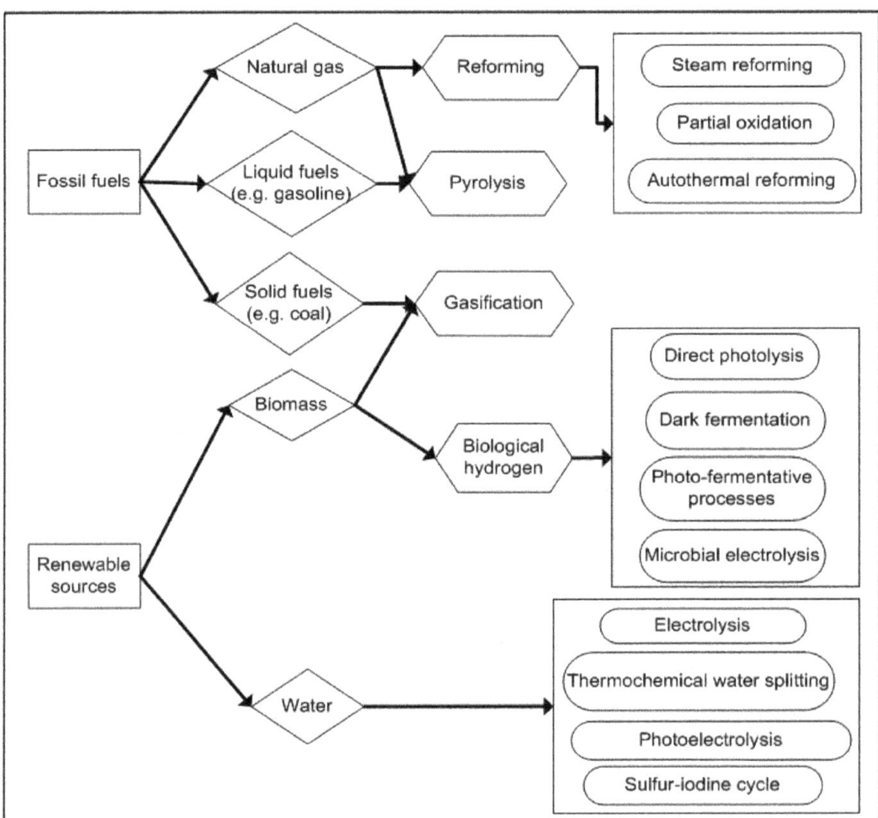

Figure 1. An overview of existing hydrogen production process from different sources.

2.1. Steam methane reforming

Steam reforming of methane (SMR) is one of the most developed and commercially used technologies. Compared to other fossil fuels, natural gas, which contains mostly methane, is a cost effective feedstock for making hydrogen. This is because methane has a high hydrogen-to-carbon ratio, meaning the yield of hydrogen is higher. Today, almost 48 % of the world's hydrogen is produced from this technology [1]. In this process, hydrogen is produced according to the following two reactions:

$$CH_4 + H_2O \rightarrow CO + 3H_2 \Delta H° = 206 \text{ kJ/mol} \quad (1)$$

$$CO + H_2O \rightarrow CO_2 + H_2 \Delta H° = -41 \text{ kJ/mol} \quad (2)$$

In the SMR, the natural gas feedstock is first reformed in the presence of steam over a catalyst at elevated temperatures (700 – 925 °C) to produce a mixture of carbon monoxide and hydrogen (syngas) as shown in Equation 1. Then, the yield of hydrogen is further increased by reacting the carbon monoxide with make up steam via the water-gas shift reaction (WGS) as shown in Equation 2. Finally, hydrogen is separated and purified by processes such as pressure swing absorption, wet scrubbing or membrane separation. SMR is currently the most cost effective hydrogen production process which offers a minimum energy efficiency of 80 – 85 % in a large scale facility if residual steam is re-used [1]. Furthermore, the process is economically viable for large scale operation [2]. According to Pardor *et al.*[3], the price of hydrogen produced from SMR ranges from \$5.97/GJ for a 25.4 million Nm^3/day plant to \$7.16/GJ for a 1.34 million Nm^3/day plant. A figure of \$11.22/GJ was estimated for hydrogen produced from a small facility (0.27 million Nm^3/day). However, the price of hydrogen varies with the price of natural gas feedstock. In general, the price of the natural gas feedstock accounts for 52 – 68 % and 40 % of the total cost for large and small SMR plants, respectively. It can be seen that decreasing the scale of operation would lead to an increase in cost of the hydrogen produced.

2.2. Partial oxidation

Hydrogen can also be produced from the partial oxidation (POX) of hydrocarbons over a catalyst at high temperatures (Equation 3).

$$CH_4 + 0.5O_2 \rightarrow CO + 2H_2 \Delta H° = -36 \text{ kJ/mol} \quad (3)$$

The reaction requires the use of high purity oxygen and is mildly exothermic. Similar to the SMR process, the yield and purity of hydrogen may be further increased by the WGS reaction and a subsequent purification process. The reported efficiency of POX is in the range of 66 – 76 % [1]. Mirabal [4] estimated the cost of hydrogen to be \$12.43/GJ for a 2.83 million Nm^3/day plant, which is higher than that produced from SMR. However, based on the use of coke off-gas and residual oil (both having a price of lower than natural gas), Pardro *et al.*[3] estimated the price to be in the range of \$6.94 – 9.83/GJ for large facilities (1.34 – 2.80 million Nm^3/day) and \$10.73/GJ for a small facility (capacity is 0.27 million Nm^3/day). Similar to SMR, the economics appears more favourable for large scale operations.

2.3. Coal gasification

Gasification can be used to convert a varied range of solid fuels such as coal and biomass into syngas (Equation 4).

$$C(s) + H_2O \rightarrow CO + H_2 \Delta H^\circ = 131 \text{ kJ/mol} \tag{4}$$

Coal gasification is a mature process and is commercially available. Although the cost of the coal feedstock is generally much cheaper than natural gas, the price of hydrogen produced from coal gasification process is estimated to be $17.45/GJ. This is higher compared to SMR ($10.26/GJ) and POX ($12.43/GJ), and this is due to the higher capital investment required for coal gasification. Coal is an economically viable option for making hydrogen in very large centralised plants where the demand for hydrogen becomes large enough to support an associated large distribution network and establishment costs. It is therefore seen that coal gasification would become more competitive than SMR and POX as the price of natural gas increases [4]. Much of the engineering experience accumulated from coal fired power plant is directly useful for coal gasification.

2.4. Other novel routes for hydrogen production

Water splitting is one of the options for producing hydrogen and has received wide attention. The current reported energy efficiency is between 10 – 27 % and the cost of hydrogen is estimated to be 3-10 times of the hydrogen produced from the SMR process [5]. Biological routes for producing hydrogen are also being considered because of the renewable nature and the mild operating conditions of these processes. These alternative routes have yet to become economically competitive with technologies in practice such as SMR and POX that use fossil fuel feedstock.

3. Hydrogen production from cyclic redox processes

There is an ongoing demand for viable processes for producing hydrogen on a small scale for decentralised distribution. For this reason, there is currently much attention being paid to the development of cyclic redox processes or commonly referred as chemical looping (CL) processes for small scale hydrogen production. In addition to the compactness of the process, another advantage is the ability to produce a near sequestration-ready stream of carbon dioxide from the process. The operating concept behind these processes resembles the well-known steam-iron process and is illustrated in Figure 2a. Some widely reported variations and applications include chemical looping combustion (CLC) for power generation and, chemical looping hydrogen production (CLH2). The schematic diagrams representing these processes are shown in Figure 2b and Figure 2c. A typical chemical looping operation consists of a reduction and an oxidation steps. During the reduction, a metal oxide is used as the oxygen carrier to oxidise carbonaceous fuels (e.g. natural gas, coal or biomass) into carbon dioxide and steam. The reduction can be optimised such that syngas (a mixture of carbon monoxide and hydrogen) can be obtained. Subsequently, the partially or fully reduced metal oxide is oxidised with air or steam to re-generate the original metal oxide and other oxidation products. When steam is used, water is split to produce hydrogen as the main product.

One of the fundamental parameters that determine the overall efficiency of many chemical looping processes is the effectiveness of the oxygen carriers. Therefore many research groups have focused on improving the activity and the stability of oxygen carrying materials. This section reports the latest developments of oxygen carrier materials for CL applications.

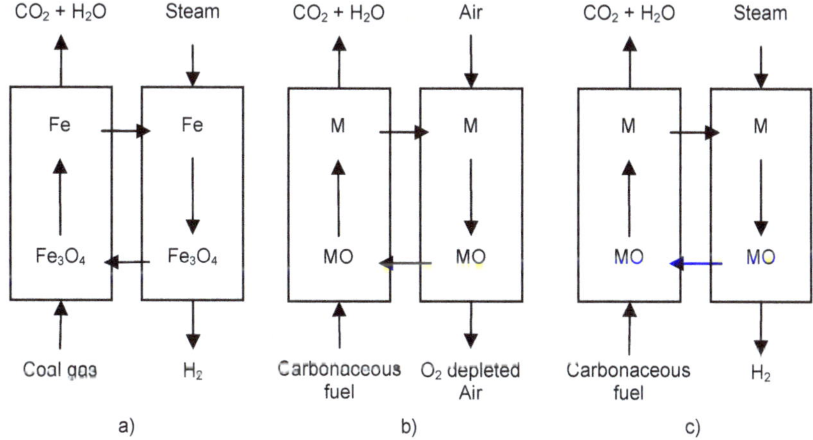

Figure 2. a) The traditional steam-iron process and chemical looping (CL) processes, b) CLC for power generation, and c) CLH2.

3.1. Thermodynamic constraints

The selection of an oxygen carrier requires comprehensive appraisal of the physiochemical properties of the material. Some properties include reaction kinetics, oxygen content, long-term recyclability and durability, attrition resistance, heat capacity, melting points, tendency to form coke, resistance to carbon deposition, cost and toxicity [6, 7]. Nevertheless, the most important requirement is the thermodynamic feasibility of oxygen transfer to and from these oxygen carriers. Figure 3 shows the changes in Gibbs free energy (ΔG) of some oxygen carriers commonly studied for CL applications. Some selected properties are provided in Table 1.

For the current topic, the oxygen carrier can be divided into two groups based on their ability to oxidise methane. The first group contains oxides that are capable of only partially oxidising methane into carbon monoxide and hydrogen. Some representative redox couples are ZnO/Zn, V_2O_5/V and CeO_2/Ce_2O_3 couples. The second group contains oxides that are able to support the complete oxidation of methane. NiO/Ni, CuO/Cu and Co_3O_4/Co are redox couples that fall into this category. In addition, the oxidations of these reduced oxides are favourable over a wide temperature range as indicated by the negative ΔG values in Figure 3. Therefore these three redox couples are often regarded as good candidates for CL applications.

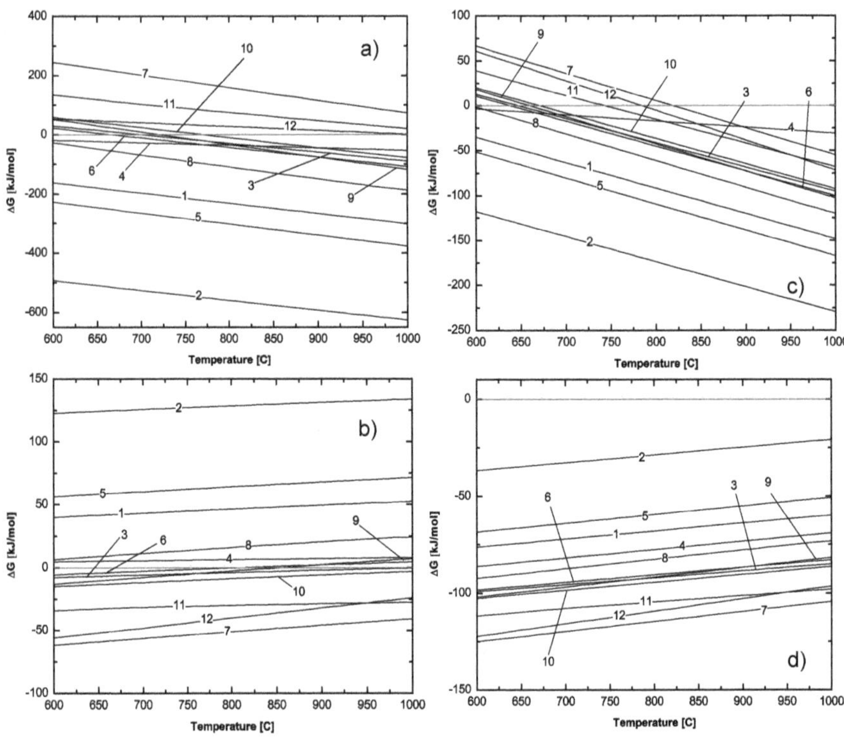

Figure 3. Variation of Gibbs free energy of reactions, a) CH_4 combustion ($CH_4 + 4/yM_xO_y \rightarrow CO_2 + 2H_2O + 4x/yM$) and b) CH_4 partial oxidation ($CH_4 + 1/yM_xO_y \rightarrow CO + 2H_2 + x/yM$), c) steam oxidation ($xM + yH_2O \rightarrow M_xO_y + yH_2$), and d) air oxidation ($xM + y/2O_2 \rightarrow M_xO_y$). See Table 1 for the legendsused.

Number	Redox couple	Melting point [°C]	Oxygen transport capacity [kg/kg-metal]	Price [USD/t]
1	NiO/Ni	1955/1455	0.27	21,800
2	CuO/Cu	1326/1084	0.25	7,680
3	Fe_3O_4/Fe	1597/1538	0.38	100
4	MnO_2/Mn	535/1267	0.58	1,500
5	Co_3O_4/Co	895/1495	0.36	39,700
6	WO_3/W	1472/3407	0.26	27,000
7	ZnO/Zn	1975/420	0.24	2,250
8	SnO/Sn	1080/232	0.13	21,000
9	In_2O_3/In	1913/157	0.21	565,000
10	MoO_2/Mo	1100/2623	0.33	34,900
11	V_2O_5/V	670/1910	0.78	25,600
12	CeO_2/Ce_2O_3	2400/2230	0.06	24,611

Table 1. Selected properties of oxygen carriers.

Compared to oxidation using molecular oxygen, the ΔG shifts to higher values when steam is used as the oxidising agent. As a result, it is not thermodynamically feasible to produce hydrogen by reacting steam with metallic Ni, Cu or Co.MnO_2/Mn and SnO/Sn couples are also not reactive when they are brought into contact with steam. ZnO/Zn and V_2O_5/V couples react with steam to produce hydrogen, however, their melting points in either the oxide or the metallic form are too low for CL applications in general. Despite the moderate ΔG values associated with Fe_3O_4/Fe, WO_3/W and CeO_2/Ce_2O_3 redox couples, the reported redox kinetics and thermo-mechanical strength have made them appealing candidates for CL processes. The Fe_3O_4/Fe couple also possesses a relatively high oxygen content, and is widely available, non-toxic and less costly. When iron oxide is used, it is only possible to oxidise the reduced state to magnetite (Fe_3O_4) due to thermodynamic limitations.

3.2. Common feedstocks for producing hydrogen

A number of studies have employed non-gaseous fuels including coal [8-13], biomass [14-17] and pyrolysis oil [18, 19]. In a syngas chemical looping (SCL) process, the fuel is first converted into syngas in a separate gasification unit. The syngas generated is then used in the reduction cycle and steam is used to regenerate the oxide and to produce hydrogen. An additional air oxidation cycle may be required to regenerate the oxygen carrier. The SCL process generally has lower efficiency for conversion, owing to the low conversions in the syngas generation step and the steam oxidation step [8]. Li et al. [9] examined the cyclic performance of a Fe-based oxygen carrier at 830 °C when a simulated syngas was used. They showed that the syngas was completely converted in the reduction half cycle giving an oxygen carrier conversion of 94.6 %. For the steam half cycle, the reduced oxygen carrier was oxidised into Fe_3O_4 producing a stream of 99.8 % pure hydrogen. In a separate study, the same group also demonstrated the feasibility of using a moving bed reactor at 900 °C for the same reaction[10]. A syngas conversion in excess of 99.5% and an oxygen carrier conversion of 50 % were recorded. A process simulation conducted by Gupta et al. [8]confirmed that the maximum efficiency for the SCL process could reach 74.2 % for hydrogen production which is comparable to or more effective than steam reforming (65-75 %), partial oxidation (50 %) and gasification (43-47 %). Considering the complexity of the SCL, it is clear that footprint of the process would be large because of the large number of unit operations involved in its design.

When coal is used as the feedstock, the solid fuel can be used to reduce oxygen carriers directly. This process is often referred as the coal direct chemical looping (CDCL) process because a gasification unit, as well as air separation and gas cleaning units, is not required [12]. The CDCL process is reported to be significantly more efficient than the SCL process for hydrogen production [6, 13].Yang et al. [11] investigated the CDCL process using a lignite-derived char in a fluidised bed reactor. The complete gasification of the char achieved a maximum carbon dioxide concentration of 90% in the presence of a K_2CO_3 catalyst. A high oxygen carrier-to-char ratio improved the complete gasification to carbon dioxide but this also led to lower hydrogen yields as a result of low conversions of the oxygen carrier. Under the optimum condition, the hydrogen production efficiency was reported to be 50.2 % at an oxygen carrier conversion of 70.2 %. The use of counter-current moving bed reactor was found to improve oxygen carrier conversion, and achieved (due to the significantly low mass required) a char conversion of > 90 % and an overall carbon dioxide capturing efficiency of > 95 %[6].

Biomass has found limited applications for SCL processes. This is because of the high water content generally associated with biomass feedstocks. Sime et al. [14] investigated the use of gases derived from woody biomass gases for SCL and reported that such process was less efficient and more costly than conventional gasification processes for producing hydrogen. Li et al. [16] pointed out that it is critical to reduce the moisture content in the biomass feedstock to less than 5 %in order to achieve a conversion of 56.6 % in gasification. Similar to other solid feedstocks, unreacted biomass must be separated before the oxygen carrier is circulated to the steam reactor. Otherwise, the unreacted biomass could be gasified and lower the purity of the hydrogen produced.

Natural gas is an efficient feedstock for CL processes since it is fed to the process in gaseous form. This minimises the need of solid handling and improves mass transfer processes [20]. Cormos (2011) recently assessed and compared hydrogen production from a natural gas CL process and a coal/lignite based SCL [21]. It was concluded that when natural gas was used to produce hydrogen, the recorded efficiency was 78.1 %. This value was higher compared to the values of 65.7 % and 63.3 % recorded for the coal- or lignite-based SCL processes, respectively. In addition, the separation and capturing of CO_2 were said to be more effective when natural gas was used. Another clear advantage of using natural gas as the feedstock is that no additional up-stream unit operations are required for producing syngas.

3.3. Oxygen carriers for cyclic redox processes

As mentioned previously, redox kinetics and thermal stability are the two main issues associated with the use of oxide-based oxygen carriers for CL processes. In order to improve their performance, support and/or promoting materials to assist in material stabilisation are often added to improve the performance of the metal oxide. A comprehensive list of oxygen carriers developed for various CL applications in the last decade can be found in an excellent review published by Adanez et al. [7]. This section highlights some recent studies on developing novel oxygen carriers.

3.3.1. Effects of metal addition to oxide carriers

Otsuka et al. [22, 23] investigated the effects of 26 different metal dopants on iron oxide. It was found that some metal dopants were more effective in preventing the iron oxide from sintering and some were more effective in facilitating the splitting of water. Among these 26 metals, Mo and Cr were found to improve the thermal stability of iron oxide in the cyclic process. The improved redox stability after the introduction of Mo metal (5 mol%) was also reported by Wang et al. [24], and Liu and Wang [25]. Despite the fact that Cr addition could improve the sintering resistance of iron oxide, temperature programmed analysis revealed that a temperature of ca. 500 °C is required to split water when compared to a temperature of 420 °C as required by iron oxide modified with Mo [22]. In addition, no oxidation of methane was observed when the temperature was lower than 700 °C [26]. It was proposed that the main role of Cr and Mo dopants was to partially transform the iron oxide into the ferrite structure ($M_xFe_{3-x}O_4$, M = Mo and Cr) [22, 26] and therefore inhabited the agglomeration of neighbouring particles.

Some metals including Ru, Rh, Pd, Ag, Ir and Pt have been shown to improve reaction kinetics by facilitating the dissociation of hydrogen, methane and water. Otsuka et al. [22] reported that the improvement on splitting of water into hydrogen by metal in a CLprocess increased in the order of Rh > Ir > Ag > Pd > Ru. Ryu et al. [27] also found that Rh was more effective than Pb, Pt and Ru in enhancing the hydrogen production step in a chemical looping process. The role of Rh was to decrease the onset temperature for the water splitting reaction. A XANES/EXAFS study on Rh-Cr-added iron oxide revealed that Rh was also able to form Rh-Fe alloy upon reductions[26]. However, Rh segregated in the alloy structure when it contacted steam and thus accelerated the sintering of iron oxide. This led to the observed deterioration in redox activity after repeated redox operation. Although Ni- and Cu-ferrites also exhibited an enhancing effect on redox kinetics, Ni and Cu were shown not to be effective in improving sintering resistance [28, 29].

The addition of a second and a third metal have been shown to further improve the redox activity [22, 24-27, 30, 31]. Common choices of metal combinations often consisted of a first metal such as Rh, Pt, Ni and Cu which is thought to catalytically activates the reducing gas (e.g. hydrogen, carbon monoxide or methane), and a second metal such as Mo and Cr which exhibits a structural stabilising effect. Otsuka et al [22] examined the addition of Rh and Mo to iron oxide for the chemical storage of hydrogen and observed an enhancement in reaction kinetics and a reduction in reaction temperature for hydrogen formation. Most importantly, the Mo provided good stabilising effect and largely mitigated the sintering of the oxygen carrier. The effect of bimetal addition on iron oxide was also investigated under methane oxidation at a temperature range of 200 – 800 °C by Takenaka et al. [30]. The methane conversion was found to increase by adding a second metal and the performance increased in the order of Rh-Cr > Ir-Cr > Pt-Cr > Ni-Cr > Pd-Cr > Cu-Cr = Co-Cr. Other research groups also reported similar findings [24, 25, 27, 31]. Despite the improvement in reactivity and thermal stability, most of the bimetallic modified oxygen carriers produce carbon upon methane oxidation. The production of carbon usually leads to a rapid deterioration of the oxygen carrier and is the source of carbon oxides (CO_x) contamination.

3.3.2. *Supported oxygen carriers*

Another approach to improve the thermal stability of oxygen carriers is to introduce inert support materials such as Al_2O_3, SiO_2, TiO_2 and ZrO_2. Adanez et al. [32] assessed the reactivity of 240 different types of oxygen carriers composed of Cu, Fe, Mn or Ni supported on SiO_2, TiO_2, ZrO_2, Al_2O_3 or sepiolite ($Mg_4Si_6O_{15}(OH)_2{\cdot}6H_2O$) over a temperature range of 950 – 1300 °C. The best Fe-based oxygen carriers were those supported on Al_2O_3 or ZrO_2. It was also found that the formation of aluminate ($NiAl_2O_4$ and $CoAl_2O_4$) lowered the oxygen transport capacity and hence reduced the redox activity [33]. SiO_2 was found to be the most suitable support for Cu-based oxygen carrier because it remained inert at high temperatures and did not form Cu-SiO_2 composites. However, Fe-based oxygen carriers showed a strong tendency to form unreactive iron silicates with SiO_2[34]. ZrO_2 and TiO_2 were suggested as the best supports for Mn- and Ni-based oxygen carriers, respectively. In terms of the cyclic redox activity, however, TiO_2 supported Ni-based oxygen carriers showed lower reactivities, compared to Ni supported on Al_2O_3. This is because NiO is more prone to react with TiO_2 and form $NiTiO_3$ which is known to be less reducible than NiO. It also exhibits a high carbon formation tendency. Therefore, Al_2O_3 supported Ni-based oxides were considered to be the most promising oxygen carrier for a large scale CLC applications.

Some metal doped iron oxide oxygen carriers were also supported on ZrO_2 for CL processes [29, 35-37]. Kodama et al. [35, 36] showed improved thermal resistance for the Ni- and Co-ferrites when ZrO_2 support was introduced. The reported methane conversion and carbon monoxide selectivity by using $Ni_{0.39}Fe_{2.61}O_2$ (33 wt%)/ZrO_2 were 46-58% and 44-48%, respectively. However, since Fe and Ni are excellent catalysts for methane decomposition, the material was severely deactivated by coke and the subsequent carbide species formed. Because Cu has lower activity for methane decomposition, $CuFe_2O_4$ was used to produce syngas from methane [29]. The results showed that no CO_x was formed during the operation. The same group also found beneficial effects of ZrO_2 and CeO_2 supports for $CuFe_2O_4$ (20 wt%) [38]. Compared to the methane conversion obtained for $CuFe_2O_4$ (34–56 %), the methane conversions achieved by $CuFe_2O_4/CeO_2$ and $CuFe_2O_4/ZrO_2$ were 89-92 % and 74-83 %, respectively. From these results, CeO_2 was found to be more active in promoting methane oxidation while ZrO_2 was considered to be a more effective stabiliser against thermal sintering. Since CeO_2 is known to be able to oxidise soot through lattice oxygen transfer [39, 40], it is thought that this property could help to minimise carbon formation when $CuFe_2O_4/CeO_2$ is used. Cha et al. [37] also confirmed that CeO_2 modified $CuFe_2O_4/ZrO_2$ was a more effective oxygen carrier than Ni-modified $CuFe_2O_4/ZrO_2$ for chemical looping syngas and hydrogen productions.

A recent study conducted by Yamaguchi et al. [41] also demonstrated the improved performance of CeO_2/ZrO_2 modified $Fe_2O_3$for producing hydrogen from methane-steam cycles. Some results obtained from temperature programmed analysis and isothermal reduction are shown in Figure 4 and are summarised in Table 2. Figure 4a shows that CeO_2 and ZrO_2 altered the redox properties of Fe_2O_3 with the most significant enhancement observed for the reducibility at low temperatures (< 600 °C) (see Table 2). The isothermal reduction analysis (Figure4b) further confirmed the accelerated reduction kinetics after the introduction of CeO_2 and ZrO_2. The observed overall enhancement was derived from the combined effects of CeO_2 and ZrO_2. CeO_2 improved the reducibility of Fe_2O_3 while ZrO_2 provided thermal stability and helped to suppress the reduction of FeO to metallic Fe. The latter was supported by the incomplete reduction of $Fe_{15}Ce_{10}Zr_{75}$ and $Fe_{40}Zr_{60}$ (Table2).Similar observations were also reported when WO_3 was modified with CeO_2 and ZrO_2[42]. The synergic effect provided by CeO_2 and ZrO_2 effectively defined the redox window of the oxygen carriers. An immediate consequence is the minimisation of carbon and carbide formation during repeated redox cycles. This can be demonstrated by the fact that CO_x free hydrogen was produced by using CeO_2-ZrO_2 modified WO_3 in a methane-steam CL process [42]. The addition of a small amount of Mo or Cr could further improve the thermal stability of this type of oxygen carrier. Galvita et al. [43] showed the addition of 2 wt% of Mo to $Fe_2O_3/Ce_{0.5}Zr_{0.5}O_2$ could maintain a stable level of hydrogen production over 100 cycles in a cyclic water-gas shift process. In this reaction, the main role of Mo is to improve the dispersion of Fe-Mo oxide material and minimise the migration of material across the boundary of adjacent particles [44].

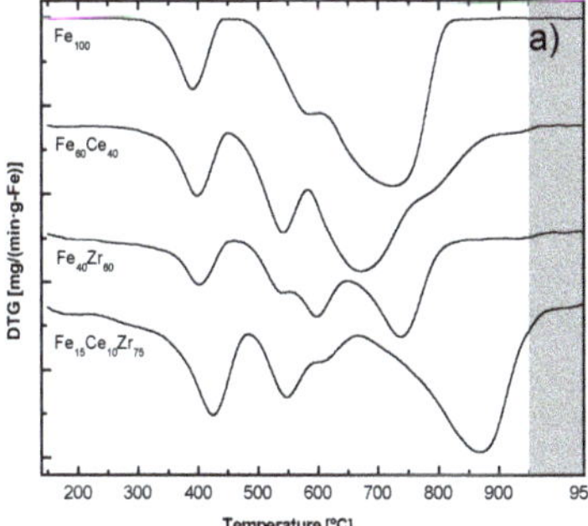

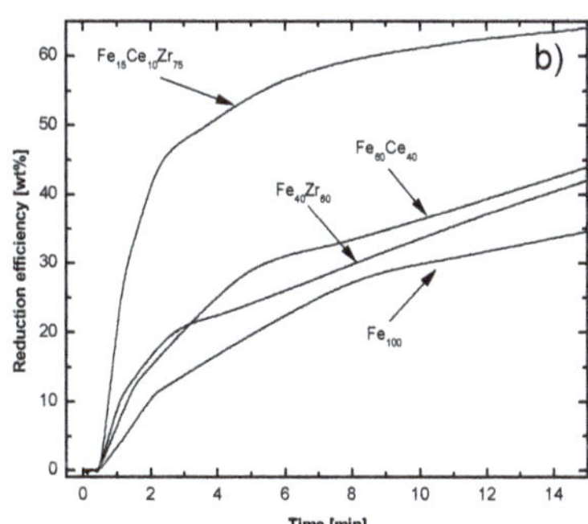

Figure 4. Effect of CeO_2 and/or ZrO_2 addition on Fe_2O_3 reducibility during a) temperature programmed and b) isothermal reduction with H_2[41].

Oxygen carrier	Oxygen removal[1] [mg-O/g-Fe]	Overall reduction efficiency[2] [wt%]	H_2 yield [μmol/g-Fe]	H_2 purity [%]
Fe_{100}	125	98.1	15	11.5
$Fe_{60}Ce_{40}$	169	97.1	368	49.1
$Fe_{40}Zr_{60}$	153	66.7	88	17.4
$Fe_{15}Ce_{10}Zr_{75}$	255	77.2	6283	97.5

Table 2. A summary of oxides used in methane-steam redox cycle [41]. [1]The oxygen removal represents a cumulative weight reduction at temperatures < 600 °C during the TPR analysis (Figure4a). [2]The overall reduction efficiency represents a final reduction efficiency obtained during isothermal reduction analysis at 750 °C for 240 min.

3.3.3. *Naturally occurring oxide materials*

Recently, many naturally occurring minerals and ashy waste produced from industry have been considered for use as oxygen carriers. These materials include natural ilmenite (Fe and Ti mixed oxide often denoted as $FeTiO_3$), iron ore, manganese ore and oxide scales. An advantage of using these materials is the low cost compared to many synthetic oxygen carriers. In addition, naturally occurring oxides usually contain Si, Al, Mg, and many other metals which have been shown to modify the physiochemical properties of the materials to various degrees. Leion et al. [45] investigated the feasibility of using ilmenite, iron ores, oxide scales from steel industry and manganese ores as oxygen carriers in a fluidised bed reactor. They concluded that many Fe based oxides, particularly ilmenite, were suitable for CLC application. However, the Mn-based oxides showed poor mechanical stability and fluidising properties, and were determined to be non-ideal candidates for this application. In a separate study, Leion et al. [46] also proved the feasibility of using ilmenite to completely capture carbon dioxide upon its reaction with syngas and reported a moderate conversion when methane is used. Adanez et al. [47] observed increases in ilmenite, and syngas and methane conversions with increasing the time on stream and the number of redox cycles. Another important finding was the enhanced activation of ilmenite when the raw ilmenite material was subjected to an oxidation pre-treatment. The authors also found the redox properties of ilmenite changed with the temperature of oxidative pre-treatment. However, the positive effect only became apparent when the ilmenite was first oxidised to pseudobrookie (Fe_2TiO_5) which is usually formed above 1000 °C.

Pre-oxidation temperature [°C]	Major crystalline phases[1]	Oxygen transfer capacity [wt%]
Raw	$FeTiO_3$, TiO_2	1.1
800	Fe_2O_3, TiO_2	1.0
1000	Fe_2TiO_5, TiO_2	1.8

Table 3. Oxygen transfer capacity and major phase of various ilmenite samples before and after pre-oxidation. [1] Phases were identified by XRD analysis

Leion et al. [46] also reported that an ilmenite sample remained active with minimum carbon formation after a continuous operation for three days at 975 °C. Furthermore, natural ilmenite is known to react just as well with petroleum coke, syngas and methane as synthetically prepared $Fe_2O_3/MgAl_2O_4$[48]. Lorente et al. [49] reported a better hydrogen storage capacity and redox stability when iron ore samples was used instead of pure Fe_2O_3. The improvement in the overall redox performance was due to the presence of impurities including SiO_2, Al_2O_3, MgO and CaO. Among these impurities, Al_2O_3 and SiO_2 are considered to be good stabilisers against sintering, while CaO and MgO are able to facilitate kinetics of water splitting.

3.4. Deactivation of oxygen carriers

The life time of the oxygen carrier is a critical factor in determining the efficiency and viability of CL processes. In general, the efficacies of oxygen carriers decrease over time because of material alternation by sintering and/or coking.

Generally, for the CLH2 application, a relatively high temperature is required for driving the reduction reaction in order to achieve satisfactory conversion and kinetics. As a result, the high temperature environment irreversibly alters the structure and the morphology of oxygen carriers, and lowers the activity during the cyclic operation. The sintering process starts as two spherical particles adhere to one another. The process involves the diffusion of metal cations between neighbouring spheres. Figure 5 shows the SEM images of a pure Fe_2O_3 sample and the same material recovered after six methane-steam redox cycles performed at

750 °C. Severe sintering is clearly evident. The heat generated from the redox reactions could accelerate the rate of sintering. When oxygen carriers sinter and agglomerate inside a fluidised bed reactor, bed defluidisation may occur. The change in solid circulation and the subsequent occurrence of gas by-pass would significantly lower the gas-solid contact and hence the overall conversion efficiency.

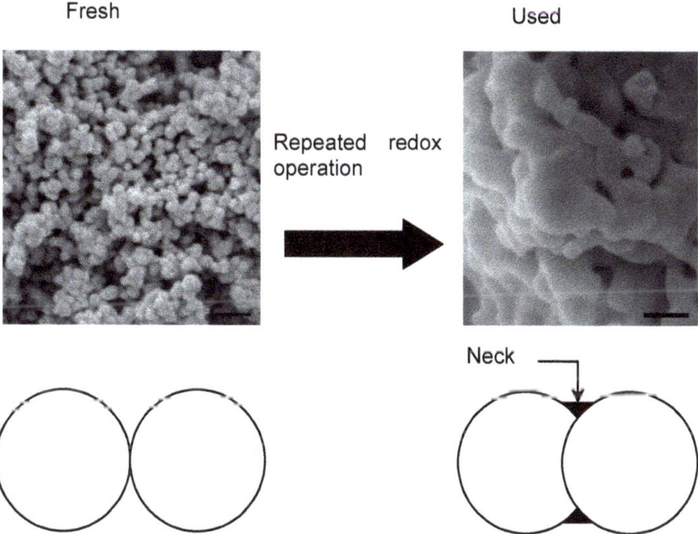

Figure 5. SEM images of Fe_2O_3 sample before and after six methane-steam redox cycles at 750 °Cand representative schematics of neck growth between two particles[41].

One of the approaches to minimise material sintering is to inhibit the diffusion in the solid particle. The complete reduction of the oxygen carrier to the corresponding zero valent metal is also a main cause of sintering since most metals agglomerates easily under elevated temperature conditions. Fukase and Suzuka [50] reported that the formation and accumulation of FeO during CL operation was mainly responsible for deactivation when iron oxide was used as the oxygen carrier. They also pointed out the importance of balancing the stoichiometry of reduction and oxidation of iron oxide and to avoid the formation of FeO by controlling reduction and oxidation temperatures. It is also important that the reduced iron species were completely oxidised to Fe_3O_4 phase. This mitigates the crystallite growth of the iron oxide and effectively prevents it from any structural changes.

Carbon is a common by-product of the CL process when a carbonaceous fuel is used as the feedstocks. Two possible routes for carbon formation are the decomposition of methane (Eq. 5) and the Boudouard reaction (Eq. 6). Methane decomposition is an endothermic reaction, and it is thermodynamically favourable at a high temperature, while the Boudourard reaction is favourable at a low temperature. These reactions could become significant in the presence of catalysts. Upon reduction, many metal oxides such as NiO, CuO and Fe_2O_3

could give rise to active metal centres which are able to rapidly produce carbon on the oxygen carrier surfaces. Once the solid carbon is formed, it will be carried over to the subsequent oxidation cycle where it is gasified to produce CO_x. When this happens, the purity of the hydrogen produced will be inevitably lowered.

$$CH_4 \rightarrow C + 2H_2 \Delta H^\circ = 74.6 \text{ kJ/mol} \quad (5)$$

$$2CO \rightarrow C + CO_2 \Delta H^\circ = -172.4 \text{ kJ/mol} \quad (6)$$

In general, as the oxygen ratio in the system decreases, there is a higher tendency towards carbon formation. The oxygen ratio is defined as the actual amount of oxygen contained in the metal oxide to the stoichiometric amount of oxygen required for complete oxidation of the fuel. It is also clear that carbon formation becomes more favourable as the oxygen in the oxygen carrier is depleted through the reaction with fuel. Cho et al. [51] reported that when more than 80 % of the available oxygen in the Ni-based oxygen carrier was consumed, the rate of carbon formation increased rapidly. This was accompanied by a drastic decrease in the fuel conversion because of the decreasing oxygen content available for oxidation. Galvita and Sundmacher [43] reported that a maximum Fe reduction of 60 % largely minimised carbon formation and a high purity hydrogen stream (< 20 ppm CO) could be obtained.

4. Process economics

In view of the lack of information on the cost of hydrogen produced from the CL process, the preliminary economic analysis and greenhouse gas footprint (GHG equivalent emissions in terms of carbon dioxide) of a methane-steam redox process will be provided in this section. A simple design for hydrogen production via a two-reactor layout was first obtained by considering the mass and energy balances as well as the overall pressure balance in order to establish a circulation of solids between the two reactors. The means of exchanging heat (direct, indirect, counter-current, available surface area, approach temperatures etc) has been considered, but has not been addressed further in this study. The pressure balance was affected by variables including the physical properties of the solid and gas, fluid velocity, solids recirculation rate as well as the geometry of the system. The pressure balance was solved using a one-dimensional model[52]. The basis of the design was a hydrogen production rate of 49 kg/h (or 547 Nm^3/h). This process considered the use of iron oxide as the oxygen carrier. Because the reduction of the iron oxide was much slower than its oxidation, a bubbling fluidised bed was chosen for the fuel reactor and a riser for the steam reactor. A particle size and density of the iron oxide particles were assumed to be 160μm and 5850kg/m^3, respectively. Other assumptions made for the operation are listed in Table 4. A high solids (i.e. the iron oxide) flow rate was required through the riser in order to meet the mass balance. This resulted in a high pressure drop across the riser, which was reduced by increasing the excess steam used for oxidation of the reduced iron oxide in the riser (at constant superficial gas velocity). The resultant mass balance is given inFigure6 and the CLH2 design is presented in Figure 5.

	Steam Reactor (riser)	Downcomer	Fuel Reactor (bubbling fluidised bed)	Loop Seal to Steam Reactor
Superficial gas velocity [m/s]	6.0	0.1	0.15	0.1
Temperature [°C]	750	700	750	700
Feed gas	Steam	Steam	Natural Gas	Steam
Feed gas temperature [°C]	240	240	500	240
Feed gas pressure [bar]	5	5	5	5
Conversion	100% FeO to Fe_3O_4	None	20% F_3O_4 to FeO 100% conversion of NG	None
Residence time required			1 minute	

Table 4. Assumptions used in the design of a CLH2process.

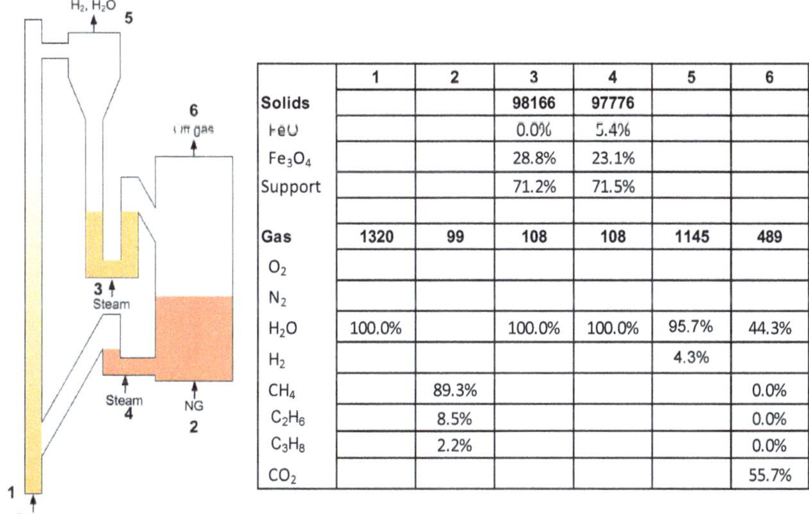

	1	2	3	4	5	6
Solids			**98166**	**97776**		
FeO			0.0%	5.4%		
Fe_3O_4			28.8%	23.1%		
Support			71.2%	71.5%		
Gas	**1320**	**99**	**108**	**108**	**1145**	**489**
O_2						
N_2						
H_2O	100.0%		100.0%	100.0%	95.7%	44.3%
H_2					4.3%	
CH_4		89.3%				0.0%
C_2H_6		8.5%				0.0%
C_3H_8		2.2%				0.0%
CO_2						55.7%

Figure 6. A schematic of CLH2processand the mass balance used for hydrogen production. Flow rates are represented in kg/hr and compositions in mass percentage.

The process flow diagram including the major peripheral equipment is shown in Figure 7. The heat from the exothermic reaction in the riser is used to raise superheated steam at 20 bar and 400 °C. This is used to generate electricity, with the steam let down to 5 bar and 240 °C. 25% of the steam is used as feed to the steam reactor and to fluidise the two loop seals. The water vapour content in the hydrogen product stream is due to the excess steam fed to the riser as well as from steam used to fluidise the loop seals. This is condensed out and returned with the water from the steam turbine to the boiler, in order to reduce the fresh water requirement. The heating required for the endothermic reaction in the fuel reactor is reduced

		Steam reactor	Steam down-comer	Fuel reactor	Units
Gas flow	Entering	957	16	159	Nm³/h
	Exiting	457	14	137	Nm³/h
Superficial gas velocity		6.0	0.1	0.15	m/s
G_s	Entering	346	346	51	kg/m²·s
Internal diameter		0.32	0.32	0.83	m
Temperature		750	700	750	°C
Pressure	Bottom	113	98	107	kPa,g
	Top	102.45	179.31	179.31	kPa,g
Height	Total internal	15	-	3.9	m
	Gas exit (from top of riser)	0.8	-	-	m
	Downcomer (not including cyclone)	-	7	-	m
	Bubbling bed /loop seal	-	0.7	1.1	m
Height relative to datum					
	Bottom	0.0	5.1	1.9	m
	Loop seal entrance to riser	1.2	-	-	m
Solids void age (ε)		0.88	0.47	0.53	

Table 5. Reactor configuration for CLH2process.

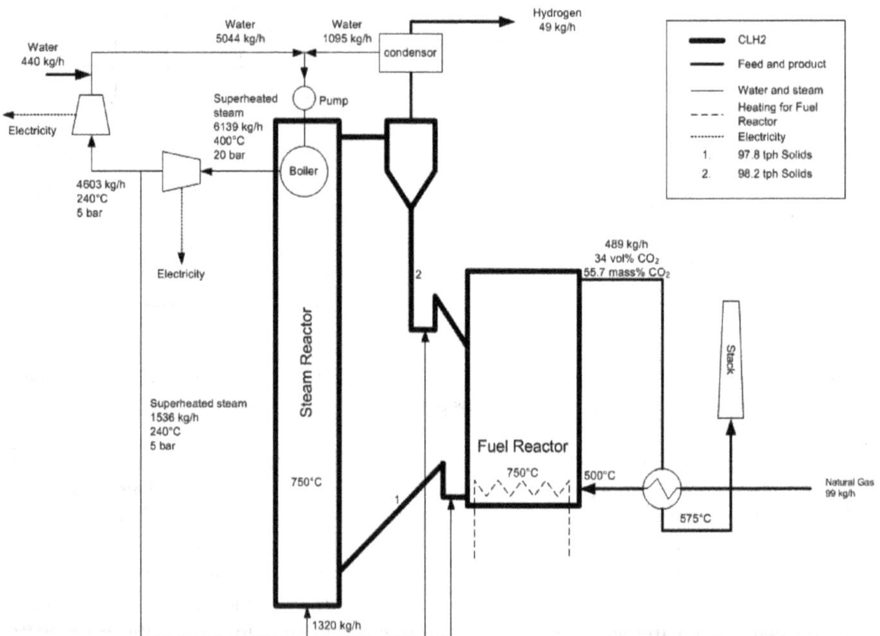

Figure 7. Proposed flow diagram of CLH2 process, showing peripheral equipment.

by pre-heating the natural gas using the waste heat from the off gas from the fuel reactor. For the current heat balance purpose it is assumed that there are different ways of supplying this remaining heat. One of the possible ways of supplying direct heat is by including a third combustion loop operated at higher temperature, which is outside the scope of this study.

The greenhouse gas emissions associated with the production of a unit of hydrogen were calculated using lifecycle assessment (LCA) techniques. Principally, LCA is a technique used to assess the environmental impacts of all stages associated with the production, use and disposal of a product or delivery of a service (product life from cradle to grave). In the case of a fossil fuel for example, this includes not only the combustion emissions associated with the fuel's use, but also includes pre-combustion or upstream emissions resulting from the extraction, production, transportation, processing, conversion and distribution of the fuel. The international standards contained in the ISO 14040 series [53] provide a basic framework in which to undertake LCA. A more general introduction to LCA may be found in Horne et al. [54]and Weidema et al. [55].I n this study, all fuel production and feedstock supply processes, as specified in Figure 7, were included in the LCA. The analysis is therefore limited to processes upstream of the refinery gate and thus does not include the delivery and combustion of hydrogen. Emission results are reported using the concept of a global warming potential (GWP), which enables different greenhouse gases to be compared and expressed using an equivalent carbon dioxide (gCO_2e) value. Data used for the analysis are summarised in Table 6 based on an hourly hydrogen production rate of 49 kg. The GHG impact of the CLH2 process under consideration is 18,690 $gCO_2e/kg\ H_2$ produced or 154 $gCO_2e/MJ\ H_2$. The impact is dominated by the need to supply process heat to the fuel reactor(redox heater emissions:9,628 $gCO_2/kg\ H_2$) as shown in Figure 8.

Inputs	Value	Units	Comments
Resources			
Natural gas	99	kg	Natural gas for reaction
Oxide material	1.26	kg	Yearly make-up (per hour)
Water	440	kg	Make-up water (reaction and cooling)
Energy			
Natural gas	151	kg	Fuel reactor heat requirement
Electricity	189	kW	Net electricity requirement
Outputs			
Hydrogen	49	kg	Compressed hydrogen output
Emissions			
H_2O	217	kg	Fuel reactor (stack emissions)
CO_2	272	kg	Fuel reactor (stack emissions)
CO_2	419	kg	Fuel reactor (heater emissions)

Table 6. LCA inputs/outputs (per hour) for the CLH2 process.

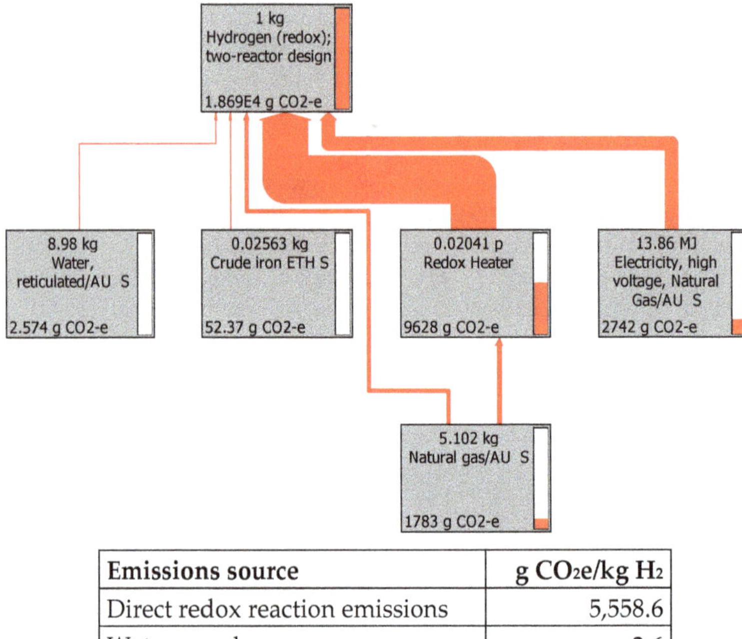

Emissions source	g CO_2e/kg H_2
Direct redox reaction emissions	5,558.6
Water supply	2.6
Gas supply (reaction feedstock)	706.1
Oxide material	52.4
Fuel reactor (process heat)	9,628.0
Electricity (from natural gas)	2,742.0
Total	18,689.7

Figure 8. Redox emissions breakdown (per kg H_2).

Preliminary results demonstrate the need to optimise the delivery of heat to the fuel reactor. The introduction of a third combustion loop operated at higher temperature is one such means to reduce upstream emissions. However, this may negatively influence total capital expenditure. The literature reports hydrogen production through current steam reforming technology produces between 9,830 gCO_2e/kg H_2 (24,000 kg H_2/day; midsized facility) and 12,130 gCO_2e/kg H_2 (480 kg H_2/day; distributed facility), and thus are higher than the direct redox process emissions [56], although significantly lower than the total CLH2 emissions. The literature only considered electricity and natural gas related emissions and thus total upstream emissions of existing technologies maybe higher than the reported values.

The commercial viability of the redox process was estimated using cost estimate practices outlined in the literature [56, 57]. Results are reported in \$/kg H_2. Material and fuel operating expenditure was calculated using the inputs identified in Figure7, as summarised in the lifecycle analysis section (Table 6). Fixed operating and maintenance costs were calculated based on the total capital expenditure. Battery limit capital expenditure (e.g. redox process) is based on the engineering judgment of the authors, with capital build-up (facilities, engineering, permitting, start-up, contingencies, working capital and land) estimated using a percentage of the battery limit cost. Capital charges are calculated using a percentage of total capital expenses. Importantly, although the estimates may look precise, they are simply estimates based on the judgment of the authors. There remains significant uncertainty about the actual cost of the redox process as it has not been commercially demonstrated. A breakdown of cost data is provided in Table 7.

Initial costing estimates show that the redox process may produce hydrogen at $8.93/kg ($9.36/kg, including carbon tax). The cost breakdown demonstrates that onsite storage of compressed hydrogen represents a significant expense. However, this arises from the conversion of stranded methane. If demand for hydrogen is identified close to a stranded gas reserve, storage costs will decrease significantly. Delivery of compressed hydrogen represents an additional cost that has not been considered in this analysis. Literature cost estimates for at

Expense	$M/yr	Comment
OpEX		Variable (fuel and materials)
Oxide material	0.55	$50/kg
Natural gas	0.20	Reaction feed and reducer heating
Electricity	0.12	Net electricity demand
Water	0.01	Make-up supply
Total OpEX	**0.88**	
CapEX	**$M**	
CLH2 reactor	10.0	
H_2 Compression	0.44	$3,000/kW capacity
H_2 Storage	5.97	$26,417/m^3 capacity; 5 days storage
Total process units	16.42	
General facilities	3.28	20 % of process unit CapEX
Engineering	2.46	15 % of process unit CapEX
Contingencies	1.64	10 % of process unit CapEX
Working capital	0.82	5 % of process unit CapEX
Total CapEX	24.62	
Balance	**$M/yr**	
OpEx (variable)	0.88	
OpEx (fixed)	0.49	2 % of total CapEX
Capital charge	2.46	10 % of total CapEX
Carbon Tax	0.18	$23/T CO_2
Total ($M/yr)	**4.02**	
Total ($/kg H_2)	**8.93**	**(ex. carbon tax)**
	9.36	**(inc. carbon tax)**

Table 7. Redox process cost estimates.

gate hydrogen production via steam reforming, using current technology, range between \$1.51/kg (midsize facility: 24,000 kg H_2/day) to \$3.68/kg (distributed facility: 480 kg H_2/day facility). Hence the hydrogen at gate cost for the CLH2 process is higher than steam reforming technology. Electrolysis production of hydrogen ranges between \$4.94 and \$6.82 per kg for a midsize and distributed facility respectively and thus is closer to CLH2production costs [56]. Experience gained through the commercialisation and deployment of the redox technology is expected to reduce costs, particularly capital build-up costs. However the stranded nature of the product may significantly increase total delivered hydrogen cost.

5. Conclusion

The feasibility of producing hydrogen from the metal/metal oxide redox process has been demonstrated in the literature. This process offers several advantages including the ability to produce hydrogen of high purity and a concentrated stream of carbon dioxide. Most importantly this process eliminates the need for a supply of high purity oxygen and a water gas shift process that are generally required by commercial processes. However, this redox process is not regarded as a fully developed technology and further R&D development is required for commercialisation.

In view of the literature, much research effort has been devoted to formulating novel oxygen carrier materials. Although several types of improved oxygen carrier materials have been identified, full appraisals of their performance and further optimisation studies are required. Iron oxides and nickel oxides appear to be attractive candidates for this application in terms of their activity. However, their thermal stabilities need further improvement. Current practices include doping, introducing a diffusional barrier provided by a second oxide, and/or adding a second oxide with higher oxygen storage capacity. There are also a limited number of studies that investigate the life time of oxygen carriers. Apart from chemical stability, the changes in the physical properties such as size and attrition of the carrier particles during fluidisation have received little attention and should be addressed in future research. It is viewed strongly that improvement in these areas would significantly increase process efficiency and economic viability of the cyclic redox process.

The lack of pilot scale studies also impedes the commercialisation of cyclic redox and chemical looping processes. Limited data are available for process design, scale-up and optimisation. For example, the transfer of the oxygen carrier particles between oxidation and reduction is a critical issue when it comes to process design. Fixed bed, moving bed and circulating fluidised bed have been proposed, and the choice of reactor will depend on the reaction kinetics and the required flow dynamics of the process. Because the cyclic redox process is considered as an unsteady process, the definition of the operation window of the process will be determined by limiting the upper and the lower oxidation states of the metal/metal oxide couple. This parameter has a direct impact on the overall conversion efficiencies, process designs and economics. Since the redox reactions usually take place at temperatures above 600 °C, most of the sensible heat stored in the gas existing from the oxidation and reduction reactors can be used to generate power with a steam generator. The co-production of excess electricity would reduce the cost of the hydrogen produced and increase overall process viability. Hence, the issue of heat management requires much closer examination when it comes to process optimisation.

Finally, the current preliminary LCA-Economic study has made the first attempt to provide an indicative price of hydrogen produced from the redox process. Although the cost of hydrogen produced from the redox process is higher than hydrogen produced from other commercial processes, several design parameters have been identified as the areas for future improvement. It is seen that the LCA techniques are valuable tools for process optimisation.

Author details

Doki Yamaguchi, Liangguang Tang, Nick Burke and Ken Chiang*
CSIRO Earth Science and Resource Engineering, Australia

Lucas Rye
CSIRO Marine and Atmospheric Research, Australia

Trevor Hadley and Seng Lim
CSIRO Process Science and Engineering, Australia

Acknowledgement

The authors acknowledge the support from CSIRO Petroleum and Geothermal Research Portfolio in conducting this study

6. References

[1] IEA (2005) Prospects for Hydrogen and Fuel Cells. OECD Publishing
[2] Mueller-Langer, F., Tzimas, E., Kaltschmitt, M., and Peteves, S. (2007) Techno-economic assessment of hydrogen production processes for the hydrogen economy for the short and medium term. Int. J. Hydrogen Energy 32: 3797-3810.
[3] Pardro, C.E.G. and Putsche, V., Survey of the Economics of Hydrogen Technologies. 1999, National Renewable Energy Laboratory: Golden, Colorado.
[4] Mirabal, S.T., (2003) An Economic Analysis of Hydrogen Production Technologies Using Renewable Energy Resources, University of Florida.
[5] Raissi, A.T. and Block, D.L. (2004) Hydrogen - automotive fuel of the future. IEEE Power Energ. Mag. November/December 40-45.
[6] Fan, L.S. and Li, F.X. (2010) Chemical Looping Technology and Its Fossil Energy Conversion Applications. Ind. Eng. Chem. Res. 49: 10200-10211.
[7] Adanez, J., Abad, A., Garcia-Labiano, F., Gayan, P., and de Diego, L.F. (2012) Progress in Chemical-Looping Combustion and Reforming technologies. Prog. Energy Combust. Sci. 38: 215-282.
[8] Gupta, P., Velazquez-Vargas, L.G., and Fan, L.S. (2007) Syngas redox (SGR) process to produce hydrogen from coal derived syngas. Energy Fuels 21: 2900-2908.

* Corresponding Author

[9] Li, F., Kim, H.R., Sridhar, D., Wang, F., Zeng, L., Chen, J., and Fan, L.S. (2009) Syngas Chemical Looping Gasification Process: Oxygen Carrier Particle Selection and Performance. Energy Fuels 23: 4182-4189.
[10] Li, F.X., Zeng, L., Velazquez-Vargas, L.G., Yoscovits, Z., and Fan, L.S. (2010) Syngas Chemical Looping Gasification Process: Bench-Scale Studies and Reactor Simulations. AIChE J. 56: 2186-2199.
[11] Yang, J.B., Cai, N.S., and Li, Z.S. (2008) Hydrogen production from the steam-iron process with direct reduction of iron oxide by chemical looping combustion of coal char. Energy Fuels 22: 2570-2579.
[12] Fan, L., Li, F., and Ramkumar, S. (2008) Utilization of chemical looping strategy in coal gasification processes. Particuology 6: 131-142.
[13] Gnanapragasam, N.V., Reddy, B.V., and Rosen, M.A. (2009) Hydrogen production from coal using coal direct chemical looping and syngas chemical looping combustion systems: Assessment of system operation and resource requirements. Int. J. Hydrogen Energy 34: 2606-2615.
[14] Sime, R., Kuehni, J., D'Souza, L., Elizondo, E., and Biollaz, S. (2003) The redox process for producing hydrogen from woody biomass. Int. J. Hydrogen Energy 28: 491-498.
[15] Kobayashi, N. and Fan, L.S. (2010) Biomass direct chemical looping process: A perspective. Biomass Bioenergy 35: 1252-1262.
[16] Li, F.X., Zeng, L., and Fan, L.S. (2010) Biomass direct chemical looping process: Process simulation. Fuel 89: 3773-3784.
[17] Hacker, V., Faleschini, G., Fuchs, H., Fankhauser, R., Simader, G., Ghaemi, M., Spreitz, B., and Friedrich, K. (1998) Usage of biomass gas for fuel cells by the SIR process. J. Power Sources 71: 226-230.
[18] Bleeker, M.F., Kersten, S.R.A., and Veringa, H.J. (2007) Pure hydrogen from pyrolysis oil using the steam-iron process. Catal. Today 127: 278-290.
[19] Bleeker, M.F., Veringa, H.J., and Kersten, S.R.A. (2010) Pure Hydrogen Production from Pyrolysis Oil Using the Steam-Iron Process: Effects of Temperature and Iron Oxide Conversion in the Reduction. Ind. Eng. Chem. Res. 49: 53-64.
[20] Galvita, V.V., Poelman, H., and Marin, G.B. (2011) Hydrogen Production from Methane and Carbon Dioxide by Catalyst-Assisted Chemical Looping, . Top. Catal. 54 907-913.
[21] Cormos, C.-C. (2011) Hydrogen production from fossil fuels with carbon capture and storage based on chemical looping systems. Int. J. Hydrogen Energy 36: 5960-5971.
[22] Otsuka, K., Kaburagi, T., Yamada, C., and Takenaka, S. (2003) Chemical storage of hydrogen by modified iron oxides. J. Power Sources 122: 111-121.
[23] Otsuka, K., Yamada, C., Kaburagi, T., and Takenaka, S. (2003) Hydrogen storage and production by redox of iron oxide for polymer electrolyte fuel cell vehicles. Int. J. Hydrogen Energy 28: 335-342.
[24] Wang, H., Wang, G., Wang, X., and Bai, J. (2008) Hydrogen production by redox of cation-modified iron oxide. J. Phys. Chem. C 112: 5679-5688.
[25] Liu, X. and Wang, H. (2010) Hydrogen production from water decomposition by redox of Fe_2O_3 modified with single- or double-metal additives. J. Solid State Chem. 183: 1075-1082.
[26] Otsuka, K. and Takenaka, S. (2004) Storage and supply of pure hydrogen mediated by the redox of iron oxides. J. Jpn. Pet. Inst. 47: 377-386.

[27] Ryu, J.C., Lee, D.H., Kang, K.S., Park, C.S., Kim, J.W., and Kim, Y.H. (2008) Effect of additives on redox behavior of iron oxide for chemical hydrogen storage. J. Ind. Eng. Chem. 14: 252-260.

[28] Kodama, T., Watanabe, Y., Miura, S., Sato, M., and Kitayama, Y. (1996) Reactive and selective redox system of Ni(II)-ferrite for a two-step CO and H_2 production cycle from carbon and water. Energy 21: 1147-1156.

[29] Kang, K.S., Kim, C.H., Cho, W.C., Bae, K.K., Woo, S.W., and Park, C.S. (2008) Reduction characteristics of $CuFe_2O_4$ and Fe_3O_4 by methane; $CuFe_2O_4$ as an oxidant for two-step thermochemical methane reforming. Int. J. Hydrogen Energy 33: 4560-4568.

[30] Takenaka, S., Hanaizumi, N., Son, V.T.D., and Otsuka, K. (2004) Production of pure hydrogen from methane mediated by the redox of Ni- and Cr-added iron oxides. J. Catal. 228: 405-416.

[31] Urasaki, K., Tanimoto, N., Hayashi, T., Sekine, Y., Kikuchi, E., and Matsukata, M. (2005) Hydrogen production via steam-iron reaction using iron oxide modified with very small amounts of palladium and zirconia. Appl. Catal., A 288: 143-148.

[32] Adanez, J., de Diego, L.F., Garcia-Labiano, F., Gayan, P., Abad, A., and Palacios, J.M. (2004) Selection of Oxygen Carriers for Chemical-looping Combustion. Energy Fuels 18: 371-377.

[33] Cho, P., Mattisson, T., and Lyngfelt, A. (2004) Comparison of iron-, nickel-, copper- and manganese-based oxygen carriers for chemical-looping combustion. Fuel 83: 1215-1225.

[34] Zafar, Q., Mattisson, T., and Gevert, B. (2005) Integrated hydrogen and power production with CO_2 capture using chemical-looping reforming-redox reactivity of particles of CuO, Mn_2O_3, NiO, and Fe_2O_3 using SiO_2 as a support. Ind. Eng. Chem. Res. 44: 3485-3496.

[35] Kodama, T., Shimizu, T., Satoh, T., Nakata, M., and Shimizu, K.I. (2002) Stepwise production of CO-RICH syngas and hydrogen via solar methane reforming by using a Ni(II)-ferrite redox system. Sol. Energy 73: 363-374.

[36] Kodama, T., Kondoh, Y., Yamamoto, R., Andou, H., and Satou, N. (2005) Thermochemical hydrogen production by a redox system of ZrO_2-supported Co(II)-ferrite. Sol. Energy 78: 623-631.

[37] Cha, K.S., Yoo, B.K., Kim, H.S., Ryu, T.G., Kang, K.S., Park, C.S., and Kim, Y.H. (2010) A study on improving reactivity of Cu-ferrite/ZrO_2 medium for syngas and hydrogen production from two-step thermochemical methane reforming. Int. J. Energy Res. 34: 422-430.

[38] Kang, K.S., Kim, C.H., Bae, K.K., Cho, W.C., Kim, S.H., Kim, W.J., Kim, Y.H., and Park, C.S. (2010) Redox cycling of $CuFe_2O_4$ supported on ZrO_2 and CeO_2 for two-step methane reforming/water splitting. Int. J. Hydrogen Energy 35: 568-576.

[39] Li, K.Z., Wang, H., Wei, Y.G., and Yan, D.X. (2009) Selective Oxidation of Carbon Using Iron-Modified Cerium Oxide. J. Phys. Chem. C 113: 15288-15297.

[40] Tang, L., Yamaguchi, D., Burke, N., Trimm, D., and Chiang, K. (2010) Methane decomposition over ceria modified iron catalysts. Catal. Commun. 11: 1215-1219.

[41] Yamaguchi, D., Tang, L., Wong, L., Burke, N., Trimm, D., Nguyen, K., and Chiang, K. (2011) Hydrogen production through methane-steam cyclic redox processes with iron-based metal oxides. Int. J. Hydrogen Energy 36: 6646-6656.

[42] Sim, A., Cant, N.W., and Trimm, D.L. (2010) Ceria-zirconia stabilised tungsten oxides for the production of hydrogen by the methane-water redox cycle. Int. J. Hydrogen Energy 35: 8953-8961.

[43] Galvita, V. and Sundmacher, K. (2005) Hydrogen production from methane by steam reforming in a periodically operated two-layer catalytic reactor. Appl. Catal., A 289: 121-127.

[44] Datta, P., Rihko-Struckmann, L.K., and Sundmacher, K. (2011) Influence of molybdenum on the stability of iron oxide materials for hydrogen production with cyclic water gas shift process. Mater. Chem. Phys. 129: 1089-1095.

[45] Leion, H., Mattisson, T., and Lyngfelt, A. (2009) Use of Ores and Industrial Products As Oxygen Carriers in Chemical-Looping Combustion. Energy Fuels 23: 2307-2315.

[46] Leion, H., Lyngfelt, A., Johansson, M., Jerndal, E., and Mattisson, T. (2008) The use of ilmenite as an oxygen carrier in chemical-looping combustion. Chem. Eng. Res. Des 86: 1017-1026.

[47] Adanez, J., Cuadrat, A., Abad, A., Gayan, P., de Diego, L.F., and Garcia-Labiano, F. (2010) Ilmenite Activation during Consecutive Redox Cycles in Chemical-Looping Combustion. Energy Fuels 24: 1402-1413.

[48] Leion, H., Mattisson, T., and Lyngfelt, A. (2008) Solid fuels in chemical-looping combustion. Int. J. Greenhouse Gas Control 2: 180-193.

[49] Lorente, E., Pena, J.A., and Herguido, J. (2011) Cycle behaviour of iron ores in the steam-iron process. Int. J. Hydrogen Energy 36: 7043-7050.

[50] Fukase, S. and Suzuka, T. (1993) Residual oil cracking with generation of hydrogen - Deactivation of iron-oxide catalyst in the steam iron reaction. Appl. Catal., A 100: 1-17.

[51] Cho, P., Mattisson, T., and Lyngfelt, A. (2005) Carbon formation on nickel and iron oxide-containing oxygen carriers for chemical-looping combustion. Ind. Eng. Chem. Res. 44: 668-676.

[52] Hadley, T.D., Chiang, K., Burke, N.R., and Lim, K.S. (2010) Multiple-loop chemical reactor design with pressure balance consideration. Fluidization XIII 463-470.

[53] ISO, ISO/DIS14040, Environmental Management Standard - Life Cycle Assessment, Principlesand Framework. International Standard. Switzerland. 2006.

[54] Horne, R., Grant, T., and Verghese, K. (2009) Life Cycle Assessment: Principles, Practice and Prospects. CSIRO

[55] Weidema, B.P., Rebitzer, G., and Ekvall, T. (2004) Scenarios in Life-cycle Assessment Society of Environmental Toxicology and Chemistry

[56] NRC (2004) The Hydrogen Economy: Opportunities, Costs, Barriers, and R&D Needs. The National Academies Press.

[57] Simbeck, D. and Chang, E. (2002) Hydrogen Supply: Cost Estimate for Hydrogen Pathways Scoping Analysis . NREL, Report: SR-540-3252. National Renewable Energy Laboratory

Chapter 5

Fermentative Hydrogen Production by Molasses; Effect of Hydraulic Retention Time, Organic Loading Rate and Microbial Dynamics

Iosif Mariakakis, Carsten Meyer and Heidrun Steinmetz

Additional information is available at the end of the chapter

1. Introduction

The generation of hydrogen by biological means is not energy intensive compared with the conventional thermochemical techniques, since the operating temperature and pressure are not very high. As raw materials organic waste streams can be used that can be considered as a renewable resource (Vijayaraghavan & Mohd Soom, 2006).The method of the dark fermentation has certain advantages compared with the other biological processes. In contrast to bio-photolysis and photo fermentation, the process needs no solar radiation, but the required energy is supplied by the organic substrates and hence the process is not interrupted during the night. Moreover, the production rate of the H_2 of the fermentative bacteria in comparison with the other biological processes is greater (Kumar et al., 2000; Nath et al., 2005).

The different process parameters that are relevant for hydrogen production have been surveyed (Li & Fang, 2007; Wang & Wan, 2009) and include the type of substrate, nutrient concentration, the inoculum, pH, reactor configuration, hydraulic retention time (HRT), organic loading rate (OLR). Carbohydrate-rich substrates are the most suitable for fermentative H_2 production systems (Hawkes et al., 2002; Kapdan & Kargi, 2006; Meherkotay & Das, 2008; Ueno et al., 2007) seeded with saccharoclastic microorganisms, They are able to break down organic substances via the Embden-Meyerhof pathway resulting to different metabolic products depending on the type of microorganism and the environmental conditions driving their catabolism (Hallenbeck, 2009).

The relevant microbial groups for the fermentative hydrogen production groups are clostridia and enterobacteria (Hallenbeck, 2005; Hawkes et al., 2007). Both groups were repeatedly experimentally confirmed as major hydrogen producers (Valdez-Vazquez & Poggi-Varaldo, 2009).

$C_6H_{12}O_6 + 2\,H_2O \rightarrow CH_3COOH + CH_3OH + 2\,CO_2 + 2\,H_2$	(1)
$C_6H_{12}O_6 + 2\,H_2O \rightarrow 2\,CH_3COOH + 2\,CO_2 + 4\,H_2$	(2)
$C_6H_{12}O_6 \rightarrow CH_3(CH_2)_2\,COOH + 2\,CO_2 + 2\,H_2$	(3)
$C_6H_{12}O_6 \rightarrow 0.6\,CH_3COOH + 0.7\,CH_3(CH_2)_2\,COOH + 2\,CO_2 + 2.6\,H_2$	(4)
$4\,H_2 + CO_2 \rightarrow CH_3COOH + 2\,H_2O$	(5)
$C_6H_{12}O_6 + 2\,H_2 \rightarrow 2\,CH_3CH_2COOH + 2\,H_2O$	(6)
$4\,H_2 + H_2SO_4 \rightarrow H_2S + 4\,H_2O$	(7)
$C_6H_{12}O_6 \rightarrow CH_3CH(OH)COOH + CH_3COOH + CO_2$	(8)
$C_6H_{12}O_6 \rightarrow 2\,CH_3CH(OH)COOH$	(9)
$3CH_3CH(OH)COOH \rightarrow 2\,CH_3CH_2COOH + CH_3COOH + CO_2 + H_2O$	(10)
$4\,CH_3OH + 2\,CO_2 \rightarrow 3\,CH_3COOH + 2H_2O$	(11)
$2\,CH_3OH + CH_3COOH \rightarrow CH_3(CH_2)_4\,COOH + 2\,H_2O$	(12)
$Fd_{red} + 2\,H^+ \rightarrow Fd_{ox} + H_2$	(13)
$CH_3CH(OH)COOH + 0.4\,CH_3COOH \rightarrow$ $0.7\,CH_3(CH_2)_2\,COOH + CO_2 + 0.4\,H_2O + 0.6\,H_2$	(14)

Table 1. Biochemical reactions relevant to hydrogen production

Enteric bacteria are gram-negative rods, facultative aerobic, with relatively simple nutrient requirements and can not form spores (Schmauder, 1992). Among the species that can produce H_2, are *Escherichia* (*E. coli*), *Proteus* (*P. vulgaris*), *Enterobacter* (*E. aerogenes*). Enteric bacteria ferment sugars to a variety of end products such as acetate, formate, lactate, succinate, ethanol, CO_2 and H_2. Hydrogen is produced according to equation 1 (Li & Fang, 2007). The maximum possible hydrogen yield by this pathway is 2 mol H_2 per mol hexose. In experiments with intestinal bacteria rather half of this value has been found (Hallenbeck, 2005).

Clostridia are spore forming, gram-positive bacteria (Schmauder, 1992). Through sporulation they can survive for example dehydration, heat and large changes in pH. Clostridial catabolism includes a variety of reactions and hence fermentation end-products such as acetate, acetone (*C. pasteurianum*), butyrate (*C. butylicum*), butanol (*C. acetobutylicum*) or caproic acid (C. *kluyveri*) (Schmauder, 1992). Hydrogen, using hexose as a substrate can be produced by two pathways with acetate and butyrate as end-products as equations 1 and 2 describe (Hallenbeck, 2005; Hawkes et al., 2007; Li & Fang, 2007). That fact that clostridia can produce higher amount of hydrogen makes them more attractive and hydrogen systems aiming at their growth must be strived. In experiments with mixed cultures yields between 1.5 mol H_2/mol hexose and 2.5 mol H_2/mol hexose were achieved (Wang & Wan, 2009). In practice the highest yield is achieved when the catabolism is driven through a mixed acetate butyrate according to equation 4 (Lengeler, 1999). This is because hexose can be also metabolized by hydrogen neutral fermentation pathways with lactic acid, ethanol, acetone, butanol as end-products, or a portion of the substrate is consumed for the production of biomass, which theoretically can be 18.5% and 14.5% of the theoretical one through the acetate and butyrate hydrogen producing pathway respectively (Aceves-Lara et al., 2008). More over, a part of hydrogen that is already produced may be consumed by certain microorganisms as homoacetogens (equation 5) with acetate as end-product (Dworkin et al., 2006), or propionic acid bacteria (equation 6) (Li & Fang, 2007), or sulfate reducing bacteria (equation 7). The co-existence of microorganisms other than hydrogen producing that compete for substrate has been observed in many hydrogen producing systems (Hawkes et al., 2007; Hung et al., 2011a, 2011b; Li et al., 2011). A major part of them belongs to the lactic acid bacteria (LAB) distinguished to heterofermentative LAB, which produce lactic acid together with CO_2 (equation 8) and minor quantities of ethanol and acetic acid and the homofermentative LAB, which produce mainly lactic acid (equation 9) (Martinko & Clark , 2009). Through the monitoring of the metabolites of carbohydrate solely is not always possible to determine the metabolic pathays used by the bacteria, since many clostridia are capable of secondary fermentation. *C. propionicum* can for instance metabolize lactic acid for the production of propionate and acetate (equation 10), while some homoacetogens utilize ethanol and CO_2 yielding acetate (equation 11). In some cases, hydrogen can be also produced by secondary fermentation *C. kluyveri* can utilize ethanol and acetate yielding hexanoic acid and molecular hydrogen by the oxidation of reduced ferredoxin (Fd) (equation 12) and *C. tyrobutyricum* can transform lactate and ethanol to butyrate (equation 14) (Martinko & Clark, 2009). It is therefore obvious that biological hydrogen production systems are complicated in terms of biological processes and microbial species involved.

Aim of this work was to experimentally study the effect of HRT and OLR on bio-hydrogen production in terms of maximization of H_2 yield, so as to optimize substrate utilization efficiency that contributes to the cost effectiveness of the process. Experiments were carried out in large-lab scale reactors of 30 L working volume. In this way, experience in the start-up procedure, selection of only H_2 producing microorganisms and the stability of long term continuous operation without methanogenesis of such a set-up could be gained. This can further be used for the scale-up of bio-hydrogen production towards the final aim of commercial implementation. An attempt towards the clarification of the possible metabolic pathways and the involved microorganisms, with along as their possible interactions was also undertaken. The way that these microorganisms behave and interract with each other and their milieu is very important for the design of effective hydrogen producing systems. The understanding of these processes can help the designer to manipulate hydrogen production by the suitable variation of the process parameters. With the use of molecular biological techniques it is possible to acquire better insights into such systems.

2. Materials and methods

2.1. Experimental set-up

Dark fermentation experiments were conducted in two identical 40 L reactors made of borosilicate glass (QVF) with a working volume of 30 L heated at 37 °C ± 2 °C by a heating pipe. The content of the fermenter was homogenized by external recirculation with eccentric screw pumps (Netsch). The pH was regulated by means of a pH glass electrode (Endress & Hauser, Orbisint CPS11) and a pH measuring transducer (Endress & Hauser, Mycom) connected to a programmable controller (Endress & Hauser, Memograph), which controlled 2 dosing pumps (Metrohm, Dosimat) for automatic addition of a sodium hydroxide solution 25% v/v and a hydrochloric acid solution 25% v/v, respectively. The Organic Loading Rate (OLR) was adjusted to the desired level by dosing (Metrohm, Dosimat) with molasses diluted 1:2 w/w and supplemented with nutrients. Every 100 mL of nutrient solution contained the following quantities in g; 1.72 $FeSO_4{\cdot}7H_2O$, 0.36 $CaCl_2{\cdot}2H_2O$, 3.78 KCl, 0.17 $MgCl_2{\cdot}6H_2O$, 11.46 NH_4Cl, 1.05 KH_2PO_4, 0.181 $FeCl_2{\cdot}4H_2O$, 0.041 $NiCl_2{\cdot}6H_20$, 0.021 $CoCl_2{\cdot}6H_2O$, 0.011 $ZnCl_2$, 0.170 KI, 0.177 $(NaPO_3)_6$, 0.0085 $MnCl_2{\cdot}4H_2O$, 0.0085 NH_4VO_3, 0.0085 $CuCl_2{\cdot}2H_2O$, 0.0061 $Al_2(SO_4)_3{\cdot}18H_2O$, 0.0085 $NaMoO_4{\cdot}2H_2O$, 0.0085 H_3BO_3, 0.0085 $Na_2WO_4{\cdot}2H_2O$, Na_2SeO_3 0.0085, 0.170 cysteine. Depending on the Hydraulic Retention Time (HRT) applied, the following quantities of this solution were added to the molasses solution; 33 mL for HRT > 2 d, 66 mL for 1 d< HRT < 2 d and 123 mL for HRT < 1 d.

The HRT was independent of substrate dosing. It was regulated by the pumping (Prominent, Gamma/L) of tap water and automatic removal of excess mixed liquor by a peristaltic pump (Ismatec, MPC Standard) controlled by a water lever sensor (Endress & Hauser, Liquiphant). The tab water was stored in containers and was daily sparged with N_2 in order to reduce the dissolved oxygen concentration bellow 1 mg/L. The produced biogas quantity was measured with a drum-type gas meter (Ritter, TG 05) and registered into the programmable controller. The produced gas was collected in gas bags (Lindte). In Figure 1 the experimental set-up is presented.

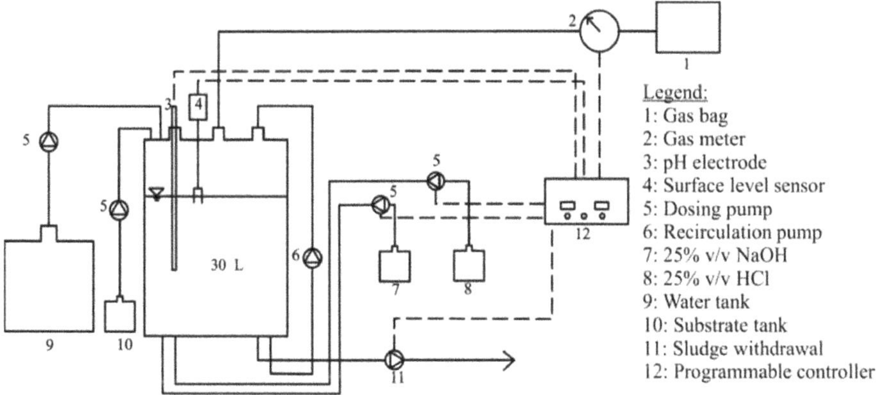

Figure 1. Experimental set-up

2.2. Reactor operation

The inoculum of the reactor has been acquired from the anaerobic digester of the Sewage Treatment Plant for Research and Education (LFKW) of the University of Stuttgart (Germany). It was diluted to 2% to 4% Total Solids (TS) concentration and sieved consecutively through 4 mm and 2 mm mesh size to prevent clogging of the tubing. It was then pretreated for 24 h at 105 °C, in order to kill the methanogenic bacteria. The dried sludge was pulverized and solved into tab water for the start-up of the system. As substrate sugar beet molasses acquired from a sugar factory in south Germany were used. In Table 2 the composition of molasses is presented. For the start-up of the system, the reactor was fed with 450 g of sucrose in a batch mode at pH 6.5 in order to enrich the biomass in H_2-producing microorganisms. Upon sucrose depletion continuous operation of the system was started. The pH was reduced to 5.5, a value that has been reported to be the optimum for continuous bio-hydrogen production (Mariakakis et al., 2011). The first phase of continuous operation aimed at the further selection of the biomass for hydrogen producing bacteria by application of high HRT and low OLR. The various experimental conditions tested during the continuous operation are presented in table 3. Their selection was based depending on the experimental progress as described in chapter 3.1. In many cases one of the two reactors had to be re-inoculated with seed sludge acquired by the other reactor. All phases had a minimum duration of 5 times the applied HRT. At phase DVII (table 3) Fe^{2+} at end-concentration in reactor of 1000 mg/L was added.

Molasses			
CODtot [mg/kg]	782000		
Ntot [mg/kg]	18610	Maltose [mg/L]	--
Ptot [mg/kg]	216	Acetate [mg/L]	1020
TS [g/kg]	848	Propionate [mg/L]	175
VS [%]	892	Butyrate [mg/L]	3062
Sucrose [g/kg]	520	D-Glucose [mg/L]	--
Lactose [mg/L]	--	Lactate [mg/L]	13736

Table 2. Composition of molasses

2.3. Analytical methods

The analyses of concern were determined according to the german standards (Deutsches Institut fuer Normung, 2002) and performed three times a week. These included; total solids (TS), volatile suspended solids (VSS), chemical oxygen demand (COD), a group parameter used for the detection of carbonaceous matter and nitrogen (in total and soluble form acquired after filtration through membrane with 0.45 μm pore diameter). Glucose, sucrose and lactic acid have been determined spectrophotometrically after enzymatic digestion by test kits according to the manufacturer's instructions (R-Biopharm). Gas Chromatography was used to analyze organic acids and alcohols. The sample was filtered through a 0.45 μm pore diameter filter and acidified with a 96% H_2SO_4 solution. Organic acids, ethanol and butanol were detected by GC (Perkin Elmer) mounted with a fused silica capillary (Varian) and using a flame ionization detector. Both the injection and capillary temperatures were set at 280 °C. Biogas composition was determined once daily after up-grading for particulate matter and water vapor removal by a gas analyzer (ABB, AO2020), equipped with an infrared detector for CH_4 and CO_2 and a thermal conductivity detector for H_2.

DNA extraction, 16S rDNA of eubacteria and clostridia PCR amplification, DGGE analysis and sequencing were performed as previously described (Mariakakis et al., 2011).

3. Results and discussion

3.1. Reactor operation

In figure 2, the reactor operation parameters, gas production and hydrogen yield performance for the experimental phases DI to DIII are presented.

After the start-up of the reactor for a week, the HRT and OLR were reduced to 4 d and 11.6 g sucrose / (L·d) respectively. After about 15 d a continuous H_2 production could be established (phase DI), which stabilized after approximately 30 d. The average soluble COD concentration in the reactor was app. 60 g/L and the average H_2 yield 2.47 mol H_2/mol hexose.

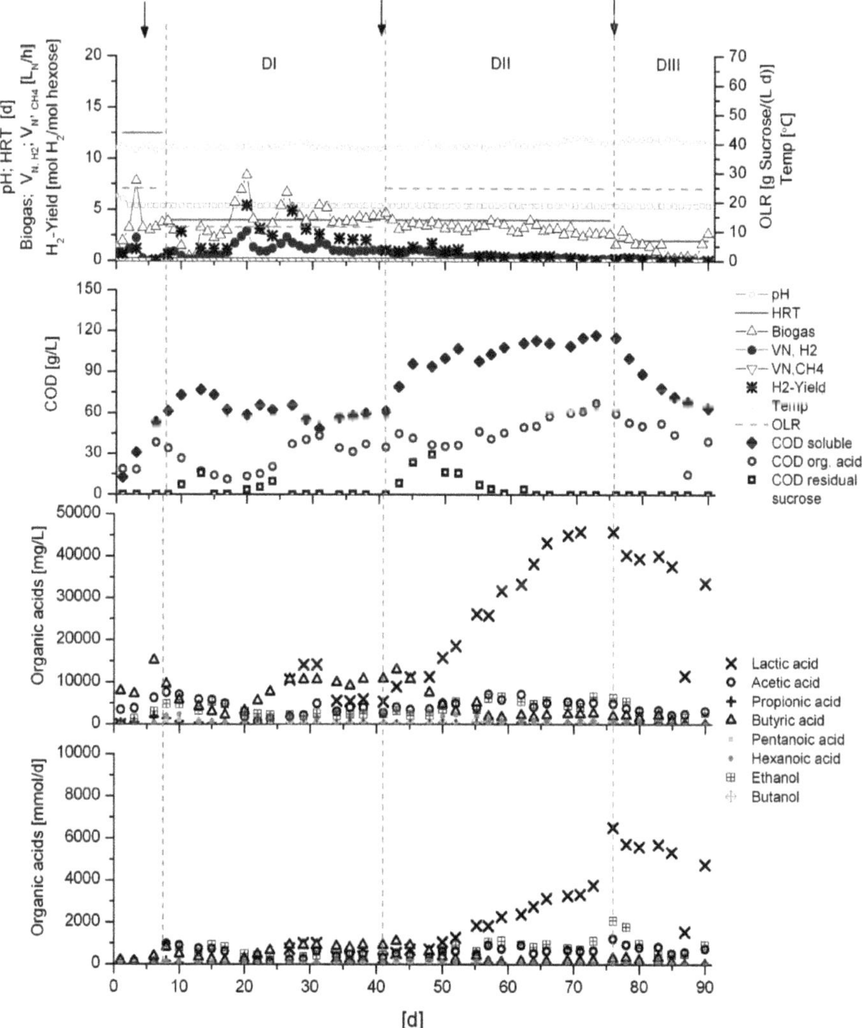

Figure 2. Operation parameters, gas production, hydrogen yield performance and production rate of relevant metabolites during phases DI to DIII. Arrows indicate the sampling dates for microbial population analyses

Phase		DI	DII	DIII	DV	DVI	DVII	D1	D2	D3	D4	D5	D6	D7
OLR [g sucrose/(L d)]		11.6	24.7	24.7	45.9	36.1	36.1	11.6	24.7	24.7	36.1	45.9	36.1	69.6
HRT [d]		4.0	4.0	2.0	1.0	2.0	0.5	3.0	3.0	1.0	1.0	0.5	0.5	0.3
Duration [xHRT]		8	11	7	20	10	47	5	6	7	20	40	41	120
Biogas [L_N/h]		3.93	3.31	1.33	3.47	6.72	8.39	2.23	2.31	2.32	9.72	6.66	6.58	5.57
H_2	[%]	32	18	8	20	16	31	22	25	18	26	26	32	35
	[L_N/h]	1.11	0.53	0.11	0.61	1.07	2.42	0.44	0.65	0.49	2.19	1.51	1.82	1.68
COD soluble [g/L]		63	101	84	61	98	27	45	97	43	49	34	22	20
Biomass	[g/L]	3.3	2.5	4.5	2.1	2.2	3.7	1.3	1.9	2.0	3.0	1.6	1.6	1.3
	[g/d]	25	19	66	64	34	161	13	19	59	89	92	94	147
Sub/te degradation [%]		93.9	92.3	100.0	92.4	100.0	99.4	99.9	71.5	100.0	99.4	95.9	97.2	83.1
Yield [mol H_2/mol hexose]		2.49	0.68	0.12	0.43	0.63	1.68	1.00	0.38	0.91	1.53	0.97	1.31	0.71
HLa	[mg/L]	9365	20252	35467	25577	36767	787	9989	13024	12309	2325	5045	2422	1451
	[mmol/d]	647	1949	5029	7447	5361	349	872	1168	3497	556	2715	1230	1276
HAc	[mg/L]	3962	4904	3334	6848	3745	4446	2581	6260	2758	5619	6183	4159	2425
	[mmol/d]	506	619	814	3495	944	4215	478	1053	1364	2788	6172	4149	4571
HPr	[mg/L]	251	82	131	41	249	0	47	31	416	277	40	0	0
	[mmol/d]	25	7	25	15	49	0	3	4	167	109	31	0	0
HBu	[mg/L]	7390	4861	1230	3678	9828	5877	4949	3683	3275	11927	5451	4064	2752
	[mmol/d]	634	393	184	1221	1674	3796	594	515	1078	3998	3646	2761	3461
HVa	[mg/L]	312	60	95	0	137	0	51	14	132	69	0	0	0
	[mmol/d]	24	5	14	0	21	0	5	1	39	20	0	0	0
HCa	[mg/L]	1805	1660	349	0	2104	30	2243	2023	131	0	0	439	44
	[mmol/d]	110	111	45	0	279	15	197	177	34	0	0	38	43
EtOH	[mg/L]	3217	4492	3194	282	1514	0	1283	1934	1468	663	194	643	0
	[mmol/d]	547	764	1042	190	513	0	287	433	963	435	255	845	0
BuOH	[mg/L]	74	281	177	0	49	0	1149	135	5	0	0	0	0
	[mmol/d]	4	30	36	0	10	0	159	19	2	0	0	0	0

Table 3. Experimental phases, operating parameters and average concentrations and yields of produced gases and metabolites

On the 41st day of operation the OLR was increased from 11.6 g sucrose / (L·d) to 24.7 g sucrose / (L·d) (phase DII), which caused an increase to the residual sucrose concentration in the reactor during the first days. The biomass was not able to metabolize the total amount of substrate. Over the total 35 d of operation of this phase H_2 in the produced biogas gradually decreased and after 70 d no H_2 was detected. The fact that biogas production was sustained, indicated biological activity, but not towards H_2 production. In phase DIII the HRT was reduced to 2 d with the OLR retained unchanged. The concentration of soluble COD decreased gradually from 100 g/L to 60 g/L due to the higher dilution rate. Nevertheless, H_2 production could not be restored and the reactor operation was terminated.

After one month of operation the excess sludge from reactor a (R-a) was used to inoculate reactor b (R-b) (phases D1 to D7). In figure 3 the reactor operation parameters, gas production and hydrogen yield performance for the experimental phases D1 to D7 together with the data from phase DI for continuity and comparison reasons are presented. Biogas and hydrogen production started immediately upon seeding of the reactor at HRT of 3 d and OLR of 11.6 g sucrose / (L·d). Hydrogen production was stable, but the yield was lower than 1 mol H_2/mol hexose. The concentration of soluble COD did not exceed 45 g/L. Increase of the OLR at 24.7 g sucrose / (L·d) lead to steep increase of the soluble COD and COD-sucrose concentration above 120 g/L and 60 g/L respectively and a decrease of the COD concentration of the metabolites. It seems that in the beginning the biomass concentration was not sufficient to metabolize the whole amount of sucrose, which on its turn accumulated in inhibitory levels as demonstrated by the cease of biogas and hydrogen production and the reduction of the COD-organic acids. It has been reported (Hafez et al., 2010) that glucose can become inhibiting for residual concentrations above 20 g/L. This problem could be overcome by the reduction of HRT to 1 d (phases D3 and on). H_2 production was restored immediately and was maintained until the termination of the experiment due to time constraints. Biogas and hydrogen production exhibited fluctuations over time. These can be generally attributed to the experimental procedure followed, which affected the actual HRT and OLR. The dosing of water and substrate was stopped everyday for different periods each time, in order to be prepared as described in Materials and methods. Stronger fluctuations were related mostly to the failure of substrate dosing due to tube clogging. In all phases, except phases D4 and D6, the H_2 yield was equal or lower than 1 mol H_2/mol hexose (table 3). In phase D4 it reached 1.53 mol H_2/mol hexose for HRT of 1 d and OLR of 36.1 g sucrose / (L·d) and in phase D6 1.31 mol H_2/mol hexose for for HRT of 0.5 d and OLR of 36.1 g sucrose / (L·d).

During the whole R-b operation, which together with the time period of phase DI reached 180 d, no methane was detected. Methanogenesis could be inhibited through the thermal pre-treatment of the seed sludge and the selected operation parameters were sufficient for hindering the proliferation of the methanogens in the system. In our previous work for which no pre-treatment of the seed sludge was carried out methanogenesis could be only be inhibited for 120 d (Mariakakis et al., 2011). Sucrose in higher concentrations could be detected only in phases D2 and D7 resulting in average substrate degradations of 71.5% and 83.1% respectively. In D2 sucrose accumulated in the beginning of the phase as results of the long HRT and the OLR of 24.7 g sucrose / (L·d). At D7 sucrose could not be degraded due to the high OLR of 69.6 g sucrose / (L·d) and the short HRT of 0.25 d.

The excess sludge from phase D3 of R-b was used to re-inoculate R-a (phases DV and DVI). In phase DVII the reactor was inoculated with excess sludge of phase D6. In Figure 4 the reactor operation parameters, gas production and hydrogen yield performance for these experimental phases together with the data from phase D3 for continuity and comparison reasons are presented.

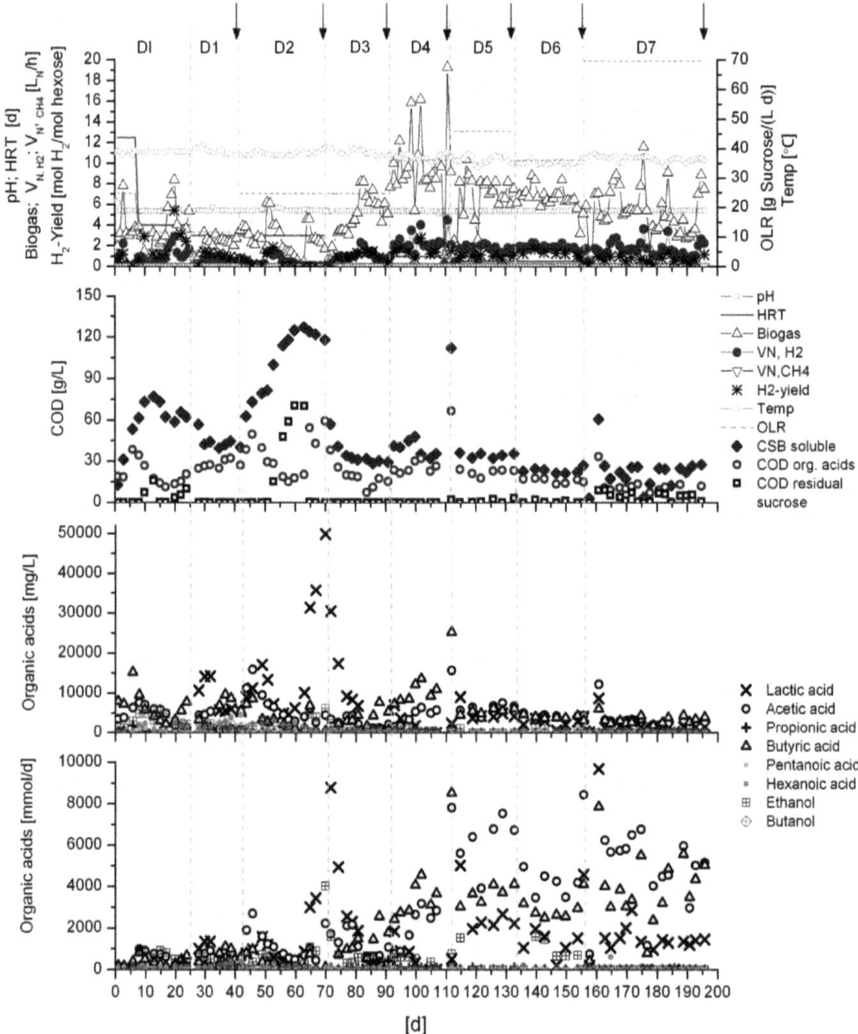

Figure 3. Operation parameters, gas production, hydrogen yield performance, HBu:HAc ratio and production rate of relevant metabolites during phases DI to D7. Arrows indicate the sampling dates for microbial population analyses

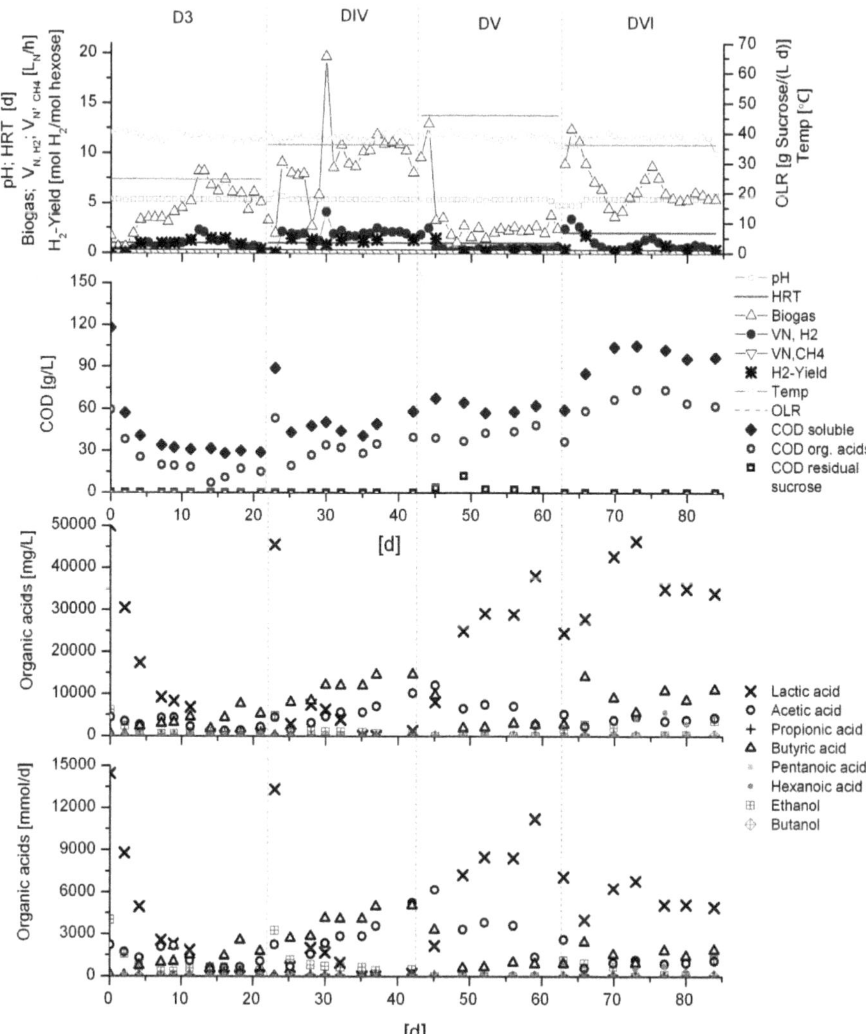

Figure 4. Operation parameters, gas production, hydrogen yield performance, HBu:HAc ratio and production rate of relevant metabolites during phases D3 to DVII. Arrows indicate the sampling dates for microbial population analyses

Biogas and hydrogen production started also in these cases immediately after seeding of the reactor. Due to the OLR set at 47.1 g sucrose / (L·d), the average soluble COD concentration reached 61 g/L. H_2 production (phase DV) yieldied 0.43 mol H_2/mol hexose. The decrease of the OLR to 36.1 g sucrose / (L·d) and increase of HRT to 2 d in phase lead to further increase of the soluble COD to an average concentration of 98 g/L. Hydrogen yield was slightly improved and reached 0.63 mol H_2/mol hexose. Biogas and hydrogen production during phase DVII were instable mainly due to malfunctions of the substrate dosing. The hydrogen yield for addition of Fe^{2+} achieved was 1.68 mol H_2 / mol hexose, which corresponds to an increase of approximately 28 % in comparison to phase D6.

The addition of Fe^{2+} enhanced hydrogen production. From table 3 it can be seen that biomass production rate reached the highest value of 161 g/d comparable only with that of phase D7 that was acquired for double as much OLR. For comparison at phase D6, biomass production rate was only 94 g/d. It seems that this nutrient that is important for H_2 production as will be explained in chapter 3.2 was limiting for the biomass growth, since it was not included in high enough concentrations in the nutrient solution with which the molasses solution was supplemented.

In Table 4 the hydrogen yield of other works in comparison to the best acquired by this work are presented. In most of the cases slightly better results were acquired by the other researchers. In a semi-scale reactor operated at HRT of 0.25 d and pH 4.5 hydrogen yield up to 1.86 mol H_2/mol hexose was achieved (Ren et al., 2006). Lay et al. (2010) maximized the yield for HRT of 0.5 d and pH of 5.5 like in this work, but achieved a somewhat lower yield of 1.35 mol H_2/ mol hexose, even though the reactor set-up was considerably smaller permitting for better control of operation. Aceves-Lara et al (2008) operated a 2 2 L CSTR at HRT of 0.25 d and pH of 5.5. The maximum yield acquired was 1.70 mol H_2/ mol hexose for an OLR of 24.2 g COD / (L·d). For OLR as high as 77.2 g COD / (L·d) the acquired hydrogen yield was much higher than that of the current work (0.84 mol H_2/ mol hexose) for the same HRT and OLR of 69.6 g sucrose/(L·d), which is equivalent to 107 g COD / (L·d) (Aceves-Lara et al., 2008). It is obvious that the process can be further optimized.

Work		OLR [g COD/ (L d)]	HRT [d]	pH	$V_{Reactor}$ [L]	Duration [d]	H_2-Yield [mol H_2/ mol Hexose]	Degradation [%]
Ren et al. (2006)		6.32	0.44	4.5	1480	10	0.34	--
		27.98	0.25			10	1.86	--
		42	0.162			10	1.81	--
Lay et al. (2010)		40	1	5.5	4	130	0.48	60.2
		80	0.5			130	1.35	64.3
		120	0.67			130	0.89	68.1
Aceves-Lara (2008)		24.2	0.25	5.5	2	17	1.70	100
		48.5	0.25			12	1.62	100
		77.2	0.25			18	1.26	98
Own	(D4)	50 (36.1)	1	5.5	30	20	1.51	99.4
	(D6)	50 (36.1)	0.5			21	1.31	97.2
	(DVII)	50 (36.1)	0.5			25	1.68	99.4

Table 4. Operating parameters, efficiencies and hydrogen yields of works with molasses as substrate for fermentative hydrogen production

3.2. Microbial metabolism

From Figures 2 to 4 and Table 3 it can be seen that the major metabolites for all tested parameters were lactic, acetic, and butyric acid together with ethanol. Propionic and hexanoic acid were produced in low quantities and only in same cases. Pentanoic acid and butanol were not detected at all. Acetate and butyrate are typical for H_2 production by mixed acid fermentation of sucrose as already mentioned in introduction. Lactate can be produced by either lactic acid bacteria or by bacteria of the genus *Clostridium* via a pathway that does not promote H_2 production (Hiligsmann et al., 2011).

During phase DI lactate was produced in less quantity than acetate and butyrate with the later being produced in almost the same proportion indicating a mixed acid fermentation, even though the theoretical yield of 2.6 mol H_2 / mol hexose was not reached (equation 4). The increase of the OLR in phases DII and DIII has lead to gradually increasing lactate production rate with simultaneous increase of acetate into a lesser extent and decrease of butyrate. The higher substrate availability forced the biomass to shift its metabolism from mixed acid fermentation to lactate fermentation. The fact that biogas production also declined is an indicator for homolactic fermentation by LAB, during which only lactate is produced (equation 9). In phases DV and DVI lactic acid production could not be solely due to homolactic fermentation, since biogas was produced. Higher lactic acid production than acetic and butyric acid production was observed in the phases for which HRT and OLR were equal or higher than 1 d and 24.7 g sucrose / (L·d) respectively. The exact mechanism of lactic acid production can be explained either by the co-existence of LAB (Hafez et al., 2009) or by the metabolic shift of the hydrogen producing clostridia (Minton & Clarke, 1989). In all cases though, high lactic acid production was combined to an increase of the OLR (Oh et al. 2004; Kim et al., 2006; Oh et al., 2004; Hafez et al., 2009) and caused a diminution of hydrogen yield. The effect of lactic acid as a metabolite in hydrogen systems has not yet been clarified. It has been reported to be promoting to hydrogen production at low concentrations and inhibiting at high concentrations. In a work (Baghchehsaraee et al., 2009), an increase of the hydrogen yield combined with the complete degradation of the externally added lactic acid in concentrations up to 3 g/L was observed. In another work, Kim et al. (2012) also observed an increase of 22% in hydrogen yield when lactic acid up to 8 g/L was added to batch fermentors operated at pH of 4.5, and a reduction when the concentration was raised at 18 g/L. The corresponding undissociated form of the lactic acid, which is the potential inhibitor (van Ginkel & Logan, 2005) was 21 mmol/L and 45 mmol/L at pH 4.5 according equation 15. In this work, the highest lactic acid concentration was 35 g/L (9 mmol/L undissociated lactic acid at pH 5.5) and was only reached temporarily in the beginning of phase D3 without any obvious long-term negative influence on the hydrogen process, like in phase DIII. It seems that lactic acid was not the inhibition factor, but another substance that was not monitored.

$$pH = pK_a + \log \frac{A^-}{HA} \tag{15}$$

For phases DI to DIII and D1 to D3 during which OLR lower or equal to 24.7 g sucrose / (L·d) was applied, acetate and ethanol were produced in almost the same rates. A ratio of EtOH:HAc equal to 1:1 has also been proposed for clostridia as described by equation 1 for enteric bacteria, when hydrogen is evolved only through the conversion of Acetyl-CoA to pyruvate, yielding 2 mol H_2 / mol hexose (Minton & Clarke, 1989). For higher loadings the observed ethanol production rate diminishes either due to a metabolic shift of the biomass or due to its consumption. Ethanol has also been detected in other hydrogen producing systems operated at pH 5.5 and various HRT and OLR (Gavala et al., 2006; Karadag & Puhakka, 2010a; Kim et al., 2006; Shen et al., 2009) as a product either of enterobacterial (Hallenbeck, 2005), clostridial hydrogen production (Akutsu et al., 2009; Lin & Lay, 2004), or heterolactic bacteria (Kandler, 1983). In the case of clostridia, it has been suggested that ethanol is produced during the late growth phase during which no hydrogen is produced, while H_2 production is favored during the exponential growth phase, during which the organic acids are produced (Nath & Das, 2004), yielding at the end of a batch fermentation a ratio of 1:1. By the adjustment of short HRT in CSTR it is possible to maintain the bacterial population in the exponential growth phase. However, the adjustment has to be suitable so that the biomass concentration in the reactor can be also maintained in concentrations that are suitable for high substrate conversion rates.

Propionic acid was produced in phases D3 and D4. Sucrose fermentation to propionic acid is a sink for hydrogen and according to equation 6 for each mol of propionic acid produced 1 mol of hydrogen is consumed. The derived hydrogen consumptions for phases D3 and D4 correspond to 15% and 19% of the total produced hydrogen respectively.

Hexanoic acid was mainly produced during phases DI, DII, DVI and D2. The HRT of all these phases was equal to or longer than 2 d. The production of hexanoic acid can only be explained by a possible secondary fermentation of *C. kluyveri* as described by equation 12 and 15.

The addition of Fe^{2+} in phase DVII did not influence the production rates of acetic acid in comparison to phase D6. On the other hand, the production rate of lactic acid was significantly reduced from 1230 mmol/d to 329 mmol/d corresponding to approximately 72% and ethanol was not produced anymore. Parallel, the production rate of butyrate increased approximately by 27%. The overall bacterial metabolism was shifted to butyrate fermentation and biomass growth as described in 3.1. This is an indication that lactic acid and ethanol production in the phases with relative short HRT (<1 d) was mainly due to the clostridial metabolism. They contributed more than 2/3 to the total lactic acid production. Iron is very important to hydrogen production, which is produced when the simple reaction of Eq. 16 takes place. This reaction is catalyzed in clostridia by a dimetallic iron only [FeFe]-hydrogenase, which receives protons by the reduced form either of ferredoxin or of NADH (Vignais & Billoud, 2007). Under iron limitation the activity of hydrogenase is also limited (Valdez-Vazquez & Poggi-Varaldo, 2009), pyruvate can not be degraded through the pathways leading to hydrogen, but fermentation is shifted towards lactic acid production (Minton & Clarke, 1989).

$$2\,H^+ + e^- \leftrightarrow H_2 \tag{16}$$

3.3. Microbial population

3.3.1. Species description and influence of HRT and OLR

The sample times for the phylogenetic analysis and the affiliated dominant species present in each experimental phase are indicated in figures 2 to 4. In figures 5 and 6 the PCR-DGGE profiles of the *Eubacteria* and *Clostridium* species are presented. The investigation on eubacterial 16S rDNA (table 5) showed mostly lactic acid bacteria of the Phylum *Firmicutes* as closest relatives to the detected DGGE bands. The analysis of 16S rDNA for clostridia showed repeating DGGE patterns (figure 5). In many cases several single bands in one lane showed the same sequence and could be affiliated to one *Clostridium* species, though it was possible to allocate several bands to different clostridia species. Most of the bands could be affiliated to the species *C. butyricum*, *C. tyrobutyricum* and *C. ljungdahlii* (table 5).

Carnobacterium sp. AT12 was the only *Eubacterium* species found in the seed sludge after the thermal pre-treatment and before start-up (table 6). It is a non-spore forming, facultative anaerobic and heterofermentative lactic acid bacterium (de Vos, 2009), which previously belonged to the genus of *Lactobacillus* (Dworkin et al., 2006). Even though this bacterium is no spore forming, it was able to survive the thermal pre-treatment. After the addition of substrate it was replaced by other species of LAB. The members of this genus are facultative anaerobic, mesophilic, non spore forming, obligately saccharoclastic with complex nutritional requirements. Their optimum pH for growth is between 5.5 and 6.2, but they can also grow for pH lower than 5. *L. fermentum* is obligately heterofermantative. It degrades hexose to equimolar quantities of lactic acid, CO_2 and acetate or ethanol via the 6-phosphogluconate pathway.

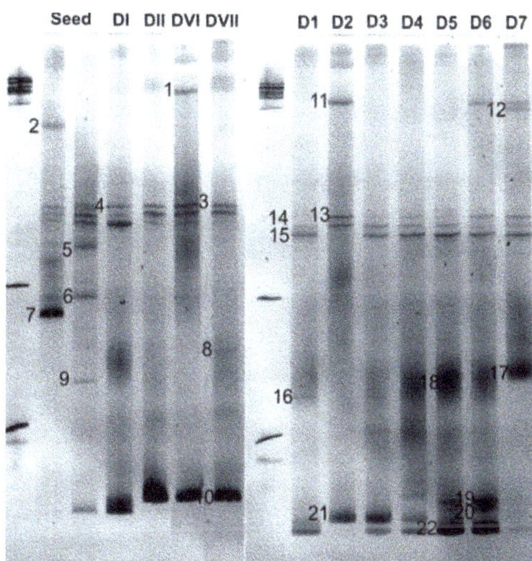

Figure 5. DGGE-profile of the *Eubacterium* species from each experimental phase. The numbers denote the bands that have been sequenced and successfully allocated. The sequencing results are presented in Table 5

L. delbrueckii subsp. *bulgaricus* on the other hand is obligately homofermentative and degrades hexose via the Embden-Meyerhoff pathway to lactic acid (de Vos, 2009). Through the phases DI to DII species of *Lactobacillus* established a stable community in the reactor. At phases DI and DII *Olsenella* species were also detected, while at phase DIII *Sporolactobacillus* has been identified (table 6). Species of the genus *Sporolactibacillus* are facultative anaerobic and obligately homofermentative. They can form spores, which are resistant to heating at 80 °C for 10 min (de Vos, 2009). They can grow only on fermentable carbohydrates at pH greater than 4.5. Their optimum growth temperature is 30 °C (Dworkin et al., 2006). For *Olsenella* sp. oral taxon 809 there are no metabolic data available, since it has not been cultivated. In general the species genus *Olsenella* is non-spore forming, homofermentative and strictly anaerobic. The end-products of sucrose fermentation consist mainly of lactic acid and to a lesser extent of acetic acid (Olsen et al., 1991). The presence of these bacteria can explain the increasing production rate of lactate and acetate indicating an overall shift of the bacterial metabolism in the system to lactic acid fermentation and more specific to homolactic fermentation as derived by the decreasing biogas production. They compete with the hydrogen producing bacteria for substrate.

In the seed sludge three species of *Clostridium* could be identified. Only *C. butyricum* could be maintained in the system until the end of phase DIII (table 6). The rest of the initially present *Clostridium* species were replaced by others indicating a constant changing population, like in the case of (Huang et al., 2010). All of them are able to produce copious amounts of hydrogen gas by mixed acid fermentation with acetate and butyrate as their major end-products as already mentioned. Even though, hydrogen producing bacteria where identified during phases DI to DII, the deterioration of hydrogen production could not be hindered. The lactic acid bacteria could proliferate in the system.

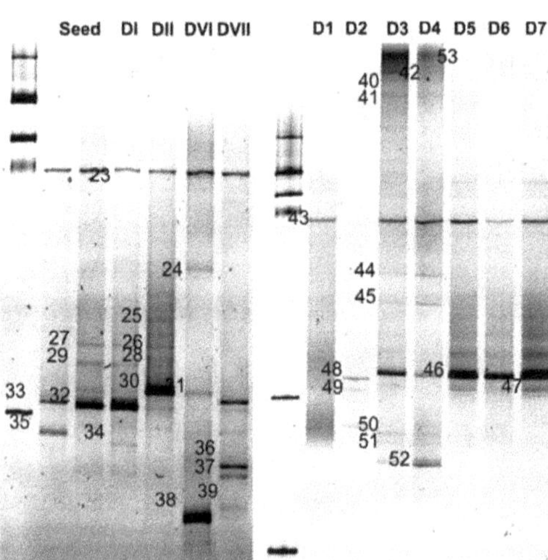

Figure 6. DGGE-profile of the *Clostridium* species from each experimental phase. The numbers denote the bands that have been sequenced and successfully allocated. The sequencing results are presented in table 5

Also in the case of phases DVI, D1 and D2 (tables 6 and 7) the symbiosis of lactic acid together with hydrogen producing bacteria could be observed resulting to high lactic acid production rates and medium to low hydrogen yields. In all cases, the employed HRT was equal or higher to 2 d. It seems that long HRT contribute to the dominance of lactic acid over hydrogen producing bacteria. In the case of phases D3 to D7, high production rates of lactic acid could still be observed, but they were lower than that of acetic and butyric acid, while medium to higher hydrogen yields could be achieved. The HRT applied was equal to or lower than 1 d. Under these conditions the hydrogen producing metabolism could dominate over the lactic acid metabolism. After phase D3 and on *C. tyrobutyricum* became one of the dominant *Clostridium* species. After phase D5 and on *Sporolactobacillus* could not be found any more in the system. It seems that HRT lower than 0.5 d does not support its growth. In the work of Fang (2002) who investigated hydrogen production by untreated secondary sludge at pH 5.5, loading rate of 25 g sucrose / (L · d), HRT 0.25 d and temperature of 26 °C, 69.1% of the microbial clones were affiliated to *Clostridium* species and 13.5% to *Sporolactobacillus racemicus*, which was able to be retained in the system due to the granular sludge reactor configuration.

Analogously, *Olsenella* disappeared from the system at phase D7 for HRT of 0.25 d. Lo et al (2008) could detect *Olsenella* and *C. butyricum* in a hydrogen producing reactor with xylose as substrate at HRT of 2 hours. They also used granular sludge. In another work (Castelló et al., 2009), the acquired low hydrogen yield was justified by the presence of *Olsenella* along with *Prevotella, Bulleidia, Mitsoukella* and *Selonomonas* species, which consumed the substrate. In this work, *C. tyrobutyricum* became visible also through the eubacteria specific primers indicating a general proliferation of *Clostridium* species for vey short HRT that comes in agreement with the low doubling times of 30 min to 3 h that have been observed for most *Clostridium* species (Dworkin et al., 2006). Despite the dominance of hydrogen producing biomass over lactic acid bacteria, the hydrogen yield was relatively low due to the low sucrose degradation efficiency. The biomass concentration in the system was not sufficient to metabolize the whole amount of substrate. *L. mucosae*, an obligate heterolactic bacterium, could maintain itself in the system even for HRT as low as 0.25 d. These findings are in agreement with the observations of (Kim et al., 2012), who investigated hydrogen production by a lactate-type fermentation in a CSTR with working volume of 4 L, glucose as substrate and digester sludge as inoculum pre-treated with acid. For HRT of 1 d and OLR equal to 10 g / (L·d) they affiliated 65% and 35% of the total bacterial population to *Eubacterium* and *Clostridium* species respectively. When the OLR was increased to 40 g / (L·d) the proportion changed in favor to *Eubacterium* species to 72% and 28% respectively. For the same OLR but HRT equal to 0.5 d a balance between the two populations with 50% each was established.

C. ljungdahlii could be detected in the system during the phases DIV, DVI, DVII and D1 to D7. More than 90% of its strains can produce H_2 during heterotrophic growth on glucose or fructose, while sucrose can not be utilized (de Vos, 2009). It can also grow autotrophically on H_2 and CO_2 or CO. It forms spores only rarely. The regulation of the diverse pathways of homoacetogens is still not understood (Dworkin et al., 2006). For the first case, extracellular enzymes released to the bulk liquid by other microorganisms that are able to degrade sucrose to glucose and fructose are required. In a mixed acid fermentation as described by

Eq. 4, a ratio of HBu:HAc should have been obtained. In most of the cases this ratio is lower than one indicating that acetic acid is also produced through a pathway other than that of hydrogen production. It seems that *C. ljungdahlii* is consuming hydrogen.

Band No.	Affiliation	Similarity	Accession Number
1	*Lactobacillus mucosae* strain TB-H32	97	AB425938
2	*Carnobacterium* sp. AT12	91	DQ027062
3	*Lactobacillus mucosae* strain FSL-04 1	87	JN092131
4	*Lactobacillus delbrueckii* subsp. *bulgaricus* strain CH3	85	JN675227
16; 18; 19; 20; 22	*Olsenella* sp. oral taxon 809	84-95	GU470903
5	*Lactobacillus fermentum* strain -O rkz-4	90	JN836490
6	*Lactobacillus delbrueckii* subsp. *bulgaricus* 2038	98	CP000156
7	*Carnobacterium* sp. AT12	93	DQ027062
8	*Lactobacillus casei* strain GIMC8:TVS-72	91	JF728260
9	Uncultured *Burkholderiales* bacterium clone CA38	96	EF434370
10; 21	*Sporolactobacillus* sp. MB-051	90; 91	AB548940
11	*Lactobacillus mucosae* strain LAB87	95	EF120376
12	*Clostridium tyrobutyricum* 5S	88	L08062
13	*Lactobacillus mucosae* strain SF1031	94	FN400925
14	Marine bacterium strain SJ-BF7	84	AM260710
17	*Clostridium tyrobutyricum* strain SCTB132	94	JN650297
23; 29; 32	*Clostridium butyricum* strain CB TO-A	94-100	AB687551
24; 39; 43; 44; 45; 50; 51; 52	*Clostridium ljungdahlii* type strain DSM13528T	87-100	FR733688
25; 26; 28; 30; 31; 40; 41; 42; 46; 47	*Clostridium tyrobutyricum* strain SCTB130	80-100	JN650295
27	*Clostridium* sp. 2NR375.1	96	JQ248567
33	*Clostridium peptidivorans* strain TMC4	100	FJ155851
34	*Clostridium intestinale*	97	AY781385
35	*Clostridium disporicum* strain NML 05A027	100	DQ855943
36;37	*Clostridium tyrobutyricum* strain S1	94; 100	JN241679
38	*Clostridium ljungdahlii* DSM 13528	89	CP001666
48; 49	*Clostridium butyricum* strain T-08B	97; 95	FR734082
53	*Clostridium tyrobutyricum* strain SCTB125	61	JN650290

Table 5. Affiliation of the DGGE bands to bacterial species after sequencing of the 16S rDNA gene

3.3.2. *Effect of Fe^{2+} addition*

The addition of Fe^{2+} at phase DVII had an influence on the bacterial population in comparison to the seed sludge (phase D6). *L. casei*, *Sporolactobacillus* from the LAB and *C. butyricum* became also dominant, while *Olsenella* was not any more detected. The presence of an extra lactic acid bacterium species though, contradicts the observed reduction of lactic acid production rate. The most probable explanation is that given in 5.3. The clostridial metabolism was shifted from lactic acid production, which is triggered under limitation of Fe^{2+} (Dürre, 2005), to butyrate. Nevertheless, LAB could still influence the system as the production of lactic acid in phase DVII indicates, also for HRT as short as 0.5 d, not like in the case of Kim et al. (2006), who discovered only one lactic acid bacterium (*Bacillus racemilacticus*) and only during the phase with OLR = 60 g COD / (L · d), while for higher or lower OLR no such species was found, although lactate has been produced in all cases, indicating that lactate in the system was a product of *Clostridium* species metabolism, which dominated the system.

In an investigation (Karadag & Puhakka, 2010b) about the influence of Fe^{2+} concentration in the range of 0.5 mg/L to 100 mg/L on a CSTR system fed with glucose and operated at HRT of 0.208 d, OLR of 43.2 g glucose /(L·d) and pH 5 an increase of 71% in hydrogen yield for Fe^{2+} concentration of 50 mg/L was achieved, followed by a fermentation shift from ethanol type to butyric acid type like in this work. In two other works (Wang & Wan, 2008) and (Lee et al., 2001) investigating the influence of Fe^{2+} concentration in batch experiments with vials, optimum concentrations of 350 mg/L and 352.8 mg/L were detected respectively. It seems that there is potential for the optimization of the quantity added to the system.

3.3.3. *Symbioses in fermentative hydrogen production systems*

There are only a few works that have investigated the bacterial population and the influence of HRT and OLR on hydrogen producing systems by molasses. Ren et al. (2007) studied the influence of pH on the microbial population structure of bio-hydrogen production by molasses in a 2.5 L CSTR seeded with sewage solids and operated at HRT of 0.25 d, OLR between 7 g COD / (L·d) and 30 g COD / (L·d) and at temperature of 35 °C. At pH between 5.5 and 6 mixed ethanol-butyrate fermentation was observed. In this work no ethanol could be detected for HRT of 0.25 d. In the reactor a co-existence of clostridia and LAB was observed like in this work. The bacterial population was dominated by *C. pasteurianum*, *Lactococcus* sp., *Desulfovibrio ferrireducens* together with uncultured species of *Actinobacterium* and *Bacteroidetes*. Chu et al. (2011) on the other hand, did not observe such a co-existence. They affiliated most clones found in a suspended sludge system treating fermented molasses with HRT raging from 0.33 d to 0.083 d to *C. butyricum*, *Megasphaera* sp. and *Corynebacterium glutamicum*. In the case of defined substrate, Kim et al. (2006) obtained the optimum hydrogen yield when a LAB (*Bacillus racemilacticus*) together with *Clostridium* species were dominant in the system.

The dominance of the lactic acid bacteria *Sporolactobacillus, Olsenella,* along with hydrogen producing *C. tyrobutyricum, C. butyricum* and *Clostridium sp* strain S6 was observed in a Continuously Stirred Tank Reactor (CSTR) at pH 5.5 and HRT of 1 d yielding 1.24 mol H_2/mol hexose (Wongtanet et al., 2007). The bacterial population examination confirmed in this case a symbiosis between the lactic acid and hydrogen producing bacteria, too. In another work (Ohnishi et al., 2010), in which kitchen waste containing lactate was used as substrate, lactic acid removal was also observed along with carbohydrate degradation during hydrogen production in an Anaerobic Sequencing Batch Reactor. The phylogenetic analysis affiliated the dominant bacteria to the genera of *Lactobacillus, Selonomonas, Veillonella* and *Megasphaera,* which belong to the phylum of *Firmicutes,* to *Prevotella* genus of the phylum of *Bacteroidetes* and to *Atopobium* and *Bifidobacterium* of the *Actinobacteria* phylum. Hydrogen production has been attributed, due to the absence of *Clostridium* species, to *Megasphaera* by simultaneous utilization of the carbohydrates and the lactate contained in the initial feed and produced by the LAB *Lactobacillus* present in the system and which in return as aerotolerant bacterium consumed residual oxygen for establishing a suitable milieu for the anaerobe *Megasphaera* and supplying it with lactate for hydrogen production.

Affiliation	Phase					
	Seed	Seed	DI	DII	DVI	DVII
Eubacterium species						
Carnobacterium sp. AT12	o					
Lactobacillus fermentum		o				
Lactobacillus delbrueckii subsp. *Bulgaricus*		o	o	o		
Uncultured *Burkholderiales* bacterium clone CA38		o				
Lactobacillus mucosae					o	o
Lactobacillus casei						o
Sporolactobacillus sp. MB-051				o	o	o
Olsenella sp. oral taxon 809		o	o			
Clostridium species						
Clostridium butyricum	+	+	+	+	+	+
Clostridium peptidivorans	+					
Clostridium disporicum	+					
Clostridium sp. 2NR375.1		+	+			
Clostridium intestinale			+			
Clostridium tyrobutyricum				+	+	+
Clostridium ljungdahlii					+/-	+/-

Table 6. Allocation of the microbial species to the seed sludge and the operation phases DI to DVIIIe. Microbial population has been simplified to species level. Symbols: "o" no hydrogen production or consumption. "+/-" some strains can produce hydrogen. "+" hydrogen production and "-" hydrogen consumption

In an other investigation (Hung et al., 2011b), it was also suggested that the facultative anaerobes of *Streptococcus* sp. and *Klebsiella* sp. found in the granular sludge of hydrogen producing fermentors seeded with untreated sewage sludge, maintained the strict anaerobic conditions required by the *Clostridium* species for hydrogen production. In the work of Kim et al. (2009), a CSTR seeded with thermally pre-treated sludge at 95 °C for 15 min was operaterd at OLR of 50 g COD / (L·d), HRT of 0.5 d, pH regulated at 5.3 and temperature at 35 °C. In this case also, a symbiosis of hydrogen producing clostridia with the lactic acid bacteria *L. delbruecki* and *Lactococcus lactis* was observed. Such an enhancing symbiosis can not be excluded that also takes place in the system of this work, so that the residual dissolved oxygen of the water added into the system could be removed. More over, the possibility of a secondary fermentation by *C. tyrobutyricum* present in most of the experimental phases and facilitated by the production of lactate by the LAB as described in our previous work (Mariakakis et al., 2011) and later demonstrated (Wu et al., 2012) has to be considered, too.

In any case, this symbiosis reduces the available substrate for hydrogen production. Furhtermore, in this work, it was not beneficial to hydrogen production in the phases with long HRTs, but deteriorated or even completely inhibited it. Lactobacilli possess the potential of inhibiting other microorganisms by different mechanisms. Their fermentation products consist mainly of lactic and acetic acid, which reduce the pH which on its side reduce their dissociation degree. In the presence of oxygen many species such as *L. lactis* and *L. bulgaricus* can produce H_2O_2, which is bacteriocidal for gram negative and bacteriostatic for gram positive bacteria (de Vos, 2009). They can produce bacteriocins, proteinaceous substances with bactericidal effect on microorganisms closely related to the producer. Most probable each *Lactobacillus* species has strains that can produce bacteriocins. Among the organisms that have been found in the present work *L. casei* and *L. fermentum* can produce bacteriocins that are inhibiting other lactobacilli (Dworkin et al., 2006). Nevertheless, it was demonstrated that bacteriocins can hinder non-closely related microorganisms, too. For instance, an inhibition effect of *L. lactis* on *C. tyrobutyricum* due to the excretion of the bacteriocin nisin Z was detected (Rilla, 2003) and of *L. paracasei* on *C. acetobutylicum* and *C. butyricum* (Noike, 2002). It can not be excluded that the lactobacilli present on this work exert a similar effect on the clostridia species. There are also some bacteriocin-like substances produced by lactobacilli, such as bulgarican produced by *L. delbrueckii* subsp. *bulgaricus* that can inhibit a wide range of non related pathogenic gram negative bacteria. *L. delbrueckii*, *L. fermentum* and *L. casei* are also capable of producing bacteriophages that can cause cell lysis (Dworkin et al., 2006). *L. casei*, *L. fermentum* and *L. bulgaricus* have been also identified in this work, so an edverse effect can not be excluded as in several other works has been suggested. It has been proposed that *L. ferintoshensis* and *L. paracasei* present along with *Clostridium* sp. and *Coprothermobacter* sp. in untreated digester sludge negatively influenced hydrogen production (Kawagoshi et al., 2005). The same was suggested for *Sporolactobacillus* sp. that was present in an anaerobic sequencing batch reactor seeded with heat pre-treated sludge and operated at pH 5.0 using sweet sorghum syrup as substrate at OLR of 25 g sugar/(L·d) (Saraphirom & Reungsang, 2011). They justified it by their ability to produce bacteriocins as Noike et al. (2002) suggested. None of these substances has been monitored, so a possible accumulation in concentrations that are inhibiting to the clostridia can only be determined by a general parameter as the COD.

Affiliation	Phase						
	D1	D2	D3	D4	D5	D6	D7
	Eubacterium species						
Marine bacterium SJ-BF7	o						
Olsenella sp. oral taxon 809	o		o	o	o	o	
Lactobacillus mucosae		o	o	o	o	o	o
Sporolactobacillus sp. MB-051		o	o	o			
Clostridium tyrobutyricum							+
	Clostridium species						
Clostridium ljungdahlii	+/-	+/-	+/-	+/-	+/-	+/-	+/-
Clostridium butyricum		+					
Clostridium tyrobutyricum			+	+	+	+	+

Table 7. Allocation of the microbial species to the seed sludge and the operation phases D1 to D7. Microbial population has been simplified to species level. Symbols: "o" no hydrogen production or consumption. "+/-" some strains can produce hydrogen. "+" hydrogen production and "-" hydrogen consumption

In most cases, heat treatment of the seed sludge was applied, but it was not sufficient to inhibit the growth of lactic acid bacteria, although they can not form spores and it has been demonstrated *in vials* that heat treatment between 50 °C and 90 °C is sufficient (Noike et al., 2002). For instance, *L. delbrueckii* and *L. fermentum* were able to be identified at batch experiments producing hydrogen seeded with activated sludge and digester sludge thermally pre-treated at 65 °C for 30 min (Baghchehsaraee et al., 2010; Hafez et al., 2010). It seems that with increasing inoculum quantity, thermal treatment becomes less effective and it can not hinder the co-dominance and activity of lactic acid bacteria.

4. Conclusion

Hydrogen production by molasses could be successfully carried out in large lab-scale reactors for a period longer than 180 d and under variable combinations of OLR and HRT. The maximum H_2 yield obtained was 1.53 mol H_2/mol hexose for HRt of 1 d and OLR of 36.1 g sucrose/(L·d). Improvement of the hydrogen production yield of 28%was achieved by the addition of Fe^{2+} to an end concentration of 1000 mg/L. In figure 7 the acquired hydrogen yields of all phases as a function of the operation parameters HRT and OLR, along with a suitable range of combination of these parameters, as determined in the current work, are presented. Combinations resulting to COD concentrations higher than 50 g/L (phase D3), was showed to be inhibitory to H_2 production. Reason was not the undissociated form of acids, but most probably the production and accumulation of bacteriocidal or bacteriostatic substances, excreted by the LAB. The second line indicates the combination for which the process becomes unfavorable in terms of substrate utilization efficiency and hence can not be considered as cost effective. The applied seed sludge pre-treatment and reactor start-up methods were successful in enriching the biomass in hydrogen producing microorganisms and killing methanogenic microorganisms that are detrimental to H_2 production. Nevertheless, H_2 production has been carried out parallel to lactic acid metabolism, which was driven either by the presence of LAB, or by the hydrogen producing clostridia due to iron Fe^{2+} limitation. A co-existence of clostridium species with lactic acid bacteria seems to be unavoidable, even for extensive pre-treatment of the seed sludge. Lactic acid bacteria influence the system primarily by consuming the substrate available for hydrogen production. The co-existence of clostridia and LAB though, seems to become beneficial to hydrogen production at HRTs in the range of 0.5 d to 1 d by supplying certain clostridia genera that are capable of performing secondary fermentation with substrate and/or by removing the residual dissolved oxygen from the system and hence establishing an appropriate milieu for the growth of clostridia. For the successful technical implementation of hydrogen production, for which process control is complex and oxygen in trace concentrations is to be expected, this symbiosis may be regarded as pre-requisite. The exact extent, to which the LAB contribution is beneficial and not adverse, requires the quantification of the biomass for the allocation of the metabolic products to specific microbial species and the monitoring of the concentrations of possible inhibiting substances.

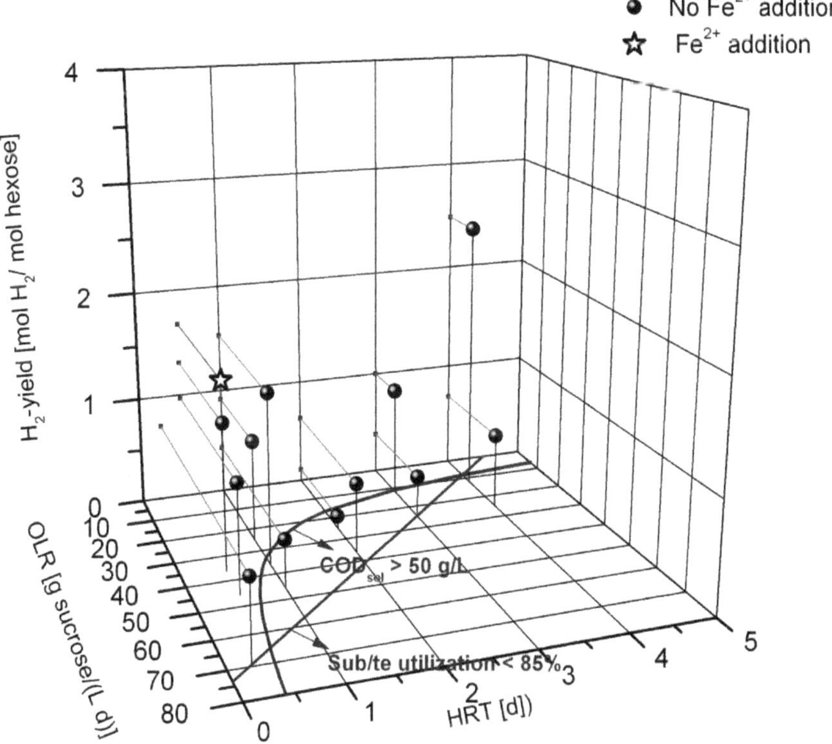

Figure 7. Suitable range of OLR-HRT combination for fermentative hydrogen production by molasses at pH 5.5

Author details

Iosif Mariakakis, Carsten Meyer and Heidrun Steinmetz
Institute for Sanitary Engineering,
Water Quality and Solid Waste Management (ISWA), University of Stuttgart, Germany

Acknowledgement

This work was supported by the German Federal Ministry of Education and Research (BMBF), EnBW AG, Purolite Deutschland GmbH and RBS wave GmbH, Grant number 03SF0351. We would like to warmly thank Ms. Kerstin Matthies of the Karlsruhe Institute of Technology for the microbial population analyses.

5. References

Aceves-Lara, C.-A., Latrille, E., Bernet, N., Buffière, P. & Steyer, J.-P. (2008). A pseudo-stoichiometric dynamic model of anaerobic hydrogen production from molasses. *Water Research.* 10-11, 2539–2550.

Akutsu, Y., Lee, D.-Y., Li, Y.-Y. & Noike, T. (2009). Hydrogen production potentials and fermentative characteristics of various substrates with different heat-pretreated natural microflora. *International Journal of Hydrogen Energy.* 13, 5365–5372.

Baghchehsaraee, B., Nakhla, G., Karamanev, D. & Margaritis, A. (2009). Effect of extrinsic lactic acid on fermentative hydrogen production. *International Journal of Hydrogen Energy.* 6, 2573–2579.

Baghchehsaraee, B., Nakhla, G., Karamanev, D. & Margaritis, A. (2010). Fermentative hydrogen production by diverse microflora. *International Journal of Hydrogen Energy.* 10, 5021–5027.

Castelló, E., García y Santos, C., Iglesias, T., Paolino, G., Wenzel, J., Borzacconi, L. & Etchebehere, C. (2009). Feasibility of biohydrogen production from cheese whey using a UASB reactor: Links between microbial community and reactor performance. *International Journal of Hydrogen Energy.* 14, 5674–5682.

Chu, C.-Y., Wu, S.-Y., Hsieh, P.-C. & Lin, C.-Y. (2011). Biohydrogen production from immobilized cells and suspended sludge systems with condensed molasses fermentation solubles. *International Journal of Hydrogen Energy.* 21, 14078–14085.

Deutsches Institut fuer Normung (2002). *Deutsche Einheitsverfahren zur Wasser-, Abwasser, und Schlamm-Untersuchungen.* Beuth, 978-3527305599, Berlin

Dürre, P. (2005). *Handbook on clostridia,* Taylor & Francis, 9780849316180, Boca Raton.

Dworkin, M., Falkow, S., Rosenberg, E., Schleifer, K.-H. & Stackebrandt, E. (2006). *The Prokaryotes,* Springer New York, 978-0-387-25492-0.

Fang, H. Liu H. and Zhang T. (2002). Characterization of a hydrogen-producing granular sludge. *Biotechnology and Bioengineering.* 78, 44–52.

Gavala, H., Skiadas, I. & Ahring, B. (2006). Biological hydrogen production in suspended and attached growth anaerobic reactor systems. *International Journal of Hydrogen Energy*. 9, 1164–1175.

Hafez, H., Baghchehsaraee, B., Nakhla, G., Karamanev, D., Margaritis, A. & El Naggar, H. (2009). Comparative assessment of decoupling of biomass and hydraulic retention times in hydrogen production bioreactors. *International Journal of Hydrogen Energy*. 18, 7603–7611.

Hafez, H., Nakhla, G., El. Naggar, M. Hesham, Elbeshbishy, E. & Baghchehsaraee, B. (2010). Effect of organic loading on a novel hydrogen bioreactor. *International Journal of Hydrogen Energy*. 1, 81–92.

Hallenbeck, P. C. (2005). Fundamentals of the fermentative hydrogen production. *Water Science & Technology*, 21–29.

Hallenbeck, P. C. (2009). Fermentative hydrogen production: Principles, progress, and prognosis. *International Journal of Hydrogen Energy*. 17, 7379–7389.

Hawkes, F., Hussy, I., Kyazze, G., Dinsdale, R. & Hawkes, D. (2007). Continuous dark fermentative hydrogen production by mesophilic microflora: Principles and progress. *International Journal of Hydrogen Energy*. 2, 172–184.

Hiligsmann, S., Masset, J., Hamilton, C., Beckers, L. & Thonart, P. (2011). Comparative study of biological hydrogen production by pure strains and consortia of facultative and strict anaerobic bacteria. *Bioresource Technology*. 4, 3810–3818.

Huang, Y., Zong, W., Yan, X., Wang, R., Hemme, C. L., Zhou, J. & Zhou, Z. (2010). Succession of the Bacterial Community and Dynamics of Hydrogen Producers in a Hydrogen-Producing Bioreactor. *Applied and Environmental Microbiology*. 10, 3387–3390.

Hung, C.-H., Chang, Y.-T. & Chang, Y.-J. (2011a). Roles of microorganisms other than *Clostridium* and *Enterobacter* in anaerobic fermentative biohydrogen production systems – A review. *Bioresource Technology*. 18, 8437–8444.

Hung, C.-H., Cheng, C.-H., Guan, D.-W., Wang, S.-T., Hsu, S.-C., Liang, C.-M. & Lin, C.-Y. (2011b). Interactions between *Clostridium* sp. and other facultative anaerobes in a self-formed granular sludge hydrogen-producing bioreactor. *International Journal of Hydrogen Energy*. 14, 8704–8711.

Kandler, O. (1983). Carbohydrate metabolism in lactic acid bacteria. *Antonie van Leeuwenhoek*, 209–224.

Kapdan, I. Karapinar & Kargi, F. (2006). Bio-hydrogen production from waste materials. *Enzyme and Microbial Technology*. 5, 569–582.

Karadag, D. & Puhakka, J. A. (2010a). Direction of glucose fermentation towards hydrogen or ethanol production through on-line pH control. *International Journal of Hydrogen Energy*. 19, 10245–10251.

Karadag, D. & Puhakka, J. A. (2010b). Enhancement of anaerobic hydrogen production by iron and nickel. *International Journal of Hydrogen Energy*. 16, 8554–8560.

Kawagoshi, Y., Hino, N., Fujimoto, A., Nakao, M., Fujita, Y., Sugimura, S. & Furukawa, K. (2005). Effect of inoculum conditioning on hydrogen fermentation and pH effect on bacterial community relevant to hydrogen production. *Journal of Bioscience and Bioengineering*. 5, 524–530.

Kim, D.-H., Kim, S.-H. & Shin, H.-S. (2009). Sodium inhibition of fermentative hydrogen production. *International Journal of Hydrogen Energy*. 8, 3295–3304.

Kim, S.-H., Han, S.-K. & Shin, H.-S. (2006). Effect of substrate concentration on hydrogen production and 16S rDNA-based analysis of the microbial community in a continuous fermenter. *Process Biochemistry*. 1, 199–207.

Kim, T.-H., Lee, Y., Chang, K.-H. & Hwang, S.-J. (2012). Effects of initial lactic acid concentration, HRTs, and OLRs on bio-hydrogen production from lactate-type fermentation. *Bioresource Technology*. 1, 136–141.

Kumar, N. Monga P.S. Biswas A.K &. Das D. (2000). Modeling and simulation of clean fuel production by *Enterobacter cloacae* IIT-BT 08. *International Journal of Hydrogen Energy*. 25, 945–952.

Lay, C.-H., Wu, J.-H., Hsiao, C.-L., Chang, J.-J., Chen, C.-C. & Lin, C.-Y. (2010). Biohydrogen production from soluble condensed molasses fermentation using anaerobic fermentation. *International Journal of Hydrogen Energy*. 24, 13445–13451.

Lee, Y.-J. Miyahara T. Noike T. (2001). Effect of iron concentration on hydrogen fermentation. *Bioresource Technology*. 80, 227–231.

Lengeler, J. W. (1999). *Biology of the prokaryotes,* Thieme [u.a.], 9780632053575, Stuttgart [u.a.].

Li, C. & Fang, H. H. P. (2007). Fermentative Hydrogen Production From Wastewater and Solid Wastes by Mixed Cultures. *Critical Reviews in Environmental Science and Technology*. 1, 1–39.

Li, R., Zhang, T. & Fang, H. (2011). Application of molecular techniques on heterotrophic hydrogen production research. *Bioresource Technology*. 18, 8445–8456.

Lin, C. & Lay, C. H. (2004). Carbon/nitrogen-ratio effect on fermentative hydrogen production by mixed microflora. *International Journal of Hydrogen Energy*. 1, 41–45.

Lo, Y.-C.; Chen, W.-M.; Hung, C.-H., Chen, S.-D. & Chang, J.-S. (2008). Dark H_2 fermentation from sucrose and xylose using H_2-producing indigenous bacteria: Feasibility and kinetic studies. *Water Research*. 42,827-842.

Mariakakis, I., Bischoff, P., Krampe, J., Meyer, C. & Steinmetz, H. (2011). Effect of organic loading rate and solids retention time on microbial population during bio-hydrogen production by dark fermentation in large lab-scale. *International Journal of Hydrogen Energy*. 17, 10690–10700.

Martinko M. & Clark D. (2009). *Biology of Microorganisms,* Pearson Publishers, 9780321536150, New York.

Meherkotay, S. & Das, D. (2008). Biohydrogen as a renewable energy resource—Prospects and potentials. *International Journal of Hydrogen Energy*. 1, 258–263.

Minton, N. P. & Clarke, D. J. (1989). *Clostridia,* Plenum Press, 9780306432613, New York.

Nath, K. & Das, D. (2004). Improvement of fermentative hydrogen production: various approaches. *Applied Microbiology and Biotechnology*. 5.

Nath, K., Kumar, A. & Das, D. (2005). Hydrogen production by Rhodobacter sphaeroides strain O.U.001 using spent media of Enterobacter cloacae strain DM11. *Applied Microbiology and Biotechnology*. 4, 533–541.

Noike, T. Takabatake H. Mizuno O. Ohba M. (2002). Inhibition of hydrogen fermentation of organic wastes by lactic acid bacteria. *International Journal of Hydrogen Energy*. 27, 1367–1371.

Oh, Y.-K., Kim, S. Hyoung, Kim, M.-S. & Park, S. (2004). Thermophilic biohydrogen production from glucose with trickling biofilter. *Biotechnology and Bioengineering*. 6, 690–698.

Ohnishi, A., Bando, Y., Fujimoto, N. & Suzuki, M. (2010). Development of a simple bio-hydrogen production system through dark fermentation by using unique microflora. *International Journal of Hydrogen Energy*. 16, 8544–8553.

Olsen, I. Johnson, J. L., Moore, V. H. & Moore, W. E. C. (1991). *Lactobacillus uli* sp. nov. and *Lactobacillus rimae* sp. nov. from the Human Gingival Crevice and Emended Descriptions of *Lactobacillus minutus* and *Streptococcus parvulus*. *International Journal of Systematic Bacteriology*. 41, 261-266

Ren, N., Li, J., Li, B., Wang, Y. & Liu, S. (2006). Biohydrogen production from molasses by anaerobic fermentation with a pilot-scale bioreactor system. *International Journal of Hydrogen Energy*. 15, 2147–2157.

Ren, N., Xing, D., Rittmann, B. E., Zhao, L., Xie, T. & Zhao, X. (2007). Microbial community structure of ethanol type fermentation in bio-hydrogen production. *Environmental Microbiology*. 5, 1112–1125.

Rilla, N. (2003). Inhibition of *Clostridium tyrobutyricum* in Vidiago cheese *by Lactococcus lactis* ssp. *lactis* IPLA 729, a nisin Z producer. *International Journal of Food Microbiology*. 1-2, 23–33.

Saraphirom, P. & Reungsang, A. (2011). Biological hydrogen production from sweet sorghum syrup by mixed cultures using an anaerobic sequencing batch reactor (ASBR). *International Journal of Hydrogen Energy*. 14, 8765–8773.

Schmauder, H.-P. (1992). Hans G. Schlegel, *Allgemeine Mikrobiologie*. Stuttgart–New York 1992. Georg Thieme Verlag. 3-13-444607-3.

Shen, L., Bagley, D. M. & Liss, S. N. (2009). Effect of organic loading rate on fermentative hydrogen production from continuous stirred tank and membrane bioreactors. *International Journal of Hydrogen Energy*. 9, 3689–3696.

Ueno, Y., Fukui, H. & Goto, M. (2007). Operation of a Two-Stage Fermentation Process Producing Hydrogen and Methane from Organic Waste. *Environmental Science & Technology*. 4, 1413–1419.

Valdez-Vazquez, I. & Poggi-Varaldo, H. M. (2009). Hydrogen production by fermentative consortia. *Renewable and Sustainable Energy Reviews*. 5, 1000–1013.

van Ginkel, S. & Logan, B. E. (2005). Inhibition of Biohydrogen Production by Undissociated Acetic and Butyric Acids. *Environmental Science & Technology*. 23, 9351–9356.

Vignais, P. M. & Billoud, B. (2007). Occurrence, Classification, and Biological Function of Hydrogenases: An Overview. *Chemical Reviews*. 10, 4206–4272.

Vijayaraghavan, K. & Mohd Soom, M. Amin (2006). Trends in bio-hydrogen generation – A review. *Environmental Sciences*. 4, 255–271.

Vos, P. de (2009). *The Firmicutes*, Springer, 0387950419, New York [u.a.].

Wang, J. & Wan, W. (2008). Effect of Fe^{2+} concentration on fermentative hydrogen production by mixed cultures. *International Journal of Hydrogen Energy*. 4, 1215–1220.

Wang, J. & Wan, W. (2009). Factors influencing fermentative hydrogen production: A review. *International Journal of Hydrogen Energy*. 2, 799–811.

Wongtanet, J., Sang, B.-I., Lee, S.-M. & Pak, D. (2007). Biohydrogen Production by Fermentative Process in Continuous Stirred-Tank Reactor. *International Journal of Green Energy*. 4, 385–395.

Wu, C.-W., Whang, L.-M., Cheng, H.-H. & Chan, K.-C. (2012). Fermentative biohydrogen production from lactate and acetate. *Bioresource Technology*. doi: 10.1016/ j.biortech. 2011.12.130

Chapter 6

Catalytic Steam Reforming of Methanol to Produce Hydrogen on Supported Metal Catalysts

Raúl Pérez-Hernández, Demetrio Mendoza-Anaya,
Albina Gutiérrez Martínez and Antonio Gómez-Cortés

Additional information is available at the end of the chapter

1. Introduction

The use of fossil fuels for energy supply in the world has caused various global environmental problems. For this reason it is becoming progressively more important to find ways of providing environmentally friendly energy. One promising alternative to fossil fuels is hydrogen, due to the importance as a clean source of energy, as well as, the increased demand in chemical industry [1; 2]. Also, fuel cells have recently attracted much attention as a potential device for energy transformation. Their performance is based on a clean process, without forming harmful by-products such as sulphur oxides and nitrogen oxides, while having a highly efficient energy transformation compared to conventional power generation processes as in heat engines. Hydrogen is a promising fuel for fuel cells and can be produced by steam reforming of natural gas, methanol and gasoline. At present, most of the world's hydrogen is produced from natural gas (~97 % CH_4) by a process called steam reforming [3-8]. The primary ways in which methane, is converted to hydrogen involve reaction with either steam (steam reforming), oxygen (partial oxidation), or both in sequence (autothermal reforming). In practice, gas mixtures containing carbon monoxide, carbon dioxide and unconverted methane. Reaction of carbon monoxide with steam (water-gas shift) over a catalyst produces additional hydrogen and carbon dioxide, and after purification, high-purity hydrogen is recovered. This reaction is highly endothermic. Although stoichiometry for the SRM suggests that only one mole of water is required for one mole of methane ($CH_4 + H_2O \rightarrow CO + 3H_2$), usually excess steam is used to reduce carbon formation. In most cases, carbon dioxide is vented into the atmosphere today, but there are options for capturing it in centralized plants for subsequent sequestration. However, steam reforming of methane does not reduce the use of fossil fuels and it still releases carbon to the environment in the form of CO_2. Thus, to achieve the benefits of the hydrogen economy, it is necessary produce hydrogen from non-fossil resources, such as water, methanol or ethanol using a renewable energy source. Among the different feedstocks available, alcohols are very promising candidates because these are easily decomposed in the presence of water and generate hydrogen-rich mixture at a relatively lower temperature. Steam reforming (SR) of methanol has been extensively studied in recent years [1–7]. Methanol has a low boiling point, a high hydrogen/carbon ratio and no C–C bonds, and can therefore be reformed at a relatively low temperature, reducing the risk of coke formation during the reaction [9]. Moreover, as methanol can be produced from renewable sources, its reforming does not contribute to a net addition of CO_2 to the atmosphere. Methanol can be converted to hydrogen by the following three reactions:

Partial oxidation of methanol

$$CH_3OH + 0.5O_2 = 2H_2 + CO_2 \qquad \Delta H° = -192 \text{ kJ mol}^{-1} \tag{1}$$

Steam reforming of methanol

$$CH_3OH + H_2O = 3H_2 + CO_2 \qquad \Delta H° = 50 \text{ kJ mol}^{-1} \tag{2}$$

Oxidative Steam Reforming of Methanol

$$CH_3OH + 1/2H_2O + 1/4O_2 = CO_2 + 5/2H_2 \qquad \Delta H° = 0 \text{ kJ mol}^{-1} \tag{3}$$

Most studies reported in the literature for the steam reforming reaction were on the application of CuO/ZnO-based and CuO/ZnO/Al_2O_3-based catalysts [10-12]. Alumina is generally added to the catalysts to improve their surface area and mechanical strength, and to prevent catalyst sintering [13]. The in situ characterization of CuO/ZnO reveals that the interaction of Cu and ZnO has a pronounced effect on the catalytic activity [14; 15]. Zinc oxide is known to improve the dispersion of Cu and the reducibility of CuO. The improvement of reducibility has been proposed as a possible cause of the good activity of CuO/ZnO-based catalysts [16]. However, some researchers have proposed that the main reason is the improvement in the adsorption properties, including the adsorption of methanol [17] and the spillover of both hydrogen from Cu to ZnO [18] and oxygen species from ZnO to Cu [19]. ZrO_2 addition to Cu-based alumina-supported catalysts has been shown to increase methanol conversion and reduce CO yields [9; 20]. However, it has been noted that the metal–support interactions in Cu/ZrO_2 are different than in the more conventional Cu/ZnO catalysts [21]. Some authors even describe a "synergy" between the Cu and ZrO_2 [13]. The higher activity of Cu-ZrO_2 catalysts has also been attributed to the stabilization of Cu_2O on the surface of the reduced catalysts or during the reaction [22; 23]. It is believed that the formation of Cu_2O leads to both more active and more durable catalysts, since Cu_2O is less susceptible to sintering compared with Cu metal [22; 23]. Cu^+ species have also been observed in CeO_2-containing Cu catalysts [24; 25] and isolated Cu^{2+} in lattice sites or in surface sites forming a nano-sized two-dimensional structure [26]. Addition of CeO_2 to Cu/Al_2O_3 catalysts has also been shown to increase methanol conversion, decrease CO selectivity and increase catalyst stability [27]. Due to this a strong effort has been directed to increase the overall efficiency of CeO_2 for different applications. CeO_2 has been widely used in purifying vehicle exhausts and became the most important rare oxides for NOx reduction with CO or hydrocarbons [28; 29]. Numerous studies of ZrO_2 or CeO_2 promoted Cu-based methanol steam reforming catalysts are available in the literature [10; 12; 21; 25; 27; 30-34]. However, comparing results between studies is challenging since the reaction evidently is very sensitive to the catalysts used and large differences in Cu loadings and catalyst compositions have been reported. For example, the copper concentrations on these types of catalysts have been varied from a few percent in some publications [35; 36] up to 70% or above in others [13]. Other studies related with nickel based catalysts on the methanol reforming process has been reported [37-39] or ethanol (SRE) as an H_2 source [40-44]. Navarro et al. [45] studied the oxidative reforming of hexadecane over Ni and Pt catalysts supported on Ce/La-doped Al_2O_3. They found for both Ni and Pt catalysts, higher specific activity when active metals were supported on alumina modified with cerium and lanthanum. However, the catalytic activity and H_2 selectivity observed on Ni-based catalysts were higher than on Pt-based catalysts. Recently, Pd-ZnO catalyst systems have been reported to be active and selective for MSR after pre-reducing with H_2 [46-48] and Pd supported on ZrO_2-TiO_2 [49]. Other catalytic systems containing highly dispersed gold have received great interest from both experimental and theoretical points of view. The role of metal oxide is to stabilize the gold nanoparticles and make the reaction take place on the gold surface reaction. Due to their high catalytic activity, particularly in CO oxidation at low temperature [50] gold-base catalysts are considered promising candidates for hydrogen production, through methanol decomposition and water-gas shift (WGS) reactions [51-54] catalytic combustion of volatile organic compounds (VOCs) [55], selective oxidation of CO in H_2-rich gas [56], adsorption of CO on Au/CeO_2 catalysts [57]. Methanol steam reforming for H_2 production has been not studied extensively with Au-base catalysts. Nevertheless, some gold-base catalysts has shown high activity for methanol oxidation at 373 K but low H_2 production as a function of time on stream [58]. On methanol decomposition was reported that gold supported on Al_2O_3 was most active than on the CeO_2, however, on the last catalyst the H_2 selectivity was better than on the former catalyst on the range temperature of 300 to 500 °C [58]. But, when water was added in the feed they observed a slight increase in the methanol conversion and, changes in the products distribution. The catalytic activity of the Au-Ag/CeO_2 catalyst and silver supported on ZnO 1D rods catalysts on the steam reforming methanol reaction for hydrogen production was reported [59-61]. The catalytic activity on Ag/ZnO sample with low Ag content showed better performance on the SRM reaction than on high silver loading catalyst. So, the sample with small Ag particle size showed best performance in methanol conversion than catalyst with big Ag particle size. Our group has previously studied the effect of nickel-copper addition to ZrO_2 by impregnation method and compared the catalytic activity of these bimetallic Cu/Ni catalysts on the oxidative steam reforming of methanol to produce H_2 [39]. The reactivity of the catalysts showed that the bimetallic samples prepared by successive impregnation had highest catalytic activity among all the catalysts studied.

The goal of this chapter is showed the effect of the metal copper or nickel addition to CeO_2 prepared by co-impregnation and sequential impregnation. Catalytic performance in oxidative steam reforming of methanol for the three Cu–Ni catalysts was compared with corresponding monometallic Cu and Ni catalysts, and Au/CeO_2 catalysts. The comparison is also made with characterization results obtained by BET (N_2 adsorption–desorption), SEM (Scanning Electron Microscopy), EDX (Energy Dispersive X-ray Spectroscopy), XRD (X-ray Diffraction), TEM (Transmission Electron Microscopy) and TPR (Temperature Programmed Reduction). In addition, the relation between the structure of bimetallic particles and catalytic performance in oxidative steam reforming of methanol is discussed.

2. Experimental

2.1. Synthesis of the catalysts

The CeO_2 synthesis was done using the precipitation method of the $Ce(NO_3)_3 \bullet 6H_2O$ (Aldrich) in NH_4OH (Fluka) at room temperature (r.t.).

$$8Ce(NO_3)_3 \bullet 6H_2O + 22NH_4OH \leftrightarrow 8Ce(OH)_4 \downarrow + 23NH_4NO_3 + 41H_2O \quad (4)$$

$$Ce(OH)_4 \xrightarrow{\Delta} CeO_2 + 2H_2O \quad (5)$$

The solid obtained was dried at 100 °C and then heated at 650°C for 5 hours in air stream. The prepared supports were impregnated with a solution of $NiCl_2 \bullet 6H_2O$, and another with a solution of $Cu(CH_3\text{-}CO_2)_2 \bullet H_2O$ at an appropriate concentration to yield 3 wt% of copper and nickel respectively. Three bimetallic samples were prepared at 50%Cu and 50%Ni respectively to obtain 3 wt. % of total metallic phase. For the first sample, CeO_2 support was successively impregnated with an aqueous solution of $Cu(CH_3COO)_2 \bullet H_2O$ (Merck), after that, the excess of water was removed at 80 °C under constant stirring and the catalyst was dried at 110 ºC and calcined at 500 °C for 2 h followed by cooling down to r.t. Then, an aqueous solution of $NiCl_2 \bullet 6H_2O$ was added and the resulting solid was calcined at the same temperature and time. The as prepared catalysts will be referred as Ni/Cu/CeO_2. For the second catalyst, the synthesis procedure was changed to the above sample mentioned. The labeling of this catalyst will be referred as Cu/Ni/CeO_2. The third sample (Cu-Ni/CeO_2) was prepared by using a simultaneous impregnation (also called co-impregnation): an aqueous solution of $Cu(CH_3COO)_2$ and $NiCl_2 \bullet 6H_2O$ were added to CeO_2 and calcined at 500 °C for 2 h. All the samples were reduced at 400 °C using a mixture of H_2 (5%)/He (50 mL/min) stream for 1 h before characterization, except for TPR technique in which the sample was calcined.

2.2. Characterization

The details of catalysts characterization have been reported in our earlier reports [26; 28; 37; 38; 49; 59; 61-64]. Nitrogen adsorption-desorption of the samples was measured at -196 C on a Belsorp-max Bel Japan equipment. Prior to the measurements the samples were degassed at 150 °C for 1 h. The surface area and pore size distribution were determined using the BET and BJH methods respectively. HRTEM and local chemical analysis of the bimetallic nanoparticles were carried out in a JEM 2200FS microscope with a resolution of 0.19 nm and fitted with an energy dispersive X-ray Spectrometer (NORAN) and a JEM 2010-HT with a point resolution of 0.19 nm fitted with an EDX microprobe Thermo-scientific. JEOL-2010 microscope with a point resolution of 0.19 nm fitted with an NORAN microprobe Thermo-scientific. The samples were dispersed in isopropanol and a drop of such a solution was placed onto copper and gold 300 mesh grids. Surface properties of the catalysts were studied by CO adsorption followed by DRIFT (Fourier Transform Infrared Spectroscopy). Experiments were done in a Nicolect Nexus 470 Spectrometer equipped with environmentally controlled Spectra Tech DRIFT (Diffuse Reflectance Infrared Fourier Transform) cell with KBr windows. For each experiment, 0.025 g of the sample was packed in the sample holder and pretreated in-situ under H_2 flow (30 mL/min) at 300 °C for 1 h. After this treatment the sample was purged with helium flow for half hour and cooled to room temperature in the same gas atmosphere before admittance for 5 min a flow (30 mL/min) of 2.5 %CO diluted in He. Afterwards, pure He was allowed to flow in the system to eliminate the residual CO gas. Spectra were collected from 128 scans with resolution of 4 cm^{-1}. For all catalysts a FTIR spectrum was obtained by making reference to the freshly reduced solid prior to CO adsorption. The spectrum of dry KBR was taken for IR single-beam background subtraction. Oxidative steam reforming of methanol was carried out at an atmospheric pressure by placing the fixed bed flow reactor (8 mm i.d.) in an electric furnace consisting of two heating zones equipped with omega temperature controllers, using a commercial flow system RIG-100-ISRI. Prior to OSRM reaction, 0.05 g of catalyst diluted in 0.150 g of SiC was reduced in situ, using a stream of H_2 (50 mL/min) increasing temperature from room to 400 °C with a heating rate of 10 °C/min and holding this temperature for 1h. A thermocouple in contact with the catalytic bed was utilized in order to monitor and control the temperature inside the catalyst. For the reaction, O_2 (5%)/He mixture (50 mL/min) and 150 mL/min of He was passed through stainless steel saturator containing methanol and water mixture (we use a hot line in the saturator in order to maintain constant the temperature ~ 25 °C). This gas was added by means of a mass flow controller (RIG-100). The total flow rate was kept at 200 mL/min. Reaction products were analyzed by Gow-Mac 580 Gas chromatograph with thermal conductivity detector equipped with two columns system (molecular sieve 5 Å and Porapack Q columns), double injector controlled by Clarity software V.2.6.04.402 and TCD. The first column was used to separate the gaseous products such as H_2, O_2, CH_4 and CO. The second column was used to separate water, methanol, methyl formate (MF) and CO_2. All the reported data were collected after a run time of 7 h. The following equations were used to determine the methanol conversion and selectivity:

$$X(\%) = \frac{C_{in} - C_{out}}{C_{in}} * 100 \quad S_{CO_2}(\%) = \frac{nCO_{2-out}}{nCO_{2-out} + nCO_{out}} * 100$$

and/or

$$S_{H_2}(\%) = \frac{nH_{2-out}}{nH_{2-out} + nCH_{4-out} + nCO_{2out} + CO} * 100$$

The subscripts in and out indicate the inlet and the outlet concentrations of the reactants or products.

3. Results

3.1. Ni/Cu/CeO_2 system

3.1.1. Textural properties of the Cu-Ni/CeO_2 system

Table 1 showed the textural properties of the catalytic materials obtained from N_2 physisorption measurement at temperature of liquid nitrogen. It showed that doping the bare CeO_2 support with copper or nickel to obtain the monometallic catalysts, results in a slight decrease on the BET surface area. The same effect was observed on the bimetallic samples, when Cu and Ni were impregnated by successive or co-impregnation method on the CeO_2. Typical SEM image with backscatter analysis of the as-synthesized Cu-Ni/CeO_2 catalyst prepared by co-impregnation method is present in Fig. 1. It showed that the sample is composed by irregular particles. It is important to mention that the bare CeO_2, as well as, the other catalysts under study had the same morphology, as the sample present on Fig. 1. This is expected because we used the CeO_2 previously stabilized at 650 °C to obtain the catalysts.

Sample	m^2/g
CeO_2	40.1
Cu/CeO_2	34.5
Cu/Ni/CeO_2	25.7
Cu-Ni/CeO_2	30.4
Ni/Cu/CeO_2	27.5
Ni/CeO_2	28.6

Table 1. Specific surface area (BET) of the Ni/Cu-base catalysts.

Figure 1. Typical SEM image of the fresh Ni-Cu/CeO_2 prepared by co-impregnation.

3.1.2. *Crystalline phases of the Cu-Ni/CeO_2 catalysts*

Fig. 2 showed the XRD patterns of the Ni/Cu/CeO_2 catalysts after thermal treatments (calcination and reduction). XRD patterns of the Ni/Cu-base catalysts yield a typical cubic fluorite structure of ceria, in addition, diffraction peak attributed to the metallic Ni was observed at 2Θ= 44.735 on the Ni/CeO_2 sample, indicating that NiO was completely reduced to metallic Ni below 500 °C. On Cu/CeO_2 sample diffraction-peaks of metallic Cu were observed at 2Θ = 43.317 and 50.449 (JCPDS 85-1326) respectively. On the bimetallic samples, diffraction peaks of Cu, Ni or Cu-Ni alloy were not observed, although the samples suffer different thermal treatments; this could be due to its low metal concentration (3.0 wt %) or because the particle size of the active phase is below of the detection limit of the technique. On samples with 3 wt. % of Cu/Ni supported on ZrO_2 [39] was observed the same effect. So, no diffraction peaks of metallic phase were observed by XRD technique.

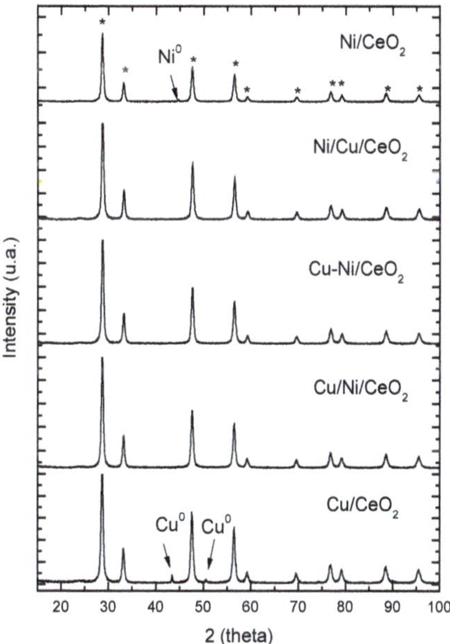

Figure 2. XRD patterns of the Ni/Cu/CeO2 catalysts. [*] cubic -CeO2

3.1.3. *Temperature-programmed reduction of the Cu-Ni-base catalysts supported on CeO_2*

Hydrogen consumption curves of the fresh bimetallic Cu/Ni-base catalysts and samples after catalytic reaction are shown in Fig. 3. Although the position of the reduction peaks strongly depends of the particle size or the interaction between metal active phase and the support, the TPR profiles of the catalytic materials are included for comparison. TPR profile of the bare CeO_2 sample showed a broad peak above 500 °C, this is assigned to reduction of surface ceria. Calcined CuO/CeO_2 catalyst showed three reduction peaks below 300 °C

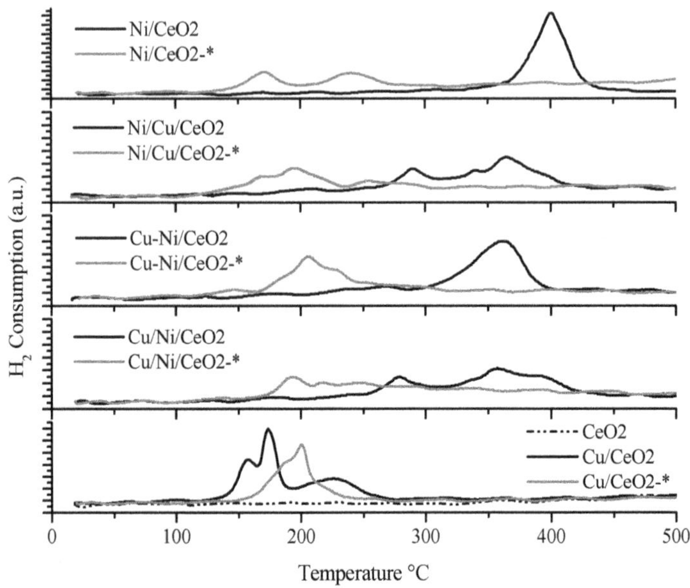

Figure 3. Temperature-programmed reduction profiles of the fresh Cu/Ni/CeO_2 catalysts (solid line) and samples after catalytic reaction (clear line).

indicating the presence of different kinds of Cu species formed during the preoxidation step [26; 63]. Peaks below 200 °C were attributed to reduction of highly dispersed CuO and the peak above 200 °C was associated with the reduction of the CuO bulk. The NiO/CeO_2 catalyst showed a sharp reduction peak at around 400 °C and may be attributed to reduction of NiO to metallic Ni crystallites. This temperature is higher than that the reported in previous work [38] for a similar catalyst. In that case the Nickel precursor was $Ni(NO_3)_2 \bullet 6H_2O$ which indicates different interaction between NiO and CeO_2. TPR profile of the bimetallic catalysts showed reduction peaks at lower temperature than Ni/CeO_2 catalyst. It has been reported that NiO supported, could be reduced at low temperatures when Cu or Pt are presented [37; 39; 65]. Because, Cu or Pt causes spillover of hydrogen onto Ni, inducing a simultaneous reduction of both, copper (platinum) oxide and NiO, causing a shift in the reduction of the active phase at low temperatures. In addition, it has been suggested that the first reduction peak observed in the TPR profile of the bimetallic catalyst, corresponded to the reduction of adjacent Cu and Ni atoms, which could be forming a bimetallic phase [37; 39; 65]. This finding indicates that the bimetallic phase had different interaction with the support and promoted the nickel reduction at lower temperatures and slows the copper reduction. In addition, it is clear that the bimetallic samples prepared by successive impregnation showed a broader reduction peak at higher temperatures than that for the Cu/CeO_2 sample, suggesting a broad particle size distribution, and slightly lower than Ni/CeO_2 sample. On the other hand, the TPR profiles of the Ni/CeO_2 and Ni-Cu/CeO_2

samples showed a sharp reduction peak than the rest of the samples. The sharp peak observed on theses samples corresponds to high uniformity in the Ni crystallite size. TPR profiles of the samples after catalytic reaction showed lower H_2 consumption, indicating that under OSRM conditions the active phase is partial oxidized. On Cu/CeO_2 samples was found by EPR technique the presence of the ion Cu^{2+} forming a nano-sized two-dimensional structure after OSRM reaction [26; 66; 67]. Oguchi et al. [22] observed a reduction peak on the CuO/ZrO_2 sample post-reaction. They concluded that the Cu_2O catalyst was stabilized during the SRM reaction. Turco et al. [68] suggested that there was a zone within the catalytic bed where the catalyst is oxidized, and another zone where it was reduced. This phenomena could be occurred on our samples, because, generally in oxidative steam reforming process, evidence suggest that the front of the catalyst bed is partially oxidized and the downstream of the catalyst bed remains in the reduced state.

3.1.4. DRIFTS of CO adsorption

In situ DRIFT spectra of the monometallic Ni/CeO_2 and Cu/CeO_2 samples and the three bimetallic Cu/Ni-base samples exposed to a 2.5%CO/He gas mixture recorded at room temperature with the aim to evaluate the influence of the metal addition to CeO_2 on the type and amount of different surface species. Fig. 4 shows an infrared spectrum in the 2200–2000 cm^{-1} regions of the Cu/Ni-base catalysts. The CO absorption band was observed at 2130 and 2100 cm^{-1} on the Ni/CeO_2 and Cu/CeO_2 samples respectively. It is generally acknowledged that carbonyl bands at wavenumbers lower than ca. 2115 cm^{-1} are due to carbonyl species adsorbed on metallic copper particles *[69]* while those at higher wavenumber correspond to carbonyls adsorbed on oxidized copper sites, so, the wavenumber increasing with the copper oxidation state. Variations in the frequency of these carbonyls have been related to changes in the nature of the exposed faces (i.e., in the degree of coordination of the copper centers). The main component at 2100 cm^{-1} was associated to CO adsorption on Cu sites of stepped particles (i.e., ‹110› plane) [70]. The band at 2135 cm^{-1} observed on the bimetallic samples was close to Ni/CeO2 catalyst, although it is slightly shifted to higher wavenumbers but it is virtually the same independently of the Cu and Ni addition to CO_2. In these bimetallic samples the absorption band corresponding to CO adsorption on Cu is totally suppressed (Fig. 4b). This finding suggests that the bimetallic samples are richer with Ni atoms in the surface of the catalysts. On the other hand, the high intensity in the CO-band observed on the bimetallic samples than monometallic catalysts could be associated to major dispersion of the metal active phase on the CeO_2. Differences in the CO-chemisorption were observed on the region of 1800–1000 cm^{-1} (Fig. 4c). CO chemisorption on the reduced surface CeO_2 and Mo/CeO_2 samples showed bidentate carbonate (as-1340, s-1680 and as-1320, s-1690 cm^{-1}), bicarbonate (1220, s-1460, s-1490 and as-1630 cm^{-1}), and bridged carbonate (as-1285 and s-1750 cm^{-1}) [71; 72]. No bands for the carbonate-like species was detected in the region of 1800–1000 cm^{-1} on the bimetallic samples prepared by successive impregnation as well for Cu/CeO_2 sample which indicates that CO does not adsorb on these materials.

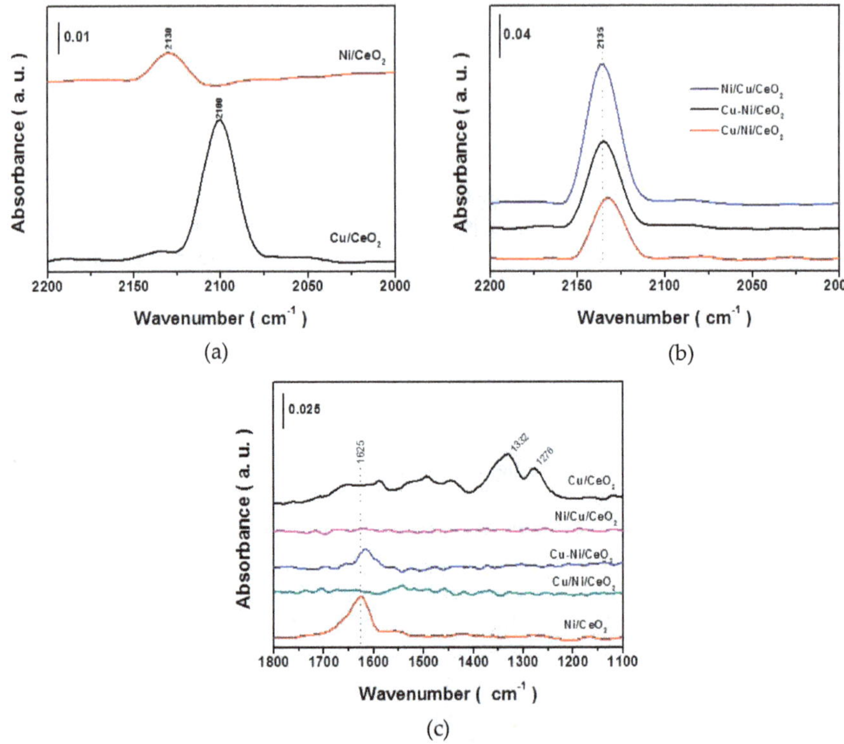

Figure 4. Diffuse reflectance FTIR spectra of CO adsorbed on monometallic-a and bimetallic-b catalysts. CO-absorption region of 1800–1000 cm^{-1} on all samples-c.

3.1.5. Catalytic activity of the Cu-Ni/CeO_2 catalysts on the OSRM reaction

The effect of the Cu and Ni addition to CeO_2 was evaluated on the oxidative steam reforming of methanol (OSRM) reaction from 200 to 400 °C. Fig. 5a-b summarizes the results of the CH_3OH conversion and H_2 selectivity over various catalysts as a function of the reaction temperature. It is clear that the bare-CeO_2 showed poor catalytic activity at the maximum reaction temperature. In general in all the samples the methanol conversion increased with an increase in the reaction temperature but, it is different when Cu and/or Ni were impregnated to CeO_2. At the beginning of the reaction the Cu/CeO_2 catalyst showed better methanol conversion than the other samples. When the temperature was raising at 300 °C the Cu/CeO_2 and the bimetallic Cu-Ni/CeO_2 (prepared by co-impregnation) catalysts had the same methanol conversion (40 %). Following by the bimetallic samples prepared by successive impregnation and the worst catalyst for methanol conversion was the Ni/CeO_2 sample. At the maximum reaction temperature the methanol conversion showed the following order: Ni/CeO_2 > Cu-Ni/CeO_2 > Ni/Cu/CeO_2 > Cu/Ni/CeO_2 > CeO_2. In previous study was observed that when Ni was supported on CeO_2 it showed better methanol conversion than Ni/ZrO_2 sample [38]. López et al. [39] reported that the Ni/Cu/ZrO_2 and Cu/Ni/ZrO_2 catalysts prepared by successive impregnation, showed high catalytic activity and H_2 selectivity than bimetallic sample prepared by simultaneous impregnation and the monometallic catalysts on the OSRM reaction. They calculated the reactivity of the model catalysts prepared by successive impregnation and observed that the band gap of the bimetallic models decreases, then, an electron transfer mechanism is favored at the interface between the bimetallic structures and the support, facilitating the redox properties of the catalysts, giving a higher OSRM activity [39]. In our case, we observed that the Ni/CeO_2 and Cu-Ni/CeO_2 (prepared by co-impregnation) samples had the best catalytic activity at the maximum reaction temperature. On these samples was observed by DRIFT technique the CO-band at 1625 cm^{-1} which was not present on the other samples. This finding can be attributed that the CO adsorption, in the carbonate species range were not favored on the other catalysts, indicating that CO does not adsorb on these materials, and so there are some blockade sites for catalytic reaction. The selectivity towards H_2 carried out at 200–400 °C on Cu-Ni-base catalysts supported on CeO_2 catalysts increased progressively by increasing the reaction temperature. It is clear that the Ni/CeO_2 and Cu-Ni/CeO_2 catalysts showed higher selectivity toward H_2 than the others samples.

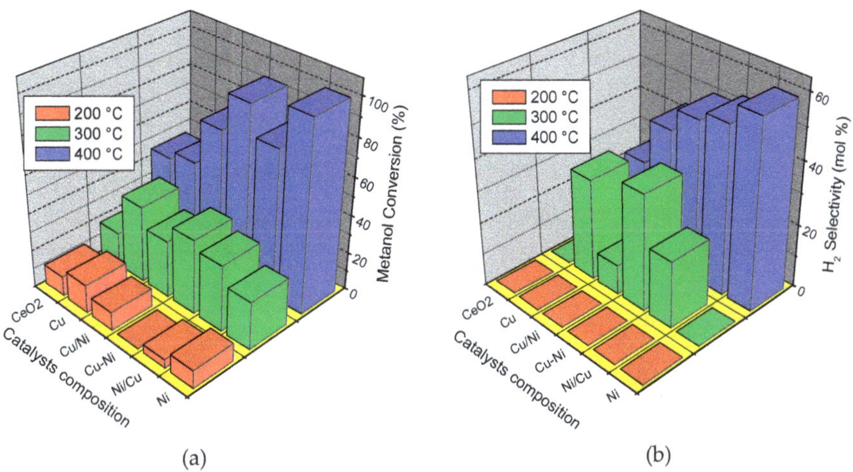

Figure 5. Methanol conversion of the Ni/Cu supported on CeO_2-(a) and H_2 selectivity-(b)

3.2. Au/CeO_2 system

3.2.1. Experimental section

CeO_2 catalyst was prepared in advance by precipitation method. NH_4OH (Baker) was added drop wise to an aqueous solution of $(Ce(NO_3)_3 \cdot 6H_2O)$. The precipitated solid was aged for 24 h and the residual liquid was removed by decanting, then the solid was dried at 100 °C for 24 h. The solid material was calcined at 100 °C 1h under an air stream and then at 500 °C for 5 h. The prepared support was impregnated with an aqueous solution of $HAuCl_4$ at an appropriate concentration to yield 1 and 3 wt % of total Au in the catalysts. The samples were dried at 100°C for 1h and then calcined at 400 °C for 2h in static air and finally reduced with a H_2(5%)/He stream at 350 °C 1h before the characterization and activity test. The labeling of different catalysts will be referred as follows: nAu/CeO_2 where n = 1 and 3 wt. % of Au in the catalyst respectively. The steady-state activity in the SRM reaction was performed in a conventional fixed-bed flow reactor (8 mm i.d.) using 0.1 g of the catalyst in a temperature range from 300 to 475 °C with steps of 25 °C with 6 h of stabilization time at each temperature and atmospheric pressure on an automatic multitask unit RIG-100 from ISR INC. The catalyst was first activated in a stream of H_2 (60 mL/min) from room temperature to 350 °C with a heating rate of 10°C/min and held at this temperature for 1h. A thermocouple in contact with the catalytic bed allowed the control of the temperature inside the catalyst was used. The sample was brought up to the reaction temperature in He and the reaction mixture was introduced. For the SRM reaction, He (60 mL/min, GHSV= 30,000 h^{-1} based on the total flow) was added by means of a mass flow controller (RIG-100) and bubbled through a tank containing mixture of water and methanol, the partial pressure of CH_3OH and H_2O was 9999.18 and 1699.86 Pa respectively. The molar ratio in the steam was CH_3OH (1.95μmol)/H_2O (1.97 μmol) = R ≈ 1.0 and the other concentration tested was CH_3OH (4.7 μmol)/H_2O (1.97 μmol) = R ≈ 2.4. The effluent gas of the reactor was analyzed by gas-chromatography (Gow-Mac 580 instrument) equipped with a two columns system (molecular sieve 5 Å and Porapack Q columns), double injector controlled by Clarity software V.2.6.04.402 and TCD. The first column was used to separate the gaseous products such as H_2, O_2, CH_4 and CO. The second column was used to separate water, methanol, methyl formate (MF) and CO_2. The GC analysis was performed in isothermal conditions (oven temperature = 100 °C). The equations used to determine the methanol conversion and selectivity was showed above.

4. Results and discussion

BET surface area calculated by the N_2 adsorption-desorption through the single point method of the 1Au/CeO_2 and 3Au/CeO_2 catalysts after thermal pretreatments were 44 and 34 m^2/g respectively. Figure 6 (a, b) showed a representative area of the 1Au/CeO_2 and 3Au/CeO_2 catalysts respectively. It showed that both samples are constituted by white spots identified as Au on large CeO_2 particles. Among these catalysts the sample with high Au loading showed big Au nanoparticles (large white spots) than on the 1Au/CeO_2 catalyst. Inset image in Figure 6a, showed an amplification of CeO_2 support using FE-SEM technique. Under this analysis we found that the CeO_2 is constituted by nanoparticles with diameters ~ 20 nm.

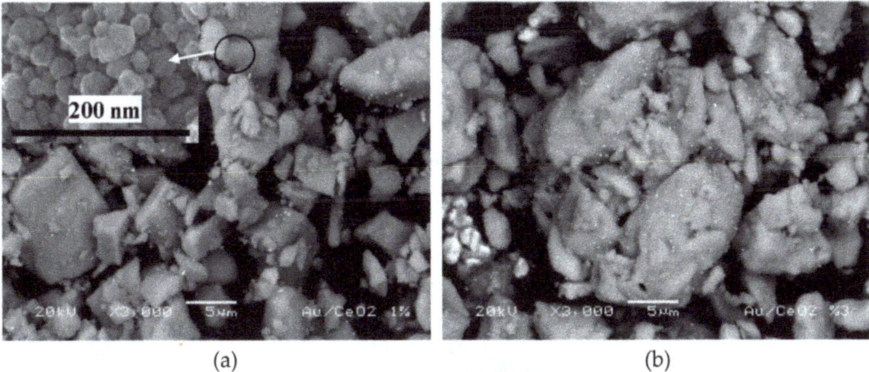

Figure 6. SEM image of: (a) $1Au/CeO_2$ catalyst. Inset image corresponds to CeO_2 support obtained by FE-SEM. (b) $3Au/CeO_2$ catalyst.

Figure 7 showed the XRD patterns of the $1Au/CeO_2$ and $3Au/CeO_2$ catalysts. It is possible to observe on the XRD pattern; characteristic peaks of the metallic gold and the others corresponding to the fluorite structure of ceria (CeO_2-cerianite). In addition, it is clear that the intensity of the diffraction peaks of the Au^0 increases proportionally as the Au was loading on CeO_2 suggesting on this sample a big Au crystallite size. The diffraction patterns of the Au/CeO_2 samples showed considerable line widths and no overlapping Au and CeO_2 diffraction peaks. We use the peaks display in the 35-45 = 2θ range to estimate the average Au crystallite size. The average value of the CeO_2 and the Au metal crystallite sizes on $1Au/CeO_2$ and $3Au/CeO_2$ samples were determined by Scherrer equation and it corresponds to 19, 23 and 33 nm respectively.

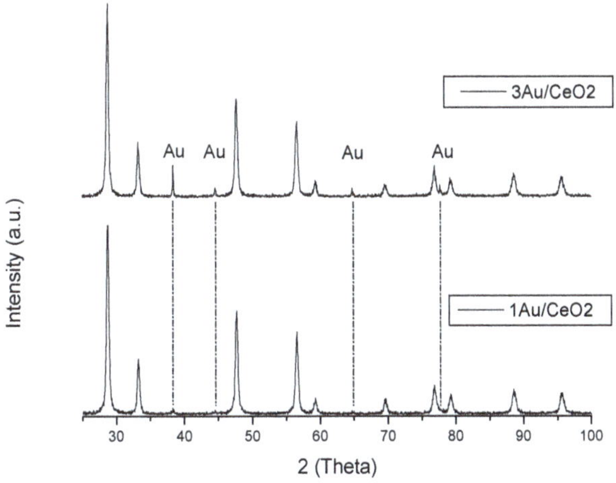

Figure 7. X ray diffraction patterns of the $1Au/CeO_2$ and $3Au/CeO_2$ catalysts. Cubic structure CeO_2 (fluorite structure) and characteristic peaks of the metallic gold.

TEM analysis of the $1Au/CeO_2$ catalyst is showed on Figure 8a; it is possible to observe the homogeneous distribution of the CeO_2 particles about 20 nm of diameter. This value is close to the FE-SEM observation and the results calculated by Scherrer equation. High resolution image of Au nanoparticle (Figure 8b) revealed that after thermal treatments, the Au nanoparticle generally had a hemi-spherical shape. The Au particle size measure was ~ 17 nm. Analysis of the electron diffraction patterns of the Au nanostructures, inset on Figure 8b, show that the crystalline structure grows as a gold fcc single crystal. These Au nanoparticles was recorded along [110] orientation. Figure 8c showed the TEM image of the $3Au/CeO_2$ catalyst, inset image on Figure 8c, showed the EDS spectra of the CeO_2 support and a gold nanoparticle the last one about 30 nm of diameter.

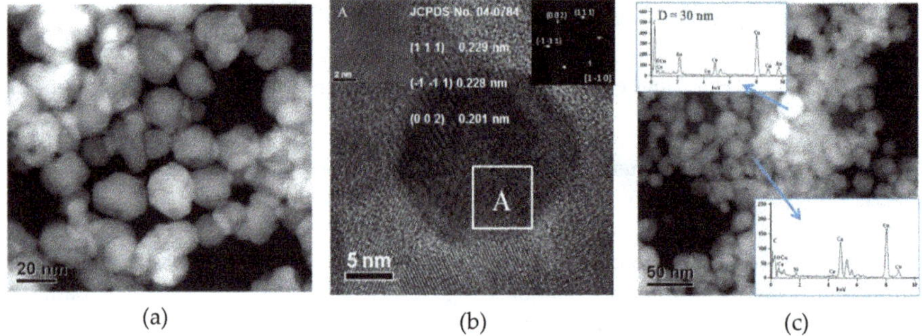

Figure 8. a) Low magnification TEM images of the $1Au/CeO_2$ catalyst. b) HRTEM image of the Au nanoparticle from $1Au/CeO_2$ catalyst which was recorded along [110] orientation. The d-spacing measured is showed inset image. The particle size was about 17 nm. c) Low magnification TEM images of the $3Au/CeO_2$ image. Inset images showed the EDS spectra of the Au nanoparticle about 30 nm of diameter and CeO_2 support.

The TPR profiles of the Au-base catalysts deposited on CeO_2 are depicted in Figure 9. This technique was also employed to found the optimal reduction temperature in the catalysts. TPR profiles showed differences in the hydrogen consumption depending of the gold content. Reduction of Au/CeO_2 samples were observed from 150 to 330 °C. For this reason, the catalysts were first activated in H_2 stream at 350 °C before all the characterization. As reported in literature, pure CeO_2 showed two reduction peaks at about 500 and 800 ° C, and were interpreted as the reduction of surface capping oxygen and bulk phase lattice oxygen, respectively [26; 73]. For Au/CeO_2 catalysts, the peak assigned to ceria surface layer reduction was reported from 120 to 178 °C [54; 74]. In our calcined $1Au/CeO_2$ catalyst, the hydrogen consumption peaks were observed within temperature range 175-325 °C, which could be deconvoluted into three components at reduction temperatures of 215, 257 and 294 °C respectively. Whereas, the calcined $3Au/CeO_2$ catalyst exhibited a broad reduction peak, which could be decompounds into three components at reduction temperatures of 184, 267 and 297 °C. It is clear that the intensity of the peak at 257 °C observed on the $1Au/CeO_2$ sample diminish significantly, and was shifted to the high-temperature region when the amount of Au was increased, because, the large crystallites tend to be reduced slower than the small ones due to their relatively lower surface area exposed to H_2. Rodriguez et al. Showed by XAFS that Au facilitates the oxide reduction of the matrix respect to pure ceria [75]. Andreeva et al. [52] observed by TPR technique two reducible species on Au-CeO_2 samples. The low-temperature peak on the TPR profile was connected with the reduction of the oxygen species on the fine gold particles, and the high-temperature peak was due to the reduction of the surface ceria. Taking into account these results and the differences in the gold loading in our samples, we assume that the former peak corresponds to the reduction of Au oxide nanoparticles, then it causes spill-over of hydrogen onto the support inducing a concurrent reduction of both the Au oxide and the surface of CeO_2 as was reported on other Ce-base catalysts [26; 28; 52].

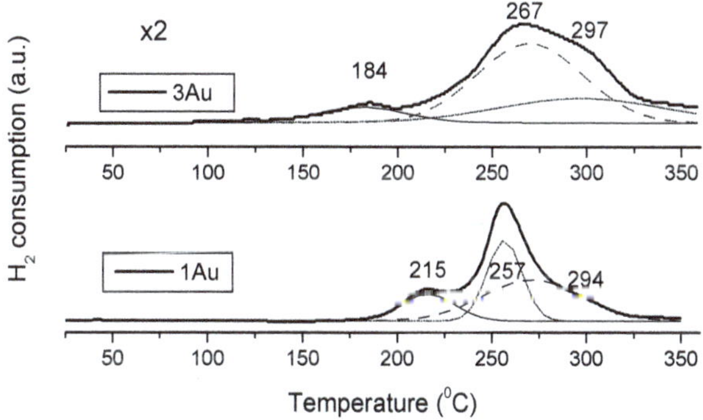

Figure 9. TPR profiles of the fresh Au/CeO_2 catalysts calcined at 400 °C.

The catalytic activity of the Au/CeO_2 catalysts as a function of the reaction temperature on the steam reforming of methanol reaction is presented on Figure 10. The light-off temperature of both samples started at ~ 350 °C at the molar CH_3OH/H_2O = 2.4 ratio, and the activity increased as temperature was rising. At the maximum reaction temperature, the methanol conversion observed on the 1Au/CeO_2 and 3Au/CeO_2 catalysts was 95 and 90 % respectively. The better catalytic activity observed on the 1Au/CeO_2 sample than on the 3Au/CeO_2 catalyst, could be attributed to differences in the Au particle size how was observed by XRD, SEM and TEM analysis. So, on the 3Au/CeO_2 catalyst the Au particle size was bigger than on the 1Au/CeO_2 catalyst. Croy et al. [76] studied the H_2 production through methanol decomposition on Pt/TiO_2 catalysts. They observed high catalytic activity on the catalysts with small particle size and diminish as the particle size increase. On methane combustion was observed better catalytic activity on the catalyst with low gold loading than the one with high gold loading, this result was associated with the dispersion of Au and the atomic ratio of Au^{3+}/Au^0 [77; 78]. However, Guzman and Gates [79] not found evidence between the Au cluster size and the catalytic activity on Au/MgO catalyst during the CO oxidation reaction with EXAFS technique. Wang et al. [80] suggested that the active sites in Au-ceria catalysts for the WGS probably contain Au nanoparticles and partially reduced ceria. It has been suggested that the presence of gold clusters weakens the bonding of the oxygen species on CeO_2 and facilitates the formation of more reactive species [53; 80; 81].

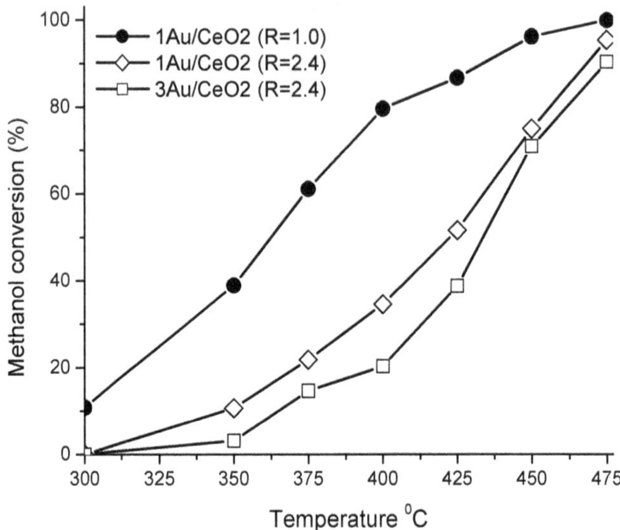

Figure 10. Temperature dependence of SRM activity of Au/CeO_2 catalysts. Partial pressure of CH_3OH and H_2O was 9999.18 and 1699.86 Pa respectively. GHVS=30,000 h^{-1}. The molar ratio in the steam was CH_3OH (1.95μmol)/H_2O (1.97 μmol) = R ≈ 1.0 and CH_3OH (4.7 μmol)/H_2O (1.97 μmol) = R ≈ 2.4.

The effect of the molar CH_3OH/H_2O ratio was evaluated on the 1Au/CeO_2 catalyst and presented on Figure 10. It showed clearly that this sample had better catalytic activity when the molar CH_3OH/H_2O ratio is close to R = 1 than on the R = 2.4. At the beginning of the reaction (300 °C) the conversion of methanol was 11 % and, increase when the temperature was rising. At 375 °C the catalyst reached near 61 % conversion whereas on the R = 2.4 ratio the methanol conversion only reached ~ 22 %. This effect is caused by the modification on the feed concentrations of the fuel as was reported on [82]. The methanol conversion observed at the maximum reaction temperature, reached almost 100 % on the molar CH_3OH/H_2O ratio R = 1, while on the R = 2.4 ratio only 95 % conversion was observed. This finding suggests that the methanol is adsorbed preferable on the surface of the catalysts than water. Table 2 showed the catalytic performance on steam reforming of methanol reaction from 300 to 475 °C range of the bare CeO_2 and 1Au/CeO_2 catalyst using the molar CH_3OH/H_2O ratio R = 1. As it can be seen, CeO_2 support showed low catalytic activity in almost all temperature range. However, at the end of the reaction the methanol conversion is almost 100 % for these samples. This finding showed the effect of the gold nanoparticles on CeO_2 during the catalytic activity. Gazsi et al. [83] suggested a cooperative effect between Au nanoparticles and CeO_2 support on the decomposition and reforming of methanol as was reported for other catalysts [9; 26; 64; 84].

Temperature (°C)	CeO_2 Methanol conversion and selectivity (%)					$1Au/CeO_2$ Methanol conversion and selectivity (%)				
	Conv.	H_2	CH_4	CO	CO_2	Conv.	H_2	CH_4	CO	CO_2
300	9.2	0	0	41.5	58.5	10.8	51.6	0	O	48.4
350	8.7	0	0	71.4	28.6	39.0	52.4	0.8	0.1	46.7
400	18.2	74.8	7.4	14.2	3.5	79.7	81.5	1.7	3.0	13.8
425	43.5	60.7	13.1	21.8	4.4	86.7	86.5	2.3	7.1	4.1
450	83.6	54.7	13.0	27.4	4.9	96.2	84.4	3.0	10.5	2.1
475	100	58.2	7.5	28.6	5.7	100	84.4	3.1	10.6	1.9

Table 2. Catalytic activity of the bare CeO_2 and $1Au/CeO_2$ catalyst. The molar ratio in the steam was CH_3OH (1.95μmol)/H_2O (1.97 μmol) = R ≈ 1.0

The reaction products observed on the steam reforming of methanol reaction of the Au/CeO_2 catalysts were H_2, CO, CO_2, CH_4 and H_2O. Small production of methyl formate as by-product of the reaction was observed in both samples. Figure 11a showed the distribution of hydrogen on the Au-base catalysts. Higher H_2 production was observed from 300 to 400 °C range on the catalysts tested with high molar CH_3OH/H_2O ratio R = 2.4 than R = 1. The drop of the hydrogen selectivity at higher temperatures could be attributed at the formation of CH_4. At the beginning of the reaction all samples showed low CO selectivity, Figure 11b. However, as the reaction temperature raise from 375 to 470 °C, the CO production increase and the selectivity toward CO_2 decrease. It is clear from Figure 11b that the $3Au/CeO_2$ sample showed high selectivity toward CO than on the $1AuCeO_2$ sample. In addition, we observed that the CO production is practically the same on the $1AuCeO_2$ catalyst (~ 10 %) independently of the molar CH_3OH/H_2O ratio. Table 2 showed the selectivity of the bare CeO_2 compared with $1Au/CeO_2$ catalyst. It showed the beneficial effect of the gold nanoparticles on the CeO_2. Because, the selectivity toward undesirable by-products such as CO and CH_4 observed on the bare CeO_2, were drop on the $1Au/CeO_2$ catalyst and the selectivity toward H_2 was improved on this catalyst.

Time on-stream studies (Fig. 12) at 350 °C in the $1Au/CeO_2$ sample with CH_3OH/H_2O ratio = 1.0, reveal high stability on the activity during the steam reforming of methanol reaction during a 65 h of reaction period, as well as high stability in the reaction products. The stability of catalysts under operating conditions is desirable for commercial applications. Fu et al. [73] found no significant change in activity after 120 h on stream on WGS. This behavior could be suggested that the catalyst maintains the same catalytic species during the SRM reaction and it does not lose by effect of the reaction conditions. So, these results showed that the CeO_2 matrix could be use to prevent the vanished of active phase during the reaction. Thus, a cooperative redox mechanism for the SRM reaction on Au–ceria is possible, similar to Cu–ceria or another kind of catalysts [26; 37; 63; 64; 73; 85] and for the WGS reaction [73].

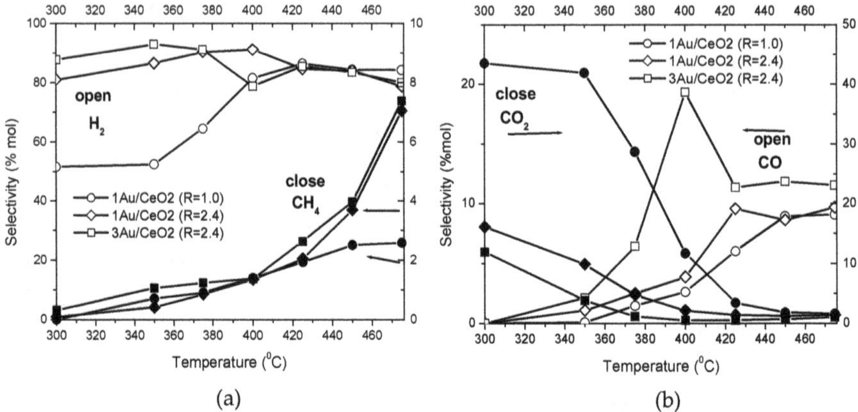

Figure 11. a) H_2 and CH_4 selectivity as a function of reaction temperature. b) CO and CO_2 selectivity as a function of reaction temperature. Partial pressure of CH_3OH and H_2O was 9999.18 and 1699.86 Pa respectively. GHVS=30,000 h^{-1}. The molar ratio in the steam was CH_3OH (1.95μmol)/H_2O (1.97 μmol) = R ≈ 1.0 and CH_3OH (4.7 μmol)/H2O (1.97 μmol) = R ≈ 2.4.

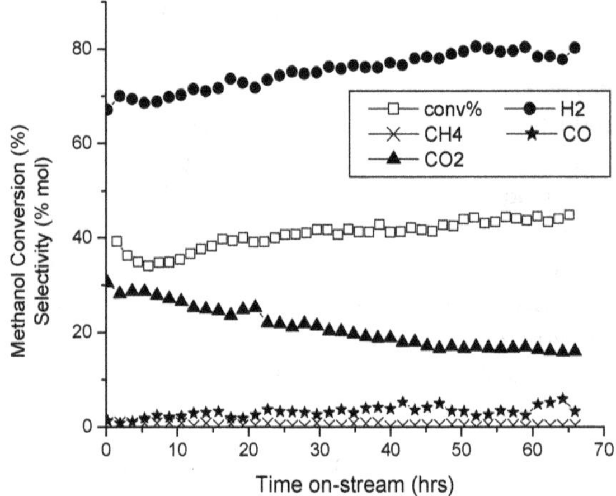

Figure 12. Stability of 1Au/CeO_2 catalyst at 350 °C. Partial pressure of CH_3OH and H_2O was 9999.18 and 1699.86 Pa respectively. GHVS=30,000 h^{-1}. The molar ratio in the steam was CH_3OH (1.95μmol)/H_2O (1.97 μmol) = R ≈ 1.0

5. Conclusion

Cu/CeO_2, Ni/CeO_2 and three bimetallic copper-nickel catalysts supported on CeO_2 were prepared by the impregnation method and tested in the OSRM reaction. The monometallic Ni/CeO_2 and the bimetallic Cu-Ni/CeO_2 (synthesized by co-impregnation) catalysts demonstrate both a higher catalytic activity in the OSRM reaction than the other catalysts. CO-chemisorption followed by DRIFT technique showed differences in the former samples. So, in the Ni/CeO_2 and Cu-Ni/CeO_2 catalysts was observed a band at 1625 cm^{-1} that was not present on the bimetallic samples prepared by successive impregnation as well for the Cu/CeO_2 samples which indicates that CO does not adsorb on these materials. This suggests that the metal active phase blockage some sites on the support or the active phase, modifies the surface of the catalyst for the adsorption of methanol and water inhibited the catalytic reaction on this system. In the case of the Au/CeO_2 system, nanosized ceria was prepared in advance with particle size of 19 nm and used as a support for gold nanoparticles. The average Au crystallite size in the 1Au/CeO_2 and 3Au/CeO_2 catalysts was 21 and 31 nm respectively. Differences in the reducibility of the Au/CeO_2 catalysts were observed depending of the Au loading. H_2-TPR results showed a shift of the reduction peaks toward high-temperature when the amount of Au was increased associated with the Au particle size present on the catalysts. The gold-base catalysts supported on CeO_2 showed catalytic activity on the SRM reaction and high selectivity toward H_2. Among these catalysts the sample with small Au particle size showed best performance in methanol conversion than on the catalyst with big Au particle size. These finding show the relationship between the Au particle size and the catalytic activity. In addition, it was demonstrate the beneficial effect of gold nanoparticles on the catalytic activity, as well as on the selectivity toward undesirable by-products such as CO and CH_4 and the selectivity toward H_2. The stability of the 1Au/CeO_2 catalyst at 350 °C was followed as a function of time on stream. This result showed high stability during the reaction. It is assumed that the interface between Au and partially reduced ceria is responsible for the high activity of Au/CeO_2 catalyst. In general we observed different behavior in the catalytic activity in our catalysts in the OSRM reaction to those reported in the literature. This difference could be attributed to the nature of the metal active phase and metal addition to the support. In this way, we suggest that the OSRM reaction could be a structure sensitive reaction according with the literature. However, further work is needed to refine and optimize the catalysts to improve the methanol conversion to produce CO-free hydrogen from the reaction under study.

Author details

Raúl Pérez-Hernández*, Demetrio Mendoza-Anaya and Albina Gutiérrez Martínez
Instituto Nacional de Investigaciones Nucleares, Carr. México-Toluca S/N La Marquesa, Ocoyoacac, México

Antonio Gómez-Cortés
Instituto de Física-Universidad Nacional Autónoma de México, D.F., México

Acknowledgement

Thanks to C. Salinas for technical support and to the projects ININ-CA-009 and CONACYT CB-2008-01-104540 for financial support. Authors would like to acknowledge Dr. Carlos Ángeles for its valuable comments and suggestions on the manuscript.

* Corresponding Author

6. References

[1] Armor, J. N. (1999) The multiple roles for catalysis in the production of H_2. Appl. Catal. A: Gen. 159-176: 159.

[2] Peña, M. A., J. P. Gómez, and J. L. G. Fierro (1996) New catalytic routes for syngas and hydrogen production. Appl. Catal. A: Gen. 144: 7-57.

[3] Pakulska, M. M., C. M. Grgicak, and J. B. Giorgi (2007) The effect of metal and support particle size on NiO/CeO_2 and NiO/ZrO_2 catalyst activity in complete methane oxidation. Appl Catal A: Gen. 332: 124-129.

[4] Sprung, C., B. Arstad, and U. Olsbye (2011) Methane Steam Reforming Over $Ni/NiAl_2O_4$ Catalyst: The Effect of Steam-to-Methane Ratio. Top Catal 54: 1063-1069.

[5] Montoya, J. A., E. Romero-Pascual, C. Gimon, P. D. Angel, and A. Monzón (2000) Methane reforming with CO_2 over Ni/ZrO_2-CeO_2 catalysts prepared by sol-gel. Catal Today. 63: 71-85.

[6] Roh, H.-S., K.-W. Jun, W.-S. Dong, J.-S. Chang, S.-E. Park, and Y.-I. Joe (2002) Highly active and stable Ni/Ce-ZrO2 catalyst for H2 production from methane. J. Mol Catal A: Chemical. 181: 137-142.

[7] Takeguchi, T., S.-n. Furukawa, and M. Inoue (2001) Hydrogen Spillover from NiO to the Large Surface Area CeO_2-ZrO_2 Solid Solutions and Activity of the NiO/CeO_2-ZrO_2 Catalysts for Partial Oxidation of Methane. J. Catal. 202: 14-24.

[8] Matsumura, Y., and T. Nakamori (2004) Steam Reforming of Methane over Nickel Catalysts. Appl. Catal. A: Gen. 258: 107-114.

[9] Agrell, J., H. Birgersson, M. Boutonnet, I. Melián-Cabreara, R. M. Navarro, and J. L. G. Fierro (2003) Production of hydrogen from methanol over Cu/ZnO catalysts promoted by ZrO_2 and Al_2O_3. J Catal 219: 389-403.

[10] Lindström, B., and L. J. Pettersson (2001) Hydrogen generation by steam reforming of methanol over copper-based catalysts for fuel cell applications. Int J Hydrogen Energy. 26: 923-933.

[11] Agrell, J., M. Boutonnet, and J. L. G. Fierro (2003) Production of hydrogen from methanol over binary Cu/ZnO catalysts: Part II. Catalytic activity and reaction pathways. Appl. Catal. A: Gen. 253: 213-223.

[12] Purnama, H., F. Girgsdies, T. Ressler, J. H. Schattka, R. A. Caruso, R. Scomäcker, and R. Schlögl (2004) CO formation/selectivity for steam reforming of methanol with a commercial $CuO/ZnO/Al_2O_3$ catalyst Catal Lett. . 94 61-68.

[13] Breen, J. P., and J. R. H. Ross (1999) Methanol reforming for fuel-cell applications: development of zirconia-containing Cu–Zn–Al catalysts. Catal. Today. 51 521-533.

[14] Günter, M. M., T. Ressler, R. E. Jentoft, and B. Bems (2001) Redox Behavior of Copper Oxide/Zinc Oxide Catalysts in the Steam Reforming of Methanol Studied by in Situ X-Ray Diffraction and Absorption Spectroscopy J. Catal. 203: 133-149.

[15] Fukahori, S., H. Koga, T. Kitaoka, A. Tomoda, R. Suzuki, and H. Wariishi (2006) Hydrogen production from methanol using a SiC fiber-containing paper composite impregnated with Cu/ZnO catalyst Appl. Catal. A: Gen. . 310 138-144.

[16] Fierro, G., M. L. Jacono, M. Inversi, P. Porta, F. Cioci, and R. Lavecchia (1996) Study of the reducibility of copper in CuO-ZnO catalysts by temperature-programmed reduction. Appl. Catal. A: Gen. . 137 327-348.

[17] Fujitani, T., and J. Nakamura (2000) The chemical modification seen in the Cu/ZnO methanol synthesis catalysts Appl. Catal. A: Gen. 191 111-129.

[18] Bowker, M., R. A. Hadden, H. Houghton, J. N. K. Hyland, and K. C. Waugh (1988) The mechanism of methanol synthesis on copper/zinc oxide/alumina catalysts J. Catal. . 109 263-273.

[19] Duprez, D., Z. Ferhat-Hamida, and M. M. Bettahar (1990) Surface mobility and reactivity of oxygen species on a copper-zinc catalyst in methanol synthesis J. Catal. . 124: 1-11.

[20] Yong-Feng, L., D. Xin-Fa, and L. Wei-Ming (2004) Effects of ZrO2-promoter on catalytic performance of CuZnAlO catalysts for production of hydrogen by steam reforming of methanol Int. J. Hydrogen Energy. 29: 1617- 1621.

[21] Szizybalski, A., F. Girgsdies, A. Rabis, Y. Wang, M. Niederberger, and T. Ressler (2005) In situ investigations of structure activity relationships of a Cu/ZrO_2 catalyst for the steam reforming of methanol. J Catal. . 233: 297-307.

[22] Oguchi, H., T. Nishiguchia, T. Matsumotoa, H. Kanaia, K. Utania, Y. Matsumurab, and S. Imamura (2005) Steam reforming of methanol over $Cu/CeO_2/ZrO_2$ catalysts. Appl Catal A: Gen. 281: 69-73.

[23] Oguchi, H., H. Kanai, K. Utani, Y. Matsumura, and S. Imamura (2005) Cu_2O as active species in the steam reforming of methanol by CuO/ZrO_2 catalysts Appl. Catal. A 293 64-70.

[24] Cheng, W.-H., I. Chen, J.-S. Liou, and S.-S. Lin (2003) Supported Cu Catalysts with Yttria-Doped Ceria for Steam Reforming of Methanol Topics Catal. 22 225-233.

[25] Liu, Y., T. Hayakawa, K. Suzuki, and S. Hamakawa (2001) Production of hydrogen by steam reforming of methanol over Cu/CeO_2 catalysts derived from $Ce_{1-x}Cu_xO_{2-x}$ precursors Catal. Commun. . 2: 195-200.

[26] Pérez-Hernández, R., A. Gutiérrez-Martínez, and C. E. Gutiérrez-Wing (2007) Effect of Cu loading on CeO_2 for hydrogen production by oxidative steam reforming of methanol. Int J Hydrogen Energy. 32: 2888-2894.

[27] Zhang, X., and P. Shi (2003) Production of hydrogen by steam reforming of methanol on CeO_2 promoted Cu/Al_2O_3 catalysts J. Mol. Catal. A: Chem. 194: 99-105.

[28] Pérez-Hernández, R., F. Aguilar, A.Gomés-Cortés, and G. Díaz (2005) NO reduction with CH_4 or CO on Pt/ZrO_2-CeO_2 catalysts. Catal. Today. 107-108: 175-180.

[29] Trovarelli, A., C. De-Leitenburg, M. Boaro, and G. Dolcetti (1999) The utilization of ceria in industrial catalysis. Catal. Today. 50: 353-367.

[30] Mastalir, A., B. Frank, A. Szizybalski, H. Soerijanto, A. Deshpande, M. Niederberger, R. Schomacker, R. Schlogl, and T. Ressler (2005) Steam reforming of methanol over $Cu/ZrO_2/CeO_2$ catalysts: a kinetic study J. Catal. 230 464-475.

[31] Ritzkopf, I., S. Vukojevic′, C. Weidenthaler, J. D. Grunwaldt, and F. Smith (2006) Decreased CO production in methanol steam reforming over Cu/ZrO_2 catalysts prepared by the microemulsion technique Appl. Catal. Gen. A. 302 215-223.

[32] Breen, J. P., F. C. Meunier, and J. R. H. Ross (1999) Mechanistic aspects of the steam reforming of methanol over a $CuO/ZnO/ZrO_2/Al_2O_3$ catalyst. Chem. Commun.: 2247-2248.

[33] Wang, L.-C., Q. Liu, M. Chen, Y.-M. Liu, Y. Cao, H.-Y. He, and K.-N. Fan (2007) Structural Evolution and Catalytic Properties of Nanostructured Cu/ZrO_2 Catalysts Prepared by Oxalate Gel-Coprecipitation Technique. J. Phys. Chem. C. 111 16549-16557.

[34] Yao, C.-Z., L.-C. Wang, Y.-M. Liu, G.-S. Wu, Y. Cao, W.-L. Dai, H.-Y. He, and K.-N. Fan (2006) Effect of preparation method on the hydrogen production from methanol steam reforming over binary Cu/ZrO_2 catalysts. Appl Catal A: Gen 297: 151-158.

[35] Lindström, B., L. J. Pettersson, and P. G. Menon (2002) Activity and characterization of Cu/Zn, Cu/Cr and Cu/Zr on g-alumina for methanol reforming for fuel cell vehicles Appl. Catal. Gen. A. 234 111-125.

[36] Jeong, H., K. Kim, T. Kim, C. Ko, H. Park, and I. Song (2006) Hydrogen production by steam reforming of methanol in a micro-channel reactor coated with $Cu/ZnO/ZrO_2/Al_2O_3$ catalyst. J. Power Sources. 159 1296-1299.

[37] Pérez-Hernández, R., G. Mondragón-Galicia, D. Mendoza-Anaya, J. Palacios, C. Angeles-Chavez, and J. Arenas-Alatorre (2008) Synthesis and characterization of bimetallic $Cu–Ni/ZrO_2$ nanocatalysts: H_2 production by oxidative steam reforming of methanol. Int J Hydrogen Energy. 33: 4569-4576.

[38] Pérez-Hernández, R., A. Gutiérrez-Martínez, J. Palacios, M. Vega-Hernández, and V. Rodríguez-Lugo (2011) Hydrogen production by oxidative steam reforming of methanol over Ni/CeO_2-ZrO_2 catalysts. Int J Hydrogen Energy. 36: 6601- 6608.

[39] López, P., G. Mondragón-Galicia, M. E. Espinosa-Pesqueira, D. Mendoza-Anaya, M. E. Fernández, A. Gómez-Cortés, J. Bonifacio, G. Martínez-Barrera, and R. Pérez-Hernández (2012) Hydrogen production from Oxidative Steam Reforming of Methanol: Effect of the Cu and Ni Impregnation on ZrO_2 and their molecular simulation studies. Int J Hydrogen Energy. 37: 9018-9027.

[40] Yang, Y., J. Ma, and E. Wu (2006) Production of hydrogen by steam reforming of ethanol over a Ni/ZnO catalyst. Int J Hydrogen Energy. 31 877-882.

[41] Biswas, P., and D. Kunzru (2007) Steam reforming of ethanol for production of hydrogen over $Ni/CeO_2–ZrO_2$ catalyst: Effect of support and metal loading Int J Hydrogen Energy 32 969-980.

[42] Mariño, F. J., E. G. Cerrella, S. Duhalde, M. Jobbagy, and M. Lombarde (1998) Hydrogen from steam reforming of ethanol. Characterization and performance of copper–nickel supported catalysts. Int J Hydrogen Energy 23 1095-1101.

[43] Mariño, F. J., M. Boveri, G. Baronetti, and M. Lombarde (2001) Hydrogen production from steam reforming of bioethanol using Cu/Ni/K/γ-Al_2O_3 catalysts. Effect of Ni. Int J Hydrogen Energy 26 665-668.

[44] Sun, J., X.-P. Qui, F. Wu, and W.-T. Zhu (2005) H_2 from steam reforming of ethanol at low temperaure over Ni/Y_2O_3, Ni/La_2O_3 and Ni/Al_2O_3 catalysts for fuel cell application. Int. J. Hydrogen energy. 30: 437-445.

[45] Navarro, R. M., M. C. Álvarez-Galván, F. Rosa, and J. L. G. Fierro (2006) Hydrogen production by oxidative steam reforming of hexadecane over Ni and Pt catalysts supported on Ce/La-doped Al_2O_3 Appl. Catal. A: Gen. 297: 60-72.

[46] Xia, B. J., J. D. Holladay, R. A. Dagle, E. O. Jones, and Y. Wang (2005) Development of highly active Pd-ZnO/Al_2O_3 catalysts for microscale fuel processor applications. Chem Eng Technol. 28(4): 515-519.

[47] Iwasa, N., S. Masuda, N. Ogawa, and N. Takezawa (1995) Steam reforming of methanol over Pd/ZnO: Effect of the formation of PdZn alloys upon the reaction. Appl Catal A: Gen. 125(1): 145-157.

[48] Karim, A. M., T. Conant, and A. Datye (2006) The role of PdZn alloy formation and particle size on the selectivity for steam reforming of methanol. J Catal 243(2): 420-427.

[49] Pérez-Hernández, R., A. D. Avendaño, E. Rosas, and V. Rodríguez (2011) Hydrogen Production by Methanol steam reforming over Pd/ZrO_2-TiO_2 catalysts. Topic Catal. . 54: 572-578.

[50] Haruta, M., T. Tsubota, T. Kobayashi, H. Kageyama, M. J. Genet, and B. Delmon (1993) Low-Temperature Oxidation of CO over Gold Supported on TiO_2, a-Fe_2O_3, and Co_3O_4. J. Catal. 144: 175.

[51] Trovarelli, C. A. (1996) Catalytic properties of ceria and CeO_2-containing materials. Catal. Rev. Sci. Eng. . 38: 439.

[52] Andreeva, D., V. Idakiev, T. Tabakova, L. Ilieva, P. Falaras, A. Bourlinos, and A. Travlos (2005) Low-temperature water-gas shift reaction over Au/CeO_2 catalysts. Catal. Today. 72: 51-57.

[53] Fu, Q., H.Saltsburg, and M. Flytzani-Stephanopoulos (2003) Active non-metallic Au and Pt species on Ceria-based Water-gas shift Catalysts. Science 301: 935-938.

[54] Sandoval, A., A. Gómez-Cortés, R. Zanella, G. Díaz, and J. M. Saniger (2007) Gold nanoparticles: Support effects for the WGS reaction. J. Mol Catal A: Chemical. 278: 200.

[55] Scire, S., S. Minnico, C. Crisafulli, C. Satriano, and A. Pistone (2003) Catalytic combustion of volatile organic compounds on gold/cerium oxide catalysts. Appl. Catal. B: Environmental. 40: 43.

[56] Panzera, G., V. Modafferi, S. Candamano, A. Donato, F. Frusteri, and P. L. Antonucci (2004) CO selective oxidation on ceria-supported Au catalysts for fuel cell application. J . Power Sources 135: 177.

[57] Manzoli, M., F. Boccuzzi, A. Chiorino, F. Vindigni, W. Deng, and M. Flytzani-Stephanopoulos (2007) Spectroscopic features and reactivity of CO adsorbed on different Au/CeO_2 catalysts. J Catal 245: 308.

[58] Chang, F.-W., H.-Y. Yu, L. S. Roselin, and H.-C. Yang (2005) Production of hydrogen via partial oxidation of methanol over Au/TiO_2 catalysts. Appl. Catal. A: General 290: 138.

[59] Pérez-Hernández, R., A. Gutiérrez-Martínez, A. Mayoral, F. L. Deepak, M. E. Fernández-García, G. Mondragón-Galicia, M. Miki, and M. Jose-Yacaman (2010) Hydrogen Production by Steam Reforming of Methanol over a Ag/ZnO One Dimensional Catalyst. Advanced Materials Research. 132: 205-219.

[60] Pérez-Hernández, R., and C. Gutiérrez-Wing (2009) Design of new Ag-Au(1-D)-CeO_2 catalysts for hydrogen production by steam reforming of methanol. EuropaCat IX, Salamanca, Spain. 1-3.

[61] Pérez-Hernández, R., A. Gutiérrez-Martínez, and C. Gutiérrez-Wing (2010) Hydrogen Production by Steam Reforming of Methanol over New Ag-Au(1-D)-CeO_2 Catalyst. Mater. Res. Soc. Symp. Proc. Vol. 127, Materials Research Society. 127: 1-4.

[62] Pérez-Hernández, R., A. Gómez-Cortés, J. Arenas-Alatorre, S. Rojas, R. Mariscal, J. L. G. Fierro, and G. Díaz (2005) SCR of NO by CH_4 on Pt/ZrO_2-TiO_2 sol-gel catalysts. Catal. Today. 107-108: 149-156.

[63] Pérez-Hernández, R., L. C. Longoria, J. Palacios, M. M. Aguila, and V. Rodríguez (2008) Oxidative steam reforming of methanol for hydrogen production over Cu/CeO_2-ZrO_2 catalysts. Energ Mater Mater Sci Eng Energ Syst. 3: 152-157.

[64] Pérez-Hernández, R., D. Mendoza-Anaya, M. E. Fernández, and A. Gómez-Cortés (2008) Synthesis of mixed ZrO_2-TiO_2 oxides by sol-gel: Microstructural characterization and infrared spectroscopy studies of NOx. J. Mol Catal A: Chemical 281: 200-206.

[65] Arenas-Alatorre, J., A. Gómez-Cortés, M. Avalos-Borja, and G. Díaz (2005) Surface Properties of Ni-Pt/SiO_2 Catalysts for N_2O Decomposition and Reduction by H_2. . J Phys Chem B. 109: 2371.

[66] Ratnasamy, P., D. Srinivas, C. V. V. Satyanarayana, P. Manikandan, R. Senthil-Kumaran, M. Sachin, and V. N. Shetti (2004) Influence of the support on the preferential oxidation of CO in hydrogen-rich steam reformates over the CuO-CeO_2-ZrO_2 system. . J Catal. 221: 455-465.

[67] Harrison, P. G., I. K. Ball, W. Azelee, W. Daniell, and F. Goldfarb (2000) Nature and surface redox properties of copper(ii)-promoted cerium(IV) oxide COoxidation catalysts. Chem Mater. 12: 3715-3725.

[68] Turco, M., G. Bagnasco, U. Costantino, F. Marmottini, T. Montanari, G. Ramis, and G. Busca (2004) Production of hydrogen from oxidative steam reforming of methanol II.

Catalytic activity and reaction mechanism on $Cu/ZnO/Al_2O_3$ hydrotalcite-derived catalysts. J. Catal. 228: 56.

[69] Hollins, P. (1992) Surf. Sci. Rep. . 16: 51.

[70] Sakakini, B. H., J. Tabatabaei, M. J. Watson, and K.C.Waugh (2000) J.Mol. Catal. A. 162 297.

[71] Nasser, H., Á. Rédey, T. Yuzhakova, and J. Kovács (2009) Thermal Stability and Surface Structure of Mo/CeO_2 and Ce-doped Mo/Al_2O_3 Catalysts. Journal of Thermal Analysis and Calorimetry. 95 69-74.

[72] Tabakova, T., F. Boccuzzi, M. Manzoli, and D. Andreva (2003) FTIR study of low-temperature water-gas shift reaction on gold/ceria catalyst. Appl Catal A: Gen. 252: 385-397.

[73] Fu, Q., W. Deng, H. Saltsburg, and M. Flytzani-Stephanopoulos (2005) Activity and stability of low-content gold–cerium oxide catalysts for the water–gas shift reaction Appl. Catal. B: Environmental. 56: 57-68.

[74] Ilieva, L., G. Pantaleo, I. Ivanov, A. M. Venezia, and D. Andreeva (2006) Gold catalysts supported on CeO_2 and CeO_2–Al_2O_3 for NOx reduction by CO Appl. Catal. B: Environmental. 65: 101-109.

[75] Rodriguez, J. A., X. Wang, P. Liu, W. Wen, J. C. Hanson, J. Hrbek, M. Pérez, and J. Evans (2007) Gold nanoparticles on ceria: importance of O vacancies in the activation of gold. Top. Catal. 44: 73-81.

[76] Croy, J. R., S. Mostafa, J. Liu, Y.-h. Sohn, and B. R. Cuenya (2007) Size dependent Study of MeOH decomposition over Size-selected Pt nanoparticles Synthesized via micelle Encapsulation. Catal. Lett. 118: 1.

[77] Zhang, Y., J. Deng, L. Zhang, W. Qiu, H. Dai, and H. He (2008) $AuOx/Ce_{0.6}Zr_{0.3}Y_{0.1}O_2$ nano-sized catalysts active for the oxidation of methane Catal. Today. 139: 29-36.

[78] Zhang, X., H. Shi, and B.Q. Xu (2007) Comparative study of Au/ZrO_2 catalysts in CO oxidation and 1,3-butadiene hydrogenation Catal. Today. 122: 330-337.

[79] Guzman, J., and B. C. Gates (2004) Catalysis by Supported Gold: Correlation between Catalytic Activity for CO Oxidation and Oxidation States of Gold J. Am. Chem. Soc. 126: 2672-2673.

[80] Wang, X., J. A. Rodriguez, J. C. Hanson, D. Gamarra, A.Martínez-Arias, and M. Fernández-García (2008) Ceria-based Catalysts for the Production of H2 Through the Water-gas-shift Reaction: Time-resolved XRD and XAFS Studies. Top. Catal. 49: 81-88.

[81] Guzman, J., S. Carrettin, J. C. Fierro-Gonzalez, Y. Hao, B. C. Gates, and A. Corma (2005) CO Oxidation Catalyzed by Supported Gold: Cooperation between Gold and Nanocrystalline Rare-Earth Supports Forms Reactive Surface Superoxide and Peroxide Species Angew. Chem. Int. Ed. 44: 4778-4781.

[82] Ahmed, S., and M. Krumpelt (2001) Hydrogen from hydrocarbon fuels for fuel cells. Int J Hydrogen Energy 26: 291.

[83] Gazsi, A., T. Bánsági, and F. Solymosi (2009) Hydrogen formation in the reaction of methanol on supported Au catalysts. Catal. Lett. 131: 33.

[84] Fisher, I. A., and A. T. Bell (1999) A Mechanistic Study of Methanol Decomposition over Cu/SiO_2, ZrO_2/SiO_2, and $Cu/ZrO_2/SiO_2$. J. Catal. . 184: 357.

[85] Burch, R. (2007) Gold catalysts for pure hydrogen production in the water–gas shift reaction: activity, structure and reaction mechanism Phys. Chem. Chem. Phys. 8: 5483-5500.

Chapter 7

Hybrid Filtration Combustion

Mario Toledo Torres and Carlos Rosales Huerta

Additional information is available at the end of the chapter

1. Introduction

Recent stringent emission regulations and depletion of energy sources have imposed special requirements on combustion technologies for the new millennium. By now it is well known that hydrocarbon sources, the bedrock of economic wealth and combustion science, are being depleted 100 000 times faster than they are being replenished. These factors will no doubt demand the development of novel combustion techniques. Among other concepts, combustion in a porous media offers a possible technological breakthrough and solutions for the near and long term. It can provide the basis for development for new combustion systems. The study of porous media phenomena itself can be a multidisciplinary field ranging from mechanical and chemical to geological and petroleum applications [1].

Porous media combustion, also known as filtration combustion, is defined as the process in which a self-sustaining exothermal reactive wave propagates over a porous reagent by means of gaseous oxidizer filtration through an inert solid matrix towards the reaction zone. As an internally self-organized process of heat recuperation, filtration combustion of gaseous mixtures in porous media differs significantly from the homogeneous flames. This difference is attributed to the following main factors: the highly developed inner surface of the porous medium results in efficient heat transfer between gas and solid; dispersion of the gas flowing through a porous media increases effective diffusion and heat transfer in the gas phase. To further elaborate, once a gas mixture is ignited inside the media, the heat release from the intense reaction zone is transferred to the solid matrix that subsequently feeds a fraction of the energy to the solid layers immediately above and below. This process facilitates a combustion process that ensures stability in a wide range of gas filtration velocities, equivalence ratios, and power loads.

Stationary and transient systems are the two major design approaches commonly employed in porous combustion [2-8]. The first approach is widely used in radiant burners and surface combustor-heaters where the combustion zone is stabilized within the finite element of the porous matrix. The second (transient) approach involves a traveling wave representing an unsteady combustion zone freely propagating in either downstream or upstream direction in the inert porous media (IPM). Strong interstitial heat transfer results in a low degree of thermal non-equilibrium between gas and solid. These conditions correspond to the low-velocity regime of filtration gas combustion, according to classification given by Babkin [9]. The relative displacement of the combustion zone results in positive or negative enthalpy fluxes between the reacting gas and the solid matrix. As a result, observed combustion temperatures can be significantly different from the adiabatic predictions and are controlled mainly by the reaction chemistry and heat transfer mechanism. The upstream wave propagation, countercurrent to filtration velocity, results in under-adiabatic combustion temperatures, while the downstream propagation of the wave leads to the combustion in a super-adiabatic regime with temperatures much in excess of the adiabatic one. Super-adiabatic combustion significantly extends conventional flammability limits to the region of ultra-low heat content mixtures.

Superadiabatic filtration combustion of rich and ultra-rich mixtures creates a situation in which partial oxidation and/or thermal cracking of hydrocarbons take place. This technology for hydrogen or synthesis gas (or syngas, a gas mixture that contains varying amounts of carbon monoxide and hydrogen) production uses an IPM [10-13]. The fuels used in porous combustion systems are basically of gaseous form due to fluidity, volumetric capacity, and shorter mixing length scale [14-21]. Liquid fuel reactors have been developed and used to a smaller extent [22-30].

According with the theory and technology of filtration combustion and the required new combustion systems, a new hybrid porous reactor can be developed changing an inert solid volume fraction by solid fuels. Now the porous medium is composed of uniformly mixed aleatory solid fuel and inert particles [31]. This change produces *hybrid filtration combustion* for simultaneous conversion of solid and gaseous fuels to energy (lean combustion) or hydrogen and synthesis gas (rich combustion).

As a result of composed porous medium, observed combustion temperatures can be lower than the IPM values and are controlled mainly by the heterogeneous reaction chemistry and heat transfer mechanism. Upstream and downstream wave propagations are present in the hybrid porous reactor. The upstream wave propagation in the composite bed is similar to the inert bed but the downstream wave propagation show a flat temperature profile with practically constant temperature along the reactor.

The combustion temperature in composite bed decreases from rich to ultra-rich gas fuel-air mixtures, and also with the increase of solid fuel volume fraction in porous medium. That suggests a change of dominant kinetic mechanism and a shift from homogeneous to heterogeneous chemistry. At high downstream propagation velocities near the stoichiometric conditions the role of gas phase kinetics is dominant. In these conditions, the upstream wave propagation suppresses the heterogeneous oxidation processes by the complete consumption of oxidizer. At higher equivalence ratios, the wave propagation is slower. This increases the role of heterogeneous kinetics. As a result, the ignition and combustion temperatures drop. In turn, this affects the wave velocity. The mechanism of heterogeneous combustion becomes dominant for downstream wave propagation. The slowly reacting solid fuel is directly exposed to oxidizer. The ignition initiated over the solid phase is further transferred to the gas phase. It is possible to suggest that the structure of the wave in this case involves a heterogeneous reaction front followed by the gas phase reaction front. These fronts are followed by a slow endothermic reaction between the formed steam and solid/gaseous fuels.

The propagation of downstream hybrid combustion waves in the fuel loaded medium has resulted in wide combustion fronts with superadiabatic combustion temperatures. Due to the wide flat temperature profiles it has been found impossible to determine the wave velocity with sufficient accuracy. The flat temperature profile also suggests that the fuel particles are burning upstream of the front.

The use of rich and ultra-rich mixtures of gas fuel with air hybrid filtration combustion for hydrogen and syngas production has big potential. Hydrogen and syngas concentration increases with an increase of equivalence ratio for gas fuel/air mixtures in the inert porous medium and composite bed. In comparison, more high concentration of hydrogen and syngas is obtained in the composite porous media. Hydrogen conversion for gas fuel/air mixtures for hybrid filtration combustion waves is 50% for higher equivalence ratio.

2. Physical and mathematical description of the hybrid process

Lean combustion results in complete burnout of the hydrocarbon fuel, with the formation of carbon dioxide and water. Thus, both the composition of the final products and the heat release are well defined. In contrast to the lean case, the partial oxidation products of rich and ultra-rich waves are not clearly defined. In such a case, fuel is only partially oxidized in the wave, and the total heat release could be kinetically controlled by the degree of partial oxidation. As a result, chemical kinetics, heat release, and heat transfer are strongly coupled in the rich and ultra-rich waves, rendering it a more complicated and challenging phenomenon than the lean wave.

In the following we describe the governing equations, derived from fundamental principles, as well as some suitable models and assumptions that allow the formulation of a mathematical model. Such a model can be tractable by numerical simulation. We consider the bulk volume V of the bed separated into the volume occupied by the solid particles, V_s, and the volume occupied by the gas V_g (i.e. $V = V_g + V_s$). As usual, we define the porosity θ as the fraction of bulk volume V occupied by the gas: $V_g = \theta V$. The composition of the gas is characterized by the molar fractions $\{y_j, j = 1,...,N_{SP}$ of the N_{SP} chemical species in the gas phase. On the other hand, the solid phase is composed of reacting fuel particles and inert particles. For the sake of conciseness, in this section we will refer to this solid fuel simply as fuel (it is implicit, by our definition of hybrid filtration combustion, that some of the species in the gas phase are also components of a gaseous fuel injected into the porous medium). To specify the composition of this solid medium we define partial (mass) densities for the fuel and inert gas as

$$\rho_f = \frac{m_f}{V_s} \ ; \ \rho_i = \frac{m_i}{V_s} \tag{1}$$

where m_f and m_i are respectively the mass of fuel and the mass of inert component in a given volume V. Note that these densities are based on the solid volume, so that $m_f = \rho_f \theta V$, and similarly for m_i. Here and in all this chapter the subscripts (g) refer to the gas phase, (s) to the solid phase, (f) to the fuel component in the solid, and (i) to the inert component in the solid.

2.1. Conservation of mass for gaseous species

For an arbitrary volume V in the porous medium, the conservation of mass for the j-th species in the gaseous mixture is given by

$$\int_V \frac{\partial}{\partial t}(\theta \rho_g y_j)\mathrm{d}V + \int_A \rho_g y_j \vec{\mathbf{u}}\cdot\vec{\mathbf{n}}\mathrm{d}A + \int_A \rho_g y_j \vec{\mathbf{w}}_j \cdot\vec{\mathbf{n}}\theta\mathrm{d}A = \int_V S_j \mathrm{d}V. \tag{2}$$

The first term on the left-hand side (LHS) is of course the local rate of variation of the molar quantity of species j in the volume, where ρ_g is the molar density of the gas. The second term is the advection of species j by the gas flowing out of V through is bounding surface A ($\vec{\mathbf{n}}$ is the outward normal vector). This flow is calculated using the filtration velocity $\vec{\mathbf{u}}$, so that the integration is taken over the whole area A. The third term is the additional transport across that bounding surface due to the diffusion velocity, $\vec{\mathbf{w}}_j$, of species j relative to the mean gas velocity. The right-hand side contains the integration of the source term, S_j, per unit of bulk volume. This term contains the net production of species j by homogeneous reaction in the gas phase, heterogeneous reaction with the solid fuel, and volatile matter release by that fuel, as explained later.

Although the diffusion velocities $\vec{\mathbf{w}}_j$ can in principle be obtained by solving a system of equations, involving the binary mass diffusion coefficients for each pair of species in the gas, such treatment is computationally difficult and expensive. Therefore, the usual simplified approach is based on Fick's law

$$y_j \vec{\mathbf{w}}_j = -D_j \nabla y_j \tag{3}$$

where D_j is the diffusion coefficient of species j into the gaseous mixture. Hence, from Eq. (2) one readily obtains

$$\frac{\partial}{\partial t}(\theta \rho_g y_j) = \nabla\cdot(\theta\rho_g D_j \nabla y_j) - \nabla\cdot(\rho_g \vec{\mathbf{u}} y_j) + S_j\ ; \qquad \text{for } j = 1,\ldots,N_{SP} \tag{4}$$

Among other parameters, the filtration velocity $\vec{\mathbf{u}}$ must be known in Eqs. (4) in order to solve them for y_j. In many applications with porous media in reactors of simple geometry (for instance, a long and narrow cylinder), the filtration velocity is determined by a global mass balance of the reactor, and the whole problem can be analyzed as one-dimensional.

By their definition, the mole fractions must satisfy

$$\sum_{j=1}^{N_{SP}} y_j = 1 \tag{5}$$

so that, one of the N_{SP} equations (4) can be eliminated, and the associated mole fraction can be obtained by difference from Eq. (5). We remark that, as shown in Ref. [32], solving (N_{SP} – 1) species equations in this way can produce numerical problems, when the exact diffusion velocities have been substituted by Fick's law, conducing to a violation in global mass conservation. The error vanishes when all diffusion coefficients are equal ($D_j = D$), and it is negligible if the species for which the equation was removed has a relatively high concentration in the mixture.

The gas volume V_g in the region V receives chemical species which come from: (i) advection by the gas flowing into V_g, (ii) enter V_g as a component of volatiles evolved from the solid fuel, or (iii) enter V_g as a product of heterogeneous reactions of the solid fuel with gaseous species. The species will react in the gaseous phase according to a homogeneous reaction mechanism. The presence of heterogeneous reactions means that simultaneously some of these species will be consumed in such reaction. The superposition of these processes determine the net source term S_j in the equation (4).

2.1.1. *Homogeneous reaction*

The homogenous reaction of the N_{SP} species in the gas is generally expressed as a system of N_{HO} reactions

$$\sum_{j=1}^{N_{SP}} \eta'_{j\alpha} \mathrm{M}_j \longleftrightarrow \sum_{j=1}^{N_{SP}} \eta''_{j\alpha} \mathrm{M}_j \quad ; \quad \text{for } \alpha = 1,\ldots,N_{HO} \tag{6}$$

where M_j represents symbolically the molecule of species j, and $(\eta'_{j\alpha}, \eta''_{j\alpha})$ are the stoichiometric coefficients of the j-th species in the α-th homogeneous reaction. These reactions are considered, in principle, reversible so that the rates of consumption and production of species j are given by

$$\sum_{\alpha=1}^{N_{HO}} \eta'_{j\alpha} r_{f\alpha} + \eta''_{j\alpha} r_{b\alpha} \text{ and } \sum_{\alpha=1}^{N_{HO}} \eta''_{j\alpha} r_{f\alpha} + \eta'_{j\alpha} r_{b\alpha}$$

respectively. Here, $r_{f\alpha}$ is the forward reaction rate (→), and similarly $r_{b\alpha}$ is the backward reaction rate (←). Thus, the net production for the j-th species by the homogeneous reaction mechanism is

$$S_{j,HO} = \sum_{\alpha=1}^{N_{HO}} (\eta''_{j\alpha} - \eta'_{j\alpha})(r_{f\alpha} - r_{b\alpha}) \tag{7}$$

The forward and backward reaction rates can be obtained from the law of mass action. Considering the mechanism (6) as composed of elementary reactions, the order of the reactions corresponds to the stoichiometric coefficients, and therefore

$$r_{f\alpha} = k_{f\alpha} \rho_g^{n'_\alpha} \prod_{j=1}^{N_{SP}} y_j^{\eta'_{j\alpha}} \text{ ; with } n'_\alpha = \sum_{j=1}^{N_{SP}} \eta'_{j\alpha} \tag{8}$$

and

$$r_{b\alpha} = k_{b\alpha}\rho_g^{n''_\alpha}\prod_{j=1}^{N_{SP}} y_j^{\eta''_{j\alpha}}; \text{ with } n''_\alpha = \sum_{j=1}^{N_{SP}} \eta''_{j\alpha} \tag{9}$$

In these expressions, $k_{f\alpha}$ and $k_{b\alpha}$ correspond to the forward and backward kinetic factors, which are usually modelled using the Arrhenius law

$$k_{f\alpha} = A_{f\alpha}T_g^{b_\alpha}\exp\left(-\frac{E_{f\alpha}}{RT_g}\right) \tag{10}$$

where the constants $A_{f\alpha}$ and b_α, as well as the activation energy $E_{f\alpha}$, are data to be provided for each reaction. These parameters constitute an essential part of the chemical reaction model (and one of the most difficult to obtain). The backward kinetic factors are computed from the forward factors using the equilibrium constants K_α of the reactions

$$k_{b\alpha} = \frac{k_{f\alpha}}{K_\alpha} \text{ with } K_\alpha = \left(\frac{p_o}{RT_g}\right)^{n''_\alpha - n'_\alpha}\exp\left(\sum_{j=1}^{N_{SP}}(\eta'_{j\alpha} - \eta''_{j\alpha})\frac{g_j^o}{RT_g}\right) \tag{11}$$

where p_o is the pressure of the standard state (p_o = 1 bar), R is the universal gas constant and g_j is the Gibbs free energy of species j ($g_j = h_j - Ts_j$). These properties depend on gas temperature, and can be obtained from available regression functions, which can be written as polynomials in T_g such as

$$h_j = \sum_{n=0}^{m} a_{nj}T_g^n \tag{12}$$

for the enthalpy, and as

$$s_j = \sum_{n=0}^{m-1} b_{nj}T_g^n + b_m \ln T_g \tag{13}$$

for the entropy (as, for example, NASA polynomials).

2.1.2. Volatiles release

The process of volatiles release from the solid fuel, by thermal decomposition and pyrolysis, can be modelled as one or several irreversible reactions, which we can represent symbolically as

$$\begin{aligned}
\{\text{Fuel}\} &\xrightarrow{k_1^{(v)}} \mu_1\text{V}_1 + (1-\mu_1)\text{S}_1 \\
\{\text{Fuel}\} &\xrightarrow{k_2^{(v)}} \mu_2\text{V}_2 + (1-\mu_2)\text{S}_2 \\
\vdots \quad & \quad \vdots \qquad \quad \vdots \\
\{\text{Fuel}\} &\xrightarrow{k_{Nv}^{(v)}} \mu_{N_v}\text{V}_{N_v} + (1-\mu_{N_v})\text{S}_{N_v}
\end{aligned} \tag{14}$$

to indicate the conversion of the solid fuel into volatiles $\{V_1, V_2, \ldots, V_{N_v}\}$,all of which have a prescribed chemical composition. The constants $\{\mu_\beta, \beta = 1, \ldots, N_v\}$, which are also parameters of the model, denote the fraction of fuel mass that is transformed into the volatile V_β via the β-th reaction in the system (14). Such transformation leaves a solid substance S_β which will react heterogeneously with the surrounding gas (as, for example, coke when coal is being burned). For each of these N_v reactions, the model postulates a kinetic factor that can be put into an Arrhenius form

$$k_\beta^{(v)} = A_\beta T_s^{b_\beta} \exp\left(-\frac{T_\beta}{T_s}\right) \text{ for } \beta = 1, \ldots, N_v \tag{15}$$

as a function of the solid particles temperature T_s. These factors are defined such that for a given mass of fuel m_f, the rate at which mass of volatile V_β is released into the gas phase is given by

$$r_\beta^{(v)} = k_\beta^{(v)} \mu_\beta m_f \tag{16}$$

The model requires the activation temperatures T_β, the pre-exponential constants, and the composition of the released volatiles. Typically for coal, the species present in the volatiles are H_2, H_2O, CO, CO_2, CH_4 and some other hydrocarbons. The composition can be defined in terms of the mass fraction of the species present in such gaseous mixtures. We use $\sigma_{j\beta}$ to indicate the mass fraction of species j in the volatile β. Therefore, superposing the effects of the N_v vias of volatilization, the source of species j in the gas phase, per unit of bulk volume, is

$$S_{j,v} = \frac{(1-\theta)\rho_f}{W_j} \sum_{\beta=1}^{N_v} k_\beta^{(v)} \mu_\beta \sigma_{j\beta} \tag{17}$$

due to the volatilization process. W_j is the molecular weight of species j and ρ_f is defined in Eq.(1).

2.1.3. *Heterogeneous reaction*

The solid substances S_β left by the volatilization mechanism can further react with several species from the surrounding gas. This heterogeneous reactions can be treated in a similar way to what is done in 2.1.1 for the homogeneous reactions. First, we need to define precisely the solids S_β. When the solid fuel is coal, the usual assumption is to consider S_β as pure carbon for all the volatilization reactions (14). Global heterogeneous reactions that can take place in such conditions are for example

$$\begin{aligned} 2C + O_2 &\longrightarrow 2CO \\ C + O_2 &\longrightarrow CO_2 \\ C + CO_2 &\longrightarrow 2CO \\ C + H_2O &\longrightarrow CO + H_2 \end{aligned} \tag{18}$$

For some other fuels, whose fixed carbon fraction is mostly C, such as wood pellets, the treatment of S_β as C could also be a good approximation. Alternatively, a more accurate representation of S_β could be a pseudo-molecule C_xH_y, that can be used to represent coke or other form of carbonized material. Since we want to consider the solid fuel as just one component of the solid phase, in addition to the inert material, the heterogeneous reactions cannot discriminate between different solid species, and therefore the heterogeneous reaction mechanism must be written as

$$\mathrm{S} + \sum_{j=1}^{N_{SP}} \nu'_{j\gamma} \mathrm{M}_j \longrightarrow \sum_{j=1}^{N_{SP}} \nu''_{j\gamma} \mathrm{M}_j \text{ for } \gamma = 1,\ldots,N_{HE} \tag{19}$$

where N_{HE} is the number of reactions in the heterogeneous mechanism and S denotes the molecular formulae for the solid material left by the pyrolysis (C or C_xH_y). As before, M_j stands for the molecule of j-th species in the gas phase, and the constants $(\nu'_{j\gamma}, \nu''_{j\gamma})$ are the stoichiometric coefficients for the j-th species in the γ-th heterogeneous reaction. A more general treatment would be to consider different solid species S_β generated by the volatilization reactions. In that case, different S molecules could appear for different γin the equations (19). The model for the kinetics could be essentially the same than the one shown below, but such an approach would require to characterize the solid phase by the content of inert material and more than one solid fuel species: ρ_i, ρ_{S1}, ρ_{S2},..., instead of ρ_i and ρ_f. Consequently, that would demand to solve mass conservation equations for ρ_{S1}, ρ_{S2},..., instead of just one for ρ_f (See 2.2). Although in conceptual terms there is no more complexity in such a procedure, we prefer do not follow that line here, in order to avoid the introduction of too many parameters, which are currently difficult to obtain and validate.

The mass rate of consumption, per unit of bulk volume, through the γ-th reaction of the fuel that has remained solid (in the S form), can be formulated as

$$r_\gamma^{(H)} = k_{f\gamma}^{(H)} \prod_{j=1}^{N_{SP}} (\rho_g y_j)^{\nu'_j} (1-\omega) F \tag{20}$$

where ω and F are defined below, and $k_{f\gamma}^{(H)}$ is a kinetic factor defined such that when multiplied by the product of the concentrations in Eq.(20), gives the rate of mass consumption per unit area of solid fuel. This kinetic energy can be postulated to follow as usual an empirical Arrhenius law

$$k_{f\gamma}^{(H)} = A_\gamma^{(H)} T_s^{b_\gamma^{(H)}} \exp\left(-\frac{E_\gamma}{RT_s}\right) \tag{21}$$

for which the activation energy E_γ and pre-exponential constants must be provided. The last two factors in Eq. (20) takes into account the surface area of solid fuel per unit of bulk volume. F is the ratio of particles surface area to the volume of porous medium. For a homogeneous porous medium with particles of characteristic mean size d, it has the value

$$F = \frac{6(1-\theta)}{d} \tag{22}$$

The factor (1-ω) is the fraction of F that corresponds to solid fuel area. The variable ω can be defined as the degree of fuel burnout, since ω = 1 corresponds to the state when all the fuel that remained in the solid phase has been consumed. The minimum value for ω depends on the initial fraction of fuel in the porous medium (see 2.2).

Taking into account the production and consumption of gaseous species by the reactions (19), the net production of species j by the whole heterogeneous reactions mechanism is then

$$S_{j,HE} = \frac{1}{W_S} \sum_{\gamma=1}^{N_{HE}} (v''_{j\gamma} - v'_{j\gamma}) r_{\gamma}^{(H)} \tag{23}$$

where W_S is the molecular weight of S. We have taken reactions (19) as irreversible, but if there were reversible reactions among them, the treatment can be done in the same way as in subsection 2.1.1, applying the equilibrium constants to obtain backward reaction rates that should be subtracted from $r_{\gamma}^{(H)}$ in Eq. (23).

In summary, the species mass conservation equations can be written more explicitly as

$$\frac{\partial}{\partial t}(\theta \rho_g y_j) = \nabla \cdot (\theta \rho_g D_j \nabla y_j) - \nabla \cdot (\rho_g \vec{\mathbf{u}} y_j) + S_{j,HO} + S_{j,v} + S_{j,HE}\,; \quad \text{for } j = 1,\ldots,N_{SP} \tag{24}$$

where the net sources of species j by homogeneous reactions, volatilization and heterogeneous reactions are given by Eqs. (7), (17) and (23), respectively.

2.2. Conservation of mass for solid fuel

To derive an equation expressing the mass conservation of solid fuel we consider first a region of volume $V(t)$, in the burning porous medium, whose boundary surface moves with the local velocity $\vec{\mathbf{v}}$ at which the solid fuel combustion propagates. In that situation the mass conservation of fuel is given by

$$\frac{\mathrm{d}}{\mathrm{d}t} \int_{V(t)} \rho_f (1-\theta) \mathrm{d}V = - \int_{V(t)} B_f \, \mathrm{d}V \tag{25}$$

where B_f is the rate of mass fuel consumption, per unit volume, by heterogeneous reaction and volatilization. The derivative on the LHS of Eq. (25) gives

$$\int_{V(t)} \frac{\partial}{\partial t}[\rho_f (1-\theta)] \mathrm{d}V + \int_{A(t)} \rho_f (1-\theta) \vec{\mathbf{v}} \cdot \vec{\mathbf{n}} \mathrm{d}A = - \int_{V(t)} B_f \, \mathrm{d}V \tag{26}$$

whence the governing equation for the partial density of fuel in the solid phase becomes

$$\frac{\partial}{\partial t}[\rho_f(1-\theta)] + \nabla\cdot[\rho_f(1-\theta)\vec{\mathbf{v}}] = -B_f \tag{27}$$

The loss of mass of fuel by volatilization is given by the summation of the $S_{j,v}$ terms in Eq. (17), for all the species. On the other hand, the loss of fuel mass due to heterogeneous reactions is given by the summation of all the consumption rates in Eq. (20), for all of such reactions. Therefore,

$$\frac{\partial}{\partial t}[\rho_f(1-\theta)] + \nabla\cdot[\rho_f(1-\theta)\vec{\mathbf{v}}] = -\sum_{j=1}^{N_{SP}} S_{j,v} - \sum_{\gamma=1}^{N_{HE}} r_\gamma^{(H)} \tag{28}$$

The evaluation of the rates $r_\gamma^{(H)}$ requires the degree of fuel burnout ω. Denoting by $\hat{\rho}_i$ and $\hat{\rho}_f$ the intrinsic densities (true densities) of the inert material and the fuel respectively, ω can be obtained from the relations

$$\rho_f + \rho_i = \hat{\rho}_f \frac{V_f}{(1-\theta)V} + \hat{\rho}_i\left(1 - \frac{V_f}{(1-\theta)V}\right) \tag{29}$$

and

$$(1-\omega)F = \frac{A_f}{V} = \frac{6}{d}\frac{V_f}{V} \tag{30}$$

where V_f represents the volume occupied by the fuel particles in a given bulk volume V, and A_f represents their surface area. Combining Eqs. (29), (30) and (22) we have

$$\omega = \frac{\hat{\rho}_f - (\rho_f + \rho_i)}{\hat{\rho}_f - \hat{\rho}_i} \tag{31}$$

so that the value of ω, at a given position, can be determined using the instantaneous local values of ρ_f and ρ_i at that position.

2.3. Conservation of energy of the gas phase

For the analysis of energy conservation in the gas, we consider the thermal and the chemical energies of the gaseous species, and neglect the kinetic energy of the flow. The energy per mole of gas is then $e_g = h_g - p/\rho_g$, with h_g being the total enthalpy per mole of gas mixture

$$h_g = \sum_{j=1}^{N_{SP}} y_j h_j \tag{32}$$

The enthalpies of the different species (h_j) include their enthalpy of formation (and they can be computed with functions like Eq. (12)). Proceeding similarly to the analysis of mass conservation, an integral balance of energy in a volume V results in

$$\int_V \left[\frac{\partial}{\partial t}(\theta \rho_g e_g) + \nabla \cdot (\rho_g e_g \vec{\mathbf{u}}) \right] \mathrm{d}V = \int_V H_V \, \mathrm{d}V + \int_A H_A \, \theta \mathrm{d}A \quad (33)$$

where H_V refers to sources of energy per unit of bulk volume, and H_A denote energy fluxes going into the volume across the boundary surface of V, per unit of gas-phase area. We have also neglected here the work done on the gas by surface forces (such as pressure and viscous stresses), and body forces (such as gravity). Those mechanical effects on the total energy are clearly negligible in comparison with the thermal effects.

The fluxes in H_A can be joined into a vector $\vec{\mathbf{q}}$ such that $H_A = -\vec{\mathbf{q}} \cdot \vec{\mathbf{n}}$, and then Eq. (33) leads to the differential equation

$$\frac{\partial}{\partial t}(\theta \rho_g e_g) + \nabla \cdot (\rho_g e_g \vec{\mathbf{u}}) = H_V - \nabla \cdot \vec{\mathbf{q}} \quad (34)$$

Generally, it is preferred to use the enthalpy as the primitive variable for the gas energy equation. In that case Eq. (34) reads as

$$\frac{\partial}{\partial t}(\theta \rho_g h_g) + \nabla \cdot (\rho_g h_g \vec{\mathbf{u}}) = \frac{\partial(\theta p)}{\partial t} + \nabla \cdot (p \vec{\mathbf{u}}) + H_V - \nabla \cdot \vec{\mathbf{q}} \quad (35)$$

The pressure terms appearing in (35) are also associated with mechanical energy of the flow. They contribute in part to the energy of deformation due to gas expansion (which is stored in the gas as thermal energy), and also to the flow acceleration due to pressure gradients (which changes the kinetic energy). As stated above, we consider these mechanical energy contributions negligible, in comparison with the thermal energy in a reactor. Consequently, the first two terms on the RHS of Eq. (35) are discarded.

The energy flux vector $\vec{\mathbf{q}}$ contains the heat diffusion, which can be expressed by Fourier's law, and the additional transport of energy due to the diffusion of species, with different enthalpies, in the multi-component gas

$$\vec{\mathbf{q}} = -\lambda_g \nabla T_g + \sum_{j=1}^{N_{SP}} \rho_g y_j \vec{\mathbf{w}}_j h_j = -\lambda \nabla T_g - \rho_g \sum_{j=1}^{N_{SP}} h_j D_j \nabla y_j \quad (36)$$

In this equation we have introduced again Fick's law for the diffusion velocities, while λ_g is the gas thermal conductivity.

According to the processes described previously, the energy source term H_V must contain (i) $H_{V,v}$: the enthalpy carried into the gas by the volatiles evolved from the solid fuel, (ii) $H_{V,HE}$: the energy released toward the gas by the heterogeneous reactions of the solid fuel and (iii) $H_{V,Q}$: the heat transfer from the solid phase (inert material and fuel) to the gas.

With regards to the enthalpy flow contributed by the volatiles, it can be calculated as

$$H_{V,v} = \sum_{j=1}^{N_{SP}} S_{j,v} h_j(T_s) \quad (37)$$

where the total molar flow of species j carried in the volatiles, $S_{j,v}$,is given by Eq. (17). We make the assumption that this volatiles gases are released at the temperature T_s of the solid phase. Note that the solid phase is characterized by a common local temperature as discussed in subsection 2.4.

The heterogeneous reactions involve the adsorption of gaseous species into the fuel particles, and the desorption of the reaction products back to the gaseous phase. Using the heterogeneous reaction model described in sub-section 2.1.3, we have that for the γ-th of such reactions, the enthalpy transport associated with those adsorbed and desorbed flows could be in principle estimated as

$$\frac{r_\gamma^{(H)}}{W_S}\sum_{j=1}^{N_{SP}} v'_{j\gamma} h_j(T_g) \text{ and } \frac{r_\gamma^{(H)}}{W_S}\sum_{j=1}^{N_{SP}} v''_{j\gamma} h_j(T''), \text{ respectively.}$$

In the first expression the enthalpies of the incoming species can be evaluated at the temperature T_g of the gas in the neighborhood of the fuel particle. In the second expression however the enthalpies of the outgoing products should be evaluated at a temperature T'' that would have to be approximated in some way. Such an estimation is a difficult task, so that a practical approach is to consider that a fraction ε of the heat reaction is retained in the solid particle, contributing to its internal energy, while the remaining fraction (1- ε) goes into the gas phase as the net flow of sensible enthalpy. More explicitly, the heat released by the γ-th heterogeneous reaction in the solid is

$$\begin{aligned} Q_{s\gamma} &= \frac{r_\gamma^{(H)}}{W_S}\left\{\sum_{j=1}^{N_{SP}}\left(v'_{j\gamma}h_j(T_g) - v''_{j\gamma}h_j(T'')\right) + h_{fS}^o\right\} \\ &= \underbrace{\frac{r_\gamma^{(H)}}{W_S}\sum_{j=1}^{N_{SP}}(v''_{j\gamma} - v'_{j\gamma})h_{fj}^o - h_{fS}^o}_{Q_{R\gamma}} + \underbrace{\frac{r_\gamma^{(H)}}{W_S}\left(\sum_{j=1}^{N_{SP}} v'_{j\gamma}\Delta h_j(T_g) - \sum_{j=1}^{N_{SP}} v''_{j\gamma}\Delta h_j(T'')\right)}_{H_{V,HE\gamma}} \end{aligned} \quad (38)$$

with $Q_{R\gamma}$ being the standard heat of reaction and Δh_j are sensible enthalpies. On the other hand, we assume that

$$Q_{s\gamma} = -\varepsilon \frac{r_\gamma^{(H)}}{W_S} Q_{R\gamma} \quad (39)$$

(By definition, $Q_{R\gamma}$ is negative for exothermic reactions, whence the minus sign in Eq. (39)). As a result, the net flow of energy toward the gas phase is

$$H_{V,HE} = \sum_{\gamma=1}^{N_{HE}} H_{V,HE\gamma} = -(1-\varepsilon)\frac{1}{W_S}\sum_{\gamma=1}^{N_{HE}} r_\gamma^{(H)} Q_{R\gamma} \quad (40)$$

We look now to the third component of H_V, namely the convective heat transfer to the gas on the surface of the solid medium. This heat flow can be calculated, in a simplified form, as

$$H_{V,Q} = \xi F(T_s - T_g) \tag{41}$$

where ζ is the heat transfer coefficient, which can be estimated from empirical relations for porous media, as for example [33]

$$\mathrm{Nu} = 2 + 1.1\,\mathrm{Re}^{0.6}\,\mathrm{Pr}_g^{\;1/3} \tag{42}$$

In this expression the characteristic length for the Reynolds and Nusselt numbers is taken as two times the particles size d, and the velocity is the effective mean velocity of the flow:

$$\mathrm{Re} = \frac{\theta|\vec{\mathbf{u}}|2d}{\nu_g}, \quad \mathrm{Nu} = \frac{\xi 2d}{\lambda_g} \tag{43}$$

Upon substituting Eq. (36) into (35), the equation for energy conservation in the gas phase becomes

$$\frac{\partial}{\partial t}(\theta\rho_g h_g) + \nabla\cdot(\rho_g h_g \vec{\mathbf{u}}) = \nabla\cdot(\theta\lambda_g \nabla T_g) + \nabla\cdot\left(\theta\rho_g \sum_{j=1}^{N_{SP}} h_j D_j \nabla y_j\right) + H_{V,v} + H_{V,HE} + H_{V,Q} \tag{44}$$

Recall that we are using total enthalpies h_j, which include the enthalpy of formation, so that the heat released by the homogeneous reactions in the gaseous phase is implicit in Eq. (44).

This equation of energy, coupled with the equations for mass conservation of species, and the equations for the solid phase, give rise to a system of simultaneous partial differential equations that must be solved by a numerical iterative scheme. In that case Eq. (44) can be solved for h_g as its primitive variable, and the gas temperature can be obtained by solving the polynomial equation

$$\sum_{n=0}^{m} c_{nj} T_g^n = h_g \ \text{ with } c_{nj} = \sum_{j=1}^{N_{SP}} a_{nj} y_j \tag{45}$$

in each iteration. This equation for T_g results from Eqs. (32) and (12).

2.4. Conservation of energy of the solid phase

The conservation of energy for the solid phase can be analyzed in the same volume $V(t)$ used in subsection 2.3, with a boundary that moves at the velocity ($\vec{\mathbf{v}}$) of propagation of solid fuel consumption. In this way we have

$$\frac{\mathrm{d}}{\mathrm{d}t}\int_{V(t)} (\rho_f e_f + \rho_i e_i)(1-\theta)\mathrm{d}V = \int_{V(t)} G_s \mathrm{d}V + \int_{A(t)} H_s \mathrm{d}A_s \tag{46}$$

or equivalently

$$\int_{V(t)} \frac{\partial}{\partial t}\left((1-\theta)(\rho_f e_f + \rho_i e_i)\right) \mathrm{d}V + \int_{A(t)} (1-\theta)(\rho_f e_f + \rho_i e_i)\vec{\mathbf{v}}\cdot\vec{\mathbf{n}}\, \mathrm{d}A = \int_{V(t)} G_s \mathrm{d}V + \int_{A(t)} H_s \mathrm{d}A_s \quad (47)$$

where e_f, e_i are the internal energies of fuel and inert particles respectively, G_s is the total source of energy for the solid medium per unit of bulk volume, and H_s is the influx of energy on the boundary surface per unit of solid area. Because of the thorough mix and contact between inert and fuel particles, we assume that they are in local thermal equilibrium. Consequently, the thermal state of the solid medium can be characterized by only one local temperature T_s. If that is not the case (for instance, if the specific heats of the inert and fuel particles were too different), separate energy equations for both the fuel and inert material can be formulated.

The source G_s is determined by the convective heat transfer between the solid and the gas, and the fraction of the heat released by the heterogeneous reactions that has remained in the solid phase. According to Eqs. (39) and (41) we have

$$G_s = -\frac{\varepsilon}{W_S} \sum_{\gamma=1}^{N_{HE}} r_\gamma^{(H)} Q_{R\gamma} - \xi F(T_s - T_g) \quad (48)$$

The influx H_s is given essentially by heat conduction through the solid medium. This can be written as $H_s = -\vec{\mathbf{q}}_s \cdot \vec{\mathbf{n}}$, with the heat flux $\vec{\mathbf{q}}_s = -\lambda_s \nabla T_s$, and λ_s being the equivalent thermal conductivity of the solid medium.

Upon substituting these expression into Eq. (47), and converting the surface integral into volume integral, the equation for energy conservation of the solid phase becomes

$$\begin{aligned}\frac{\partial}{\partial t}\left((1-\theta)(\rho_f C_f + \rho_i C_i)T_s\right) + \nabla\cdot\left((1-\theta)(\rho_f C_f + \rho_i C_i)\vec{\mathbf{v}}T_s\right) &= \nabla\cdot\left((1-\theta)\lambda_s \nabla T_s\right) \\ &\quad - \frac{\varepsilon}{W_S}\sum_{\gamma=1}^{N_{HE}} r_\gamma^{(H)} Q_{R\gamma} - \xi F(T_s - T_g)\end{aligned} \quad (49)$$

2.5. Summary

In this section, we have presented the formulation of a mathematical model for hybrid filtration combustion, along with some model assumptions that would allow a computational approach. Such treatment will depend of course on the boundary and initial conditions for particular reactor designs. As we have shown, a crucial factor for the modelling of these complex phenomena is the availability of model parameters such as ε, the composition of the volatiles V_β, etc. The parameterization of a model of this nature requires extensive experimental work. Our purpose is not to derive here all these parameters, but to show some experimental results that highlight the potential of ultra-rich hybrid filtration combustion for hydrogen and syngas production.

3. Hydrogen production by hybrid filtration combustion

In this section, we show experimental results for wave velocities, combustion temperature, and H_2 and CO concentrations for hybrid combustion of gas and solid fuels in porous media.

3.1. Natural gas and coal particles

The porous medium is composed of uniformly mixed aleatory coal and alumina spheres with varying volume fractions. This section provide the data on natural gas-air flames stabilized inside hybrid porous media with a volumetric coal content from 0 to 75% for an equivalence ratio $\phi = 2.3$ and a filtration velocity of 15 cm/s.

The solid temperatures recorded for hybrid filtration combustion in coal-alumina beds, decreased from 1537 K at 15 % of coal in porous media to 1217 K at 75 % (Fig. 1a). The wave velocity is reduced with increase of coal volume fraction. At high coal contents the wave velocity is limited by the oxygen availability and the bed displacement resulting from the coal consumption. The latter factor suggests that the wave velocity in the packed bed with 100% coal content will be close to zero.

The hydrogen concentration increased with an increase of coal content with the maximum of 22% at a coal content of 75%. The result shows that high combustion temperatures facilitate hydrogen production in hybrid filtration combustion. The carbon monoxide also increased with an increase of coal content.

The maximum hydrogen conversion reached 55% at a coal content of 75% compare to 35.1% for natural gas flames propagating in the inert porous bed (Fig. 1b).

3.2. Butane and wood pellets

The porous media was composed of uniformly mixed aleatory wood pellets and alumina spheres. The equal volumes of 5.6 mm solid alumina balls and wood pellets were mixed resulting in the packet bed with a porosity of ~40%. Experimental data were collected for a slightly increasing filtration velocity in a range of equivalence ratios from stoichiometry (ϕ = 1.0) to $\phi = 2.6$.

The solid temperature of butane/air mixtures in the inert porous medium increases from 1510 K at $\phi = 1.45$ to 1610 K at $\phi = 2.6$ (Fig. 2a). The solid temperatures recorded for hybrid filtration combustion in wood pellets-alumina spheres porous media, decrease from 1477 K at $\phi = 1.45$ to 1216 K at $\phi = 2.6$. For butane/air mixtures in the inert and composite porous media the maximum absolute velocity value of ~0.012 cm/s is observed for $\phi = 1.45$ (Fig. 2b).

Hydrogen concentration increases with an increase of equivalence ratio for butane/air mixtures in the inert porous medium. In comparison, more than four times higher concentration of hydrogen is measured for butane/air mixtures in the composite porous media made of alumina and wood pellets. For butane/air mixtures in the inert porous medium the concentration of carbon monoxide increased with equivalence ratio and for an-inert porous media the concentration of CO remains almost constant with the equivalence ratio.

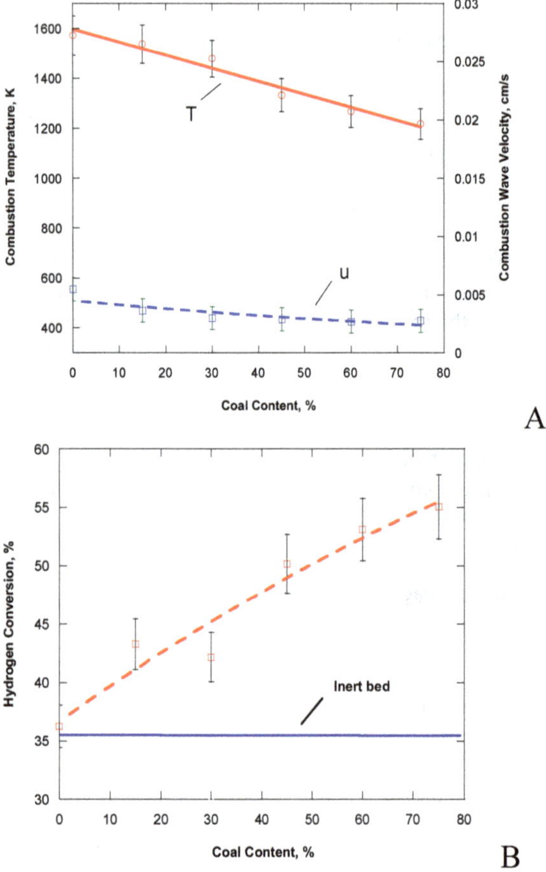

Figure 1. Combustion temperatures and wave velocities (A) and degree of conversion to hydrogen (B) for rich and ultra-rich waves for natural gas mixtures with air varying the volume of coal in the porous media from 0 to 75% [34].

The hydrogen yield is calculated using the initial hydrogen content in butane for the case of inert bed and the initial hydrogen content in butane and wood pellets for the case of the composite bed. The maximum yield recorded for the inert porous medium is close to 10% at $\phi = 2.6$ (Fig. 2c). The maximum yield recorded for the composite pellets is ~48% at $\phi = 2.4$.

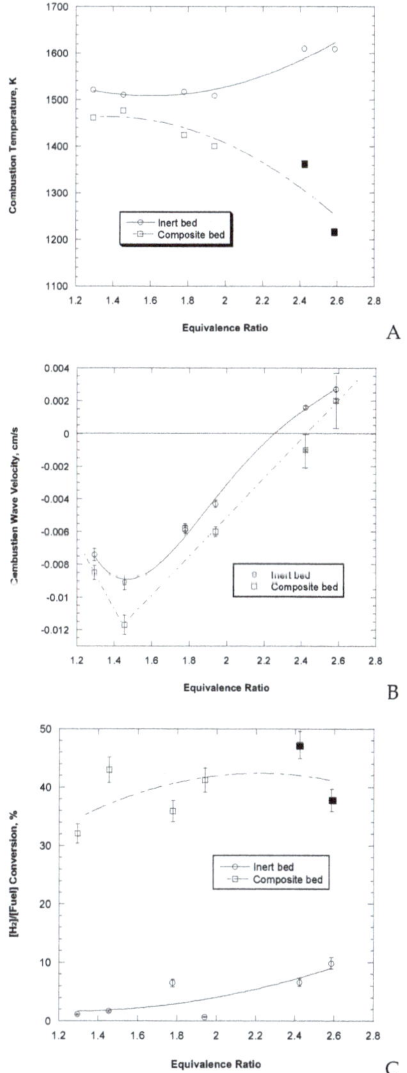

Figure 2. Combustion temperatures (A), wave velocities (B) and degree of conversion to hydrogen (C) for rich and ultra-rich waves for butane mixtures with air in the inert bed and composite porous media [35].

3.3. Propane and polyethylene pellets

The reactor is filled with a uniformly mixed aleatory polyethylene pellets and alumina spheres whose diameter is 5.6 mm. The geometry of polyethylene pellets are 2.4x5.0x4.4

mm. Experimental data were collected at a range of equivalence ratios (ϕ) from ϕ =1.0 to ϕ =1.7. The air/propane flow rate was maintained constant at 5.7 L/min yielding a filtration velocity of 11.3 cm/s.

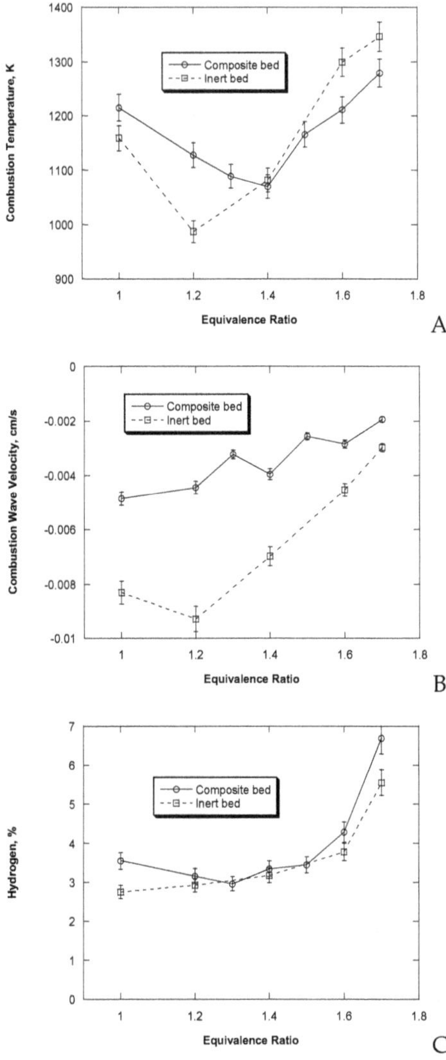

Figure 3. Combustion temperatures (A), wave velocities (B) and hydrogen concentration (C) in rich mixtures of propane with air for gaseous and hybrid filtration combustion waves.

The solid temperature of propane/air mixtures in the inert porous medium decreases from 1159 K at ϕ = 1.0 to 987 K at ϕ = 1.2 (Fig. 3a). Then the temperature increase to 1346 K at ϕ = 1.7. The solid temperatures recorded for hybrid filtration combustion in polyethylene pellets-alumina spheres porous media, shows similar behaviour and decrease from 1215 K at ϕ = 1.0 to 1070 K at ϕ = 1.4. Then the temperature increase to 1279 K at ϕ = 1.7. The experimental results for propane/air mixtures in the inert and composite porous media upstream wave were recorded in all the range of equivalence ratios studied (Fig. 3b).

Hydrogen concentration increases with an increase of equivalence ratio for propane/air mixtures in the inert porous medium and composite porous media made of alumina and polyethylene pellets. The maximum generated mole fraction of hydrogen is 6.7% for hybrid propane. Carbon monoxide is also generated in both porous media reactor.

4. Conclusion

In this chapter we have defined the hybrid filtration combustion process. The physical phenomena involved as well as their mathematical description by the governing equations enforcing mass and energy conservation for both the gas and solid phases are discussed.

We presented experimental evidence that such a process of hybrid filtration combustion can actually be achieved in practice. In particular we showed applications for reformation of gaseous and solid fuels into hydrogen and syngas. This was shown for several combinations of gaseous and solid fuels.

Author details

Mario Toledo Torres* and Carlos Rosales Huerta
Department of Mechanical Engineering, Technical University Federico Santa Maria, Valparaiso, Chile

Acknowledgement

The authors wish to acknowledge the support by the CONICYT-Chile (FONDECYT 1121188 and BASAL FB0821 - FB/05MT/11).

5. References

[1] Kaviany M (1995) Principles of heat transfer in a porous media. Springer-Verlag. New York.
[2] Toledo M, Bubnovich V, Saveliev A, Kennedy L (2009) Hydrogen production in ultrarich combustion of hydrocarbon fuels in porous media. Int. J. Hydrogen Energy 34: 1818-1827.

* Corresponding Author

[3] Dhamrat RS, Ellzey JL (2006) Numerical and experimental study of the conversion of methane to hydrogen in a porous media reactor. Combust. Flame 144: 698-709.

[4] Babkin V, Korzhavin A, Bunev V (1991) Propagation of premixed gaseous explosion flames in porous media. Combust. Flame 87: 182-190.

[5] Babkin V. (1993) Filtrational combustion of gases. Present state of affairs and prospects. Pure Appl. Chem. 65: 335-344.

[6] Vogel BJ, Ellzey JL (2005) Subadiabatic and superadiabatic performance of a two-section porous burner. Combust. Sci. Technol. 177: 1323-1338.

[7] Abdul M, Abdullaha M, Abu Bakarb M (2009) Combustion in porous media and its applications - a comprehensive survey. J. Environ. Manage 90: 2287-2312.

[8] Bingue JP, Saveliev AV, Fridman AA, Kennedy LA (2002) Hydrogen production in ultra-rich filtration combustion of methane and hydrogen sulfide. Int. J. Hydrogen Energy 27: 643-649.

[9] Babkin V (1993) Filtrational combustion of gases. Pure and Applied Chemistry 65: 335-344.

[10] Dobrego KV, Zhdanok SA, Khanevich EI (2000) Analytical and experimental investigation of the transition from low velocity to high-velocity regime of filtration combustion. Exp. Therm. Fluid. Sci. 21: 9-16.

[11] Foutko SI, Shabunya SI, Zhdanok SA, Kennedy LA (1996) Superadiabatic combustion wave in a diluted methane–air mixture under filtration in a packed bed. Proc. Combust. Inst. 25: 1556-65.

[12] Kennedy LA, Bingue JP, Saveliev AV, Fridman AA, Foutko SI (2000) Chemical structures of methane–air filtration combustion waves for fuel-lean and fuel-rich conditions. Proc. Combust. Inst. 28: 1431-1438.

[13] Drayton MK, Saveliev AV, Kennedy LA, Fridman AA, Li Y- E (1998) Syngas production using superadiabatic combustiom of ultra-rich methane-air mixtures. Proc Combust. Inst. 27: 1361-7.

[14] Weinberg FJ, Bartleet TG, Carleton FB, Rimbotti P, Brophy JH, Manning RP (1988) Partial oxidation of fuel rich mixtures in a spouted bed combustor. Combust Flame 72: 235-9.

[15] Ytaya Y, Oyashiki T, Hasatani M (2002) Hydrogen production by methane-rich combustion in a ceramic burner. J. Chem. Eng. Jpn. 35: 46-56.

[16] Kennedy LA, Saveliev AV, Fridman AA (1999) Transient filtration combustion. Mediterr. Combust. Symp. 1: 105-39.

[17] Kennedy LA, Saveliev AA, Bingue JP, Fridman AA (2002) Filtration combustion of a methane wave in air for oxygen enriched and oxygen-depleted environments. Proc. Combust. Inst. 29: 835-41.

[18] Howell JR, Hall MJ, Ellzey JL (1996) Combustion of hydrocarbon fuels within porous medium. Prog. Energy Combust. Sci. 22: 121-45.

[19] Gavrilyuk VV, Dmitrienko YM, Zhdanok SA, Minkina VG, Shabunya SI, Yadrevskaya NL, et al. (2001) Conversion of methane to hydrogen under superadiabatic filtration combustion. Theor. Found. Chem. Eng. 35: 589-96.

[20] Gerasev AP (2008) Hybrid autowaves in filtration combustion of gases in a catalytic fixed bed. Combust. Explos. 44: 123-32.

[21] Schoegl I, Newcomb SR, Ellzey JL (2009) Ultra-rich combustion in parallel channels to produce hydrogen-rich syngas from propane. Int. J. Hydrogen Energy 34: 5152-63.

[22] Dixon MJ, Schoegl I, Hull CB, Ellzey JL (2008) Experimental and numerical conversion of liquid heptane to syngas through combustion in porous media. Combust. Flame 154: 217-31.

[23] Pedersen-Mjaanes H, Chan L, Mastorakos E (2005) Hydrogen production from rich combustion in porous media. Int. J. Hydrogen Energy 30: 579-92.

[24] Tarun K. Kayal, Mithiles Chakravarty (2007) Modeling of a conceptual self-sustained liquid fuel vaporization–combustion system with radiative output using inert porous media, International Journal of Heat and Mass Transfer 50: 1715–1722.

[25] Sumrerng Jugjai, Nopporn Polmart (2003) Enhancement of evaporation and combustion of liquid fuels through porous media, Experimental Thermal and Fluid Science 27: 901-909.

[26] Trimis D, Wawrzinek K, Hatzfeld O, Lucka K, Rutsche A, Haase F, Kruger K, and Kuchen C (2001) High modulation burner for liquid fuels based on porous media combustion and cool flame vaporization. Proceedings of the Sixth International Conference on Technologies and combustion for clean environment, v.2, Porto, Portugal, 9-12 July (Ed. M. G. Carvhalho), pp. 1-8.

[27] Tarun K. Kayal, Mithiles Chakravarty (2005) Combustion of liquid fuel inside inert porous media: an analytical approach. International Journal of Heat and Mass Transfer 48 : 331-339.

[28] Haack DP (1993) Mathematical analysis of radiatively enhanced liquid droplet vaporization and liquid fuel combustion within a porous inert medium. MSc Thesis, University of Texas, Austin.

[29] Kaplan M and Hall MJ (1995) The combustion of liquid fuels within a porous media radiant burner. Exp. Thermal. Fluid Sci. 11: 13-20.

[30] Tseng CJ and Howell JR (1994) Liquid fuel combustion within inert porous media. Heat Trans. Combined Modes, ASME, HTD 299 : 63-69.

[31] Salganskii EA, Fursov VP, Glazov SV, Salganskaya MV, Manelis GB (2006) Model of vaporeair gasification of a solid fuel in a filtration mode. Combust. Explos. 42: 55-62.

[32] Poinsot T, and Veynante D (2001) Theoretical and Numerical Combustion. R. T. Edwards Inc. Philadelphia.

[33] Salgansky EA, Kislov VM, Glazov SV, Zholudev AF, Manelis GB (2008) Filtration Combustion of a Carbon–Inert Material System in the Regime with Superadiabatic Heating. Combustion, Explosion, and Shock Waves 44: 273-280.

[34] Toledo MG, Utria KS, González FA, Zúñiga JP, Saveliev AV (2012) Hybrid filtration combustion of natural gas and coal. Int. J. Hydrogen Energy 37 : 6942-6948.
[35] Toledo M, Vergara E, Saveliev A (2011) Syngas production in hybrid filtration combustion. Int. J. Hydrogen Energy 36: 3907-3912.

Chapter 8

Use of Hydrogen-Methane Blends in Internal Combustion Engines

Bilge Albayrak Çeper

Additional information is available at the end of the chapter

1. Introduction

In today's modern world, where new technologies are continually being introduced, transportation energy use is increasing rapidly. Fossil fuel, particularly petroleum fuel, is the major contributor to energy production[1]. Fossil fuel consumption is steadily rising as a result of population growth in addition to improvements in the standard of living. It can be seen from Figure 1 that the world's population has been increasing steadily over the last 5 decades, and this trend is expected to continue [2]. As a result, total energy consumption has grown by about 36% over the last 15 years [3]. Energy consumption is expected to increase further in the future, as the world's population is expected to grow by 2 billion people in the next 30 years [2]. These energy trends can be seen in Figure 2. Increased energy demand requires increased fuel production, thus draining current fossil fuel reserve levels at a faster rate. In addition, about 60% of the world's current oil reserves are in regions that are in frequent political turmoil [3]. This has resulted in fluctuating oil prices and supply disruptions.

Rapidly depleting reserves of petroleum and decreasing air quality raise questions about the future. As world awareness about environmental protection increases so too does the search for alternatives to petroleum fuels [1].

Alternative fuels such as CNG, HCNG, LPG, LNG, bio-diesel, biogas, hydrogen, ethanol, methanol, di-methyl ether, producer gas, and P-series have been tried worldwide. The use of hydrogen as a future fuel for internal combustion (IC) engines is also being considered. However, several obstacles have to be overcome before the commercialization of hydrogen as an IC engine fuel for the automotive sector. Hydrogen and CNG blends (HCNG) may be considered as an automotive fuel without requiring any major modification in the existing CNG engine and infrastructure [4].

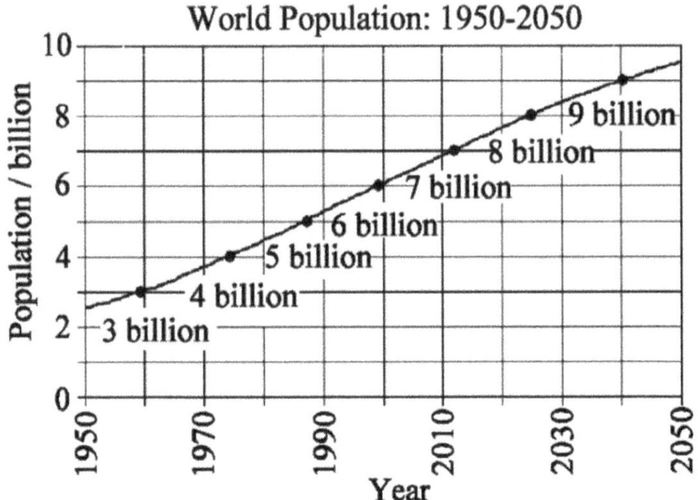

Figure 1. World population 1950-2050 [2].

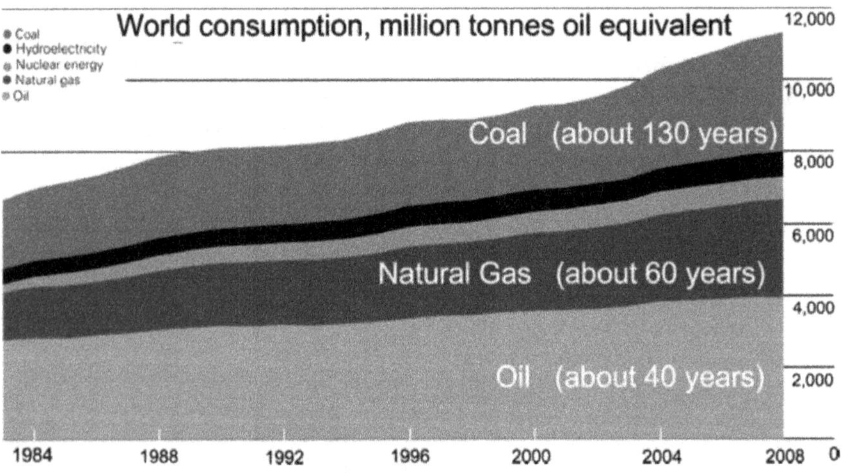

Figure 2. Fossil fuel consumption from 1983 to 2008 with approximate current reserves-to-production ratios in remaining years [3].

Alternative fuels are derived from resources other than petroleum. The benefit of these fuels is that they emit less air pollutants compared to gasoline and most of them are more economically viable compared to oil and they are renewable [5]. Figure 3 shows the percentages of alternative fuels used according to total automotive fuel consumption in the world as a futuristic view.

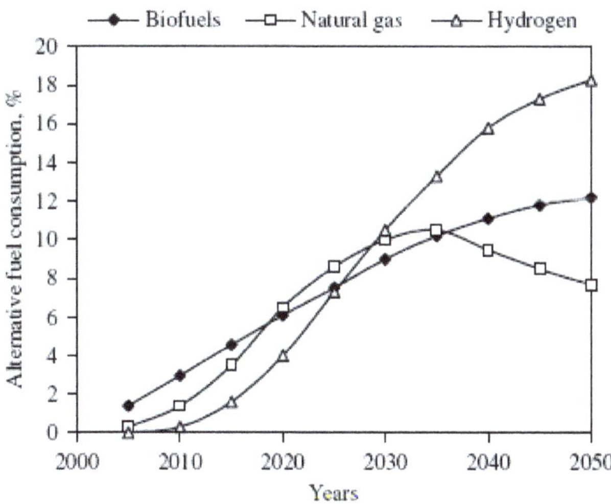

Figure 3. Percentages of alternative fuels compared to total automotive fuel consumption in the world [6]

2. Hydrogen specifications

Hydrogen is acknowledged to offer great potential as an energy carrier for transport applications. A number of technologies can use hydrogen as an energy carrier, with the internal combustion engine being the most mature technology [7]. Currently, 96% of hydrogen is made from fossil fuels. Based on 2004 data, in the United States 90% is made from natural gas, with an efficiency of 72%. Only 4% of hydrogen is made from water via electrolysis. Currently, the vast majority of electricity comes from fossil fuels in plants that are 30% efficient and from electrolysis which means that electricity is run through water to separate the hydrogen and oxygen atoms. Using renewable energy is much more effective than using fossil fuel to produce hydrogen. Current wind turbines perform at 30-40% efficiency, producing hydrogen at an overall efficiency rate of 25%. The best solar cells available have an efficiency rate of 10%, leading to an overall efficiency rate of 7%. Algae can be used to produce hydrogen at an efficiency rate of about 0.1% (see Figure 4)[8].

The use of hydrogen as an automotive fuel appears to promise a significant improvement in the performance of spark-ignition engines [9]. The self-ignition temperature of the hydrogen/air mixture is greater than that of other fuels and, therefore, hydrogen produces an antiknock quality of fuel. The high ignition temperature and low flame luminosity of hydrogen makes it a safer fuel than others, it is also non-toxic. Hydrogen is characterized by having the highest energy–mass coefficient of all chemical fuels and in terms of mass energy consumption it exceeds conventional gasoline fuel by about three times, and alcohol by five to six times [10]. Therefore, the results clearly establish that hydrogen fuel can increase the effective efficiency of an engine and reduce specific fuel consumption. A small amount of hydrogen mixed with air produces a combustible mixture, which can be burned in a conventional spark-ignition engine at an equivalence ratio below the lean flammability limit of gasoline/air mixture. The resulting ultra lean combustion produces a low flame temperature and leads directly to lower heat transfer to the walls, higher engine efficiency and lower NO_x exhaust emissions [11–13].

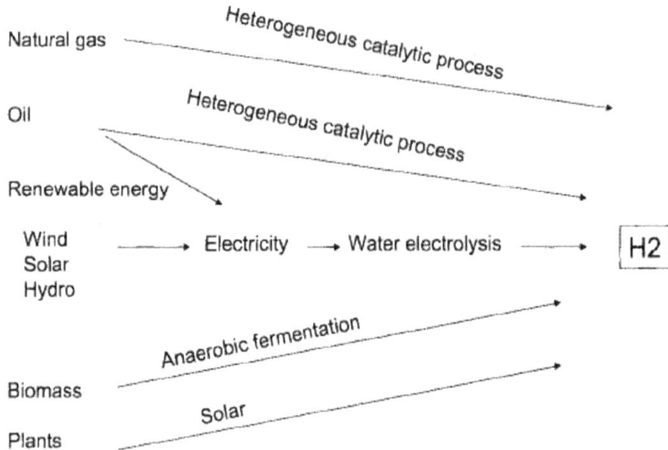

Figure 4. Various processes for the production of hydrogen[8]

The burning velocity of hydrogen/air mixture is about six times higher than that of gasoline/air mixtures. As the burning velocity rises, the actual indicator diagram is nearer to the ideal diagram and a higher thermodynamic efficiency is achieved [14,15]. Figure 5 plots the laminar burning velocities against the equivalence ratio for hydrogen–air mixtures at normal pressure and temperature (NTP)[7]. The solid symbols in Figure 5 denote stretch-free burning velocities (or rather, burning velocities that were corrected to account for the effects of the flame stretch rate), as measured by Taylor [16], Vagelopoulos et al. [17], Kwon and Faeth [18] and Verhelst et al. [19]. The empty symbols denote other measurements that did not take stretch rate effects into account, as reported by Liu and MacFarlane [20], Milton and Keck [21], Iijima and Takeno [22] and Koroll et al. [23]. These experiments result in consistently higher burning velocities, with the difference increasing for leaner mixtures.

Hydrogen is a clean fuel with no carbon emissions; the combustion of hydrogen produces only water and a reduced amount of nitrogen oxides. Conversely, combustion products from fossil fuels, such as CO, CO_2, nitrogen oxides, or other air pollutants, cause health and environmental problems. Hydrogen will help reduce CO_2 emissions as soon as it can be produced in a clean way either from fossil fuels, in combination with processes involving CO_2 capture and storage technologies, or from renewable energy. These features make hydrogen a potentially excellent fuel to meet the ever increasingly stringent environmental controls regarding exhaust emissions from combustion devices, including the reduction of green house gas emissions [24–27].

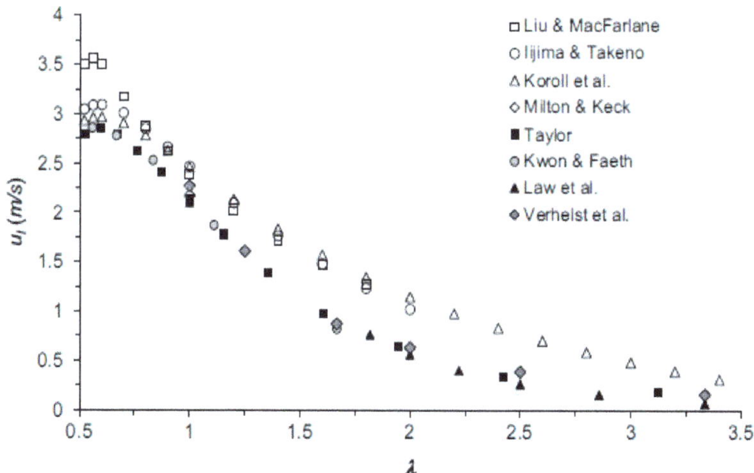

Figure 5. Laminar burning velocities plotted against air-to-fuel equivalence ratio, for NTP hydrogen–air flames[7]. The experimentally derived correlations are from Liu and MacFarlane [20], Milton and Keck [21]. Iijima and Takeno [22] and Koroll et al. [23]. Other experimental data are from Taylor [16], Vagelopoulos et al. [17], Kwon and Faeth [18] and Verhelst et al. [19].

3. Methane specifications

Natural gas (CNG) is considered as an alternative vehicle fuel because of its economical and environmental advantages [28]. CNG, which is a clean fuel with methane as its major component, is considered to be one of the most favorable fuels for engines, and the utilization of CNG has been realized in spark-ignition engines. However, due to the slow burning velocity of CNG and its poor lean-burn capability, the CNG spark-ignition engine still has some disadvantages like low thermal efficiency, large cycle-by-cycle variation, and poor lean-burn capability, and these decrease engine power output and increase fuel consumption [29]. The advantages of CNG compared to petrol are as follows: unique combustion and suitable mixture formation; due to the high octane number of CNG, the engine operates smoothly with high compression ratios without knocking; CNG with lean burning quality leads to the lowering of exhaust emissions and fuel operating cost; CNG has a lower flame speed; and engine durability is very high. CNG is produced from gas wells or related to crude oil production. CNG is made up primarily of methane (CH_4) but frequently contains trace amounts of ethane, propane, nitrogen, helium, carbon dioxide, hydrogen sulfide, and water vapor. Methane is the principal component of natural gas [30].

CNG has many other advantages as well. It has a high octane number of 130, which enables an engine to operate with little knocking at a high compression ratio. In addition, gasoline and diesel engines can be easily converted into CNG engines without major structural changes [31]. Not only does the CNG engine have good thermal efficiency and high power, but its combustion range is also broad. This is an advantage when striving for lean combustion resulting in low fuel consumption and less NO_x production [32]. The CNG engine also yields very low levels of PM emissions when compared with other conventional engines. These facts are supported by an experimental study performed to explore the combustion and emission characteristics of both gasoline and CNG fuels using a converted spark-ignition engine [33]. In light of these advantages, the number of CNG vehicles is continuously growing, and old vehicles are being converted into CNG vehicles through engine modifications [34].

4. Hydrogen-methane mixtures for internal combustion engines

Traditionally, to improve the lean-burn capability and flame burning velocity of natural gas engines under lean-burn conditions, an increase in flow intensity is introduced in the cylinder, and this measure always increases the heat loss to the cylinder wall and increases the combustion temperature as well as the NO_x emission [35]. One effective method to solve the problem of the slow burning velocity of natural gas is to mix natural gas with fuel that possesses fast burning velocity. Hydrogen is regarded as the best gaseous candidate for natural gas due to its very fast burning velocity, and this combination is expected to improve lean-burn characteristics and decrease engine emissions [36]. The hydrogen blends in CNG can range from 5 to 30% by volume. Hythane is a 15% blend of hydrogen in CNG by energy content, which was patented by Frank Lynch of Hydrogen Components Inc, USA [37]. A typical 20% blend of hydrogen by volume in CNG is 3% by mass or 7% by energy. An overall comparison of the properties of hydrogen, CNG, and 5 % HCNG blend by energy and gasoline is given in Table 1. It is to be noted that the properties of HCNG lie in between those of hydrogen and CNG [4].

Properties	H_2	CNG	HCNG	Gasoline
Stoichiometric volume fraction in air,(vol %)	29.53	9.43	22.8	1.76
Limits of flammability in air, (vol %)	4-75	5-15	5-35	1.0-7.6
Auto ignition temp. K	858	813	825	501-744
Flame temp in air K	2318	2148	2210	2470
Maximum energy for ignition in air, mJ	0.02	0.29	0.21	0.24
Burning velocity in NTP air, cm s^{-1}	325	45	110	37-43
Quenching gap in NTP air, cm	0.064	0.203	0.152	0.2
Diffusivity in air cm^2 s^{-1}	0.63	0.2	0.31	0.08
Percentage of thermal energy radiated	17-25	23-33	20-28	30-42
Normalized flame emissivity	1.00	1.7	1.5	1.7
Equivalence ratio	0.1-7.1	0.7-4	0.5-5.4	0.7-3.8

Table 1. Overall comparison of properties of hydrogen, CNG, HCNG and gasoline[4].

Hydrogen also has a very low energy density per unit volume and as a result, the volumetric heating value of the HCNG mixture decreases (Table 2) as the proportion of hydrogen is increased in the mixture [38].

Properties	CNG	HCNG 10	HCNG 20	HCNG 30
H_2 [vol %]	0	10	20	30
H_2 [mass %]	0	1.21	2.69	4.52
H_2 [energy %]	0	3.09	6.68	10.94
LHV [$MJkg^{-1}$]	46.28	47.17	48.26	49.61
LHV [$MJNm^{-3}$]	37.16	34.50	31.85	29.20
LHV stoichiometric mixture [$MJNm^{-3}$]	3.376	3.368	3.359	3.349

Table 2. Properties of CNG and HCNG blends with different hydrogen content [39]

Many researchers have studied the effect of the addition of hydrogen to natural gas on performances and emissions in the past few years[40-65]. Blarigan and Keller investigated the port-injection engine fueled with natural gas–hydrogen mixtures [40]. Bauer and Forest conducted an experimental study on natural gas–hydrogen combustion in a CFR engine [41]. Wong and Karim analytically examined the effect of hydrogen enrichment and hydrogen addition on cyclic variations in homogeneously charged compression ignition engines. The results indicated that the addition of hydrogen can reduce cyclic variations while extending the operating region of the engine [42]. Karim et al. theoretically studied the addition of hydrogen on methane combustion characteristics at different spark timings. The theoretical results showed that the addition of hydrogen to natural gas could decrease the ignition delay and combustion duration at the same equivalence ratio. It indicated that the addition of hydrogen could increase the flame propagation speed, thus stabilizing the combustion process, especially the lean combustion process [43]. Ilbas et al. [44] experimentally studied the laminar burning velocities of hydrogen–air and hydrogen–methane–air mixtures. They concluded that increasing the hydrogen percentage in the hydrogen–methane mixture brought about an increase in the resultant burning velocity and caused a widening of the flammability limit (Figure 6).

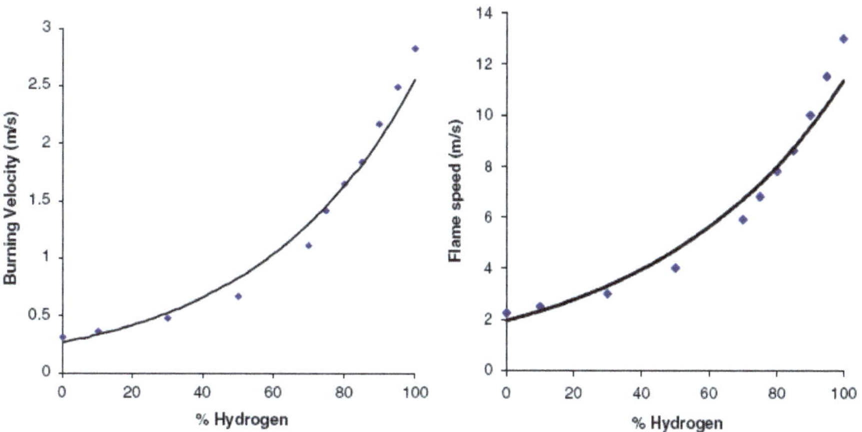

Figure 6. Burning velocities and flame speed for different percentages of hydrogen in methane ($\phi = 1.0$)[44].

Shudo et al., analyzed the characteristics combustion and emission of a methane direct injection stratified charge engine premixed with hydrogen lean mixture [45]. Their results showed that the combustion system achieved higher thermal efficiency due to higher flame propagation velocity and lower exhaust emissions. An increase in the amount of premixed hydrogen stabilizes the combustion process to reduce HC and CO exhaust emission, and increases the degree of constant volume combustion and NO_x exhaust emission. The increase in NO_x emission can be maintained at a lower level with retarded ignition timing without reducing the improved thermal efficiency. Nagalingam et al. [46] investigated hydrogen enriched CNG (hythane). He noted that the power was reduced due to the lower volumetric heating value of hydrogen compared with methane. However, since the flame speed of hydrogen was significantly higher than that of CNG, less spark advance was required to produce maximum brake torque (MBT). Wallace and Cattelan experimentally studied natural gas and hydrogen mixtures in a combustion engine. The experiments were conducted by studying the emissions of an engine fueled with a mixture of natural gas and approximately 15% hydrogen by volume [47].

Raman et al. [48] carried out an experimental study on SI engines fueled with HCNG blends from 0% to 30% of H_2 in a V8 engine. The authors observed a reduction in NO_x emissions using 15%-20% hydrogen blends with some increase in HC emissions as a result of ultra-lean combustion. The experiments were performed using a Chevrolet Lumina, which has six cylinders, four stroke cycles, is water cooled, with a total engine cylinder volume of 3.135 l, bore of 89 mm, stroke of 84 mm and compression ratio of 8.8:1. In their study, the BSFC of an 85/15 CNG/H_2 mixture was less than that of natural gas. The BSFC values decreased for both natural gas and the 85/15 CNG/H_2 mixture while spark timing (BTDC) values increased. The BSHC of CNG was higher than that of the fuel mixture. However, the $BSNO_x$ emission values of the 85/15 CNG/H_2 mixture were higher than that of CNG. If a catalytic converter is used, the $BSNO_x$ values are decreased drastically. Larsen and Wallace [49] conducted experimental tests on heavy-duty engines fueled by HCNG blends. The authors found that HCNG blends improve efficiency and reduce CO, CO_2 and HC emissions. Collier et al. examined the untreated exhaust emissions of a hydrogen-enriched compressed natural gas (HCNG) production engine [50]. They used variable composition hydrogen/NG mixtures and drew the following conclusions: the addition of hydrogen increases NO_x emission for a given equivalence ratio while it decreases total HC emissions which is in good agreement with Akansu's results [51]. They also found that as the hydrogen percentage increases, the lean limit of combustion is significantly extended. Hoekstra et al. [52] observed a reduction in NO_x for hydrogen percentages up to 30%, beyond this limit no improvement was observed. An important point was the higher flame speed and a consequent reduction of the spark advance angle to obtain the maximum brake torque, as already indicated by Nagalingam et al. [46]. Wang et al. investigated the combustion behavior of a direct injection engine operating on various fractions of NG–hydrogen blends [53]. The results showed that the brake effective thermal efficiency increased with the increase of hydrogen fraction at low and medium engine loads. The rapid combustion duration decreased, and the heat release rate and exhaust NO_x increased with the increase of hydrogen fraction in the blends. Their study suggested that the optimum hydrogen volumetric fraction in NG–hydrogen blends is around 20% to achieve a compromise in both engine performance and emissions.

Ceper [54] studied different CH_4/H_2 mixtures experimentally and numerically. Her experimental study was performed with a four-stroke, four-cylinder, water cooled, Ford 1.8-liter internal combustion engine. CH_4/H_2 (100/0, 90/10, 80/20, 70/30) gas fuel mixtures of fuels were tested at different engine speeds and excess air ratios. Kahraman et al. [55] experimentally researched the performance and exhaust emissions of a spark-ignition engine fueled with methane-hydrogen mixtures (100% CH_4, 10% H_2 + 90% CH_4, 20% H_2 + 80% CH4, and 30% H_2 + 70% CH_4) at different engine speeds and different excessive air ratios. The results demonstrated that while the speed and excess air ratio increased, CO emission values decreased. Furthermore increasing the excess air ratio also decreased the maximum peak cylinder pressure. Çeper et al. [56] experimentally analyzed the performance and the pollutant emissions of a four-stroke spark-ignition engine operating on natural gas-hydrogen blends of 0%, 10%, 20% and 30% at full load and 65% load for different excess air ratios. The results showed that while the excess air ratio increased, CO and CO_2 emission values decreased. In addition, increasing the excess air ratio led to a decrease in peak pressure values and by increasing the H_2 amount, peak pressure values were close to TDC, and the brake thermal efficiency values increased.

Sierens and Rossel [57] determined that the optimal HCNG composition to obtain low HC and NO_x emissions should be varied with engine load. Huang et al. [58] conducted an experimental study for a direct-injection spark-ignition engine fueled with HCNG blends under various ignition timings and lean mixture conditions. The ignition timing is an important parameter for improving engine performance and combustion. Dimopoulos et al. [59] optimized a state of the art passenger car natural gas engine for hydrogen–natural gas mixtures and high exhaust gas recirculation (EGR) rates in the major part of the engine map. Increasing the hydrogen content of the fuel accelerated combustion leading to efficiency improvements. Well-to-wheel analysis revealed paths for the production of the fuel blends still having overall energy requirements slightly higher than a diesel benchmark vehicle but reducing overall green house gas emissions by 7%.

Based on the results of an experimental test campaign carried out in ENEA labs, Ortenzi et al [60], aimed at identifying the potential of using blends of natural gas and hydrogen (HCNG) in existing ICE vehicles. The tested vehicle was an IVECO Daily CNG, originally fueled with natural gas and the tests were made on an ECE15 driving cycle to compare the emission levels of the original configuration (CNG) with the results obtained with different blends (percentage of hydrogen in the fuel) and control strategies (stoichiometric or lean burn). Dulger investigated an 80% CNG and 20% H_2 mixture burning SI engine numerically [61]. Swain et al. [62] and Yusuf [63] investigated the same mixture with a different engine. Yusuf used a Toyota 2TC type engine with the following specifications: year 1976 1:6 l, 1588 cc, maximum HP 88 and maximum speed of 6000 rpm, bore 85 mm, stroke 70 mm, compression ratio of 9.0:1 and four cylinder engine. The engine was tested at 1,000 rpm, using best efficiency spark advance and light loading conditions. When the methane–hydrogen mixture was compared to pure methane operation with the same equivalence ratios, the methane and hydrogen mixture increased BTE and NO_x emissions while decreasing the best efficiency spark advantage, unburned HCs and CO. Moreover, the lean limit combustion of natural gas was reduced from 0.61 to 0.54. The lean limit of combustion was defined as an operation with at least 38% of the cycles not completing combustion. By hydrogen addition, the equivalence ratios could be reduced by about 15% without increasing combustion duration and ignition delay.

Ma and Wang [64], experimentally investigated the extension of the lean operation limit through hydrogen addition in an SI engine which was conducted on a six-cylinder throttle body injection natural gas engine. Four levels of hydrogen enhancement were used for comparison purposes: 0%, 10%, 30% and 50% by volume. Their results showed that the engine's lean operation limit could be extended through adding hydrogen and increasing load level (intake manifold pressure). The effect of engine speed on lean operation limit is smaller. At a low load level an increase in engine speed is beneficial in extending the lean operation limit but this is not true at high load level. The effects of engine speed are even weaker when the engine is switched to hydrogen enriched fuel. Spark timing also influences the lean operation limit and both over-retarded and over-advanced spark timing are not advisable. Road tests on urban transport buses were performed by Genovese et al. [65], comparing energy consumption and exhaust emissions for NG and HCNG blends with hydrogen content between 5% and 25% in volume. The authors found that average engine efficiency over the driving cycle increases with hydrogen content and NO_x emissions were higher for blends with 20% and 25% of hydrogen, despite the lean relative air fuel ratios and delayed ignition timings adopted. Having reviewed the main experimental papers published in the past, we conclude that numerical analysis also plays a fundamental role in research activities, allowing a better design of the experimental tests in terms of cost savings and time reduction[66-70].

4.1. Emissions

Air pollution is fast becoming a serious global problem arising from an increasing population and its subsequent demands. This has resulted in increased usage of hydrogen as fuel for internal combustion engines. Hydrogen resources are vast and it is considered as one of the most promising fuels for the automotive sector. As the required hydrogen infrastructure and refueling stations do not currently meet demand, the widespread introduction of hydrogen vehicles is not feasible in the near future. One of the solutions for this hurdle is to blend hydrogen with methane. Such types of blends take benefit of the unique combustion properties of hydrogen and at the same time reduce the demand for pure hydrogen. Enriching natural gas with hydrogen could be a potential alternative to common hydrocarbon fuels for internal combustion engine applications [71].

When experimental or simulation studies on reciprocating engines are carried out, much attention is paid to pollutant CO, HC and NO_x emissions. Nevertheless, although CO_2 is one of the most important greenhouse gases, these emissions are not usually taken into account, and measurements and calculations of CO_2 emissions are omitted from many studies [72].

Fuel costs and their relationship to equivalent CO_2 emissions are represented in Figure 7 for several types of fuel ([73] and data from the authors). As observed, the global CO_2 emissions associated with CNG and their costs are lower than those produced by gasoline or diesel. Hydrogen produces lower CO_2 emissions than CNG, gasoline or diesel, but hydrogen always originates from renewable sources. Due to the high price of crude oil, in some cases the cost of H_2 is lower than that of gasoline or diesel. In any case, these data have been prepared without taking into account the possible effects of an increase in demand or mass production [72].

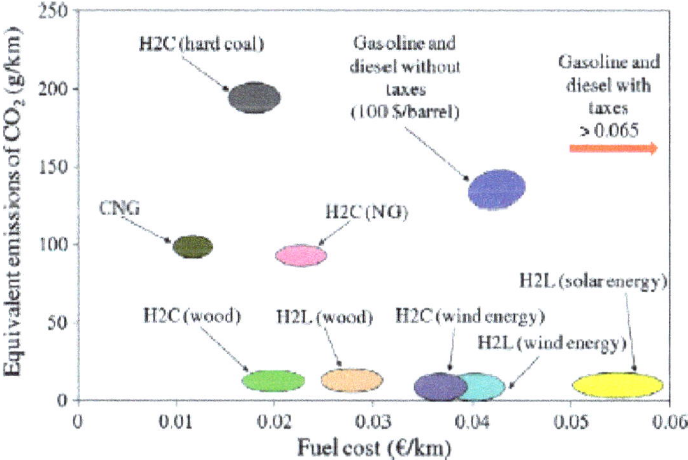

Figure 7. Cost and CO_2 emissions for several fuels [72].

All these performance parameters have a direct relationship with the exhaust emissions produced, often with contradictory effects. For instance, while higher compression ratios are favored in order to increase thermal efficiency, they also result in higher NO_x emissions because of the resultant higher combustion chamber temperatures. This is also the case when running stoichiometric fuel-air mixtures, as seen in Figure 8 (which is applicable to gasoline engines, but the general trends are similar for natural gas engines as well). In addition, while the combustion of lean fuel-air mixtures ($\phi < 1$) results in low NO_x emissions (as seen from Fig. 7) this can also result in lower power output. However, running an engine on fuel-rich mixtures ($\phi > 1$) is also undesirable and this results in high unburnt HC and CO emissions. Knock limits are also a factor when deciding ideal operating parameters. For instance if an engine is running too high a compression ratio, resistance to knock is lowered. This would require the need for spark retardation with respect to combustion TDC (which can affect thermal efficiency and therefore power output as well as exhaust emissions)[74].

Figure 9 illustrates the $BSNO_x$ (g/kWh) values versus equivalence ratios from different studies [75]. As seen in this figure, according to studies, with increasing H_2 percentage, $BSNO_x$ values increase or decrease. According to refs [62,49,57] and Bauer and Forest [41] (there is no data value in graphics), with increasing H_2 percentage, the $BSNO_x$ values increase. However, in the experiments performed by Raman et al. [48], with increasing H2 percentage, the $BSNO_x$ values decrease. Moreover, if the equivalence ratios decrease, the BSNOx values reach a low value. It is interesting to note that Hoekstra et al. [52], as well as Larsen and Wallace [49], obtained extremely low NO_x emission.

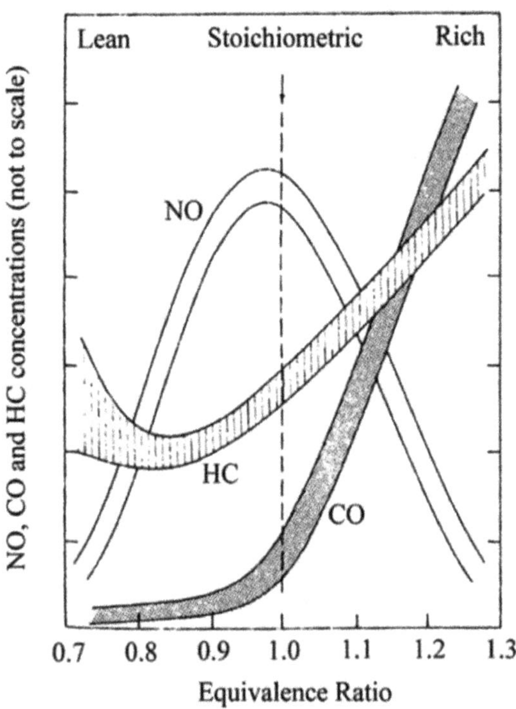

Figure 8. Typical NO, HC and CO trends with equivalence ratio in an SI engine, adapted from [74].

Figure 10 shows the BSHC (g/kW h) values in different studies [75]. As seen in this figure, with increasing H_2 percentage and equivalence ratio, the BSHC values decrease. If fuel is to be 100% H_2 fuel, the BSHC value will be zero. We can say that BSHC values decrease as the amount of H_2 increases. By increasing the equivalence ratios Swain et al.[62] obtained the highest BSHC values in these studies. The maximum value is about 64 g/kW h, for a 20% H_2 and 80% CH_4 mixture with $\phi = 0.60$. However, hydrocarbon emissions of 20% H_2 and 80% CH_4 mixture are less than those of pure methane [62]. In this figure, the BSHC values of Ref. [49] are at their highest value. BSHC values increase with increasing engine load.

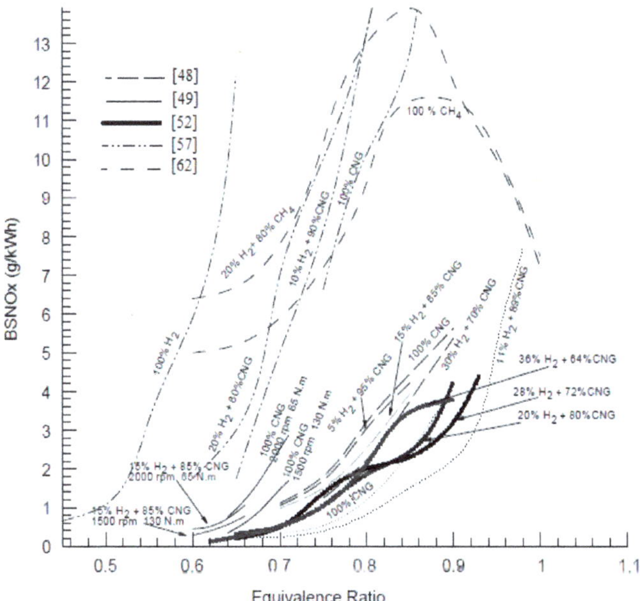

Figure 9. BSNOx (g/kWh) values of different studies versus equivalence ratios[75].

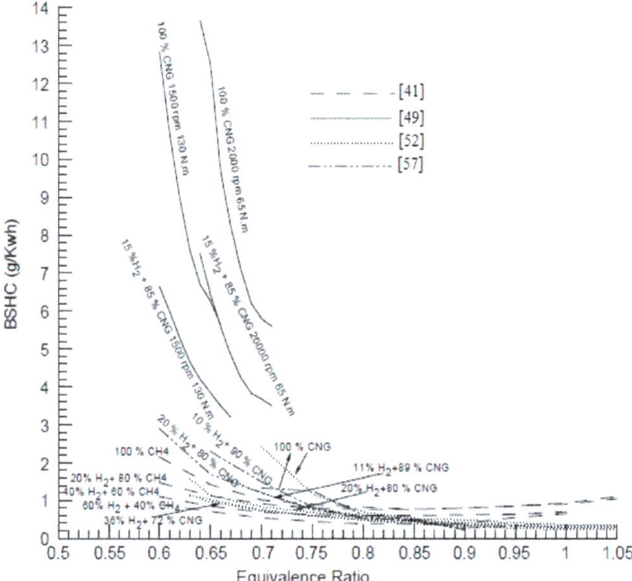

Figure 10. Brake specific hydrocarbons (BSHC g/kw h) values in different studies[75].

Larsen and Wallace obtained 1.65 and 2:41 g/kW h CO values at 1500 rpm, and ϕ = 0.65 equivalence ratio, using an 85/15 CNG/H_2 and 100% CNG, respectively [49]. Yusuf measured all engine/fuel configurations performed similarly over normal operating ranges. An important variation occured with rich mixtures. In addition, the 80/20 CH_4/H_2 mixture showed a small but significant reduction in BSCO output [62,63]. Bauer and Forest's experiments demonstrated that production of CO was highly dependent on combustion stoichiometry and less so on the engine. They obtained a general reduction in BSCO with the addition of hydrogen because of the reduction of carbon in the fuel. They added up to 60% hydrogen by volume and found that BSCO decreased up to 20 g/kW h (60/40 CH_4/H_2) at ϕ=1.0. In the ultra lean region (ϕ<0.4), an increase in BSCO was noted, due to incomplete combustion combined with sharply dropping power [41]. Figure 11 shows the BSCO emission values of some studies[75]. As seen in this figure, a ϕ value between 0.65 and 0.8 placed BSCO values at a dramatically low level.

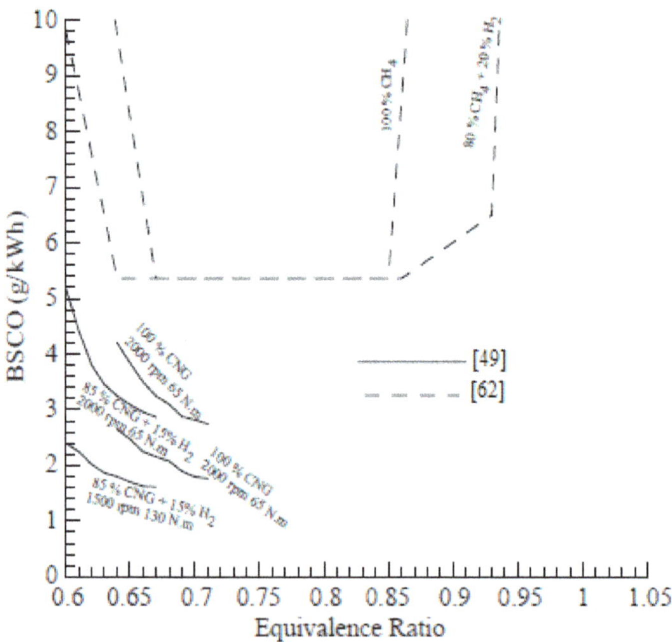

Figure 11. BSCO (g/kw h) values versus equivalence ratio in different studies[75].

Figure 12 gives the brake NOx, HC, CO and CO_2 emission versus hydrogen fraction at various injection timings[76]. Brake NOx emission increases with increasing hydrogen fraction when the hydrogen fraction is less than 10%, and it decreases with the increase of hydrogen fraction when the hydrogen fraction is larger than 10% at various injection timings. The comprehensive effects of in-cylinder temperature, excess air ratio and combustion duration contribute to this. As excess air ratio in this experiment is larger than

1.0 and combustion duration is slightly decreased with increasing hydrogen fraction, the effect of cylinder gas temperature plays an important part, thus the trend of brake NO_x emission is consistent with that of the maximum mean gas temperature. Brake HC emission decreases with the increase of hydrogen fraction. This is because the quench distance of the fuel blends is decreased and the lean flammability limit of the natural gas-hydrogen fuel blends is extended with hydrogen addition. Meanwhile, combustion is improved with the increase of hydrogen fraction, and this enhances the post-flame oxidation of the already formed HC. Furthermore, the C/H ratio decreases with increasing hydrogen fraction and this also contributes to the decrease of brake HC emission with the increase of hydrogen fraction.

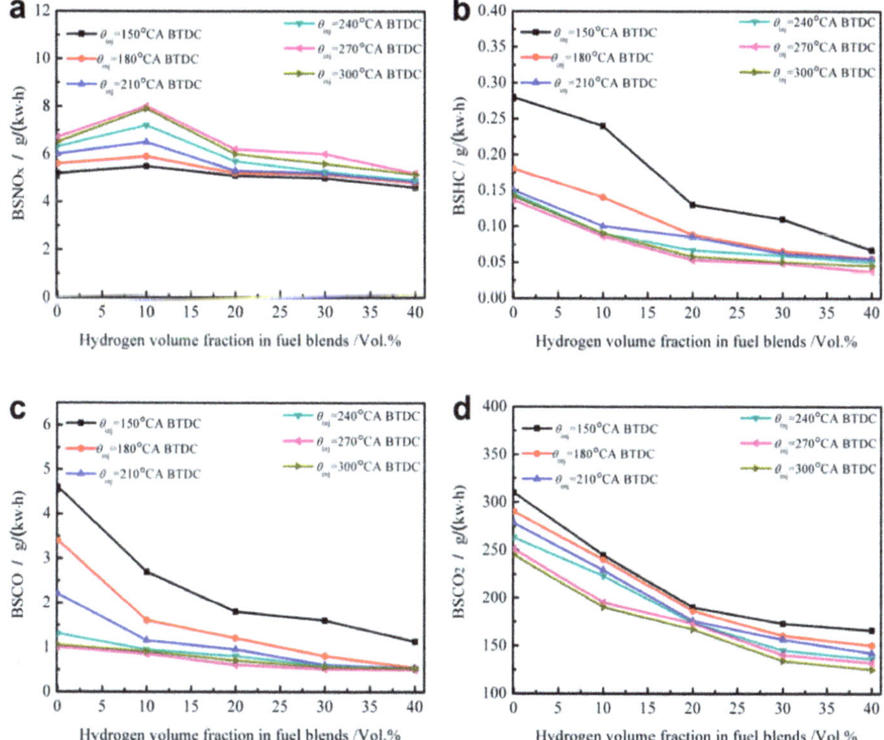

Figure 12. Brake NO_x, HC, CO and CO_2 emission versus hydrogen fractions. (a) Brake NOx emission versus hydrogen fractions. (b) Brake HC emission versus hydrogen fractions. (c) Brake CO emission versus hydrogen fractions. (d) Brake CO_2 emission versus hydrogen fractions[76]

Brake CO emission decreases with increasing hydrogen fraction. As overall excess air ratio in the cylinder increases with hydrogen addition, and CO is strongly related to the air-fuel ratio, the sufficiency of oxygen in the cylinder makes the CO emission low. Also, combustion is improved with the increase of hydrogen fraction, and this enhances the post-flame oxidation of the already formed CO. Furthermore, the C/H ratio decreases with increasing hydrogen fraction in the fuel blends and this also contributes to the decrease of brake CO emission with the increase of hydrogen fraction. Brake CO emission achieves its minimum value at a fuel-injection timing of 270 °CA BTDC. Brake CO_2 emission decreases with the increase of hydrogen fraction. The decrease in the C/H ratio of the mixtures with the increase of hydrogen fraction is responsible for this. A low carbon fraction produces low CO_2 concentration [76].

4.2. Cylinder pressure

Figure 13 shows the cylinder pressure at engine speeds of 2000 and 3000 rpm for different values of H_2 percentages (0, 3%, 5% and 8% in ref 77; 0, 10%, 20% and 30% in ref 54) and λ=1.0. For all cases, the cylinder pressure increased with the increase in the amount of H_2. The maximum pressures for the 8% H_2, 5% H_2, 3%H_2 and pure CNG occurred at 11, 12, 12.5, and a 13.5° crank angle ATDC respectively [77]. The maximum pressures for the 30% H_2, 20% H_2, 10%H_2 and pure CNG occurred at about 53, 48, 44, and a 36° crank angle ATDC respectively [54]. At an engine speed of 3000 rpm, the maximum cylinder pressures occurred at a 13.5° crank angle ATDC with their magnitudes being the highest of all values of H_2 percentage [77]. In Ref [54], the maximum cylinder pressure occurred at a 30° crank angle ATDC. In Ref [77], the compression ratio of the engine was 14:1 and in ref [54] the compression ratio of the engine was 10:1. So the maximum cylinder pressure values were obtained at 8% H_2 in both figures. For all the previous cases, the cylinder pressure increased with the increase in the amount of H_2. The explanation for this phenomenon is mainly due to fact that the flame speed of hydrogen is faster than the flame speed of CNG. Therefore, burning CNG in the presence of a small amount of hydrogen will result in faster and more complete combustion. This will result in higher peak pressure closer to TDC and it will produce a higher effective pressure [77].

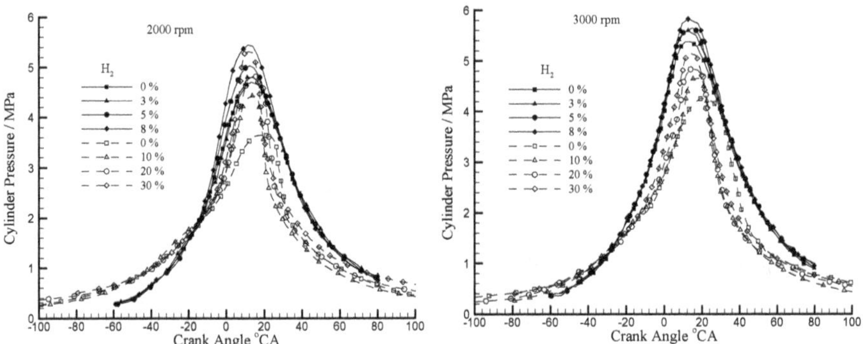

Figure 13. Cylinder pressure values versus the crank angle for different engine speeds and different H_2 fractions (solid ref [77] and dashed ref [54])

Figure 14 shows the in-cylinder pressure curve under various λ for different fuels: pure CNG, 30% HCNG, 55% HCNG[78]. From Figure 14(a), as the mixture is leaner, the maximum in-cylinder pressure is smaller. Figure 14(b,c,d) shows further that the position of the maximum in-cylinder pressure is later before λ=1.5. On the other hand, when λ > 1.5, the maximum in-cylinder pressure is nearer the TDC.

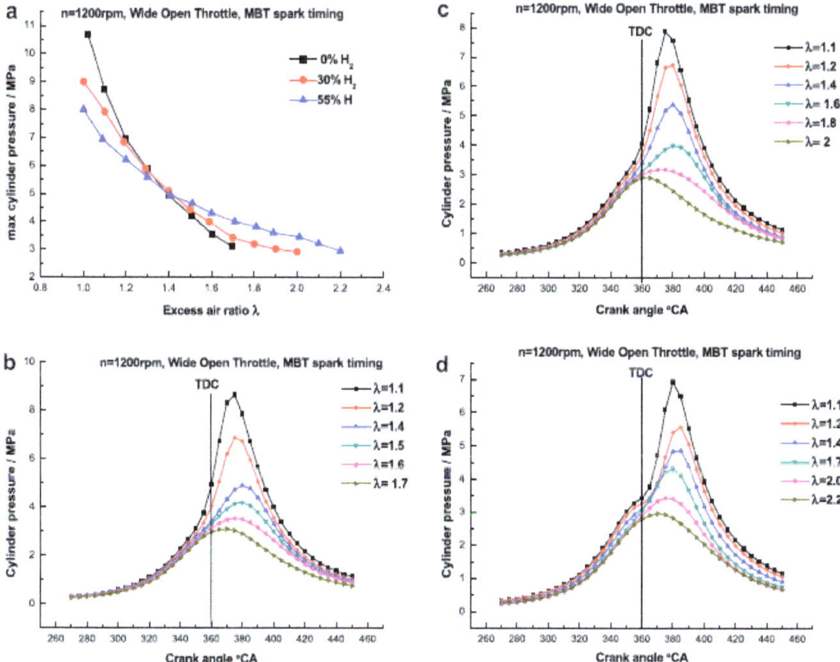

Figure 14. (a) Max cylinder pressure versus excess air ratio. (b) In-cylinder pressure for CNG. (c) In-cylinder pressure for 30% hydrogen volumetric ratio. (d) In-cylinder pressure for 55% hydrogen volumetric ratio[78].

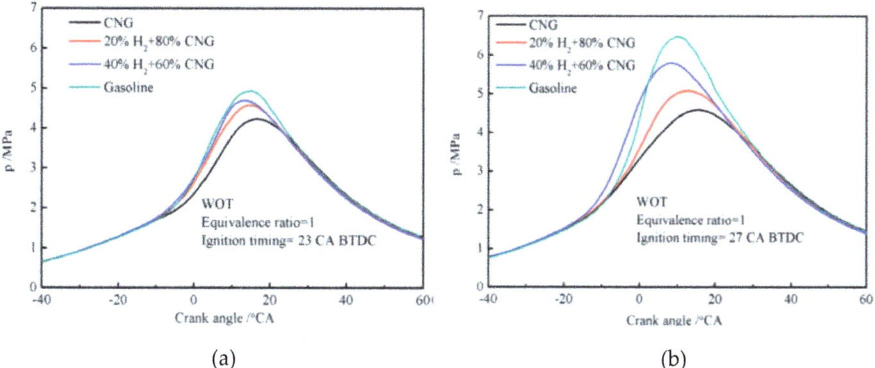

Figure 15. Cylinder pressure versus crank angle for 2000 and 3000 rpm in different fuels[79].

Figure 15 shows cylinder pressure versus crank angle for 2000(a) and 3000(b) rpm respectively [79]. As shown in these figures, the timing of the maximum cylinder pressure fueled with natural gas is postponed compared with that fueled with gasoline, and it advances as hydrogen is added.

4.3. Brake thermal efficiency

Figure 16 depicts BTE versus equivalence ratio [75]. As seen in this figure, the BTE of a 20% H_2 +80% CH_4 mixture is higher than that of 100% CH_4 [57,62]. Since only one cylinder was used in the experiment it is expected that efficiency be lower compared to an experiment using a four-cylinder engine. According to the experiments in Ref[41], the BTE values decreased, while the H_2 percentage increased. The highest efficiency values were between 0.7 and 0.9 equivalence ratios. According to the study in Ref. [55], the maximum efficiency was at about ϕ=0.75–0.8 for a 30%H_2+70%CH_4 mixture.. Also, effective efficiency had about a ϕ= 0.75–0.8 equivalence ratio [41,57,62].

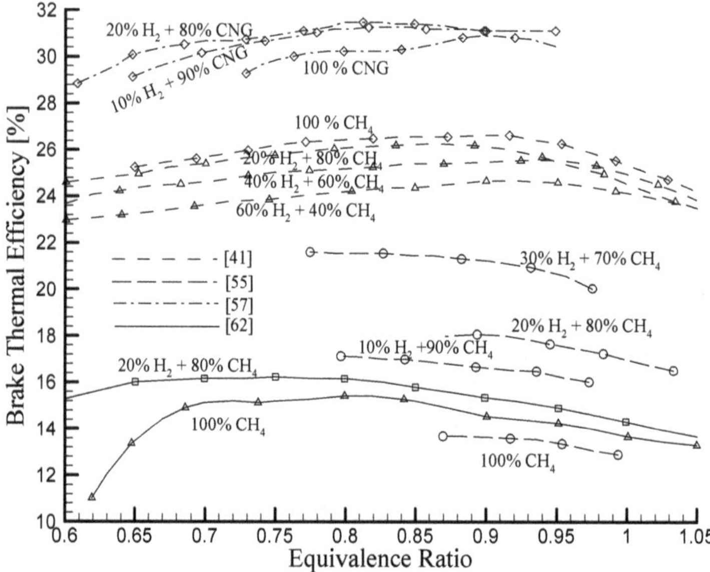

Figure 16. Brake thermal efficiency versus equivalence ratio

5. Conclusions and perspectives for further development

The results in this study can be summarized as follows:

- The ultimate goal of hydrogen economy is to displace fossil fuels with clean burning hydrogen and CNG is the best route to ensure the early introduction of hydrogen fuel into the energy sector.

- The lean-burn capability and flame burning velocity of the natural gas engine was improved by blending it with fast burning velocity fuel such as hydrogen.
- HCNG engines are superior to CNG engines from a fuel economy, power, and torque point of view due to better combustion.
- The addition of hydrogen to natural gas increases BMEP compared with that of natural gas combustion. This is due to the increased burning velocity of the mixture by hydrogen addition which shortens combustion duration and increases the cylinder gas temperature.
- The HCNG engine improves power by 3 - 4 % and torque by about 2 - 3 % compared to the CNG engine. The HCNG engine operates on the leaner side than the CNG engine which reduces fuel consumption by about 4% compared to CNG engine.
- The HCNG fuel reduces CO emissions and NO_x emissions more than the neat CNG operation. Thus the blended HCNG fuel is more environmentally friendly.
- Engine operating parameters have to be carefully chosen by the designer, taking into account their effect on engine performance and emission.
- Any attempt to control emissions by operating the engine with leaner mixtures has to take into account the effect on other variables like power.
- Compression ratio and equivalence ratio have a significant effect on both the performance and emission characteristics of the engine and have to be carefully designed to achieve the best engine performance characteristics.
- Higher engine rotational speeds can be used in lean mixtures to increase the power output of an engine operating on hydrogen while maintaining high efficiency and pre-ignition free operation.
- The variation in spark timing with hydrogen is very effective in controlling the combustion process.
- Higher compression ratios can be applied satisfactorily to increase power output and efficiency, mainly because of the relatively fast burning characteristics of hydrogen–air mixtures.
- The addition of hydrogen to methane gives a good alternative fuel to hydrocarbon fuels as it gives good flame stability, wide flammable regions and relatively higher burning velocity.
- NO_x emission values generally increase with increasing hydrogen content. However, if a catalytic converter, an EGR system or lean-burn technique are used, NO_x emission values can be reduced to extremely low levels.
- HC, CO_2 and CO emission values decrease with increasing hydrogen percentage.
- The addition of H_2 (up to 20-30% vol.) to NG may constitute an effective short-term solution for the green-house gases problem and at the same time to introduce H_2 into the fuel market without requiring changes in current engine technology.
- In conjunction with new and advanced technologies, hydrogen-methane mixture gases can provide a large part of the rapidly growing need for clean and affordable energy services in the world.

Future research of the hydrogen enriched compressed natural gas fuel include continuous improvement on performance and emissions, especially to reduce the hydrocarbon emissions (including methane if necessary) which are currently not heavily regulated but will probably be more closely regulated in the future. Although the exhaust emissions from hydrogen-enriched natural gas are already very low, further refinement must be done in order to further reduce emissions and to achieve Enhanced Environmentally Friendly Vehicle (EEV) standards. Therefore finding the optimal combination of hydrogen fraction, ignition timing and excess air ratio along with other parameters that can be optimized is certainly a large hurdle. It is not only a challenge to locate the ideal combination of hydrogen fraction, ignition timing, and excess air ratio, but it can also be a large challenge to control these parameters. This requires sufficient control system to be developed for the HCNG engine to maximize the performance simultaneously minimizing the exhaust emissions. Other potential improvements include the reduction of emissions which can be transpire with the addition of a catalytic converter or by implementing an exhaust gas recycle system, lastly there is potential for performance improvements with an increase in the compression ratio[80].

As a result, today are faced with environmental problems, tomorrow hydrogen will solve all environmental problems due to road transports: Natural gas-hydrogen blends may be a potential bridge from today to tomorrow.

Abbreviations

AFR	Air-fuel ratio
ATDC	After top dead center
BSCO	Brake specific carbon monoxide
BSFC	Brake specific fuel consumption
BSHC	Brake specific hydrocarbon
$BSNO_x$	Brake specific nitrogen oxide
BTDC	Before top dead center
BTE	Brake thermal efficiency
CA	Crank angle (°)
CFR	Co-operative fuel research
CI	Compression ignition engine
CNG	Natural gas
CO	Carbon monoxide
CO_2	Carbon dioxide
ECE15	European driving cycle
EGR	Exhaust gas recirculation
ENEA	Italian national agency for new technologies
HC	Hydrocarbon
HCNG	Hydrogen-natural gas blend
H_2	Hydrogen
IC	Internal combustion engines
LNG	Liquid natural gas
LPG	Liquid petroleum gas

MBT	Maximum brake torque
NEDC	New European driving cycle
NG	Natural gas
NOx	Nitrogen oxides
NTP	normal pressure and temperature
rpm	Revolutions per minute
SI	Spark-ignition engine
TDC	Top dead center
THC	Total unburned hydrocarbon
u_L	Laminar burning velocity
WOT	Wide open throttle

Greek symbols

ϕ	Equivalence ratio
λ	Excess air ratio

Author details

Bilge Albayrak Çeper
Erciyes University Faculty of Eng., Dept. of Mech Eng., Kayseri, Turkey

Acknowledgement

Bilge Albayrak Ceper would like to thank Professor Nafiz Kahraman and Assoc. Prof. Selahaddin Orhan Akansu at Erciyes University for their encouragement on this study.

6. References

[1] Rahman M.M., Mohammed M.K., and Bakar R.A., Effect of engine speed on performance of four-cylinder direct injection hydrogen fueled engine, Proceedings of the 4th BSME-ASME International Conference on Thermal Engineering 27-29 December, 2008, Dhaka, Bangladesh
[2] Population Division United States Census Bureau International Database. World population: 1950-2050. http://www.census.gov/ipc/www/idb/worldpopgraph.php.
[3] British Petroleum, BP statistical review of world energy, BP Annual Review; June 2009.
[4] Patil K.R., Khanwalkar P.M., Thipse S.S., Kavathekar K.P., Rairikar S.D., Development of HCNG Blended Fuel Engine with Control of NOx Emissions, International Journal of Computer Information Systems and Industrial Management Applications (IJCISIM), ISSN: 2150-7988, 2010;2:087-095.
[5] Pourkhesalian A.M., Shamekhi A.H., Salimi F., Alternative fuel and gasoline in an SI engine: A comparative study of performance and emissions characteristics, Fuel 2010;89: 1056–63.

[6] Adeeb Z., Glycerol delignification of poplar wood chips in aqueous medium, Energy Educ Sci Technol 2004;13:81–8.
[7] Verhelst S., Maesschalck P., Rombaut N., Sierens R., Increasing the power output of hydrogen internal combustion engines by means of supercharging and exhaust gas recirculation, Int J Hydrogen Energy 34 (2009) 4406–4412
[8] Doble M., Kruthiventi A.K., Alternate Energy Sources, Green Chemistry and Engineering, 2007, Pages 171-192.
[9] Veziroglu T.N., Barbir F., Solar-hydrogen energy system: the choice of the future. Environ. Conserv. 1991;18(4):304–12.
[10] Veziroglu T.N., Petkov T., Sheffield J.W., An outlook of hydrogen as an automotive fuel. Int J Hydrogen Energy 1989;14(7):449–74.
[11] North D.C., Investigation of hydrogen as an internal combustion fuel. Int. J. Hydrogen Energy 1992;17(7):509–12.
[12] Asadiq A-B.M., Ashahad A-J. H., A prediction study of the effect of hydrogen blending on the performance and pollutants emission of a four stroke spark ignition engine. Int. J Hydrogen Energy 1999;24(4):363–75.
[13] Al-Baghdadi Maher A.S., Al-Janabi Haroun A.S., A prediction study of a spark ignition supercharged hydrogen engine. Energy Conversion Manage. 2003;44(20):3143–50.
[14] Desoky A.A., El-Emam S.H., A study on the combustion of alternative fuel in spark ignition engines. Int. J. Hydrogen Energy 1985;10(8):497–504.
[15] Sher E., Hacohen Y., Measurements and predictions of the fuel consumption and emission of a spark ignition engine fueled with hydrogen-enriched gasoline. Proc. Instan. Mech. Engrs1989;203:155–62.
[16] Taylor S.C., Burning velocity and the influence of flame stretch. PhD thesis, Leeds University, 1991.
[17] Vagelopoulos C.M., Egolfopoulos F.N., Law C.K., Further considerations on the determination of laminar flame speeds with the counterflow twin-flame technique, 25th Symp. (Int.) on Combustion 1341–1CK347, 1994.
[18] Kwon O.C., Faeth G.M., Flame/stretch interactions of premixed hydrogen-fueled flames: measurements and predictions, Combust Flame 2001;124:590–610.
[19] Verhelst S., Woolley R., Lawes M., Sierens R., Laminar and unstable burning velocities and Markstein lengths of hydrogen–air mixtures at engine-like conditions, Proc Combust Inst 2005;30:209–16.
[20] Liu D.D.S., MacFarlane R., Laminar burning velocities of hydrogen–air and hydrogen–air–steam flames, Combust Flame 1983;49:59–71.
[21] Milton B., Keck J., Laminar burning velocities in stoichiometric hydrogen and hydrogen–hydrocarbon gas mixtures, Combust Flame 1984;58:13–22.
[22] Iijima T., Takeno T., Effects of temperature and pressure on burning velocity, Combust Flame 1986;65:35–43.
[23] Koroll G.W., Kumar R.K., Bowles E.M., Burning velocities of hydrogen–air mixtures, Combust Flame 1993;94:330–40.
[24] Ghazi A.K., Hydrogen as a spark ignition engine fuel, Int. J. Hydrogen Energy 2003;28(5):569–77.

[25] Jehad A.A., Yamin H.N., Gupta B.B.B., Srivastava O.N., Effect of combustion duration on the performance and emission characteristics of a spark ignition engine using hydrogen as a fuel,Int. J Hydrogen Energy 2000;25(6):581–9.

[26] Das L.M., Gupta R., Gupta P.K., Performance evaluation of a hydrogen-fuelled spark ignition engine using electronically controlled solenoid-actuated injection system, Int. J. Hydrogen Energy 2000;25(6):569–79.

[27] De Ferrières S., El Bakali A., Lefort B., Montero M., Pauwels J.F., Experimental and numerical investigation of low-pressure laminar premixed synthetic natural gas/O2/N2 and natural gas/H2/O2/N2 flames, Combustion and Flame 2008;154: 601–23.

[28] Kim K., Kim H., Kim B., Lee K. and Lee K., Effect of Natural Gas Composition on the Performance of a CNG Engine, Oil & Gas Science and Technology – Rev. IFP, 2009;64(2); 199-206.

[29] Hu E., Huang Z., He J., Zheng J., Miao H., Measurements of laminar burning velocities and onset of cellular instabilities of methane–hydrogen–air flames at elevated pressures and temperatures, Int J Hydrogen Energy 2009;34: 5574 –84.

[30] Semin, R.A.B., A Technical Review of Compressed Natural Gas as an Alternative Fuel for Internal Combustion Engines, American J. of Engineering and Applied Sciences 2008;1(4): 302-311.

[31] Kubesh J., King S.R., Liss W.E., Effect of Gas Composition on Octane Number of Natural Gas Fuels, *SAE* 922359 1992.

[32] Lee Y., Kim G., Effect of Gas Compositions on Fuel Economy and Exhaust Emissions of Natural Gas Vehicles, *KSAE* 7, 1999;8: 123-31.

[33] Aslam M.U., Masjuki H.H., Kalam M.A., Abdesselam H., Mahlia T.M.I., Amalina M.A. An Experimental Investigation of CNG as an Alternative Fuel for a Retrofitted Gasoline Vehicle, Fuel 2006; 85; 5-6, 717-24.

[34] Ryu K., Kim B., Development of Conversion Technology of a Decrepit Diesel Vehicle to the Dedicated Natural Gas Vehicle, *KSAE* 14, 2006;6:73-81.

[35] Cho H.M., He B.Q., Spark ignition natural gas engines – a review, Energy Convers Manage 2007;48(2):608–18.

[36] Tunestal P., Christensen M., Einewall P., Andersson T., Johansson B., Jonsson O., Hydrogen addition for improved lean burn capability of slow and fast burning natural gas combustion chambers, SAE paper 2002-02-2686; 2002.

[37] Thipse S.S., et al, Development of a CNG injection engine compliant to Euro-IV norms and development strategy for HCNG operation, SAE Paper 2007-26-029.

[38] Bell, S.R., Gupta, M., Extension of a Lean Operating Limit for Natural Gas Fuelling of a Spark Ignition Engine Using Hydrogen Blending., *Combustion Sciences and Technology*, 1997;123: 1-6, 23-48.

[39] Xu J., Zhang X., Liu J., Fan L., Experimental Study of a Single Cylinder Engine Fueled with Natural Gas – Hydrogen Mixtures, *International Journal of Hydrogen Energy*, 2010;35(7): 2909-14.

[40] Blarigan P.V., Keller J.O., A hydrogen fuelled internal combustion engine designed for single speed/power operation. International Journal of Hydrogen Energy 2002; 23(7):603–9.

[41] Bauer C.G., Forest T.W., Effect of hydrogen addition on the performance of methane-fueled vehicles. Part I: effect on S.I. engine performance. International Journal of Hydrogen Energy 2001;26(1):55–70.
[42] Wong Y.K., Karim G.A., An analytical examination of the effects of hydrogen addition on cyclic variations in homogeneously charged compression-ignition engines. International Journal of Hydrogen Energy 2000;25(12):1217–24.
[43] Karim G.A., Wierzba I., Al-Alousi Y., Methane–hydrogen mixtures as fuels. International Journal of Hydrogen Energy 1996;21(7):625–31.
[44] Ilbas M., Crayford A.P., Yılmaz I., Bowen P.J., Syred N., Laminar-burning velocities of hydrogen–air and hydrogen–methane–air mixtures: An experimental study, Int J Hydrogen Energy 31;2006:1768 – 1779.
[45] Shudo, T., Shimamura, K., Nakajima, Y., Combustion and emissions in a methane DI stratified charge engine with hydrogen pre-mixing, JSAE Review, 2000; 21:3-7(5).
[46] Nagalingam B., Duebel F. and Schmillen K., Performance study using natural gas, hydrogen-supplemented natural gas and hydrogen in AVL research engine, Int. J. Hydrogen Energy, 1983; 8(9): 715-20.
[47] Wallace J.S., Cattelan A.I., Hythane and CNG fuelled engine exhaust emission comparison, Proceedings 10th World Hydrogen Energy Conference, Cocoa Beach, USA, June 20–24;1994:1761–1770.
[48] Raman V., Hansel J., Fulton J., Lynch F., Bruderly D., Hythane-an ultraclean transportation fuel. Procs. of 10th World Hydrogen Conference, Cocoa Beach, Florida, USA, 1994;3: 1797–806.
[49] Larsen J.F., Wallace J.S., Comparison of emissions and efficiency of a turbocharged lean-burn natural gas and hythane-fueled engine. J Eng for Gas Turbines Power 1997;119:218-26.
[50] Collier K., Hoekstra R.L., Mulligan N., Jones C., Hahn D., Untreated exhaust emissions of a hydrogen-enriched CNG production engine conversion., SAE Paper No. 960858; 1996.
[51] Akansu S.O., Kahraman N., Çeper B., Experimental study on a spark ignition engine fuelled methane-hydrogen mixtures, Int. J. Hydrogen Energy, 2007; 32(17): 4279-84.
[52] Hoekstra R.L., Collier K., Mulligan N., Chew L., Experimental study of a clean burning vehicle fuel. Int J Hydrogen Energy 1995;20:737-45.
[53] Wang J., Huang Z., Fang Y., Liu B., Zeng K., Miao H., et al., Combustion behaviors of a direct-injection engine operating on various fractions of natural gas hydrogen blends, Int J Hydrogen Energy 2007; 32(15):3555–64.
[54] Ceper B.A., Usability of hydrogen–natural gas mixtures in internal combustion engines, Erciyes Unversity, Institute of Natural Sciences; Phd thesis, 2009. Turkey.
[55] Kahraman N., Ceper B., Akansu S.O., Aydin K., Investigation of combustion characteristics and emissions in a spark-ignition engine fuelled with natural gas–hydrogen blends. Int J Hydrogen Energy 2009;34: 1026-34.
[56] Ceper B.A., Akansu S.O., Kahraman N., Investigation of cylinder pressure for H2/CH4 mixtures at different loads. Int J Hydrogen Energy 2009;34: 4855-61.

[57] Sierens R., Rosseel E., Variable composition hydrogen/natural gas mixtures for increased engine efficiency and decreased emissions. J Eng for Gas Turbines Power 2000;122:135-40.

[58] Huang Z., Wang J., Liu B., Zeng K., Yu K., Jiang D., Combustion characteristics of a direct-injection engine fueled with natural gas-hydrogen blends under different ignition timings. Fuel 2007;86:381-7.

[59] Dimopoulos P., Bach C., Soltic P., Boulouchos K., Hydrogen–natural gas blends fuelling passenger car engines: Combustion, emissions and well-to-wheels assessment, Int J Hydrogen Energy 2008;33: 7224–36.

[60] Ortenzi F., Chiesa M., Scarcelli R., Pede G., Experimental tests of blends of hydrogen and natural gas in light-duty vehicles, Int J Hydrogen Energy 2008;33: 3225–29.

[61] Dulger Z., Numerical modeling of heat release and flame propagation for methane fueled internal combustion engines with hydrogen addition. PhD thesis, University of Miami, 1991.

[62] Swain M.R., Yusuf M.J., Dulger Z., Swain M.N., The effects of hydrogen addition on natural gas engine operation. SAE paper 932775, 1993.

[63] Yusuf M.J., Lean burn natural gas fueled engines: engine modification versus hydrogen blending. PhD thesis, University of Miami, 1993.

[64] Ma F., Wang Y., Study on the extension of lean operation limit through hydrogen enrichment in a natural gas spark-ignition engine, Int J Hydrogen Energy 2008;33: 1416–24.

[65] Genovese A., Contrisciani N., Ortenzi F., Cazzola V., On road experimental tests of hydrogen/natural gas blends on transit buses. Int J Hydrogen Energy 2011;36:1775e83

[66] Mariani A., Morrone B., Unich A., Numerical evaluation of internal combustion spark ignition engines performance fuelled with hydrogen- Natural gas blends, Int J Hydrogen Energy 2012;37: 2644-54.

[67] Mariani A., Morrone B., Unich A., Numerical modelling of internal combustion engines fuelled by hydrogen-natural gas blends. ASME Int Mech Eng Congress Exposition; 2008.

[68] Morrone B., Unich A., Numerical investigation on the effects of natural gas and hydrogen blends on engine combustion. Int J Hydrogen Energy 2009;33:4626-34.

[69] Tinaut F.V., Melgar A., Gime´nez B., Reyes M., Prediction of performance and emissions of an engine fuelled with natural gas/hydrogen blends. Int J Hydrogen Energy 2011;36:947-56.

[70] Park J., Cha H., Song S., Min Chun K., A numerical study of a methane-fueled gas engine generator with addition of hydrogen using cycle simulation and DOE method. Int J Hydrogen Energy 2011;36:5153-62.

[71] Nanthagopal, K., Subbarao R., Elango T., Baskar P., Annamalaı K., Hydrogen Enriched Compressed Natural Gas-A Futuristic Fuel for Internal Combustion Engines, Thermal Science, 2011; 15(4): 1145-54.

[72] Navarro E., Leo T.J., Corral R., CO_2 emissions from a spark ignition engine operating on natural gas–hydrogen blends (HCNG), Applied Energy (2012), http://dx.doi.org/10.1016/j.apenergy.2012.02.046

[73] Wurster R., H2-based road transport in comparison. In: Proceedings of DIME international conference "Innovation, Sustainability and Policy". Bordeaux, France; 2008. http://www.lbst.de/ressources/docs2008/DIME-Conference_H2 AutomotiveFuel_LBST_11SEP2008_V4.pdf

[74] Heywood J.B., Internal combustion engine fundamentals. McGraw-Hill; 1988.

[75] Akansu S.O., Dulger Z., Kahraman N., Veziroglu T.N., Internal combustion engines fueled by natural gas–hydrogen mixtures. International Journal of Hydrogen Energy 2004;29:1527–9.

[76] Zheng J., Hu E., Huang Z., Ning D., Wang J., Combustion and emission characteristics of a spray guided direct-injection spark-ignition engine fueled with natural gas-hydrogen blends, Int J Hydrogen Energy 2011;36: 11155-63.

[77] Mohammed S.E.L., Baharom M.B., Aziz A.R.A., Analysis of engine characteristics and emissions fueled by in-situ mixing of small amount of hydrogen in CNG, Int J Hydrogen Energy 2011;36: 4029–37.

[78] Ma F., Wang M., Jiang L., Deng J., Chen R., et al., Performance and emission characteristics of a turbocharged spark-ignition hydrogen-enriched compressed natural gas engine under wide open throttle operating conditions, Int J Hydrogen Energy 2010;35: 12502-509.

[79] Gao Z., Wu X., Gao H., Liu B., Wang J., et al., Investigation on characteristics of ionization current in a spark-ignition engine fueled with natural gas-hydrogen blends with BSS de-noising method, Int J Hydrogen Energy 2010;35: 12918-29.

[80] Ma F., Naeve N., Wang M., Jiang L., Chen R., Zhao S., Hydrogen-enriched compressed natural gas as a fuel for engines,http://cdn.intechopen.com/pdfs/11490/InTech-Hydrogen_enriched_compressed_natural_gas_as_a_fuel_for_engines.pdf

Section 3

Hydrogen Storage

Chapter 9

Mg-Based Thin Films as Model Systems in the Search for Optimal Hydrogen Storage Materials

Małgorzata Norek

Additional information is available at the end of the chapter

1. Introduction

There is a growing necessity of reducing greenhouse gases emissions to the atmosphere. A renewable energy system based on the use of electricity and hydrogen as energy carriers does not result in harmful pollutants being released to the natural environment. Hydrogen and fuels cells satisfy all requirements for an environmentally friendly vehicle. However, on-board hydrogen storage remains a key problem. It still remains to understand better the mechanisms involved in the interaction of hydrogen with matter. The search of an optimal storage material is a complicated process, were an interplay of many variables need to be considered.

In recent years, magnesium hydride (MgH_2) has attracted considerable attention as hydrogen storage material because of its large gravimetric density of 7.7 wt%, reversible hydrogen storage and low cost. The research has been focused on improving poor hydrogen sorption kinetics and lowering high dissociation temperature (~ 673 K). Strategies to reduce the stability of MgH_2 include alloying with various elements and nanostructuring. It was shown that mechanical ball milling can reduce particle size up to ~10 nm, introduce defects, large surface areas, and grain boundaries and in turn decrease the strength of the metal-hydrogen bond. Milling with different catalysts has improved noticeably hydrogen sorption kinetics.

Although the ultimate goal is the production of large amounts of hydrogen storing materials to be used in the transport sector, thin film processing is an alternative method that provides the opportunity to synthesize nanostructured materials in specific compositions, well-defined microstructures and dimensions. This, in turn, offers the possibility of more rational approach to the problem which might contribute to progress in hydrogen technology. Upon the size reduction to the nanoscale, the analysis of the finite-size effects, such as: surface or interface contributions to the hydrogenation thermodynamics and quantum-size effects, is possible. Moreover, cooperative phenomena (elastic interaction within the interfacial region) can be introduced through the synthesis of multilayer films and kinetic limitations can be minimized, leading to novel materials with unique properties. Thin films do not suffer as much from embrittlement and/or decrepitation as bulk materials, allowing to study cyclic absorption and desorption. Buckling of the film due to hydrogen loading occurs when the elastic energy stored in the film exceeds the adhesion energy. In case of weak adhesive forces between the film and the substrate the mechanical stress generated during the hydride formation relaxes by a buckle-and-cracks network formation. If the stress is high, but the adhesion still strong enough to hinder detachment of the film, the film will remain intact but the stress relaxes via plastic deformations (e. g. dislocations etc.). As a result of the dominant elastic out of plane expansion of a strongly clamped film, the morphology of the film does not alter. Thanks to a well-defined structure, composition and dimension of thin films, the thin film approach allows to study: the reaction pathways, the role of catalysts, the phase segregation and diffusion phenomena occurring during hydrogen absorption and release, which are of great importance in the rational design of the hydrogen storage materials for practical applications.

The chapter will briefly review the techniques and methods used to investigate the interaction of hydrogen with matter. Next, the hydrogen properties of Pd-capped Mg thin films will be presented. The most important achievements in the research of the Mg-based alloy thin films will be described in detail. Alloying Mg with small quantity of 3d transition metals (Sc, Ti, V, Cr, Ni, Fe, Cu, Co) can tailor the hydrogen reaction enthalpy and consequently reduce the operating temperatures during hydrogen absorption and desorption. High-throughput screening methods have allowed to select many promising Mg-based alloys with specific chemical compositions which are characterized by very good kinetics and high hydrogen capacity.

2. Methods and measurements

The techniques applied to research the hydrogen interaction with matter in bulk materials are usually not applicable in thin films. Classical methods such as volumetric and gravimetric techniques, in which the hydrogen is measured via the pressure drop or by the increased weight of the sample, are difficult to apply to thin films due to the small quantities of hydrogen involved. Hydrogen concentration within the thin films can be measured by: electrochemical loading, nuclear reaction analysis (NRA), elastic recoil detection analysis (ERDA), or neutron reflectivity (NR) [1].

In the electrochemical loading the metal film is used as a negative electrode. The electrochemical reaction corresponds to the reduction of water and involves one electron per absorbed hydrogen atom. As a result of the reaction hydrogenation of the thin film occurs. For a material able to absorb a quantity of hydrogen C_{sg} by solid–gas reaction, an equivalent electrochemical capacity, C_{el}, can be calculated according to the Equation:

$$C_{el}(mAh/g) = \frac{C_{sg}F}{3.6M} \quad (1)$$

where F is the Faraday constant, C_{sg} is expressed in H atoms per formula unit (H/f.u.), and M is the molecular weight of the alloy in g/f.u. [2].

The ion beam analytical methods, such as NRA or ERDA, are nondestructive, straightforward and completely quantitative. The high accuracy of the techniques is mainly due to the precision with which cross-sections of the involved atomic and nuclear processes are known. During the bombardment the interaction of the particle beam with a material (elastic and inelastic scattering, nuclear reactions and electromagnetic excitation) takes place. The material composition can be deduced from the number of observed events per incident beam particle.

In NRA, the ion beams of MeV (up to ~50 MeV) is applied for materials analysis. When the light projectiles impinge on light to medium heavy atoms, nuclear reactions in the target nuclei can occur. The yield of the reaction products (γ, p, n, d, ^{3}He, ^{4}He, etc.) is proportional to the concentration of the specific elements in the sample. Absolute concentrations can be calculated easily with the help of simple standards (e.g. bulk material or compounds of the analyzed elements). NRA is most often applied for the analysis of H, Li, Be, B, C, N, O, F, Na, Al, P with detection limits range from 10^{-3} to 10^{-7}. Reactions with narrow resonances (100 eV to 1 keV) can be found for many of the aforementioned elements. By stepping up the accelerator energy and thus shifting the depth (d) within the target at which the reaction takes place, the depth profiling with a resolution of the order of 10 nm is measured (Figure 1).

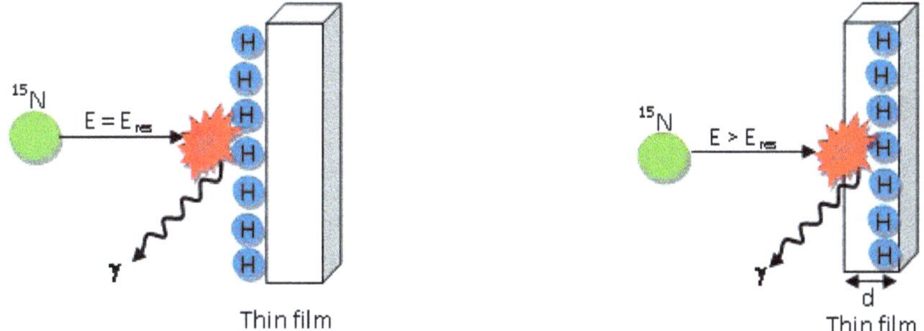

Figure 1. A scheme of hydrogen depth profiling with the ^{15}N ions.

Within the film the ^{15}N ions are slow down to the resonant energy E = 6.4 MeV and react with hydrogen according to the formula: ^{1}H(^{15}N,^{12}C)$\alpha\gamma$. The detected number of γ quants, of a characteristic energy of 4.965 MeV, is proportional to the hydrogen concentration at a certain depth.

ERDA is one of the few fully quantitative hydrogen profiling methods. Elements lighter than the incident beam particles (2 MeV ^{4}He beam for hydrogen detection) are analyzed (e.g.

hydrogen profiling) by detection of the recoiling target atoms in a grazing angle geometry (Figure 2). Absorber foils or mass discriminating detectors are used in order to discriminate between forward scattered projectiles and different types of recoiling particles. The depth profiles of all light target elements can be obtained simultaneously well separated from each other.

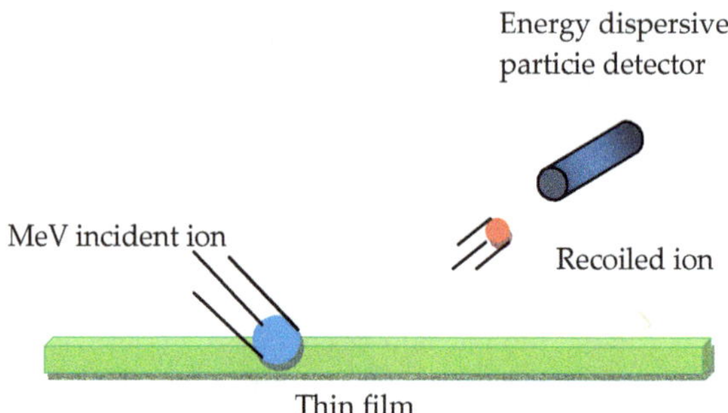

Figure 2. A scheme for ERDA experiment

Neutron reflectivity (NR) method is based on the fact the neutrons obey the same laws as electromagnetic waves and as such display reaction and refraction on passing from one medium to another. The sample is exposed to a beam of neutrons and the reflectivity is measured as a function of the momentum transfer q. A specular reflectivity experiment measures the scattered intensity as a function of $q_z = 2k_z$ (perpendicular to the interface). As such, the reflectometry experiment provides information about structure perpendicular to the interface. For specular reflection (the reflection which is defined as reflection in which the angle of reflection equals the angle of incidence, Figure 3) the critical angle is given by Equation 2:

$$\theta_c = \lambda\sqrt{\frac{\delta}{\pi}} \tag{2}$$

where λ is the neutron wavelength 4.75 Å, δ is the scattering length density, which is defined by the Equation 3:

$$\delta = \frac{\Sigma_i^n b_i}{V} \tag{3}$$

where b_i is the scattering length of the relevant atom and V is the volume containing the n atoms [3]. The hydrogen concentration can be, thus, calculated directly from a measure of the critical angle [4].

There are many indirect methods exploiting hydrogen induced changes of other physical properties of the film. The hydrogen uptake has often been monitored by measuring the sample resistance. To study the hydrogen loading and the thermodynamic properties four point probe measurement is commonly used [5]. Under applied hydrogen pressure at constant temperature, the resistance of metal thin film increases due to a decrease in electron mobility with H acting as scattering centers. Also the changes of the elastic, the magnetic and the optical propertied could be used to follow the hydrogen uptake. The conversion to hydrogen concentrations must then be carried out by another independent calibration measurement.

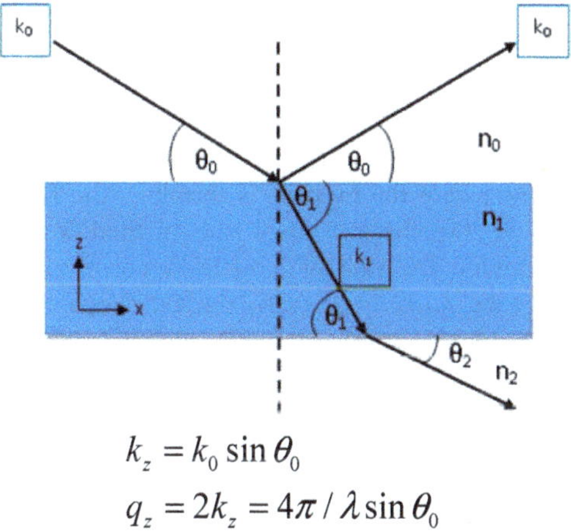

Figure 3. A thin film of refractive index n_1 between two bulk media of refractive indices n_0, n_2.

Thin films open new fields of research such as screening for new composite materials. Matrix samples or samples with defined chemical composition allow the simultaneous investigation of hydrogen sorption by means of combinatorial approaches. Two of them rely on changes in the optical response, as measured by the emissivity in the IR, or the optical transmission in the visible region, during the change from the metal to hydride, or hydride to metal, phase. The third one is carried out by X-ray microdiffraction, a sequential characterization which aimed at identification of structural phases rather the decomposition temperatures and storage capacities.

IR screening technique is based on the change of the apparent temperature due to the surface emissivity variations. According to Stefan-Boltzmann law (Eq. 4) the radiation power of the material is proportional to its emissivity (ε):

$$W = \sigma\varepsilon T^4 \tag{4}$$

Emissivity, in turn, depends on the resistivity of the material (ρ), as defined by the Hagen-Rubens low (Eq. 5):

$$\varepsilon = 2(2\varepsilon_0\omega\rho)^{1/2} \tag{5}$$

Since metal hydrides are characterized by higher resistivity than their corresponding pure metals, it is possible to measure hydrogenation/dehydrogenation process by monitoring the emissivity of a given material. During the measurement a large part of the radiation emitted by the material is partially reabsorbed by the atmosphere either inside or outside the chamber where the experiment is being performed and is reflected by the IR camera lenses. Thus, the apparatus measures in fact the effective radiation (*W'*) which is automatically converted into an apparent temperature (T_a) [6]:

$$T_a = \left(\frac{W'}{\sigma}\right)^{1/4} = T(\varepsilon)^{1/4} \tag{6}$$

The metal thin film blocks IR radiation and appears black in the IR camera. Upon hydrogenation the material becomes optically transparent and the IR camera will see the bright region at the point where the hydride is forming. The IR screening method was introduced by Olk [7, 8]. Oguchi et.al used the method, combined with structural characterization, to determine the rate of a hydride phase growth [9]. Although the hydrogen content cannot be quantify directly, the IR screening is a valuable tool for studying the kinetics of the hydrogen sorption or fast identification of catalyst.

The method based on the change of optical transmission upon hydrogen loading is called "Hydrogenography" and was developed by Gremaud and coworkers [10]. The amount of light transmitted through a thin film is measured by means of a 3CCD camera as a function of hydrogen pressure at constant temperature. Moreover, the optical transmission can be related to the hydrogen concentration via Lambert-Beer's law [11]. Therefore, the hydrogenography allows for a detailed quantitative study of both kinetics and the thermodynamics of hydrogenation/dehydrogenation process. The concentration depends linearly with the logarithm of the optical transmission T [12]:

$$\ln\left(\frac{T}{T_0}\right) \propto c \tag{7}$$

Pressure vs. $\ln(T/T_0)$ isotherms (PTI's which are equivalent to the standard PCI's) of thousands of different samples can be measured simultaneously. Based on these measurements performed at various temperatures Van't Hoff plots can be built and the thermodynamic parameters, such as enthalpy of the hydrogen-induced phase transitions, can be determined (Figure 4). An advantage of the method is that it can works equally for both materials which perform the metal-to-insulator transition upon hydrogen uptake and for materials which remain metallic after the process [13]. The stresses associated with the clamping of a thin film to a substrate must be taken, however, into account when transferring the experimental data to bulk samples.

The other combinatorial methods which make use of the stresses was developed by Ludwig and coworkers [14]. Thin films are deposited onto the array of 24 Si cantilevers and are exposed to H_2 gas. As a result of the in-plane stresses and out-of-plain strains induced by the volume expansion occurring during the H_2 loading, the bending of the cantilevers takes place. The cantilever bending is monitored by an optical approach: a laser diode beam is split into parallel lines by the optics and the lines are projected on the cantilevers. The individual reflections from each cantilever are projected on a semitransparent screen and are recorded by CCD camera (Figure 5).

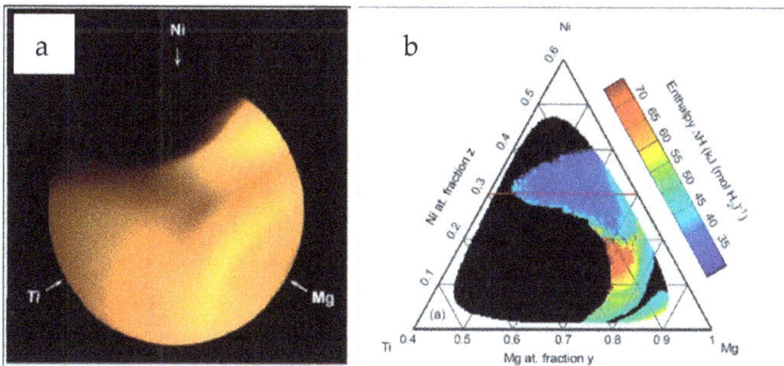

Figure 4. a) Optical transmission image of a $Mg_yNi_zTi_{1-y-z}$ gradient film during loading at 333K and b) Hydride formation enthalpies of the Mg-Ni-Ti-H system, determined from the Van't Hoff plots for each pixel of the $Mg_yNi_zTi_{1-y-z}$ gradient film (from Ref. [10]).

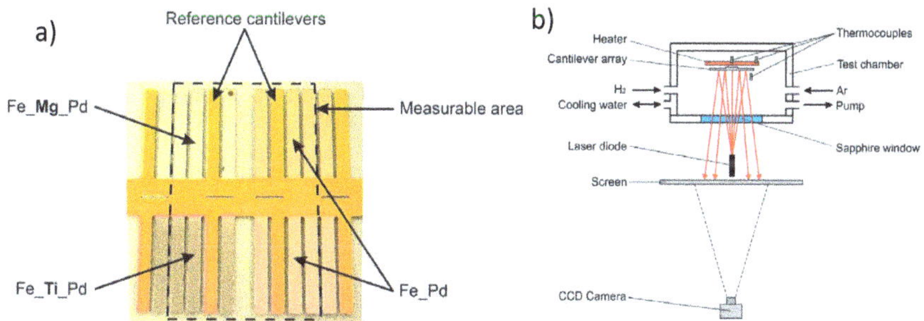

Figure 5. a) A cantilever material library: 24 cantilevers are arranged in 4 quadrates, 16 cantilevers are simultaneously observable; uncoated cantilevers serve as reference, b) schematic of the gas phase apparatus (from Ref. [14]).

The method was successfully applied to Mg-Ni system: the hydrogen induced stress was correlated with the composition and microstructure of the films [15]. Similar technique based on measurement of the surface curvature of a dense array of 2500 MEMS-fabricated cantilevers was developed by Woo et al. [16].

A different screening method was presented by Guerin and coworkers [17]. The technique uses a silicon microfabricated MEMS array with 49 heaters, independently controllable, which allow both temperature programmed desorption and infrared thermography measurements. The integrals of the TPD (Temperature Programmed Desorption) provide a direct measure of the hydrogen storage capacity as a function of composition. Due to the application of high heating rates to detect hydrogen with sufficient sensitivity, the measurements are performed far away from the equilibrium conditions. Therefore, the method is complementary to the techniques which allow for the determination of thermodynamic parameters.

3. Pd-capped Mg thin films

Hydride uptake by metal is governed by a thermodynamic process as described by pressure - concentration isotherm (Figure 6) [18]. At the beginning of the process the host metal dissolves some portion of hydrogen to form solid state solution (α-phase). Upon the H concentration increase in the metal, the H-H interaction become dominant and the nucleation of the hydride (β-phase) occurs. The region where the two phases coexist (the plateau pressure in the isotherm) defines the amount of hydrogen that can be stored reversibly at a given pressure and temperature. The H content decreases with increasing H_2 pressure and temperature [19]. The equilibrium pressure, p_{eq}, (the plateau pressure) depends strongly on the temperature. The relation is expressed by the Van't Hoff equation (Eq. 8):

$$\ln\left(\frac{p_{eq}}{p_{eq}^{0}}\right)=\left(-\frac{\Delta H}{R}\right)\left(\frac{1}{T}\right)+\frac{\Delta S}{R} \tag{8}$$

where R is the gas constant, the ΔH and the ΔS are the changes of enthalpy and entropy, respectively, and $p_{eq}{}^{0}$ = 1.013 × 10^5 Pa is the standard pressure. Thus, the slope of the van't Hoff curve determines the enthalpy, whereas the intercept gives the values of the entropy of the process.

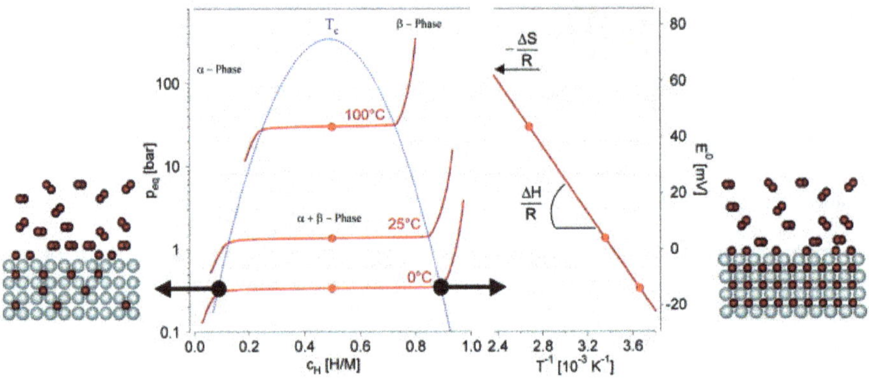

Figure 6. Pressure-concentration-temperature plot and a van't Hoff curve (logarithm of the equilibrium or plateau pressure vs. the reciprocal temperature) for an intermetallic compound (from Ref. [18]).

MgH_2 decomposes into Mg bulk and H_2 gas, the reaction being endothermic (ΔH>0), while the reaction of the formation of MgH_2 from Mg bulk and H_2 gas is exothermic (ΔH<0). The value of the enthalpy for MgH_2 formation/decomposition is ~ 75 kJ/mol, which means that at ambient pressure (~1 bar) the operating temperature is of ca. 600K [20, 21]. This is much too high for the practical application. It is not easy to lower the enthalpy of hydrogenation (or dehydrogenation) process since it involves the binding energy between the hydrogen atom and a metal. Recent DFT and HF calculations made for the Mg/MgH_2 system clearly show that considerable changes in the thermochemistry of hydrogen absorption/desorption can be achieved only if the particle size is reduced below 2 nm. A steep decrease in the enthalpy of MgH_2 formation was found within the ultrasmall MgH_2 particle region (0.62-0.92 nm) [22]. The desorption enthalpies were calculated to be -20.64, 34.54, and 61.86 kJ/mol H_2 for the MgH_2 nanowires with diameter of: 0.68 nm, 0.85 nm, and 1.24 nm, which were derived from the Mg nanowires with diameter of 0.32 nm, 0.71 nm, and 1.04 nm, respectively (Figure 7) [23]. The phenomena was explained by the fact that in the case of such an extreme downsizing the vast majority of Mg and H atoms are exposed to the surface, where hydrogen atoms occupy the less stable top and bridges sites. In larger crystals the hydrogen resides in the more stable, three-coordinate sites. Consequently, surface Mg atoms are left uncoordinated and the hydrogen atoms are more easily absorbed/desorbed from these energetically less favored sites. The same trend was observed for the Mg/MgH_2 thin films: the enthalpy reduction of 5kJ/mol as compared to the bulk value was calculated for the film thickness of 2 unit cell (0.16 nm Mg/ 0.30 nm MgH_2) [24]. It is not surprising, thus, that in thin films the ΔH is still very close to the value for bulk materials. Giving an example, the ΔH for hydrogen uptake by Pd capped 1 μm thick Mg thin films, as determined by resistance measurements, was similar to the value for bulk Mg [25]. The desorption enthalpy of ~ 67 kJ/mol was estimated for 600 nm -thick iron-doped Mg thin film [26]. Almost the same value of 68 kJ/mol was determined for MgH_2 formation in sandwiched Pd/Mg/Pd thin film delaminated from the substrate (to get rid of any effect related to the film clamping) [27]. The small change of ΔH in the latter two cases was probably due to the additional lattice defects introduced by doping with Fe and Pd metals.

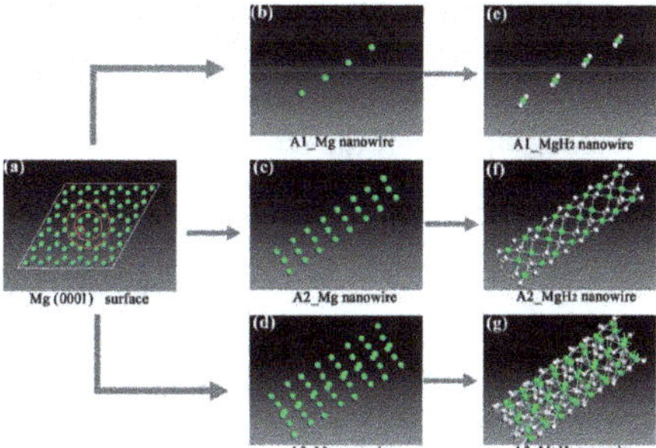

Figure 7. (a) Mg(0001) surface. Magnesium nanowires of infinite length along [0001] and the diameter size increase in the order: (b) 0.32 nm Mg nanowire; (c) 0.71 nm Mg nanowire and (d) 1.04 nm Mg nanowire. Optimized structure for (e) 0.68 nm MgH_2 nanowire; (f) 0.85 nm MgH_2 nanowire and (g) 1.24 nm MgH_2 nanowire (from Ref. [23]).

On the other hand, Baldi and coworkers showed, using hydrogenography, that it is possible to tune the thermodynamics of hydrogen absorption in Mg by means of elastic clamping [28]. They prepared the block layer 2x[Ti(10nm)/Mg(20nm)]/Pd(10nm) and proved that Pd is able to increase the hydrogen plateau pressure more than 200 times with respect to bulk Mg via the elastic constrain exerted on the expanding Mg layer. Furthermore, it was demonstrated the Mg-alloy-forming elements, such as Pd and Ni, increase substantially the hydrogen absorption plateau pressure, whereas the elements such as Ti, Nb and V, which are immiscible with Mg, are elastically disconnected from Mg and have little effect on thermodynamics properties of Mg (Figure 8). The model, based on the elastic interaction, predicted also the increase of an equilibrium hydrogen pressure with increasing Mg thickness.

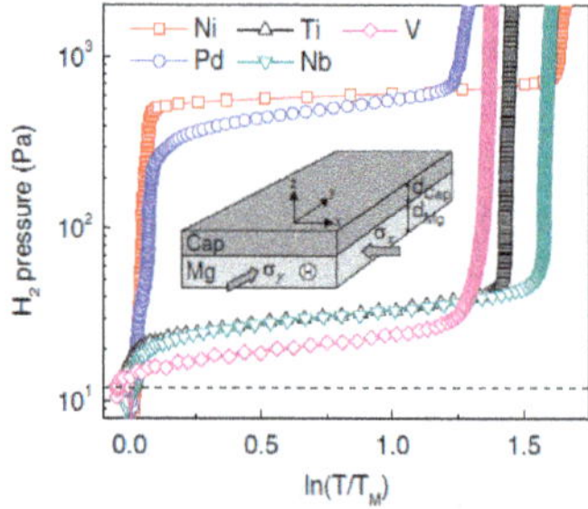

Figure 8. Effect of cap layer: PTI's measured at 333 K for Ti(10 nm)Mg(20 nm)X(10 nm)Pd(10 nm) samples deposited on glass with X = Ni, Pd, Ti, Nb, and V. The dashed line is the pressure at which coexistence of α and β phases begins to appear upon hydrogen absorption in bulk Mg. σ_x and σ_y are compressive stresses in the x and y direction due to the cap layer (from Ref. [28]).

The change in the equilibrium pressure plateaus may be also caused by clamping to the substrate which creates severe in-plane stresses in thin films during hydrogen uptake. The large impact on the thermodynamics was observed for rigid substrates which cause resistance toward volume expansion [29, 30]. The increase of p_{eq} due to the stresses caused by capping layer or a substrate would be very desirable with respect to practical application if these stresses would not relax after few hydrogen absorption/desorption cycles drawing back the p_{eq} to its original value. Moreover, the stresses relax by the formation of buckle-to-crack network, changing the structure of the thin films [31, 32]. This feature impedes the systematic studies of the interaction of the hydrogen with a metal. Recently, it was demonstrated that thin films built up on the porous, regular hexagonal AAO template, contains enough free space to allow for metals lattice unhindered expansion and, therefore, a reduction of the mechanical stress in the layer [33]. After hydrogenation, some prominent differences between two films deposited on the non-porous and porous substrate are observed (Figure 9). While in the layer deposited on the rigid glass many cracks and bulges can be observed after hydrogen absorption, the same layer deposited on the porous AAO template remains completely smooth without any cracks (compare Figure 9 (A) with (E) and (B) with (F)). In the thin film deposited on the glass locally some protrusions have appeared as an effect of swelling of the layer (Figure 9 (C)). Based on these observations it can be concluded that a porous substrate may offer an effective way to release the stress without layer deterioration.

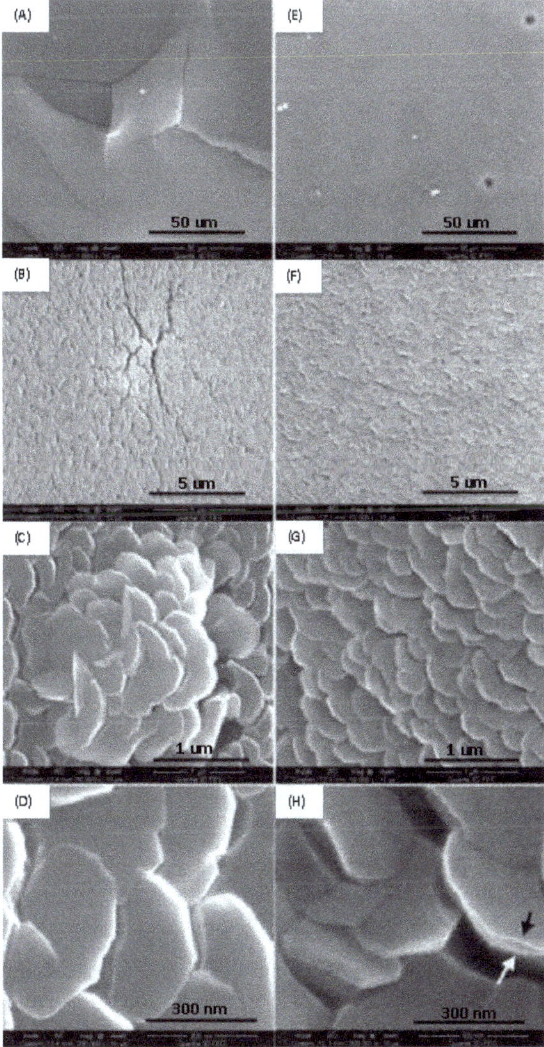

Figure 9. SEM images of Pd/Mg/Pd films on (A, B, C, D) non-porous glass substrate and (E, F, G, H) porous AAO template after hydrogenation at room temperature. Mg develops a plate-like structures of a hexagonal shape, suggesting a preferential growth along c-axis. The arrows in Figure 9(H) demonstrate that each platelet, in fact, consists of two identical hexagonal crystals stacked perfectly along their basal planes (from Ref. [33])

Another serious problem related with the hydrogenation of pure Mg is poor kinetics of process and an easiness of Mg to oxidation. The kinetics of a metal to hydride transformation is typically described by an Arrhenius type expression (Eq. 9):

$$k(T) = A\exp\left(-\frac{E_a}{k_B T}\right) \tag{9}$$

where k is the rate of the hydrogen absorption or desorption process, k_B is the Boltzmann constant, T is temperature, A the apparent pre-exponential factor and E_a – the apparent activation energy. The plot ln(k) vs. $1/k_BT$ gives, thus, the E_a value. The lower the activation energy, the faster the process. The E_a is 115-122 kJ/mol for hydrogen absorption and 126-160 kJ/mol for hydrogen desorption, as calculated for pure Mg [34]. The relation between E_a and ΔH is demonstrated graphically in Figure 10.

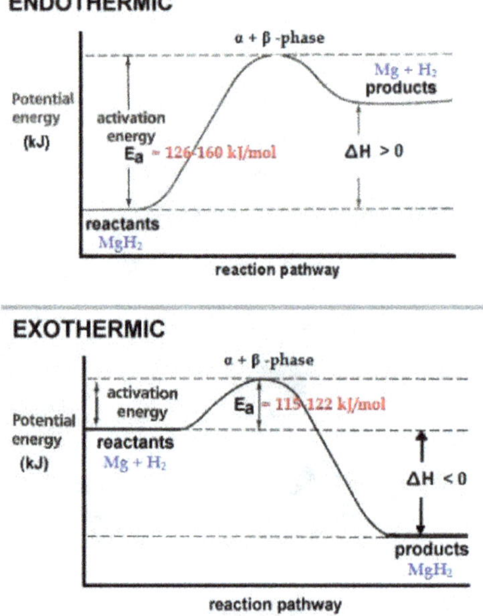

Figure 10. The relation between E_a and ΔH for Mg/MgH_2.

Reducing distances in nanoscale materials translate into faster overall reaction kinetics. Pd coating is usually applied to facilitate hydrogenation and to avoid oxidation of Mg. Hydrogen uptake for Pd-capped Mg thin films can occur at room temperature thanks to the high H_2 dissociation rate and high hydrogen diffusivity in the Pd outermost layer [35]. MgH_2 formation in thin films occurs mostly at the Pd/Mg interface where the Pd-Mg interaction is maximized. The thicker the Pd-Mg intermixing region the higher the number of Pd/Mg interfaces and the higher the hydrogen uptake can be expected [36-38]. The lattice mismatch at the interface creates defect sites with a low local electron density which weakens the hydrogen bonding and therefore, within this interface hydrogenation of Mg occurs very fast. It was shown that the reaction at the interface control the growth of MgH_2 for the hydride layer thickness smaller than 60 nm [39]. Above the thickness hydrogenation is limited by hydrogen diffusion in the hydride layer. The overall diffusion coefficient for the Mg-to-MgH_2 transition, including nucleation and growth, was determined to be as low as $D = 1.1 \times 10^{-20}$ m^2/s [40]. Due to the extremely slow diffusion, the hydride nucleation and growth above a certain film's thickness will be practically stopped. In other words, there is a limited amount of hydrogen that can be loaded into the Mg layer above a certain thickness. It was demonstrated that the thinner Mg layer the better the hydrogen uptake rate and the higher hydrogen content at the beginning of the process. The Mg layer of 20 nm thickness has shown the best absorption kinetics with saturated hydrogen content of 5.5 wt% at 298 K and 0.7 bar H_2 [41]. The E_a for desorption process for Pd capped 100 nm thick Mg layer was determined to be 80 kJ/mol [42]. Beneficial effect of Pd was studied by hydrogenography: the rate of hydrogen absorption by Mg layer increased with Pd doping [43]. The Pd-doped Mg can absorb hydrogen at room temperature and under less than 1 bar pressure in few minutes. On the other hand, no significant influence of Pd was observed on hydrogen desorption at room temperature in air. The hydrogen release from the layer was no completed after 5h.

The enhancement of the H_2 absorption process was achieved by an electric current. The 100 nm thick Mg/Pd film, exposed to 1 bar of H_2 pressure and simultaneously to a voltage of 20-30 V, was easily hydrogenated without an external heat source [44]. Better hydriding kinetics was also observed for the films heated up to 473 K prior to hydrogenation. The annealing optimized the morphology and structure of the film in terms of finer particles size and better crystallinity [45]. No influence of the crystallization degree on absorption kinetics was, however, found by Higuchi and coworkers [46]. Yet, the crystallinity affected the hydrogen desorption kinetics which was manifested by lower desorption temperature. The lower the degree of Mg crystallization, as estimated from the intensity of Mg(002) peak in X-ray diffraction spectra, the lower the temperature at which the hydrogen was releasing. The most amorphous Mg in Pd/Mg films desorbed H_2 at a temperature lower than 463 K in vacuum of 7.0×10^{-1} Pa.

The annealing at elevated temperature can cause the alloying of Mg and Pd elements. The Mg/Pd multilayered films activated at 474 K for 2 h under 30 bar of H_2 pressure transformed completely into the Mg_6Pd phase [47]. After three cycles the films consisted of a mixture of MgPd, Mg_5Pd_2, and Mg_6Pd intermetallics phases. Qu et al. annealed the Pd/Mg film at various temperature up to 473 K in vacuum for 2 h and did not find the presence of the intermetallics [34]. It seems that in the metals interdiffusion and alloying occur dependent on whether the annealing is performed under H_2 pressure or in vacuum. To avoid the formation of the Pd-Mg intermetallics phases the tantalum (Ta) layer can be used [48]. The Ta is well-known as an excellent diffusion barrier for metals [49]. Moreover, it possesses a very high hydrogen permeability ($1{,}3 \times 10^{-7}$ mol/ms $Pa^{1/2}$ at 500 °C) [50].

In order to increase the Pd-Mg intermixing region, the sandwiched Pd/Mg/Pd thin films were prepared. Such systems demonstrate optimal hydrogen sorption properties not only thanks to the extended Pd/Mg interactions but also due to the cooperative phenomena. The cooperative phenomena is an elastic interaction between the two metals [51]. Upon the hydriding/dehydriding cycles both metals experience the lattice expansion/contraction. During desorption the hydrogen is first released from the Pd layers. The stress induced on the top and down surface of the Mg film, force desorption of hydrogen from Mg (Figure 11).

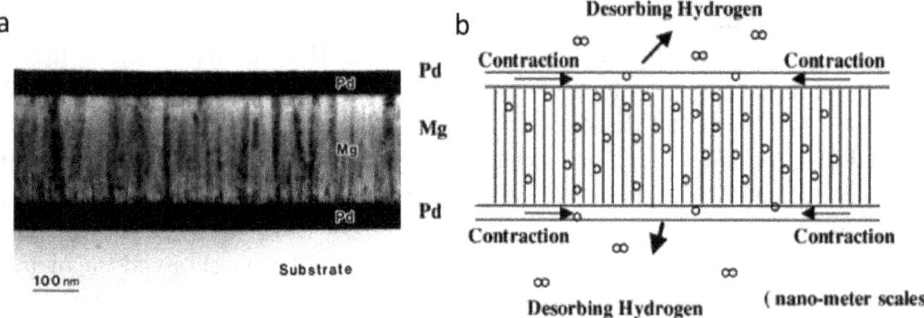

Figure 11. a) TEM micrograph for the cross section of Pd/Mg/Pd film before hydrogenation; b) scheme of hydrogenated Pd/Mg/Pd film demonstrating the cooperative effect (from Ref. [51]).

The sandwiched Pd/Mg (100 nm)/Pd thin films were fully hydrogenated at room temperature for 4h under 0.04 bar and could dehydrogenate completely and rapidly in air [52]. Hydrogen desorption process took only 20 min at 338 K, while at room temperature it lasted approximately 300 min. The overall activation energy for desorption process was estimated to be 48 kJ/mol, thus, significantly smaller than the value for pure Mg. Similar Pd/Mg/Pd films, but hydrogenated at 353K for 4h under 1 bar of H_2 pressure, exhibited the E_a of ca. 60 kJ/mol [53]. The discrepancy between the E_a values suggests that the hydrogenation conditions influence the desorption process. Faster kinetics and lower activation energy was attributed both to more extended Pd/Mg interface and to the cooperation effect.

The Mg crystallographic patterns of as-deposited Mg thin films exhibits usually strong (002) and (004) reflections implying a preferential growth along the c-axis, i.e. a single (001) out-of-plane orientation [33, 54]. MgH_2 transforms epitaxially relative to Mg according to the relation: Mg(001)//MgH_2(110) for the glass substrate and for the Al_2O_3 (001), whereas for the $LiGaO_2$ (320) substrate the relation Mg(110)//MgH_2(200) was found [55]. Due to a preferential growth along c-axis the Mg develops into a shape of hexagonal platelets (Fig. 9) forming closely packed columns (Figure 11) containing a number of polycrystalline and grain boundary defects [56, 57]. It was observed that the hydrogenation leads to a reduction of the defects. Since the defects provide the sites where hydrogen bonds are weakened and, consequently, the hydrogen migration energy is lowered leading to faster diffusion, the reduction of the defects is rather disadvantageous with respect to the hydriding/dehydriding properties. The evolution of the lattice defects in Mg during the hydriding/dehydriding cycles was also studied by the positron annihilation spectroscopy (PAS) [58]. After the very first decomposition of MgD_2 the increase of void-like defects was observed accompanied by the decrease in deuterium storage capacity. Upon subsequent absorption and desorption cycles the concentration of these defects was reduced without, however, restoring the original storage capacity. The activation energy of desorption process decreased with the film thickness which was ascribed to a higher concentration of void-like defects in thinner layers [34].

First principle calculations show that hydrogen diffusion in MgH_2 is dominated by motions of charged defects such as positively charged vacancies at H sites, and negative H interstitials. The data are supported by some experimental observations (like for instance the enhancement of H_2 absorption by an electric current [44]). The latter type of defects dominates diffusion because the activation energy for motion of this defect is small [59]. Therefore, it should be possible to improve H diffusion by doping of Mg with charged elements which are able to increase the concentration of the diffusion defects. The DFT calculations demonstrate that only a small number of dopants have these properties [60]. In particular, the diffusion in MgH_2 can be significantly enhanced only by Co, which act as a n-type dopant that increase the concentration of negatively charged H interstitials. Nevertheless, the overall kinetics of H_2 uptake or release is also affected by lattice defects accumulated around grain boundaries and surfaces or by isolated dislocations. And the dissociation or recombination process taking place around these defects may be enhanced by other dopants.

The hydrogen sorption kinetics was significantly improved by addition of Cr-Ti bimetallic compounds [61]. At 200 °C Mg-Cr-Ti nanocomposites absorb 5 wt% hydrogen in several seconds, and desorb in 10-20 min. The influence of Ti and Fe dopants on the hydride formation was studied by infrared emission imaging of wedge-shaped thin films during hydrogen loading [62]. Ti addition did not influence Mg hydride growth rate, however, it resulted in the formation of thicker hydride layer on top of the films as compared to the un-doped Mg films. The addition of atomic fraction of 3.1% of Fe increased the hydrogenation rate by an order of magnitude and twice the MgH_2 layer thickness as compared to pure Mg. A beneficial effect of Fe on Mg hydrogenation process was thoroughly studied at temperature ranging from 363 K to 423 K by Tan et al. [26] The results suggest that Mg-Fe powders of 1-2 μm size can be fully hydrogenated within 1 min at temperature of 150 °C (423K). Catalytic effect of other transition metals was also investigated. Molybdenum plays a significant role in Mg hydrogenation [63, 64]. Mo and Ti act mainly as catalysts. The catalytic affect of NbO_x as a function of the oxygen concentration was studied [65]. The lowest activation energy for hydrogen absorption was found for the highest oxygen content (Nb_2O_5). Pure Nb does not display a considerable catalytic effect [66]. Hydrogenation kinetics was also speed up by vanadium (V) implementation into magnesium films [67]. The effect was, however, small as compared to the addition of Fe or Ti. Better effect was achieved for Mg-Cu multilayered films: Mg was fully converted to MgH_2 at temperature of 473K [68]. As will be seen in the next paragraph, a significant effect on the hydrogenation process of Mg may be exerted by the 3D transition metals which may form binary hydrides with Mg, including Co, Ni or Mn [69-71].

4. Binary and ternary Mg-based alloys

The attempts to destabilize MgH_2 (lower the desorption enthalpy) and to improve its kinetics by alloying it with transitions metals has led to discovery of many binary systems, including Mg-Ni, Mg-Fe, Mg-Co or Mg-Mn. Among these systems the magnesium – nickel hydride phases seem to be very promising candidates for hydrogen storage materials owing to a relatively high hydrogen storage capacity and low cost. The stoichiometric Mg_2Ni, due to the existence of the parent metal alloy, can be easily and reversibly hydrogenated even at room temperature. Mg_2NiH_4 has a storage capacity of 3.6 wt% and the desorption enthalpy of ~ 64 kJ/mol [72]. However, because hydrogen loading capacity increases with the Mg content, other, non-stoichiometric Mg-Ni compositions are of a special interest with respect to hydrogen storage and were intensively investigated.

An interesting hydriding properties was demonstrated for amorphous $Mg_{1.2}Ni_{1.0}$ films deposited on molybdenum [73]. It undergoes reversible hydriding/dehydriding reaction at about 150 °C and under 3.3 MPa of H_2 pressure. Moreover, based on the DSC measurement, the enthalpies for hydrogen absorption and desorption process were estimated to be -39.9 and 42.0 kJ/mole, respectively, and therefore, substantially lower than for pure Mg. The TPS (Thermal Desorption Spectroscopy) measurement for an amorphous, 1.7 nm thick, Mg_5Ni_1 thin films deposited on molybdenum, shows that hydrogen desorption starts already at 77 °C and has its maximum at around 147 °C, similarly to the $Mg_{1.2}Ni_{1.0}$ films [74]. Furthermore, the PES (Photoelectron Spectroscopy) demonstrated that no Mg segregation or desorption occurs when hydrogen is desorbed from the film, rendering the system reversible. The PCI measurement of amorphous MgNi/Pd multilayer thin films deposited on Ni substrate by magnetron sputtering revealed that up to 4.6 wt% of hydrogen can be loaded to the films at room temperature [75]. Approximately the same, reversible hydrogen content of 4.45 wt% at 224 °C has been obtained for the 1160 nm thick $Mg_{2.9}Ni$ film [76]. The $Mg_{2.9}Ni$ thin film was composed of nanocrystalline Mg_2Ni and Mg phase being of about 20-50 nm in size. The lower absorption/desorption temperature, as compared to pure Mg_2Ni and Mg systems, was attributed to the extra free energy in the interfacial region between the two nanocrystalline phases and the stress exerted on the MgH_2 phase by the adjacent Mg_2Ni, which decomposes from its hydride prior to Mg. Structural and optical properties of the Mg_yNi_{1-y} system in the composition range $0.5<y<0.95$ was studied by Gremaud and coworkers [77]. For $0.6<y<0.8$ crystalline Mg_2Ni coexists with amorphous Mg and/or Ni. After hydrogenation mostly Mg_2NiH_4 phase is present and some MgH_2 on the Mg-rich side. The abrupt structural changes is observed around the Mg-Mg_2Ni eutectic point ($y = 0.886$) which are accompanied by the drastic change in optical properties in the hydride state (Figure 12). Above the eutectic point amorphous Ni is embedded in crystalline Mg and hydrogen absorption is hampered by slow diffusion in the MgH_2 phase. Consequently a large part of Mg remains metallic. This is followed by a sudden drop in transmittance for $y > 0.886$, which one should expect to be high upon hydrogenation (Figure 12). Moreover, owing to the intimately mixed microstructure around the eutectic point, hydrogenation reaction is destabilized (the reaction occurs at lower temperature) with respect to the Mg_2Ni – Mg_2NiH_4 reaction, and the hydrogen sorption mechanism is changed from being governed by hydrogen dissociation at the Pd capping layer to being limited by diffusion in $Mg_yNi_{1-y}H_x$ [78].

Optical properties of Pd-capped Mg_yNi_{1-y} thin films for the Mg-rich side of the Mg-Ni system (0.7<y<0.9) was also investigated by Yoshimura et al.[79]. The films showed much better optical transmittance upon hydrogen exposure compared to the stoichiometric Mg_2Ni compound (indicated in Figure 12 by dashed, vertical line for y = 0.667), contrary to the results obtained by Gremaud et al. which showed rather constant transmittance in the 0.7<y<0.9 region (see Figure 12 [77]). The reason for this discrepancy may lay in the difference between the films thickness or in subtle dissimilarities between their microstructure. In agreement with the results given in Figure 12, the decrease of transmittance for Mg-poor side of the Mg-Ni system was observed by Johansson and coworkers [80]. The change in the effective optical band gap from 3.6 eV for Mg-rich to 2.4 eV for Mg-poor hydrogenated Mg-Ni alloys, was determined. Furthermore, almost linear increase of hydrogen capacity with increasing Mg content was observed.

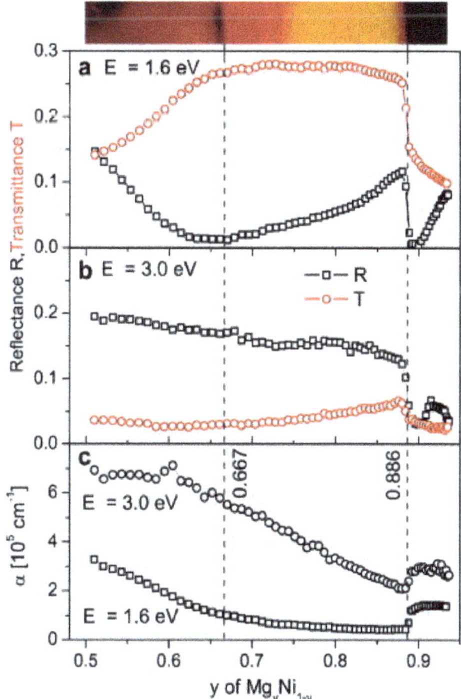

Figure 12. $Mg_yNi_{1-y}H_x$ compositional dependence of the optical reflectance (R) and transmittance (T) for two photo energies (a) $\hbar\omega = 1.6$ eV and (b) $\hbar\omega = 3.0$ eV. (c) Corresponding absorption coefficient α for both photon energies. At the top the optical appearance (in transmission) of a $Mg_yNi_{1-y}H_x$ gradient thin film in the hydrogenated state is shown. For y > 0.883 the film appears to be black as a result of a sudden drop in transmission (from ref. [77])

The dependence of optical properties of $Mg_yNi_{1-y}H_x$ on the alloy composition, slightly deviated from the stoichiometric Mg_2Ni, was investigated by Baldi et al. [81]. A large difference in optical behavior upon hydrogenation was explained by different hydrogen uptake mechanisms, namely, different nucleation and growth process of the hydride. Apart from a self-organized double layer structure (so-called black state): a transparent Mg_2NiH_4 layer at the substrate - film interface and a top layer of $Mg_2NiH_{0.3}$, which is always present at the beginning of hydrogenation process irrespective of the film composition, the following hydrogenation depends critically on the composition of the parent metal alloy. In slightly Mg-rich $Mg_{2\pm\delta}Ni$ films the hydrogenation proceeds by random nucleation and growth of the Mg_2NiH_4 phase within the upper metallic layer (slow process). For slightly Mg-poor $Mg_{2\pm\delta}Ni$ films hydrogenation occurs via the growth of the transparent Mg_2NiH_4 layer formed at the substrate interface (fast process). The reason for the slow hydrogenation in Mg-rich Mg_yNi_{1-y} films was explained by the presence of MgH_2, which forms prior to Mg_2NiH_4, and acts as blocking impurities for the growth of the Mg_2NiH_4 phase.

The black-state, present at a low hydrogen concentration, is characterized by low reflectance and zero transmittance in the whole visible region and suggests the application of Mg-Ni hydrides as a termochromic device for temperature control of hybrid solar collectors [82, 83]. The origin of the black state comes from preferred nucleation of Mg_2NiH_4 near the substrate/film interface after a solid solution $Mg_2NiH_{0.3}$ is formed. For achieving the black state, only a change in the first 30 nm of the films is necessary. The mixed double layer at the substrate/film interface consists of approximately 20 vol% of $Mg_2NiH_{0.3}$ and 80 vol% of Mg_2NiH_4. It was shown that the unusual hydrogen loading sequence, which starts near the substrate and not close to the catalytic Pd layer capping layer as one would expect, is due to locally enhanced kinetics [84]. Microstructural analysis revealed that up to film thickness of 50 nm, the film is built up of small grains located near the substrate, which most probably lower the activation energy, favoring nucleation of Mg_2NiH_4 [85]. The grains size increases upon further deposition, until a columnar microstructure is developed (Figure 13). After the nucleating layer is fully loaded to Mg_2NiH_4, subsequent hydrogenation process, slow or fast dependent on the composition of $Mg_{2\pm\delta}Ni$, takes place, until the entire film is loaded to semiconducting, transparent Mg_2NiH_4 (Figure 13).

The similar hydrogenation mechanism was observed for Mg-Co system [86]. Likewise the Mg-Ni, the nucleation of hydrogen rich phase (Mg_2CoH_5 in this case) initiates at the film-substrate interface. Consequently, a double layer is formed leading to the appearance of the optical black state at intermediate hydrogen concentration. In contrast, Mg-Fe showed more homogeneous hydrogen absorption. In fully loaded state in Mg-Ni and Mg-Co systems the ternary hydrides: Mg_2NiH_4 and Mg_2CoH_5, respectively, were detected, while in Mg-Fe system MgH_2 is mainly formed. The difference in hydrogen absorption mechanism is probably caused by the absence of any intermediate compound in the Mg-Fe system which does not favor the formation of Mg_2FeH_6. The presence of the black states was also observed in Mg-Mn alloys [87, 88].

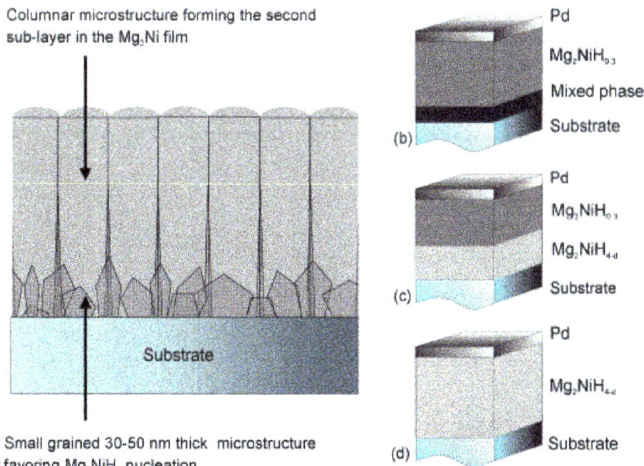

Figure 13. (a) Schematic representation of the Mg_2Ni thin film microstructure and hydrogenation mechanism for Mg-poor $Mg_{2\pm\delta}Ni$ film (fast process). The first 30–50 nm thick layer consists of small grains. From this layer, a columnar microstructure develops. At the first stage after the exposure to hydrogen a homogeneous $Mg_2NiH_{0.3}$ solid solution is formed. Nucleation of the Mg_2NiH_4 phase starts in this 30–50 nm sub-layer resulting in the optical black state (b). This layer hydrogenates to Mg_2NiH_4 and increases in thickness upon further hydrogenation resulting in a bilayer system (c). This continues until the whole film is loaded to semiconducting, transparent Mg_2NiH_4 (d) (from ref. [85]).

Mg was also alloyed with Cu to get an improved hydrogen storage properties. It was demonstrated that Pd-capped $Mg_{90}Cu_{10}$ amorphous thin film can reversibly store 5.8 wt% hydrogen in near ambient condition and could desorb it at temperature around 100 °C [89]. A comprehensive study of several metastable, crystalline single-phase $Mg_{80}X_{20}$ (X = Sc, Ti, V, Cr) thin film alloys was presented by Niessen and Notten [90]. These compounds showed high reversible hydrogen storage capacities. The best hydrogen absorption and desorption kinetics, with respect to pure Mg, was obtained in the case of Sc and Ti doping. By means of electrochemical loading $Mg_{80}Sc_{20}$ and $Mg_{80}Ti_{20}$ can reversibly store 6.7 wt% and 6.53 wt% hydrogen, respectively [91, 92]. Furthermore, it was noticed that after hydrogenation a homogeneous fcc - structured hydride is formed in both $Mg_ySc_{(1-y)}$ and $Mg_yTi_{(1-y)}$ alloys.

The structural, optical and electrical properties of the $Mg_yTi_{1-y}H_x$ thin films was thoroughly studied by Borsa et al. [93]. In metallic state all the films have zero transmission and a relatively high reflection that decreases with increasing Ti content. After hydrogen absorption under 10^5 Pa H_2 at room temperature, the reflection is low for all compositions, whereas the transmission decreases continuously with the metal ratio in the parent alloy: for $y \geq 90$ the transmission is significant, for $y < 90$ it becomes very low. Gremaued et al has presented the hydrogenography results in a form of the change of optical transmission vs $p(H_2)$ Pa for continuous gradient in the alloy composition, hydrogenated at 363K (Figure 14, ref [10]). There, a somewhat different trend was observed: the transmission is low at $p(H_2) < 10^3$ Pa but is getting higher upon decreasing y, while for the pressure approximately within the range of 1x10³ Pa < $p(H_2)$ < 4x10³ Pa the transmission becomes high for all compositions studied (0.60 < y < 0.89).

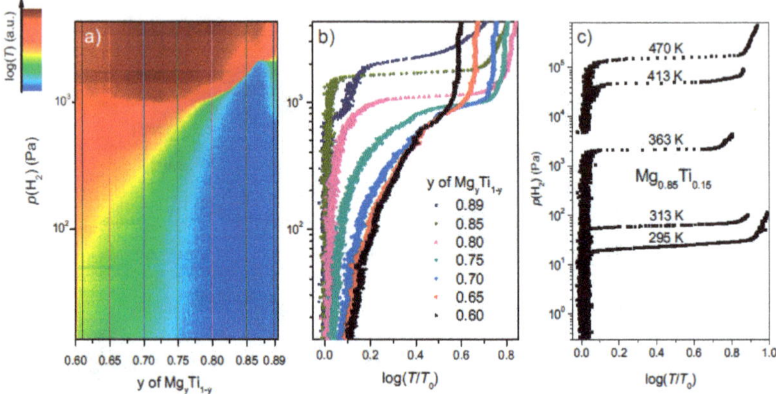

Figure 14. PTI's of Mg_yTi_{1-y} thin films with a continuous gradient in an alloy composition. a) Logarithm of the optical transmission T plotted in false colors as a function of the hydrogen pressure $p(H_2)$ and Mg atomic fraction y during hydrogenation at 363 K. Red corresponds to a high transmission and blue to a low transmission. As each column of pixels corresponds to one composition value y, the color map contains the isotherms of about 500 compositions. b) PTI's for the representative Mg–Ti alloy compositions indicated by colored vertical lines in a). c) PTI's for $Mg_{0.85}Ti_{0.15}$ at five different temperatures. The transmission T is normalized to the transmission in the metallic state T0 (from ref. [10]).

The combination of low reflection and low transmission in the hydrogenated state contributes to the highly absorbing state. Such a highly absorbing state was accounted for the presence of a mixed double layer: metallic TiH_2 and semiconducting MgH_2 phase. Based on the experimental data it was suggested that the phases do not form a composite material consisted of independent TiH_2 and MgH_2, but constitute rather a coherent structure. Hydrogenation process of the Mg_yTi_{1-y} thin films was described as follows: at very low hydrogen pressure a solid solution is formed in the Mg_yTi_{1-y} alloy, causing a small expansion of the host lattice; with increasing pressure Ti-related sites are hydrogenated at first, giving rise to the internal lattice strains release, thanks to the equality of the molar volume of TiH_2 and Mg; upon further hydrogen pressure increase, Mg is hydrogenated (Figure 15). Due to the coherent coupling between Mg and TiH_2 and the local stress induced, the Mg crystallizes to a cubic, fluorite MgH_2 structure, for y < 0.87. The complete reversibility of the system suggested that this ordering is robust and preserves upon hydrogen cycling most probably owing to the accidental equality of the molar volume of Mg and TiH_2.

Structurally, as-prepared films for all compositions have one crystalline phase that corresponds to a hexagonal Mg-Ti alloy. In the hydrogenated states the structure differs depending on the metal ratio in the parent alloy. Three regimes where identified: (1)

single fluorite phase for y < 0.87; (2) single phase rutile for y > 0.9 and (3) two phase coexistence for 0.87 < y < 0.90. The presence of a chemically partially segregated but structurally coherent metastable phase in Mg-Ti-H thin films was further confirmed by Extended X-ray Absorption Fine Structure (EXAFS) spectroscopy and by the positron Doppler broadening depth profiling method [94, 95]. Positron depth-profiling was also applied to monitor the effects of hydrogenation on thin films [96]. The analysis revealed a single homogenous layer for most metal and metal hydride films, except for the $Mg_{0.9}Ti_{0.1}H_x$ film, where a double layer was detected: a thin unloaded $Mg_{0.9}Ti_{0.1}$ or Mg-Ti-Pd alloy layer on top of a hydrogenated $Mg_{0.9}Ti_{0.1}H_x$. X-ray diffraction pattern confirmed the presence of two phases in the hydrogenated $Mg_{0.9}Ti_{0.1}$ film: one, identified as hexagonal $Mg_{0.9}Ti_{0.1}$ and a second, recognized as the rutile $Mg_{0.9}Ti_{0.1}H_2$ phase. For Ti fraction larger than about 15% only a single, broad peak in the X-ray pattern was detected, the position of which was typical for the fluorite metal hydride phase. The crystallographic analysis was confirmed by in situ recording diffraction patterns at various tilt angles, which allowed a precise identification of the crystal structures of the Mg_yTi_{1-y} thin films [97]. In the as-deposited state the film alloys have a hexagonal closed packed crystal structure. Hydrogenation under 10^5 Pa H_2 transformed the structures of $Mg_{0.7}Ti_{0.3}$ and $Mg_{0.8}Ti_{0.2}$ films into a rhombohedrally distorted unit cell with face-centered cubic (fcc) symmetry, whereas hydrogenated $Mg_{0.9}Ti_{0.1}$ has a body-centered tetragonal (bct) structure. Moreover, it was noticed that the hydrogen desorption kinetics changes along with the crystal structure from rapid for fcc-structures hydrides to sluggish for hydrides with a bct symmetry.

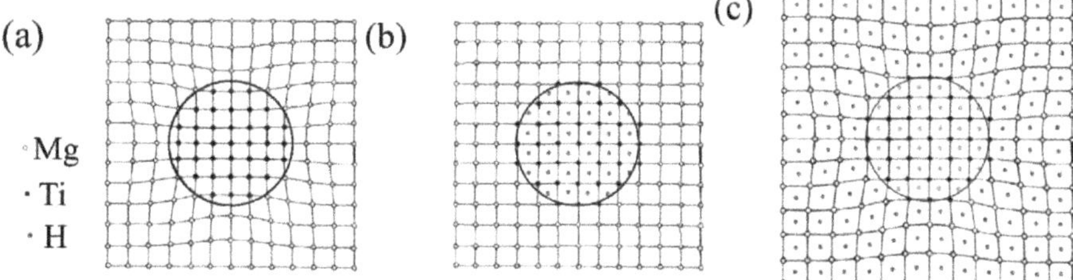

Figure 15. (a) Schematic representation of a coherent crystalline grain consisting of a Mg and Ti region. (b) The same crystalline grain after hydrogen uptake in the Ti-related sites. The accidental equality of the molar volumes of TiH_2 and Mg leads to an almost perfect crystal in situation (c). Fully hydrogenated state (from ref. [93]).

The enthalpy of hydrogenation process was established by hydrogenography [10]. For y<0.87, when the Mg-Ti hydrides have a fcc symmetry, the ΔH=-65 kJ/mol and does not differ much from the enthalpy for a bct symmetry (ΔH=-61 kJ/mol). The advantage of the fcc crystal is, however, a favorable kinetics of hydrogen diffusion within the fcc structure. The properties of the Mg-Ti thin films are very attractive for the design of hydrogen sensors, solar absorbers, or color-neutral switchable mirrors [98, 99].

The appealing kinetics of the Mg_yTi_{1-y} material has entailed the research to improve the thermodynamic of the hydrogenation process. The plateau pressure of the Mg-Ti-H system is very low (~ 0.1 Pa at room temperature). To increase the equilibrium pressure, the alloy has to be destabilized by adding additional elements. Substitution of Mg and Ti by lightweight Al or Si resulted in a shift of the plateau to higher pressures, while remaining at room temperature and maintaining a high gravimetric storage capacity of approximately 6 wt% [100]. By means of hydrogenopraphy an optimal composition in Mg-Ni-Ti system was found, $Mg_{69}Ni_{26}Ti_5$, with a relatively high hydrogen capacity (3.2 wt%) and the formation enthalpy value of -40 kJ/mol [12]. Adding Fe to the Mg-Ti system has resulted in excellent kinetics and reversible hydrogen sorption [101]. At 200 °C the Mg-Fe-Ti thin films absorbed ~ 5 wt% hydrogen in seconds and desorbed in minutes. However, rather no effect on thermodynamics of Mg-Fe-Ti was observed. The high-throughput screening technique was applied to search for an optimal material in Mg-Ti-B system [102]. The material with the composition $Mg_{0.36}Ti_{0.06}B_{0.58}$ was identified, with 10.6 wt% H_2 capacity. However, only partial reversibility was observed for the compound in the thin film.

In order to keep high hydrogen capacity of the Mg-based systems, Mg has to be destabilized with light metals, which posses a lower affinity for hydrogen. Aluminum was very often the element of choice. Beside its potential beneficial effect on kinetics and thermodynamics, the addition of Al to Mg may lead to the formation of complex hydrides, such as $Mg(AlH_4)_2$ with hydrogen storage capacities of 9.3 wt%. The magnesium alanate releases hydrogen in two steps upon heating. In the first step it desorbs 7 wt% H_2 leading to the formation of MgH_2 and Al.

Combinatorial synthesis and hydrogenation of Mg/Al libraries was prepared by electron beam physical vapor deposition at room temperature and was studied by wavelength dispersive spectroscopy (WDS) and micro-X-ray diffraction [103]. As-prepared thin films did not contain any Mg_yAl_{1-y} intermetallics or solid solution. For all Mg_yAl_{1-y} compositions full hydrogenation was obtained. MgH_2 was the main phase observed for all compositions above 20 at.% Mg. It was found that Al can act as a catalyst for hydrogenation reaction of magnesium. The hydrogenation/dehydrogenation of Pd/(Fe, Ti)/Mg-Al/(Fe, Ti)/Pd thin films was investigated [104]. The films absorb hydrogen at low temperature (~50 °C) and with excellent kinetics (few minutes). The addition of Al has improved the H_2 sorption properties and hydrogen capacities, and the catalytic effect of Ti and Fe was acknowledged.

Absorption and desorption properties of Pd-covered Mg_yAl_{1-y} alloy thin films was studied as a function of temperature and alloy composition by means of neutron reflectivity technique (NR) [105]. Hydrogen absorption was performed at 430K for 20h at 6.8 MPa. Hydrogen content and distribution in the Mg_yAl_{1-y} thin films was determined in situ. For all compositions the hydrogen was uniformly distributed throughout the MgAl film thickness (of about 52 nm for the MgAl layer and ~ 10 nm for Pd coating), whereas no hydrogen was found in the Pd layer. The $Mg_{0.7}Al_{0.3}$ film could store 4.1 wt% hydrogen with complete H_2 desorption at a temperature of 448 K. A superior absorption and desorption properties of $Mg_{0.7}Al_{0.3}$ over those of $Mg_{0.6}Al_{0.4}$ was observed. Except lower hydrogen content (3.1 wt%),

$Mg_{0.6}Al_{0.4}$ released the hydrogen at temperature around 25 K higher than $Mg_{0.7}Al_{0.3}$. Moreover, along with the higher hydrogen desorption temperature, the Pd interdiffusion into the MgAl film was noticed. A systematic study of structural change of the films during hydrogen desorption was performed later by the same group of researchers [106]. It was concluded that the Pd interdiffusion into hydrogenated MgAl film occurs for both $Mg_{0.6}Al_{0.4}$ and $Mg_{0.7}Al_{0.3}$ alloy thin films, leading to a complete destruction of the films structure. In contrast, for the as-prepared hydrogen-free Pd-coated $Mg_{0.7}Al_{0.3}$ thin film, the Pd layer remained almost intact and only small zone in the Pd/MgAl interface was infiltrated by Pd.

To prevent the interdiffusion, the Ta layer was added between the Pd and the $Mg_{0.7}Al_{0.3}$ layers (Figure 16) [107]. Tantalum was found to improve significantly hydrogen absorption kinetics due to lowering of the nucleation barrier for the formation of the hydride phase in the $Mg_{0.7}Al_{0.3}$ layer. Hydrogenation of the alloy could occur at pressure 10 time lower than for the Ta-free sample, without reducing the storage capacity. It was observed that between the catalysts (either Pd or Ta/Pd) and the $Mg_{0.7}Al_{0.3}$ layer a non-absorbing region appears, which was attributed to interfacial stress. According to the measurement ~5 wt% H can be stored in the $Mg_{0.7}Al_{0.3}$ alloy thin film under mild conditions (at room temperature and under 0.13 MPa).

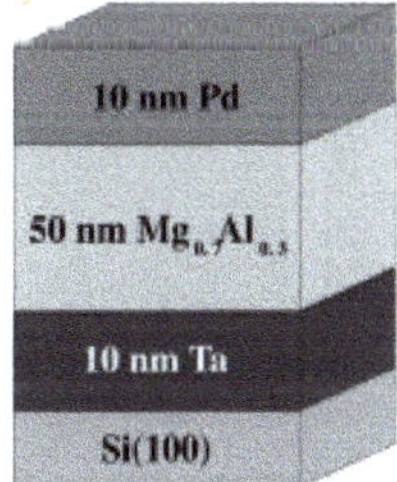

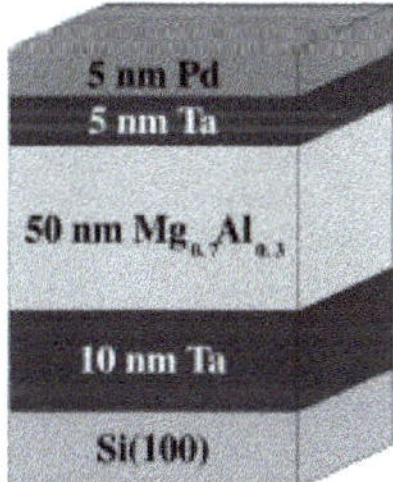

Figure 16. The structure of the film with the single Pd catalyst layer (left) and the Ta/Pd catalyst bilayer (right) (from ref. [107]).

Structure and electrochemical hydrogen storage properties of sandwiched $Pd/Mg_{1-x}Al_x/Pd$ (x=0, 0.13, 0.1, 0.39) thin films were investigated [108]. The improvement of both thermodynamics and kinetics of Mg-Al alloys as compared to pure Mg was confirmed. X-ray analysis showed that the layer was constituted of a single phase Mg(Al) solid solution. All films demonstrates a globular surface structure (Figure 17). The surface roughness of the films depends on the Al concentration. In the absence of Al (x=0) the surface was very rough and consisted of large particles. Upon the Al increase the surface was becoming smoother. Also porosity depends on the Al concentration. The best hydriding properties was detected for the $Pd/Mg_{0.79}Al_{0.21}/Pd$ thin films (storage capacity of ~ 1.9 wt%), which was related to a high porosity of the film (28%).

The complex $Mg(AlH_4)_2$ hydride was experimentally observed by Jain et al. [109]. A sandwiched-like Pd/Al/Mg/Pd thin film was hydrogenated at 150 °C under 0.2 MPa H_2 pressure for 2h. The formation of the $Mg(AlH_4)_2$ along with MgH_2, formed mostly at the

Pd/Mg interface, has eliminated the generation of Mg_xPd_y intermetallics and reduced the oxygen content in the film. Moreover, an increment in hydrogen storage capacity was noticed. The $Mg(AlH_4)_2$ was also obtained under high-flux, low-energy hydrogen ion irradiation of Mg/Al bilayer films [110, 111]. Its synthesis takes place under intensive intermixing of Mg and Al atoms and with continuous supply of hydrogen. The 2 keV hydrogen ions pass the naturally formed 2 – 4 nm thick Al_2O_3 layer without destroying it. Moreover, it was suggested that the magnesium alanate was protected by the Al_2O_3 layer. The experimental results have confirmed a two-step decomposition mechanism of the $Mg(AlH_4)_2$ and demonstrated that dehydrogenation is controlled by transport process through the oxide layer.

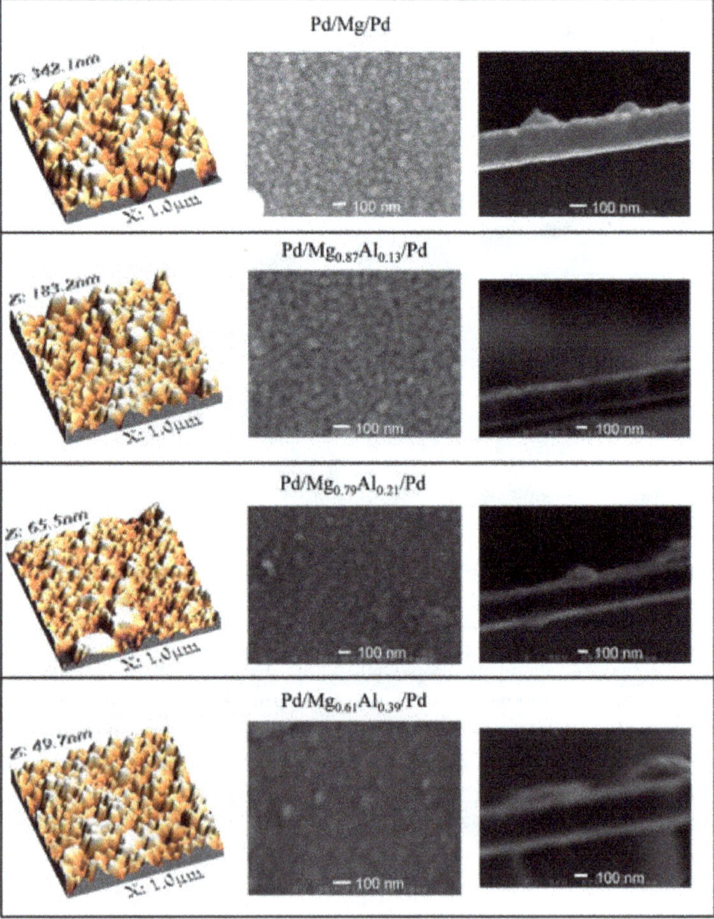

Figure 17. 3D AFM images, surface and cross-section SEM images of the $Pd/Mg_{1-x}Al_x/Pd$ films with x = 0, 0.13, 0.21 and 0.39 from top to bottom (from ref. [108]).

5. Conclusion

Notwithstanding the remarkable achievements in hydrogen absorption–desorption attained for thin films, it remains difficult to transfer the results to bulk materials (powders), which are thought to be used in transport sector. The difficulties originates from the fact that the hydrogen interaction with mater depends strongly on a subtle tuning between parameters such as: a number and type of defects, microstructure, size and morphology of crystallites, which are not easy controllable by the application of the powders preparation methods, like for instance mechanical milling, as compared to the thin films deposition techniques. As an example, mechanically milled Mg-Ti powders display much slower kinetics, reduced storage capacities and lower structural stability as compare to the Mg-Ti thin films [112]. The thin films synthesis methods can lead to the formation of non-equilibrium phases, which are rather not accessible by bulk preparation methods. In contrast to bulk preparation techniques, in a thin film approach it is relatively easy to control the size and morphology of material. The size reduction to the nanoscale leads to the energy contributions which are typically negligible in bulk, including the interaction of the sample with the substrate, surface or interface energies and quantum size effects. These contributions allows one to overcome the hydrogenation kinetics limitations and to study the intrinsic thermodynamic properties of hydrogen absorption. Thin films are particularly suited for combinatorial techniques: allowing one to perform combinatorial research on thousands of different concentrations of elements at once. It helps to identify an optimal alloy's composition and to find the best catalyst for hydrogen sorption. Moreover, other technological applications for thin films, such as in batteries, coatings, or solar collectors, have been proposed due to discovered spectacular changes in the optical properties of metal–hydride films near their metal–insulator transition ("switchable mirror" behavior). The transition is reversible and can be simply induced by changing the surrounding hydrogen gas pressure or electrolytic cell potential.

Author details

Małgorzata Norek
Department of Advanced Materials and Technology, Military University of Technology, Warsaw, Poland

Acknowledgement

Thanks are due to Prof. Leszek R. Jaroszewicz from Military University of Technology in Warsaw, Poland, and to the Polish Ministry of Science and Higher Education, Key Project POIG.01.03.01-14-016/08 for the financial support.

6. References

[1] Remhof A, Borgschulte A (2008) Thin-Film Metal Hydrides. ChemPhysChem 9: 2440-2455.

[2] Cuevas F, Joubert J-M, Latroche M, Percheron-Guegan A Intermetallic compounds as negative electrodes of Ni/MHbatteries (2001) Appl. Phys. A 72: 225-238.

[3] More information can be found at the website: http://www.ncnr.nist.gov/programs/reflect/

[4] Remhof A, Song G, Sutter Ch, Schreyer A, Siebrecht R, Zabel H (1999) Hydrogen and deuterium in epitaxial Y (0001) films: Structural properties and isotope exchange. Phys. Rev. B 59: 6689.

[5] Ingason A S, Olafsson S (2005) Thermodynamics of hydrogen uptake in Mg films studied by resistance measurements. J. Alloys and Compd. 404-406: 469-472.

[6] Demènech-Ferrer R, Rodrìguez-Viejo J, González-Silveira, Garcia G (2011) In situ infrared thermographic screening of compositional spread Mg-Ti. J. Alloys and Compd. 509: 6497-6501.

[7] Olk CH, Tibbetts GG, Simon D, Moleski JJ (2003) Combinatorial preparation and infrared screening of hydrogen sorbing metal alloys. J. Appl. Phys 94: 720.

[8] Olk CH (2005) Combinatorial approach to material synthesis and screening of hydrogen storage alloys. Meas. Sci. Technol. 16: 14-20.

[9] Oguchi H, Tan Z, Heilweil EJ, Bendersky LA (2010) In-situ infrared imaging methodology for measuring heterogeneous growth process of a hydride phase. Int. J. Hydrogen Energy 35: 1296-1299.

[10] Gremaud R, Broedersz CP, Borsa DM, Borgschulte A, Mauron P, Screuders H, Rector JH, Dam B, Griessen R (2007) Hydrogenography: an optical combinatorial method to find new light-weight hydrogen-storage materials. Adv. Mater. 19: 2813-2817.

[11] Borgschulte A, Lohstroh W, Westerwaal RJ, Schreuders H, Rector JH, Dam B, Griessen R (2005) Combinatorial method for the development of a catalyst promoting hydrogen uptake. J. Alloys and Compd. 404-406: 699-705.

[12] Dam B, Gremaud R, Broedersz C, Griessen R (2007) Combinatorial thin films methods for the search of new lightweight metal hydrides. Scripta Mater. 56: 853-858.

[13] Gremaud R, Slaman M, Schreuders H, Dam B, Griessen R (2007) An optical method to determine the thermodynamics of hydrogen absorption and desorption in metals. Appl. Phys. Lett. 91: 231916.

[14] Ludwig A, Cao J, Savan A, Ehmann M (2007) Hugh-throughput characterization of hydrogen storage materials using thin films on macromachined Si substrates. J. Alloys and Compd. 446-447: 516-521.

[15] Ludwig A, Cao J, Dam, B, Gremaud R (2007) Opto-mechanical characterization of hydrogen storage properties of Mg-Ni thin film composition spreads. Appl. Surf, Sci. 254: 682-686.

[16] Woo NC, Ng BG, van Dover RB (2007) High-throughput combinatorial study of local stress in thin films composition spreads. Rev. Sci. Instrum. 78: 072208.

[17] Guerin S, Hayden BE, Smith DCA (2008) High-throughput synthesis and screening of hydrogen – storage alloys. J. Comb. Chem. 10: 37-43.

[18] Züttel A (2003) Materials for hydrogen storage. Materials Today 6: 24–33.

[19] Krozer A, Kasemo B (1987) Unusual kinetics due to interface hydride formation in the hydriding of Pd/Mg sandwich layers. J Vac Sci Technol A 5: 1003-1005.

[20] Pozzo M, D. Alfè D (2008) Structural properties and enthalpy of formation of magnesium hydride from quantum Monte Carlo calculations. Phys. Rev. B 77: 104103.
[21] Aguey-Zinsou KF, Ares-Fernandez JR (2008) Synthesis of Colloidal Magnesium: A Near Room Temperature Store for Hydrogen. Chem. Mater. 20: 376-378.
[22] Cheung S, Deng WQ, van Duin ACT, Goddard III WA (2005) ReaxFFMgH Reactive Force Field for Magnesium Hydride Systems. J. Phys. Chem. A 109: 851-859.
[23] Li L, Peng B, Ji W, Chen J (2009) Studies on the Hydrogen Storage of Magnesium Nanowires by Density Functional Theory. J. Am. Chem. Soc. 113, 3007–3013
[24] Liang JJ (2005) Theoretical insight in tailoring energetic of Mg hydrogen absorption/desorpion through nano-engineering. Appl. Phys. A 80: 173-178.
[25] Ingason AS, Olafsson S (2005) Thermodynamics of hydrogen uptake in Mg films studied by resistance measurements. J Alloys Compd 404-406: 469-472.
[26] Tan Z, Chiu C, Heilweil EJ, Bendersky LA (2011) Thermodynamics, kinetics and microstructural evolution Turing hydrogenation of iron-doped magnesium thin films. Int. J. Hydrogen Energy 36: 9702-9713.
[27] Barcelo S, Rogers M, Grigoropoulos CP, Mao SS (2010) Hydrogen storage property of sandwiched magnesium hydride nanoparticle thin film." International Journal of Hydrogen Energy 35: 7232 – 7235.
[28] Baldi A, Gonzalez-Silveira M, Palmisano V, Dam B, Griessen R (2009) Destabilization of the Mg-H system through elastic constrains. Phys. Rev. Lett. 102: 226102.
[29] Nicolas M, Dumoulin L, Burger JP (1986) Thickness dependence of the critical solution temperature of hydrogen in Pd films. J. Apl. Phys. 60: 3125.
[30] Vermeulen P, Ledovskikh A, Danilov D, Notten HL (2006) The impact of the layer thickness on the thermodynamics properties of Pd hydride thin film electrodes. J. Phys. Chem. B 110: 20350-20353.
[31] Siviero G, Bello V, Mattei G, Mazzoldi P, Battaglin G, Bazzanella N, Checchetto R, Miotello A (2009) Structural evolution of Pd-capped Mg thin films under H2 absorption and desorption cycles. Int J Hydrogen Energy 34: 4817-4826.
[32] Gremaud R, Gonzalez-Silveira M, Pivak Y, de Man S, Slaman M, Schreuders H, Dam B, Griessen R (2009) Hydrogenography of PdHx thin films: Influence of H-induced stress relaxation processes. Acta Materialia 57: 1209-1219.
[33] Norek M, Stępniowski WJ, Polański M, Zasada D, Bojar Z, Bystrzycki J (2011) A comparative study on the hydrogen absorption of thin films at room temperature deposited on non-porous glass substrate and nano-porous anodic aluminum oxide (AAO) template. Inter. J. Hydrogen Energy 36: 11777-11784.
[34] Norberg NS, Arthur TS, Fredrick SJ, Prieto AL (2011) Size-Dependent Hydrogen Storage Properties of Mg Nanocrystals Prepared from Solution. J. Am. Chem. Soc. 133: 10679–10681.
[35] Yoshimura K, Yamada Y, Okada M (2004) Hydrogenation of Pd capped Mg thin films at room temperature. Surf Sci 566-568: 751-754.
[36] Paillier J, Bouhtiyya S, Ross GG, Roué L (2006) Influence of the deposition atmosphere on the Pd/Mg interface characteristics of hydriding Pd-Mg thin films prepared by pulsed laser deposition. Thin Solid Films 500: 117-123.

[37] Zhdanov VP, Krozer A, Kasemo B (1993) Kinetics of first-order phase transitions initiated by diffusion of particles from the surface into the bulk. Phys Rev B 47: 11044-11048

[38] Paillier J, Roue (2005) Hydrogen electrosortpion and structural properties of nanostructured Pd/Mg thin films elaborated by pulse laser deposition. J. Alloys Compd 404-406: 473-476.

[39] Kelly ST, Clemens BM (2010) Moving interface hydride formation in multilayered metal thin films. J. Appl. Phys. 108: 013521.

[40] Spatz P, Aebischer HA, Krozer A, Schlapbach L (1993) The Diffusion of H in Mg and the Nucleation and Growth of MgH2 in Thin Films. Z. Phys. Chem. 181: 393-397

[41] Qu J, Sun B, Yang R, Zhao W, Wang Y, Li X. (2010) Hydrogen absorption kinetics of Mg thin films under mild conditions. Script. Mater. 62: 317-320.

[42] Qu J, Wang Y, Xie L, Zheng J, Liu Y, Li X, (2009) Superior hydrogen absorption and desorption behavior of Mg thin films. J Power Sources 186: 515-520.

[43] Pasturel M, Slaman M, Schreuders H, Rector JH, Borsa DM, Dam B, Griessen R (2006) Hydrogen absorption kinetics and optical properties of Pd-doped Mg thin films. J. Appl. Phys 100: 023515.

[44] Borgschulte A, Rector JH, Schreuders H, Dam B, Griesssen R (2007) Electrohydrogenation of MgH2-thin films. Appl. Phys. Lett. 90: 071912.

[45] Qu J, Wang Y, Xie L, Zheng J, Liu Y, Li X (2009) Hydrogen absorption–desorption, optical transmission properties and annealing effect of Mg thin films prepared by magnetron sputtering. Int J Hydrogen Energy 34: 1910-1915.

[46] Higuchi K, Kajioka H, Toiyama K, Fujii H, Orimo S, Kikuchi Y (1999) In situ study of hydriding–dehydriding properties in some Pd/Mg thin films with different degree of Mg crystallization. J Alloys Compd 293-295: 484-489.

[47] Ye SY, Chan SLI, Ouyang LZ, Zhu M (2010) Hydrogen storage and structure variation in Mg/Pd multi-layer film. J Alloys Compd 504: 493-497.

[48] Tan X, Harrower CT, Amirkhiz BS, Mitlin D (2009) Nano-scale bi-layer Pd/Ta, Pd/Nb, Pd/Ti and Pd/Fe catalysts for hydrogen sorption in magnesium thin films. Int. J. Hydrogen Energy 34: 7741-7748.

[49] Hoogeveen R, Moske M, Geisler H, Samwer K (1996) Texture and phase transformation of sputter-deposited metastable Ta films and Ta/Cu multilayers. Thin Solid Films 275: 203-206.

[50] Ockwig NW, Nenoff TM (2007) Membranes for hydrogen separation. Chem Rev. 107: 4078-4110.

[51] Higuchi K, Yamamoto K, Kajioka H, Toiyama K, Honda M, Orimo S, Fujii H (2002) Remarkable hydrogen storage properties in three-layered Pd/Mg/Pd thin films. J Alloys Compd 330-332: 526-530.

[52] Qu J, Sun B, Zheng J, Yang R, Wang Y, Li X (2010) Hydrogen desorption properties of Mg thin films at room temperature. J. Power Sources 195: 1190-1194.

[53] Qu J, Sun B, Liu Y, Yang R, Li Y, Li X (2010) Improved hydrogen storage properties in Mg-based thin films by tailoring structures. Int J Hydrogen Energy 35: 8331 – 8336.

[54] [54]Özgit Ç, Akyıldız H, Öztürk T (2010) Isochronal hydrogenation of textured Mg/Pd thin films Thin Solid Films 518: 4762 – 4767

[55] Kelekar R, Giffard H, Kelly ST, Clemens BM (2007) Formation and dissociation of MgH2 in epitaxial Mg thin films. J. Appl. Phys. 101: 114311.

[56] Singh S, S. Eijt WH, Zandbergen MW, Legerstee WJ, Svetchnikov VL, (2007) Nanoscale structure and the hydrogenation of Pd-capped magnesium thin films prepared by plasma sputter and pulsed laser deposition. J. Alloys Compd. 441: 344-351.

[57] Yamamoto K, Higuchi K, Kajioka H, Sumida H, Orimo S, Fujii H (2002) Optical transmission of magnesium hydride thin film with characteristic nanostructure. J Alloys Compd 330-332: 352-356.

[58] Checchetto R, Bazzanella N, Miotello A, Brusa R, Zecca A, Mengucci A (2004) Deuterium storage in nanocrystalline magnesium thin films. J. Appl. Phys. 95: 1989-1995.

[59] Hao S, Sholl DS (2008) Hydrogen diffusion in MgH2 and NaMgH3 via concerted motions of charged defects. Appl. Phys. Lett. 93: 251901.

[60] Hao S, Sholl DS (2009) Selection of dopants to enhance hydrogen diffusion rates in MgH2 and NaMgH3. Appl. Phys. Lett. 94: 171909.

[61] Zahiri B, Amirkhiz BS, Mitlin D (2010) Hydrogen storage cycling of MgH2 thin film nanocomposites catalyzed by bimetallic Cr Ti. Appl. Phys. Lett. 97: 083106.

[62] Tan Z, Heilweil EJ, Bendersky LA (2010) In-situ kinetics studies on hydrogenation of transition metal (=Ti, Fe) doped Mg films. Mater. Res. Soc. Symp. Proc. 1216: 148-152.

[63] Stolz SE, Popovic D (2007) A high-resolution core-level study of Ni-catalyzed absorption and desorption of hydrogen in Mg-films. Surf. Sci. 601:1507-1512.

[64] Farangis B, Nachimuthu P, Richardson TJ, Slack JL, Meyer BK, Perera RCC, Rubin MD (2003) Structural and electronic properties of magnesium-3D transition metal switchable mirrors. Solid State Ionics 165: 309-314.

[65] Borgschulte A, Lohstroh W, Westerwaal RJ, Schreuders H, Rector JH, Dam B, Griessen R (2005) Combinatorial method for the development of a catalyst promoting hydrogen uptake. J. Alloys and Compd. 404-406: 699-705.

[66] Mosaner P, Bazzanella N, Bonelli M, Checcetto R, Miotello A (2004) Mg: Nb films produced by pulsed laser deposition for hydrogen storage. Mater. Sci. Eng. B 108: 33-37.

[67] Leon A, Knystautas EJ, Huot J, Russo SL, Koch CH, Schulz R (2003) Hydrogen sorption properties of vanadium- and palladium-implanted magnesium films. J. Alloys and Compd. 356-357: 530-535.

[68] Akyıldız H, Özenbas M, Öztürk T (2006) Hydrogen absorption in magnesium based crystalline thin films. Int. J. Hydrogen Energy 31: 1379-1383.

[69] Richardson TJ, Farangis B, Slack JL, Nachimuthu P, Perera R, Tamura N, Rubin M (2003) X-ray absorption spectroscopy of transition metal-magnesium hydride thin films. J. Alloys and Compd. 356-357: 204-207.

[70] Farangis B, Nachimuthu P, Richardson TJ, Slack JL, Perera R, Gullikson EM, Lindle DW, Rubin M (2003) In situ x-ray-absorption spectroscopy study of hydrogen absorption by nickel-magnesium thin films. Phys. Rev. B 67: 085106.

[71] Richardson TJ, Slack JL, Farangis B, Rubin MD (2002) Mixed metal films with switchable optical properties. Appl. Phys. Lett. 80: 1349-1351.

[72] Reilly JJ, Wiswall RH (1968) Reaction of hydrogen with alloys of magnesium and nickel and the formation of Mg2NiH4. Inorg. Chem. 7: 2254-2256.

[73] Chen J, Yang H-B, Xia Y-Y, Kuriyama N, Xu Q, Sakai T (2002) Hydriding and dehydriding properties of amorphous magnesium-nickel films prepared by a sputtering method. Chem Mater. 14: 2834-2836.

[74] Stoltz SE, Stoltz D (2007) Spectroscopic evidence for reversible hydrogen storage in unordered Mg5Ni1 thin films. J. Phys.: Condens. Matter 19: 446010.

[75] Ouyang LZ, Wag H, Chung CY, Ahn JH, Zhu M (2006) MgNi/Pd multilayer hydrogen storage thin films prepared by dc magnetron sputtering. J. Alloys and Compd. 422: 58-61.

[76] Quyng LZ, Ye SY, Dong HW, Zhu M (2007) Effect of interfacial free energy on hydriding reaction of Mg-Ni thin films. Appl. Phys. Lett. 90: 021917.

[77] Gremaud R, van Mechelen JLM, Schreuders H, Slaman M, Dam B, Griessen R (2009) Structural and optical properties of MgyNi1-yHx gradient thin films in relation to the as-deposited metallic state. Int. J. Hydrogen Energy 34: 8951-8957.

[78] Gremaud R, Broedersz CP, Borgschulte A, van Setten MJ,, Schreuders H, Slaman M, Dam B, Griessen R (2010) Hydrogenography of MgyNi1-yHx gradient thin films: interplay between the thermodynamics ad kinetic of hydrogenation. Acta Mater. 58: 658-668.

[79] Yoshimura K, Yamada Y, Okada M (2002) Optical switching of Mg-rich Mg-Ni alloy thin films. Appl. Phys. Lett. 81: 4709-4711.

[80] Johansson E, Chacon C, Zlotea C, Andersson Y, Hjorvarsson B (2004) Hydrogen uptake and optical properties of sputtered Mg-Ni thin films. J. Phys.: Condens. Matter 16: 7649-7662.

[81] Borsa DM, Lohstroh W, Gremaud R, Rector JH, Dam B, Wijngaarden RJ, Griessen R (2007) Critical composition dependence of the hydrogenation of Mg2±δNi thin films. J. Alloys and Compd 428: 34-39.

[82] Lokhorst AC, Dam B, Giebels IAME, Welling MS, Lohstroh W, Griessen R (2005) Thermochromic metal-hydride bilayer devices. J. Alloys and Compd. 404-406: 465-468.

[83] Mechelen JLM, Noheda B, Lohstroh W, Westerwaal RJ, Rector JH, Dam B, Griessen R (2004) Mg-Ni-H films as selective coatings: Tunable reflectance by layered hydrogenation. Appl. Phys. Lett. 84: 3651-3653.

[84] Lohstroh W, Westerwaal RJ, Mechelen JLM, Chacon C, Johanssson E, Dam B, Griessen R (2004) Structural and optical properties of Mg2NiHx switchable mirrors upon hydrogen loading. Phys. Rev. B 70: 165411.

[85] Westerwaal RJ, Borgschulte A, Lohstroh W, Dam B, Kooi B, Brink G, Hopstaken MJP, Notten PHL (2006) The growth-induced microstructural origin of the optical black states of Mg2NiHx thin films. J. Alloys and Compd. 416: 2-10.

[86] Lohstroh W, Westerwaal RJ, Lokhorst AC, Mechelen JLM, Dam B, Griessen R (2005) Double layer formation in Mg-TM switchable mirrors (TM: Ni, Co, Fe). J. Alloys and Compd. 404-406: 490-493.

[87] Richardson TJ, Slack JL, Farangis B, Rubin MD (2002) Mixed metal films with switchable optical properties. Appl. Phys. Lett. 80: 1349-1351.
[88] Jangid MK, Singh M (2012) Hydrogenation and annealing effect on electrical properties of nanostructured Mg/Mn bilayer thin films. Int. J. Hydrogen Energy 37: 3786-3791.
[89] Akyıldız H, Öztürk T (2010) Hydrogen sorption in crystalline and amorphous Mg-Cu thin films. J. Alloys and Compd. 492: 745-750
[90] Niessen RAH, Notten PHL (2005) Electrochemical Hydrogen Storage Characteristics of Thin Film MgX (X=Sc, Ti, V, Cr) Compounds. Electrochem. Solid-State Lett. 8: A534-A538.
[91] Niessen RAH, Notten PHL (2005) Hydrogen storage in thin film magnesium-scandium alloys. J. Alloys and Compd. 404-406: 457-460.
[92] Vermeulen P, Niessen RAH, Notten PHL (2006) Hydrogen storage in metastable MgyTi(1-y) thin films. Electrochem. Comm. 8: 27-32.
[93] Borsa DM, Gremaud R, Baldi A, Schreuders H, Rector JH, Kooi B, Vermeulen P, Notten PHL, Dam B, Griessen R (2007) Structural, optical, and electronical properties of MgyTi1-yHx thin films. Phys. Rev. B 75: 205408.
[94] Baldi A, Gremaud R, Borsa DM, Balde CP, Eerden AMJ, Kruijzer GL, Jongh PE, Dam B, Griessen R (2009) Nanoscale composition modulations in MgyTi1-yHx thin films alloys for hydrogen storage. Int. J. Hydrogen Energy 34: 140-1457.
[95] Leegwater H, Schut H, Egger W, Baldi A, Dam B, Eijt SWH (2010) Divacancies and the hydrogenation of Mg-Ti films with short range chemical order. Appl. Phys. Lett. 96: 151902.
[96] Eijt SWH, Leegwater H, Schut H, Anastasopol A, Egger W, Ravelli L, Hugenschmidt C, Dam B (2011) Layer-resolved study of the Mg to MgH2 transformation in Mg-Ti films with short-range chemical order. J. Alloys and Compd. 509S: S567-S571.
[97] Vermeulen P, Graat PCJ, Wondergem HJ, Notten PHL (2008) Crystal structures of MgyTi100-y thin film alloys in the as-deposited and hydrogenated state. Int. J. Hydrogen Energy 33: 5646-5650.
[98] Slaman M, Dam B, Pasturel M, Borsa DM, Schreuders H, Rector JH, Griessen R Fiber optic hydrogen detectors containing Mg-based metal hydrides. Sensor Actuat. B Chem. 123: 538-545.
[99] Bao S, Tajima K, Yamada Y, Okada M, Yoshimura K (2007) Color-neutral switchable mirrors based on magnesium-titanium thin films. Appl. Phys. A 87: 621-624.
[100] Vermeulen P, Thiel EFMJ, Notten PHL (2007) Ternary MgTiX-alloys: a promising route towards low-temperature high-capacity, hydrogen storage materials. Chem. Eur. J. 13: 9892-9898.
[101] Zahiri B, Harrower CT, Amirkhiz BS, Mitlin D (2009) Rapid and reversible hydrogen sorption in Mg-Fe-Ti thin films. Appl. Phys. Lett. 95: 103114.
[102] Amieiro-Fonseca A, Ellis SR, Nuttall JC, Hayden BE, Guerin S, Purdy G, Soulie JP, Callear SK, Culligan SD, David WIF, Edwards PP, Jones MO, Johnson SR, Pohl AH (2011) A multidisciplinary combinatorial approach for tuning promising hydrogen storage materials towards automotive applications. Faraday Discuss. 151: 369-384.

[103] Garcia G, Domenech - Ferrer R, Pi F, Santiso J, Rodriguez-Viejo J (2007) Combinatorial synthesis and hydrogenation of Mg/Al libraries prepared by electron beam physical vapor deposition. J. Comb. Chem. 9: 230-236.

[104] Domenech - Ferrer R, Sridharan MG, Garcia G, Pi F, Rodriguez-Viejo J (2007) Hydrogen properties of pure magnesium and magnesium-aluminium thin films. J. Power Sources 169: 117-122.

[105] Fritzsche H, Saoudi M, Haagsma J, Ophus C, Luber C, Harrower CT, Mitlin (2008) Neutron reflectometry study of hydrogen desorption in destabilized MgAl alloy thin films. Appl. Phys. Lett. 92: 121917.

[106] Fritzsche H, Saoudi M, Haagsma J, Ophus C, Luber C, Harrower CT, Mitlin (2009) Structural changes of thin MgAl films during hydrogen desorption. Nucl. Instr. And Meth. A 600: 301-304.

[107] Harrower C, Poirier E, Fritzsche H, Kalisvaart P, Satija S, Akgun B, Mitlin D (2010) Early deuteration steps of Pd- and Ta/Pd-catalyzed Mg70Al30 thin films observed at room temperature. Int. J. Hydrogen energy 35: 10343-10348.

[108] Bouhtiyya S, Roue L (2010) Structure and electrochemical hydrogen storage properties of Pd/Mg1-x Alx/Pd thin films prepared by pulsed laser deposition. J. Mater. Sci. 45: 946-952.

[109] Jain P, Jain A, Vyas D, Kabiraj D, Khan SA, Jain IP (2012) Comparative study on hydrogenation properties of Pd capped Md and Mg/Al. films. Int. J. Hydrogen Energy 37: 3779-3785.

[110] Pranevicius LL, Milcius D (2005) Synthesis of Mg(AlH4)2 in bilayer Mg/Al thin films under plasma immersion hydrogen ion implementation and thermal desorption process. Thin Solid Films 485: 135-140.

[111] Pranevicius L, Templier C, Pranevicius LL, Milcius D (2005) Influence of surface barriers on hydrogen storage in MgAl films on permeable stainless steel membranes. Vacuum 78: 367-373.

[112] Liang G, Schulz R (2003) Synthesis of Mg-Ti alloy by mechanical alloying. J. Mater. Sci.38: 1179 – 1184.

Section 4

Hydrogen Power

Chapter 10

Electrode/Electrolyte Interphase Characterization in Solid Oxide Fuel Cells

Analía Leticia Soldati, Laura Cecilia Baqué,
Horacio Esteban Troiani and Adriana Cristina Serquis

Additional information is available at the end of the chapter

1. Introduction

The increasing global energy requirements and the growing environmental conscience demand high performance technology for energy production, storage and transport. Fuel cells (FCs) have reached a recognized place within the potentially efficient devices to convert chemical energy into electricity. First developed by William Grove in 1835 [1], FCs had a profound impact in the aerospatial industry and more recently gained special attention for wider application, because they are considered as environmentally friendly devices. Some advantages of FCs include, that their efficiency is independent of the Carnot cycle for thermal machines [2], and that they can be designed for mobile and stationary applications. These devices also produce only water as residue when Hydrogen is used as fuel, without undesired residual emissions and consequently has no impact on the greenhouse effect.

Contrary to the conventional energy conversion devices, FCs reduce the emission of contaminating gases such as CO_2, Sulfur, Nitrogen dioxide and hydrocarbons. Between them, Carbon dioxide is thought as one of the principal sources of the global warming effect that forces the climate change we are experiencing for the past decades. A technology which would reduce these emissions is in agreement with the actual international political framework and is ethically desirable.

Another great advantage of the FC technology is that it allows the distributed energy generation. FCs have minimal installation and maintenance costs, allowing on-site generation and reducing production and transport costs when compared to conventional central power stations (fossil fuel, nuclear, hydroelectric, etc.). For example, FCs are a suitable opportunity of energetic autonomy (at fair costs) for small communities, hospitals or schools, which are not connected to national distribution grids.

Finally, Hydrogen has been chosen historically to supply FC devices. This element is considered the clean fuel of the future because its combustion only generates water vapor as a product. However, recent investigations show that the FC can also be operated with conventional hydrocarbon fuels such as natural gas or bio-gas [3]. Compared to the gas-fired power stations, for the same quantity of raw material, FCs show higher efficiency in the energy generation and decrease the contaminants released into the atmosphere. Thus, FCs are an alternative and clever way for using the resources which allows for a stepwise approach towards a future hydrogen-based economy.

2. Materials for solid oxide fuel cells

2.1. The Solid Oxide Fuel Cell (SOFC)

Between the different types of FCs, those based on solid oxides (SOFC) have found a place in a wide range of powers, between the large mobile (cars) and small to intermediate stationary (hospitals, schools) applications (Figure 1). These cells are operated between high (1000ºC-800ºC) and intermediate (600ºC-400ºC) temperature ranges and the principal characteristic is that their fundamental parts consist on ceramic materials. Some advantages are the reversibility of the electrode reaction, low internal resistance, high tolerance to catalytic poisons, and high quality residual heat production; which can be used in other applications such as internal fuel reforming or heating.

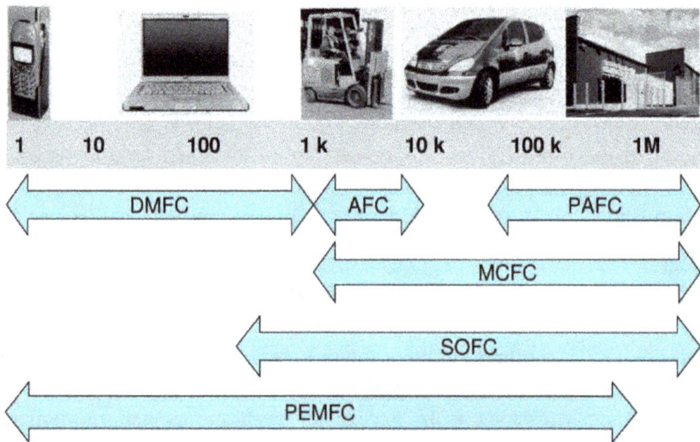

Figure 1. Different fuel cells and their power range. Abbreviations refer to the type of fuel cell: DMFC: Direct Methanol, AFC: Alkaline, PAFC: Phosphoric Acid, MCFC: Molten Carbonate, SOFC: Solid Oxide, PEMFC: Proton Exchange Membrane.

The core of a conventional SOFC design involves three main parts: a cathode, an electrolyte and an anode (Figure 2). Additional components such as sealants and connectors are also needed. The reduction of the Oxygen molecule (O_2) takes place in the cathode material, incorporating two electrons from the external circuit; afterwards the O^{2-} ion travels through the electrolyte to the anode material, where it oxidizes the hydrogen molecule. In this way a water molecule and two electrons are released, closing the electrical circuit.

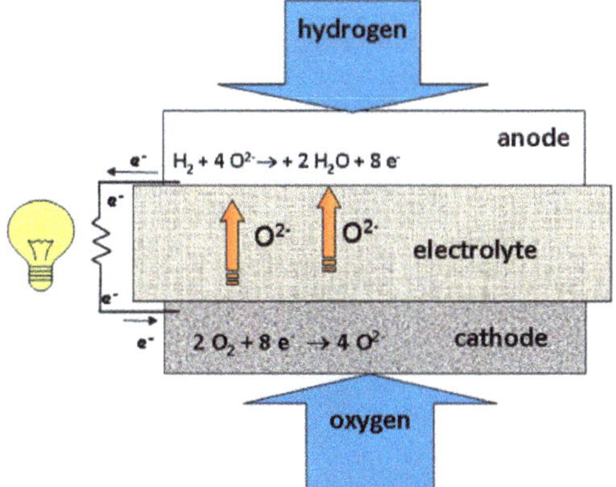

Figure 2. Sketch of a conventional SOFC array, using Hydrogen as fuel.

The output voltage (E) of a SOFC can be expressed as:

$$E = E_0 - (\Delta U_{Cathode} + \Delta U_{Electrolyte} + \Delta U_{Anode}) \quad (1)$$

Where E_0 is the open circuit voltage of the cell and $\Delta U_{Cathode}$, $\Delta U_{Electrolyte}$, and ΔU_{Anode} are the cathode, electrolyte and anode overpotentials respectively [4]. SOFC output voltage diminishes when reducing the temperature because all component (cathode, electrolyte and anode) overpotentials increase. Cathode overpotential is the one which augments more than the others, limiting the cell performance.

The overpotential is a measure of the energy required for the occurrence of the reactions and processes involved in the cell operation. For that reason one of the main challenges in SOFC design is to minimize all component overpotentials and specifically the cathodic overpotential. Good gas permeability and high electron (and Oxygen ion) conductivities of electrodes, as well as low inter-material reactivities, are essential requirements for a successful operation independently of the electrode/electrolyte nature. On one hand, combinations of mixed conductor cathodes such as rare earth doped perovskites with Ceria or Ytria based electrolytes have shown promising results. On the other hand, improving the electrode morphology or reducing cathode particle size, have been shown as good alternatives to decrease the cathode overpotential thus improving the general efficiency of the device. However, after improving the performance of each component, the bottle neck is to optimize the assembly electrode/electrolyte, with special attention in the interfacial zone.

2.2. Electronic and ionic conductors

Regardless of the electrode nature, the working mechanism of a SOFC can be understood as follows [5]: the porous cathode material (α) is in close contact with a ceramic electrolyte phase (γ) along an interphase (α/γ), which is exposed to a gas phase (β) rich in Oxygen. The Oxygen diffuses from an external source through the cathode pores and is reduced in some place of the α/β/γ interphase, called the tree-phase boundary or TPB. The α phase is connected in a point far away from this interphase to a current collector which provides a conduction path for the electrons. On the other hand, the electrolyte γ provides an ionic conduction path to transport the O^{2-} ions from the Oxygen reaction zone to the anode, where they can oxidize the fuel and be recombined into water.

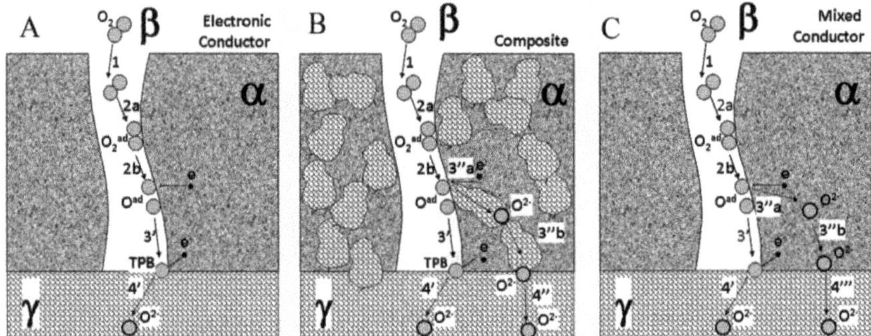

Figure 3. Sketch showing the Oxygen reduction reaction steps in pure electronic conductor (A), composite (B) and mixed conductor (C) cathodes.

Historically the first choice for cathode materials was pure electronic conductor oxides (Figure 3A). A good example of this kind of ceramics is the family of $(La,Sr)MnO_{3-\delta}$ (LSM) manganites. Some structural improvements to achieve better performance comprised decreasing the electrode grain size to nanometric levels in order to increase the active surface, or increasing its porosity to allow a better diffusion of the Oxygen molecules. However, in these materials the Oxygen reduction reaction is limited to the TPB zone, where the Oxygen ion can be transferred to the electrolyte in the same place where the reduction reaction occurs.

Another excellent way to increase the SOFC device performance is enlarging the area where the electrode reaction takes place. Different ways were explored and successful alternatives were detected. One of the most used solutions was enlarging the TPB area by incorporating to the electrode an ionic conductor with the composition of the electrolyte (Figure 3B). When using Ytria doped Zirconia (YSZ) or Gadolinium or Samarium doped Ceria (CGO and SDC) as electrolyte materials, LSM/YSZ, LSM/CGO and LSM/SDC combinations are often found as composite cathodes, respectively [6].

The discovery of ceramic materials which present mixed conductivity opened new perspectives in the SOFC design. Mixed conductors (MC) are able to conduct both Oxygen ions and electrons, and are therefore excellent as SOFC cathodes. Good examples are ABO_3 perovskites with rare-earth cations in the A site and transition metals in the B site. To them belong the emerging MC family of rare earth - iron doped cobaltites of composition $La_{1-x}Sr_xCo_{1-y}Fe_yO_{3-\delta}$ (LSCF) and related families. The B ions in these structures are surrounded by Oxygen octahedral, and the ionic conductivity is achieved by the diffusion of Oxygen vacancies. This property bypasses the need of a TPB in this sense: the MC cathode material allows the Oxygen reduction reaction occurrence over the entire cathode/gas interphase (α/β), at the same time that provides a conduction path through which the O^{2-} ions can travel towards the electrolyte γ (Figure 3C).

2.3. Oxygen reduction reaction and cathode performance

A strategy to reduce cathode overpotential is to facilitate the Oxygen reduction reaction (ORR) at the cathode. Besides the different reaction paths proposed in the literature, there is a general consensus about that ORR involves the following steps [5] (see Figure 3):

1. O_2 gas-phase diffusion through cathode pore and/or boundary layer.
2. (a-b) O_2 adsorption / desorption and successive dissociation. This stage can also occur in one single step of dissociative adsorption. In addition, Oxygen can be partially reduced thus forming electroactive species.
3. Transport of Oxygen and/or electroactive species to cathode/electrolyte interphase.

   (3′) This stage consists in surface diffusion of adsorbed Oxygen (O^{ad}) and/or electroactive species (see Figures 3A, B and C).

   (3′′a-b) Only for MC materials, this stage can also involve the following steps (see Figures 3B and 3C): (3′′a) charge transfer (Oxygen reduction) and ionic incorporation into cathode material, and (3′′b) O^{2-} diffusion through bulk cathode.

   It is worth to note that in the case of cathodes composed of a mixture of pure electronic and ionic conductor materials, this stage can occur at the points where the gas and the ionic conductor are in contact only if the ionic conductor is also in contact with the electrolyte (see Figure 3B).
4. Incorporation of electroactive species into the electrolyte

   (4′) For pure electronic materials, this stage can only take place at the TBP (see Figure 3A). (4′′) In the case of cathodes composed of a mixture of pure electronic and ionic conductor materials, this stage can occur at the TBP and at the points where the ionic conductor is in contact with the electrolyte (see Figure 3B).

   (4′′′) In the case of MC materials, this step can occur in the whole cathode/electrolyte interphase including the TPB (see Figure 3C).

Because of the variety and complexity of the involved processes, it is difficult to know which of the steps mentioned above actually happens during the ORR and if they occur simultaneously or successively. However, in real systems some stages are slower than others and, hence, the ORR rate is defined by them. The ORR limiting step or steps depend on several parameters such as temperature, Oxygen partial pressure, cathode/electrolyte interphase and cathode microstructure, composition and electronic nature (i.e. if the material is a pure electronic conductor, a pure mixed conductor or a composite).

Electrochemical impedance spectroscopy (EIS) is a useful technique for studying component performance as well as for determining the limiting steps of the reactions occurring in a SOFC [7]. It consists in applying an electrical signal (voltage or current) of small amplitude and recording the response (current or voltage). From the relation between the voltage and the current it is possible to obtain the impedance $Z = |Z|\, e^{j\omega t+\phi} = Z' + j\, Z''$ (where $|Z|$ is the amplitude, ϕ is the phase, Z′ is the real component and Z″ is the imaginary component). As the input frequency signal (ω) is varied, an impedance spectrum is obtained which can be represented in a Nyquist (Figure 4A) or in a Bode plot (Figure 4B).

Several microscopic phenomena happen when an electrical signal is applied to an electrochemical system such as a SOFC. The variation of applied frequency allows discriminating processes characterized by different time constants. Generally speaking, the selection of frequency range allows focusing on the phenomena taking place in certain component of the cell. The flow of charged particles (current density) depends on the ohmic resistances of electrodes and electrolyte and on the rate of the reactions occurring in the system. The overpotential (ΔU) of the system or cell component is related to the current j by $\Delta U = j \cdot ASR$, where ASR stands for area specific resistance. This value can be estimated as the difference between the low frequency ($\omega \rightarrow 0$) and the high frequency ($\omega \rightarrow \infty$) intersection of the spectrum with the Z′ axis (see Figure 4A). A reasonable value of maximum ASR is 0.45 Ωcm^2 for the entire fuel cell [2], yielding to an average maximum value of 0.15 Ωcm^2 for each cell component (i.e. cathode, anode and electrolyte).

EIS is also a useful technique for identifying the rate limiting mechanisms of the reactions involved in a fuel cell. In order to do so, several impedance spectra must be recorded as a function of operation variables as temperature, oxidant or fuel gas partial pressure, carrier gas, and bias potential or current. Then, all EIS spectra must be fitted with an equivalent circuit using a Complex Nonlinear Least Squares (CNLS) procedure. Impedance expressions of the element generally used in equivalent circuits can be found in reference [7]. The equivalent circuit is proposed according to the shape of spectra and the nature of the system in study. The identification of the reaction limiting steps is made by analyzing the variation of the fitted parameters as a function of operation conditions.

Figure 4 shows an example of cathode spectrum fitted using an equivalent circuit [8]. This spectrum corresponds to the Oxygen reduction reaction at a nanostructured $La_{0.4}Sr_{0.6}Co_{0.8}Fe_{0.2}O_{3-\delta}$ cathode. Points in Figures 4A and B represent experimental data, continuous lines represent equivalent circuit fitting and dotted lines represent the individual contributions. The equivalent circuit used for fitting is shown in Figure 4C. The analysis of the circuit parameter as function of temperature, Oxygen partial pressure and gas carrier allows identifying the ORR limiting steps. These are O_2 gas-phase diffusion and Oxygen ion transport into cathode bulk (steps 1 and 3″b in Figure 3C, respectively). In Figure 4C, the sub-circuit composed of a resistance (R) in parallel with a constant phase element (CPE) is related to Oxygen gas-phase diffusion whereas the finite-length Warburg element (W) represents the Oxygen ion transport into cathode bulk.

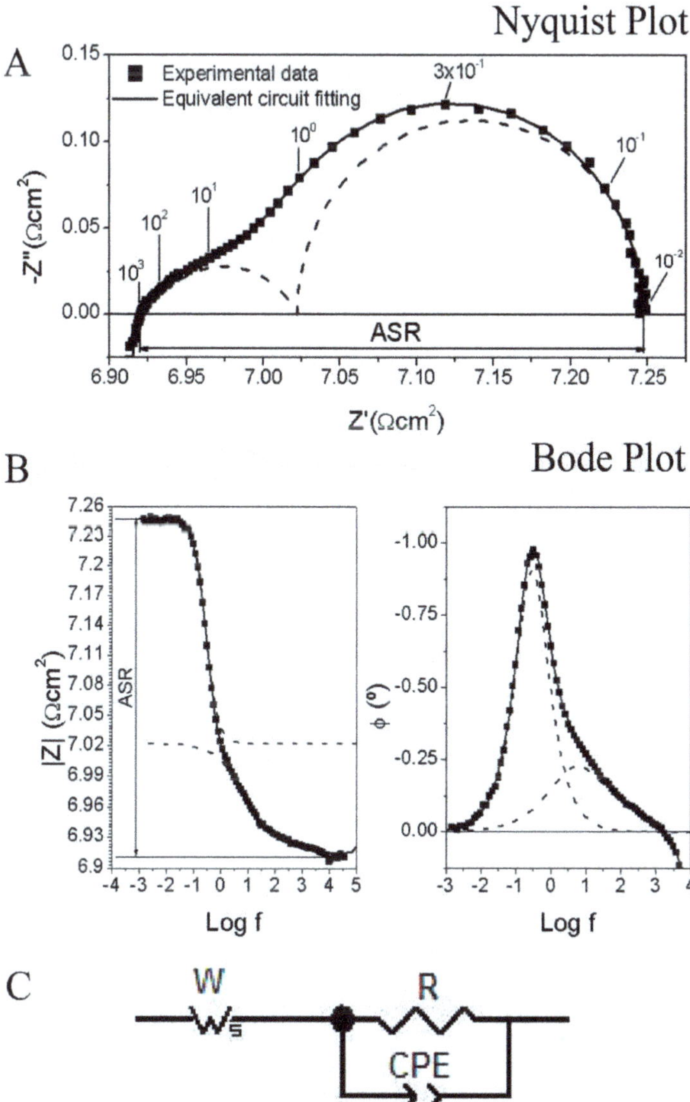

Figure 4. Example of an impedance spectrum [8]. (A) Nyquist and (B) Bode plots. Numbers in Figure A indicate the measuring frequency. Points are experimental data, continuous lines represent equivalent circuit fitting and dotted lines represent the individual contributions. (C) Equivalent circuit used for fitting.

3. Interphases and interfaces in SOFC

The region where two systems meet is known as "interface" or alternatively as "interphase". The former word is often used when two different crystalline orientations meet at a specific face (or surface), while the latter is used to describe the convergence of two different phases. Both concepts are involved in the study of fuel cells. Inside a SOFC core there are two main interphases that, as stated before, influence the performance of the cell: they are the cathode/electrolyte and the anode/electrolyte interphases.

3.1. The role of the electrode/electrolyte interphase in a SOFC

The Oxygen reduction reaction at the cathode/electrolyte boundary or the hydrogen oxidation reaction at the anode/electrolyte boundary are principally determined by the charge transfer capacity at the interphase and by those structural parameters which affect the gas diffusion in the solids and its concentration in the adsorption surfaces [9, 10]. Although these reduction and oxidation mechanisms are not fully understood, it is generally accepted that the reactions follow different stages and involve more than one reaction path. Therefore, the reaction kinetics at the interphases strongly depends on chemical and structural properties of the cell components [6, 11-13], the electrode morphology (particle size, porosity, thickness, etc.) [14], the characteristics at the interfacial boundary [15, 16] and other working parameters as the operation temperature and the gas partial pressure [17].

3.2. Sample preparation and interphases characteristics types

The three primordial SOFC parts (cathode, electrolyte or anode) can be assembled in several ways. They differ in which part of them constitutes the mechanical supporting element and in the different possible deposition methods used to mount the assembly. Thus, the mechanical properties and characteristics of the interphase zone strongly depend on the selected technique.

Two designs established a precedent in the SOFC market: cells supported by the anode and cells supported by the electrolyte. In both cases the supporting part is pressed as a dense pellet of some millimeters thickness and the other two parts are deposited as layers on its surfaces.

Painting, spraying and coating are the cheapest and less complicated available techniques to deposit layers between 200 nm and 20 μm on dense pellets. The interphases obtained are generally incoherently grown and often contain zones of no adherence (interphase pores or voids). Such an interphase could be an advantage for electronic conductor cathodes for example, because they need higher abundance of TPB points. However, it is clearly a disadvantage for MC materials, because the presence of interphase pores decreases the area where the Oxygen can be effectively transferred towards the electrolyte. In all these slurry deposition methods the material to be deposited must be transformed into an ink of controlled viscosity. They differ in the way the ink is placed on the pellet. In dip coating methods for example, the dense pellet is immersed in the ink, taken out at a controlled rate and let dry [18]. On the other side, in spin coating techniques, the ink is dropped into the pellet surface, while it spins at a certain angular velocity [19]. The desired layer thickness is obtained after repeating the processes a number of times. After that, thermal treatments are applied to improve the adherence to the base material. Presence of undesired micro-fractures, pores, delaminating zones, etc. can be controlled by changing deposition parameters such as characteristic times, velocities, ink composition, viscosity and thermal treatments.

Screen printing (also known as serigraphy) is other mechanical process used to deposit an electrode (or electrolyte) layer on the supporting surface [20]. A paste made of the electrode powder is pressed through a stencil with open zones. As a consequence, the material deposits only in these open zones. The advantage of the method is that the electrode can be painted thus forming the desired intricate patterns to tailor the cell design. These layers have a thickness of a several μm . Electrophoretic deposition is another method that can be used to deposit dense uniform coatings with thicknesses in a wider range. In this case the charged electrode particles are dispersed in a fluid and migrate to the substrate forced by a potential gradient [21, 22]. Besides, spray pyrolysis method involves atomizing a solution of the material's precursors salts onto a heated substrate, using either air pressure or high voltage [23].

A refined deposition technique is Pulsed Laser Deposition (PLD) [11]. Usually this technique is used to grow the electrode on an electrolyte supported cell. The cathode is synthesized as a dense core, which is after ablated by Laser pulses at controlled energy and frequency. The sputtered cathode material is deposited on the electrolyte pellet in atomic-size layers. This methodology allows not only depositing ultra-thin layers and forming coherent electrode/electrolyte interfaces, but also building composite cathodes by changing the ablating core successively between two different compositions. Besides, it allows shaping the cathode with special nano-structured morphologies that improve the element performance. An excellent example are the vertically aligned nano-pores (VANP), which enhances Oxygen-gas phase diffusivity [24, 25].

3.3. The effect of different interphases and micro-nanostructures in the SOFC performance

It is well known that, for the same compositions of the trio cathode/electrolyte/anode, the performance of a SOFC can be improved or deteriorated by modifying the micro/nano-structure of the components and the nature of their interphases. Jorgensen et al. reported for example, that a less dense microstructure with smaller grain sizes in LSM/YSZ composite cathodes enlarged the TPB active zone [6]. Beckel et al. characterized $La_{0.6}Sr_{0.4}Co_{0.2}Fe_{0.8}O_{3-\delta}$ and $Ba_{0.25}La_{0.25}Sr_{0.5}Co_{0.8}Fe_{0.2}O_{3-\delta}$ cathodes on YSZ electrolytes and found that modifying the cathode composition or the microstructure the assembly showed a significant enhancement in performance [13]. The authors reported that reducing the grain size and introducing a dense layer between the porous cathode and the electrolyte were favorable structure modifications. On the other side, Chen et al modified the microstructure of a NiO/YSZ composite anode instead of the cathode materials [26]. In this way, they improved the cell performance by reducing the particle size to the sub micrometric level and increased the material porosity by adding organic substances (to nucleate pore formation) in a mixture 5:3. The same effect was reported by Li et al, who demonstrated that the SOFC performance could be enhanced by modifying the TPB size and density and thus optimizing the microstructural properties of a Ni/YSZ composite cermet anode [27].

Regarding the influence of the composition in the cell performance using electronic conductors or composites, Barbucci et al. reported the electrochemical characteristics of rare earth doped manganites and Yttria stabilized Zirconia (LSM/YSZ) composite cathodes, an excellent material for intermediated temperature (IT-) SOFCs [28]. Using potentiometric and complex impedance analyses they found that a volume ratio of 1:1 enlarges the electrochemical activity, because it extends the TPB region inside the electrode volume. At the same time, Chen et al. compared the performance achieved by cathodes of pure LSM, LSM/YZS composite and LSM/SDC composite using YSZ electrolytes. In that study the LSM/SDC cell reached the highest performance at an operating temperature of 700ºC, and the efficiency difference was related to both the gas diffusion rate and the TPB size in each of the three arrays [29].

On the other side, the electrode/electrolyte interphase using MC cathodes is limited only to the small area where both materials make contact. Nowadays, the challenge with these promising materials tends both, to maximize the contact area improving the deposition methods and to increase the mixed conductivity by modifying the material composition and microstructure. Following the idea of composite cathodes, Dusastre et al. investigated mixtures of $La_{0.6}Sr_{0.4}Co_{0.2}Fe_{0.8}O_{3-\delta}/Ce_{0.9}Gd_{0.1}O_{2-\delta}$ and found that the electrochemical properties of the composite cathode mounted on pure CGO electrolytes were highly dependent on the electrode microstructure and its composition [30]. The authors found that the porous electrode performance is affected not only by the mixed conduction transport but also by catalytic properties inherent to the TPB and the gas transport to and from the TPB. Afterwards, Baqué et al synthesized $La_{0.6}Sr_{0.4}Co_{0.8}Fe_{0.2}O_{3-\delta}$ powders with nanometric sized grain and deposited it by different methods on CGO dense electrolytes [18]. The authors demonstrated that this array reached ASR values of 0.4 Ω cm^2 at 450°C and 0.18 Ω cm^2 at 550°C in air [31]. This excellent electrochemical performance is even better than many of the cathodes and composite cathodes known at the moment for these operating temperatures. This especially attractive behavior is explained by two main factors: (i) the increase in the cathode active zones. This is due to the nanometric particle sizes, the increase in porosity and improved connectivity. (ii) The improvement of the electrode/electrolyte interface. This is a consequence of the combination of the used deposition method and the optimized electrode/electrolyte composition [15].

3.4. Reactivity at the interphase

A good electrode/electrolyte interphase in SOFC assemblies must fulfill two main requirements: first of all it must facilitate the transfer of the O^{2-} ions in an efficient way from the cathode to the anode and second, the interfacial reactivity between the components must be minimized or avoided at the working temperatures and under the presence of reductive/oxidative atmospheres, for the entire cell lifetime.

However, due to the SOFC hard operating conditions, the presence of electrode/electrolyte reactivity is not rare. In a recent review of solid state chemistry applied to SOFC materials Backhaus-Ricoult [10, 32] studied different solid oxide cathodes with Zirconia based electrolytes. The author found that formation of reaction products, diffusion, segregation and modification of the electronic structure in the cathode/electrolyte interphase affected the Oxygen exchange rate and had an undesirable impact on the SOFC performance. Bevilacqua et al. studied, by different X-ray absorption spectroscopy techniques, the composition and speciation of Ni and Fe in the cathode material $LaNi_{0.6}Fe_{0.4}O_3$ deposited on YSZ or SDC electrolytes [33]. Under operation conditions the authors found that the $LaNi_{0.6}Fe_{0.4}O_3$ /YSZ assemblies formed an isolating layer of $La_2Zr_2O_7$ at the interphase, while LNF/SDC showed no parasite phases and excellent electrochemical properties, together with an enlargement of the TPB. On the other side, Grosjean et al. studied the behavior of $La_{0.8}Sr_{0.2}MnO_3$ on YSZ electrolytes and found a decrease in the electrical performance of one order of magnitude, together with a quick deterioration after a long term treatment at 850ºC [34]. The authors correlated the efficiency loss to both an inter-diffusion of Mn and the presence of the resistive $La_2Zr_2O_7$ phase. This conditions lead to a rise in the electronic conductivity and a decrease in the ionic conductivity, deteriorating the cell performance. Besides, Izuki et al. studied a $La_{0.8}Sr_{0.2}Co_{0.2}Fe_{0.8}O_x$ cathode deposited by PLD on a CGO substrate [35]. After long term operation at 1000-1200 ºC the authors observed inter-diffusion of La, Ce and Gd across the interphase, and presumed the formation of Lanthanum doped Ceria in the vicinity of the interface. Recently, Montenegro Hernandez et al. (2012) investigated the reaction of the rare earth niquelates La_2NiO_4 , Nd_2NiO_4 and Pr_2NiO_4 deposited by spin coating on CGO and YZS electrolytes [36]. After 1000 hours operation the authors found by X-ray diffraction analysis and complex impedance measurements significant performance deterioration. This could be in part associated with the formation of new phases, product of decomposition of cathode and electrolyte materials at the interphase.

4. Techniques for characterization

4.1. Scanning electron microscopy

In a SOFC assembly the change from one phase into the other occurs in the nanometer scale. Therefore, for a precise characterization of the interfacial zone, high quality sample preparation and high spatial resolution techniques are required. A first appropriate tool to find out morphology and chemical composition at both sides of the interphase is the Scanning Electron Microscope (SEM). High magnified images can be acquired using a Field Emission Gun (FEG-SEM) device instead a conventional Tungsten filament. The bright source and the wide field depth allow producing a topographic image of high magnification. A coarse view of the material at both sides of the interphase can be thus be achieved by detecting secondary electrons emitted from the sample (Figure 5A). Besides, these microscopes are equipped with backscattering electron detectors, which provide an intensity profile depending on the atomic number of the target atom. A complementary Energy or Wavelength Dispersive System (EDS or WDS, respectively) coupled to the microscope makes possible to map the composition by analyzing the characteristic X-ray spectra.

4.2. Transmission electron microscopy

A powerful technique to investigate (even at the atomic level) crystal structure and chemical composition is the Transmission Electron Microscope (TEM). In this case, the electron beam is allowed to pass through the sample and interacts with the material. The intensity of transmitted and/or diffracted electrons is afterwards converted into images or diffraction patterns, whose contrast can be (carefully) related to the material properties [37]. In addition, EDS systems can also be coupled to this microscope, with the advantage that the interaction volume is very small as compared to the SEM situation. However, the difficulty with this method is that, to let transmission happen, the sample must have thicknesses of about 0.1 μm or less in the region of interest. How to get those thin samples from an electrode/electrolyte interphase can be very challenging: these are regions of superior reactivity, which often contain boundaries between different microstructures, chemical compositions, densities, hardness, grain sizes, etc. Traditional TEM preparation methods as Ar-milling or electrochemical polishing, which may affect differently both phases, are therefore inappropriate. A solution to that problem arrived with Focused Ion Beam (FIB) technology, which is one of the most versatile tools to prepare thin samples of site-specific areas for microscopy and spectroscopy analyses [38].

4.3. Focused ion beam

A desired area can be selected with micrometric precision with help of a dedicated SEM installed in the same chamber as the FIB (Figure 5A). A thin Pt-layer is deposited on the surface to protect the underlying material from the sputtering process and a Ga-ion beam is used to cut a slab of the bulk material (Figures 5B and 5C). This slab can be further polished to a final thickness of about 100 nm or less. After, the foil is deposited with a micromanipulator in an adequate substrate and is ready for analysis [39]. This procedure is known as FIB/lift-out preparation, and is often found related with SEM and TEM *ex-situ* analysis [40]. On the other hand, the FIB technique can be also used for site-specific *in-situ* analyses. FIB-SEM nano-tomography for example, is more and more being used for a detailed three dimensional (3D) reconstruction of the micro- and nanostructure in a material. In this case, the FIB is used stepwise to mill a slice in the selected area. The sample is then alternatively micrographed with SEM or EDS and further milled in slices. The 3D reconstruction is possible using computer programs to merge the images together. Porosity, tortuosity, phase composition, percentage of adherence, etc. can be obtained from these reconstructions.

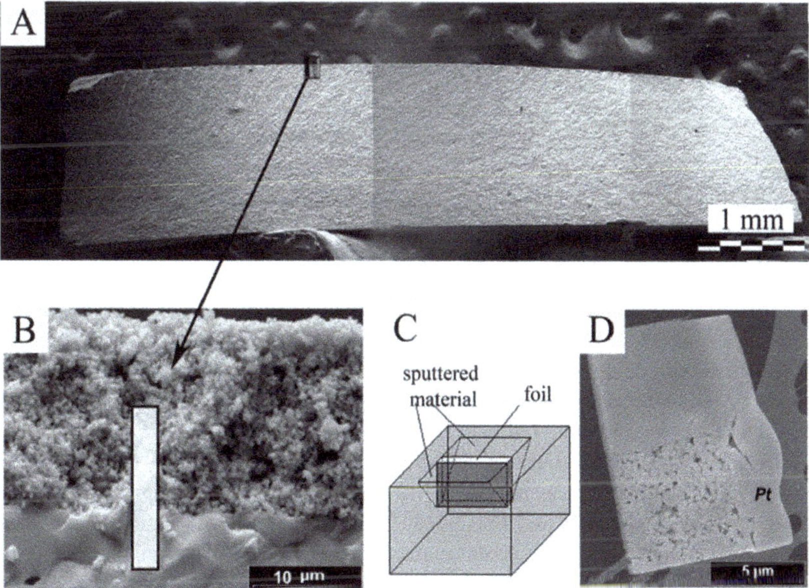

Figure 5. (A and B) SEM images showing the cathode/electrolyte interphase in an electrolyte-supported SOFC; the images were obtained from a fractured piece. (B) A Pt layer is deposited on the surface of interest to protect the underlying material. (C) Schema of a FIB cut: a Ga-ion beam is used to sputter the material in front and back of the region of interest. (D) TEM image of the obtained foil: a 10×15×0.1 µm slice is extracted from the bulk and deposited in an appropriate substrate using a micromanipulator.

Since 2005 the increasing number of publications using FIB to study SOFC materials shows that the methodology's usefulness in this field has been recognized. The fact that the beam homogeneously mills the whole sample makes it an attractive method to prepare thin specimens of the interfacial SOFC regions, where two very different materials converge in a sensible and reactive area [41]. Liu and Jiao (2005) reported for the first time FIB/lift-out prepared TEM samples of the anode/electrolyte interface from a long-term tested Ni/YSZ-YSZ half-cell [42]. One year later Grosjean et al. (2006), investigated the diffusion and reactivity processes at the interphase of a planar SOFC, between an YSZ electrolyte and a $La_{0.8}Sr_{0.2}MnO_3$ cathode [34]. Using FIB/lift-out samples the authors analyzed the microstructure and made site-specific nano-scaled chemical analysis at both sides of the interfacial zone. Recently, Soldati et al. (2011) studied by FEG-SEM, TEM and EDS, FIB/lift out foils of the interfacial region between nano-sized LSCF cathodes and dense CGO electrolytes [15]. The authors compared the effect of different synthesis routes in the interfacial characteristics and found correlations with the cell performance.

Regarding the 3D-FIB/SEM profiles, the first report applied to SOFC materials corresponds to the group of Wilson et al., who studied with this technique a thin YSZ electrolyte layer cast onto a thick NiO–YSZ anode support, with a LSM-YSZ composite cathode; 3D-microstructural features such as phase volume fractions, total phase boundary areas, TPB lengths and density, connectivity, tortuosity, etc. could be related to cathode polarization resistances and cell performance [43]. These authors analyzed also the properties of composite LSM-YSZ cathodes [44, 45]. Smith et al. used the same technique to find the relationship between the cathode microstructure and the electrochemical performance in symmetric LSM-YSZ-LSM cells [46]. Afterwards, Holzer et al. statistically characterized size, shape and topology of complex granular cermets, while Jiao et al. reported the 3D microstructure of a NiO-YSZ anode and $72.5ZnO–27MnO_2–0.5Al_2O_3$ cathode [47] and Chae et al. studied the internal microstructure of Mg-, Sr- and Co- doped $LaGaO_3$ powders pellets used as SOFC electrolyte [48].

Secondary effects due to the FIB preparation must be evaluated in *in-situ* as well as in *ex-situ* samples. Shearing et al. for example, commented on some problems associated with charging, shadowing and re-deposition of sputtering material observed in FIB-SEM profiles that could be overcome using coupled FIB/lift-out preparation and *ex-situ* SEM imaging [49-51]. On the other side, Soldati et al. (2012) compared two FIB/lift-out prepared foils of the same Nd_2NiO_4 / $Ce_{0.9}Gd_{0.1}O_{1.95}$ interphase: in one case the foil was extracted from the "as prepared" bulk material and in the second case Pt was deposited prior to the extraction filling all material pores; TEM and FEG/SEM-EDS analyses showed no differences between both samples, demonstrating that no detectable re-deposition or structural change occurred because of the FIB preparation [36, 52].

5. Some particular case studies[1]

5.1. The case of LSCF/CGO adherence

When evaluating a SOFC design, the electrode/electrolyte adherence and the nature of its interphase are relevant points to take into account to improve the performance. A well connected porous interphase would be desirable for electronic conductor cathodes, but a full covered contact area would be preferred in case of a mixed conductor cathode. In the former case, the existence of pores at the interphase plane would enlarge the TPB zones where the electrolyte, the cathode and the gas phases coexist, thus enlarging the points where Oxygen can be reduced and transferred. In the latter case, a TPB is not mandatory. A porous interphase would reduce instead the cell performance by reducing the cathode/electrolyte contact area for Oxygen transfer.

Figure 6 shows TEM images of two FIB foils extracted from the interfacial region of two different SOFC assemblies. In both cases the dense electrolyte is a $Ce_{0.9}Gd_{0.1}O_{2-\delta}$ polished pellet and the cathode corresponds to nanostructured $La_{0.4}Sr_{0.6}Co_{0.8}Fe_{0.2}O_{3-\delta}$ synthesized by a modified acetate route [31]. The cathodes were deposited by spin coating forming a 5 μm or a 15 μm layer (upper and bottom figures respectively). No micro-fractures, delamination or damaged boundaries were observed, supporting a good electrode/electrolyte attachment. In addition, no amorphous or re-deposited material was found inside the pores indicating the

[1] In this section the acronnysms LSCF and CGO are used specifically for the compounds $La_{0.4}Sr_{0.6}Co_{0.8}Fe_{0.2}O_{3-\delta}$; and $Ce_{0.9}Gd_{0.1}O_{1.95}$.

absence of secondary effects due to the FIB preparation [34, 42]. However, although both samples were thermally treated and prepared in the same way, the interphase presented a different microstructure: The 5 μm sample (Figures 6A to 6C) showed a partially open interphase, presenting many regions where the cathode layer did not cover the electrolyte (i.e. interfacial pores), while the 15 μm sample (Figures 6D to 6F) presented an interfacial boundary with almost 100% coverage factor.

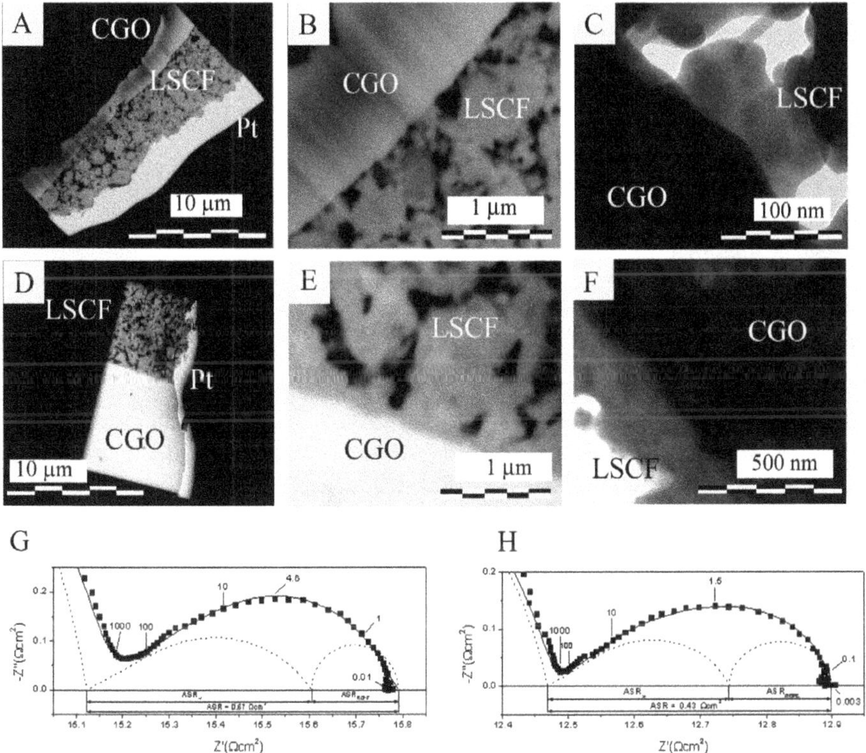

Figure 6. (A-D) FEG-SEM and TEM images of two LSCF/CGO assemblies showing the cathode/electrolyte interphase. The cathodes were deposited by spin coating. In the (A-C) images the interphase observed in the foil presents regions of no adherence called "interfacial pores" or "voids". The (D-F) images show a totally covered interphase. Impedance measurements of both arrays are shown in (G and H). LSCF=$La_{0.4}Sr_{0.6}Co_{0.8}Fe_{0.2}O_{3-\delta}$; CGO=$Ce_{0.9}Gd_{0.1}O_{1.95}$

EIS was used for evaluating the electrochemical performance of these assemblies as a function of temperature, Oxygen partial pressure and gas carrier [8]. The obtained results point out that, in both cases the ORR is limited by Oxygen gas-phase diffusion and Oxygen ion transport into the cathode bulk. Figures 6G and 6H show the impedance spectra of the two arrays measured at 500ºC in air and fitted with the equivalent circuit of Figure 4C. As it can be observed, the total area specific resistance (ASR) resulted lower for the thickest cathode ($0.43\Omega cm^2$, Figure 6H) than for the thinnest one ($0.67\Omega cm^2$, Figure 6G). ASR_w values related to Oxygen ion transport into cathode bulk are 0.27 Ωcm^2 and 0.48 Ωcm^2 for the thickest and the thinnest cathodes, respectively; while ASR_{RCPE} values related to Oxygen gas-phase diffusion are similar for both cathodes (i.e. 0.16 Ωcm^2 for the first cathode and 0.19 Ωcm^2 for the latter), as expected because of the similar pore structures. Accordingly, the difference in the total ASR can be mainly attributed to different efficiencies in the Oxygen ion transport into the cathode bulk. The ASR related to this step can be expressed as follows [53]:

$$\mathrm{ASR_w} = A\frac{RT}{4F^2}\frac{l_\delta}{SC_v|_{x=0}D}\frac{th\sqrt{j\omega\frac{l_\delta^2}{D}}}{j\omega\frac{l_\delta^2}{D}} \tag{2}$$

where A is the cathode geometric area, R and F are the ideal gas and Faraday constants, T is the temperature, l_δ is the effective diffusion length, S is the electrode/electrolyte interface area, $C_v|_{x=0}$ is the Oxygen vacancy concentration at the electrode/electrolyte interface, D is the vacancy diffusion coefficient and ω is the measuring angular frequency.

Both cathodes, which have exactly the same composition, were prepared in the same way and presented the same morphology and nano/microstructure. As a result, A, $C_v|_{x=0}$, and D in Eq. 2 are equal for both materials. In addition, a higher ASR_w is expected for a thicker cathode [54], contrary to the observed in Figure 6. Consequently, the ASR_w difference can be mainly attributed to the difference in interfacial contact area (S).

The dissimilar interphases in both samples may be due to some lack of reproducibility in the substrate polishing procedure or in the ink preparation, but they could also be inherent to the spin coating deposition. Similar results were also observed by other authors. Murray et al., for example, reported variations in the impedance results of LSCF spin coated samples, sometimes as large as a factor of 3, while attempting to reproduce the processing conditions from cathode to cathode [19]. Indeed, a systematic study including the characterization of substrate roughness and a comprehensive ink rheology analysis is needed to optimize the reproducibility of the electrode/electrolyte interphase.

5.2. The case of LSCF nano-crystals

The strategy generally adopted for enhancing the Oxygen ion transport into the cathode bulk is modifying its composition. Nevertheless, outstanding results can be also achieved by using nanostructured cathodes. The study of cathode nanostructure even at atomic level is very important not only for characterizing it accurately, but also for detecting detrimental impurities at grain boundaries which could block the Oxygen ion transport. This task can be done by combining FIB/lift-out and TEM techniques.

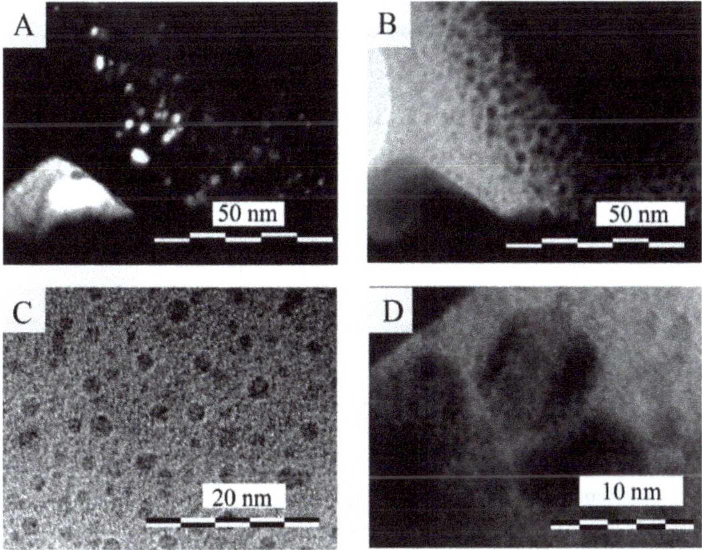

Figure 7. Dark Field (A) and Bright Field (B) TEM images of an $La_{0.4}Sr_{0.6}Co_{0.8}Fe_{0.2}O_{3-\delta}$ cathode showing nanocrystallites. (C and D) High Resolution TEM images detailing the nano-sized crystals.

Figure 7 shows TEM images from nanostructured $La_{0.4}Sr_{0.6}Co_{0.8}Fe_{0.2}O_{3-\delta}$ cathodes [8, 15, 31]. Bright spots reveal diffracting nanometric domains in the dark field (DF) image of Figure 7A. The inverted phenomenon is applied in the bright field (BF) image of Figure 7B and hence the nanocrystalline domains appear obscure. Figure 7C displays high resolution (HR) images where several nanocrystallites of 10 nm or less are surrounded by zones with some degree of crystalline disorder. Atomic planes inside the nanocrystallites can be clearly distinguished in the HR image shown in Figure 7D.

ORR reaction occurring at these nanostructured cathodes was studied by EIS as a function of temperature, Oxygen partial pressure and carrier gas [8]. The limiting steps are O^{2-} transport into the cathode bulk and Oxygen gas-phase diffusion. This reaction path was previously proposed by Grunbaum et al. for a $La_{0.6}Sr_{0.4}Co_{0.8}Fe_{0.2}O_{3-\delta}$ cathode with sub-micrometric grains [55]. The ASR_{RCPE} related to Oxygen gas-phase diffusion is similar for both cathodes [8], while ASR_{W} related to Oxygen ion transport into cathode bulk is more than two orders of magnitude lower for the nanostructured cathode (see Figure 8) at temperatures lower than 600°C. This significant difference cannot solely be explained by the different compositions [56]. Thus, this behavior can be attributed to an Oxygen diffusion enhancement as a result of the advantageous nanostructure exhibited by our cathodes. Nanostructured materials exhibit a considerable larger grain boundary volume than those with sub-micrometric grains. In particular, grain boundaries are zones with high density of defects and disorder. Hence, the diffusion coefficient is higher in these zones and the effective diffusion length is shorter yielding to lower ASR_{W} values (see Eq. 2).

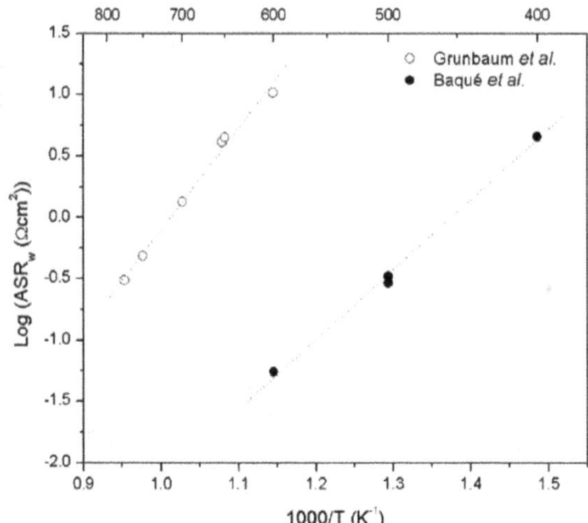

Figure 8. Area Specific Resistance (ASR_w) in nanostructured $La_{0.4}Sr_{0.6}Co_{0.8}Fe_{0.2}O_{3-\delta}$ cathodes (Baqué et al. [31]) compared to that reported for sub-micrometric grains with the same composition (Grunbaum et al.[55])

5.3. The case of a semi-coherent LSCF/CGO interphase

Cathode/electrolyte symmetric cells were assembled coating a dense $Ce_{0.9}Gd_{0.1}O_{1.95}$ pellet with an ink of nano-sized particles of composition $La_{0.4}Sr_{0.6}Co_{0.8}Fe_{0.2}O_{3-\delta}$ [18, 31]. The final thicknesses of the cathode layers were between 2 and 10 μm. The FIB/lift-out technique was used to sample the interphase area. Although the cathode porosity was high, and the cathode part was expected to be mechanically fragile, the foils were extracted and manipulated, keeping the original granular structure. Afterwards, the slices were studied with FEG-SEM and TEM, and the images obtained from two of them are shown in Figure 9 and 10.

Figures 9A and 9B show an LSCF/CGO foil that resulted in a slice of 10×15μm and ~100 nm thickness. The images were obtained with FEG-SEM by detecting backscatter electrons. Therefore a Z contrast is observed. The very light layer in the right side of the foil in Figure 9A corresponds to the Pt layer used to prepare the sample. A contrast of light and dark-gray tones is obtained for CGO and LSCF materials. CGO is observable as a dense pellet, while LSCF grains can be clearly recognized in the porous part.

A TEM view of the electrode/electrolyte interphase is shown in Figure 9F. The higher magnification achieved with TEM allows noting that the change from one phase to the other occurs in less than 200 nm. Diffraction in the electrolyte side (Figure 9C) showed a mono-crystalline pattern, corresponding to the CGO crystals. On the other side, the diffraction pattern of the LSCF cathode well far from the interphase shows a poly-crystalline pattern (Figure 9E). This is expected for random-oriented nano-sized particles. Conversely, the diffraction pattern of LSCF near the interphase (see star in Figure 9D) corresponds to a single crystal pattern. The spots coincide with that of the CGO diffraction. This means that the LSCF nano-grains near the interphase are not randomly oriented. Instead, they repeat the orientation of the underlying CGO grains.

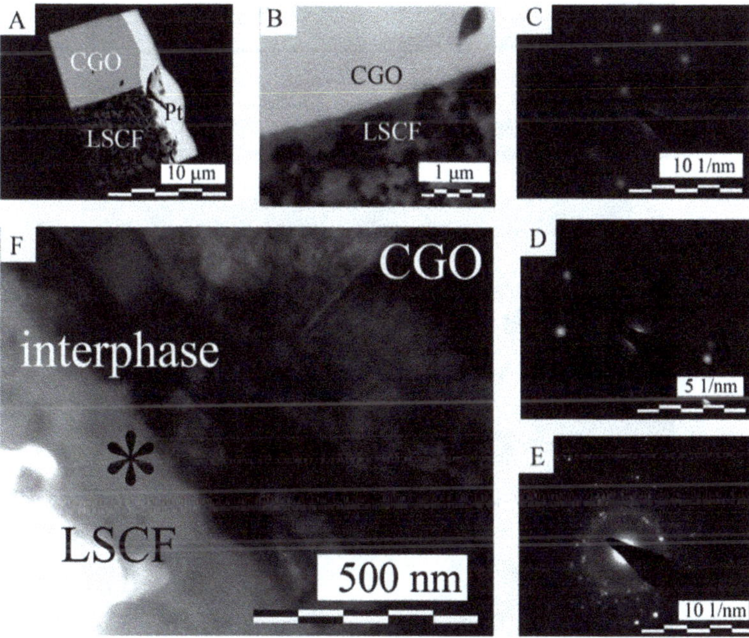

Figure 9. (A) FEG-SEM images of a thick FIB foil detecting backscattered electrons. The gray contrast is due to differences in the atomic number Z; Pt is observed as a very bright layer, while CGO grains are light gray and LSCF nano-grains are dark colored. (B) Detail of the interphase plane. Electron diffraction patterns of the electrolyte (C), the interphase region (D) and the cathode far from it (E). (F) TEM image of the LSCF/CGO interphase; the star indicates the position where the diffraction pattern shown in (D)·was obtained. LSCF=$La_{0.4}Sr_{0.6}Co_{0.8}Fe_{0.2}O_{3-\delta}$; CGO=$Ce_{0.9}Gd_{0.1}O_{1.95}$

Figure 10 shows a LSCF/CGO foil that resulted thinner than the one observed in Figure 9. In this case, no Z contrast can be detected using backscattered electrons in the FEG-SEM (Figure 10A). However, exactly because this foil is very thin, it is excellent for doing high resolution (HR) TEM images.

Indeed, a close view of the interphase (Figure 10B) at higher magnification (Figure 10C) using the high resolution (HR)-TEM mode demonstrated that the crystal planes of CGO and $La_{0.4}Sr_{0.6}Co_{0.8}Fe_{0.2}O_{3-\delta}$ are semi-coherently oriented along the interphase [15]. The fact that LSCF may growth semi-coherently with the substrate seems to be the result of using an ink prepared with powders at temperature slightly below of the final phase formation temperature. [31].

Figures 9 and 10 demonstrate that the LSCF and the CGO atoms near the interphase are not randomly oriented. Furthermore, it could be proved using two different microscopic techniques (electron diffraction in a thicker foil and HR-TEM in a thinner foil) that the interphases are semi-coherently oriented. This ordered transition implies that the Oxygen

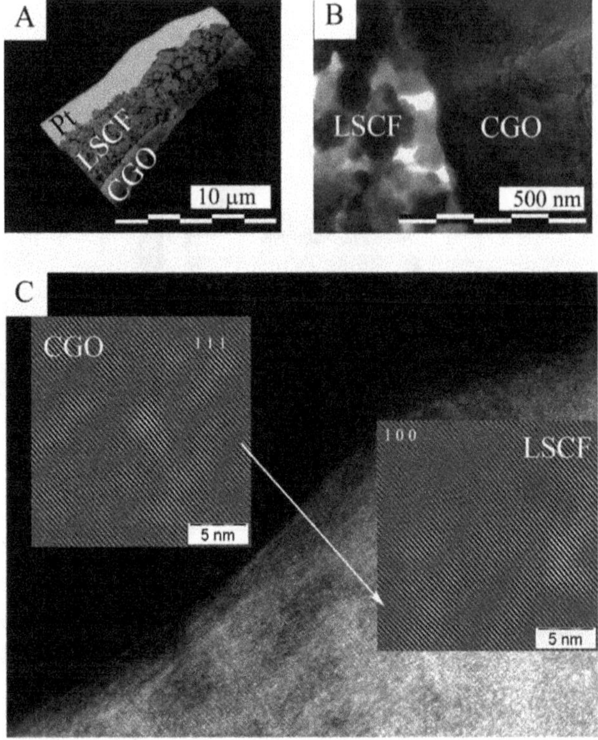

Figure 10. (A) FEG-SEM image of a FIB foil detecting backscattered electrons. Pt is observed in light colors. This sample was extremely thin and therefore no Z contrast is observed between the LSCF and the CGO materials. (B) TEM image showing a detail of the interphase plane; the nano-grains of the cathode and the micro-grains of the electrolyte are observed. (C) High resolution TEM image of the LSCF/CGO interphase, showing a semi-coherent interphase; the two insets are Fourier filtered reconstructions using only the 111 crystal plane in CGO and the 100 plane in LSCF. LSCF=$La_{0.4}Sr_{0.6}Co_{0.8}Fe_{0.2}O_{3-\delta}$; CGO=$Ce_{0.9}Gd_{0.1}O_{1.95}$

atoms can be easily transferred from the cathode to the electrolyte by Oxygen vacancy mobility between the components. These structural properties contribute thus, to increase the cell performance by enhancing the Oxygen ionic transfer through the cathode/electrolyte interphase and thus, lowering the ASR values [5].

6. Overview

There are numerous examples in the SOFC field that illustrate how dependent the cell performance of the electrode/electrolyte interphase characteristics is. The combination of site-specific sampling techniques, with high spatial resolution microscopy and electrochemical impedance spectroscopy, is an excellent way to evaluate a given design for improving the cell performance. Several cases are given in the different sections of this text and can be resumed in two main groups:

- Studies correlating the cell performance with the interphase micro-structure:

An interesting case of this group was reported in reference [57]. The study comprises two SOFC assemblies, where $La_{0.4}Sr_{0.6}Co_{0.8}Fe_{0.2}O_{3-\delta}$ cathodes were deposited on $Ce_{0.9}Gd_{0.1}O_{2-\delta}$ electrolytes by spin coating methods. The cathodes and the electrolytes presented the same morphology and composition, and the sintering treatments applied to favor adherence were similar. However, electrochemical impedance studies demonstrated that the efficiency of one cell was very low in comparison to the other. A microscopic and spectroscopic characterization of FIB foils obtained from the cathode/electrolyte region, from both assemblies, showed that the interphase morphologies were different. In one case the electrode/electrolyte interphase presented interfacial pores and voids, and in the other case it was completely covered which explained the better performance. Other examples where the interphase micro-structure was related to the cell performance was reported in reference [15]. This study reported enhanced electrochemical performance in LSCF cathodes synthesized by the novel HMTA route [31]. High resolution TEM images demonstrated that the CGO and LSCF atoms at both sides of the interphase were semi-coherently oriented. The ordered structure was thus responsible of facilitating the Oxygen ion transfer by Oxygen vacancy diffusion. Consequently, a decrease in the array's ASR value produced an increase in the cell electrochemical performance.

- Studies correlating the cell performance with presence of reactivity at the interphase:

Studies that comprise the characterization of the interphases after long term operation conditions are excellent examples of this group. A SOFC operates at intermediate to high temperatures and reducing/oxidizing conditions. In addition, to make these devices economically competitive, a long lifetime span with minimal maintenance is desirable. Thus, chemical reaction or diffusion between the components are problematic because they alter the cell integrity and directly affect the performance [58]. A well known case is that of $La_{0.8}Sr_{0.2}MnO_3$ cathodes mounted on ZrO_2-based electrolytes (YSZ) [34]. Cells operated at 850ºC showed one order of magnitude less electrical performance than expected, and deteriorated quickly under operating conditions. The causes were diffusion of Mn into the cathode and the electrolyte, and the formation of parasite phases. Another example regards a $La_{0.8}Sr_{0.2}Co_{0.2}Fe_{0.8}O_x$ cathode deposited as thick film onto sintered $Ce_{0.9}Gd_{0.1}O_{1.95}$ [35]. The inter-diffusion of La into CGO, and of Ce and Gd into LSCF together with the formation of lanthanum doped ceria near the interface, were observed after long term treatments at 1000-1200ºC. Of course, this behavior is undesirable because it deteriorates the cells integrity and consequently the electrochemical properties.

7. Conclusion

The cathode, the electrolyte, and the anode materials form the core of a SOFC assembly; therefore, to improve the cell performance it is necessary to tailor each of these three components. Microstructure, composition, and thermal treatments were largely recognized as potentially useful parameters to change. Common strategies involved were, adding traces of other elements to achieve mixed conduction in the cathode, reducing the particles grain size to increase the surface area, or increasing the quantity of pores (and their connectivity) to enlarge the gas diffusion across the electrodes. Once these parameters are optimized for each of the three components, the bottle neck is the optimization of the contact regions between them.

The electrode/electrolyte interphases are the most relevant because they influence the transfer of ions and electrons from one material to the other and thus directly affect cell performance. Good contact area, improved mechanical/thermal behavior, absence of micro-fractures or delamination, and lack of inter-material reactivity after long-term operation conditions are desired properties. Some of these parameters can be engineered by optimizing, for example, the composition at both sides of the interphase, or through the methods used for deposition and/or the sintering treatments.

However, SOFC interphases occur at the nanometric scale and thus their characterization needs high spatial resolution techniques such as TEM. Preparing a sample of these interphases is not trivial. Two materials of very different microstructures, compositions, and mechanical and thermal properties converge in this zone and make it very reactive and mechanically weak. One of the most versatile tools to prepare samples of selected areas for microscopy and spectroscopy analyses is the Focused Ion Beam (FIB). With help of an electron microscope, a desired area can be selected with micrometric precision and a Ga-ion beam is used to extract a thin sample of the bulk. This foil, with only some nm thickness, can then be analyzed afterwards with the scanning and the transmission electron microscopes and related techniques. Z-contrast, Energy dispersive spectroscopy (EDS) and Energy Loss Spectroscopy (EELS) are common choices for composition analyses. Electron Diffraction (ED) produces information about the atomic order and crystal structure in the materials. Electron Backscatter Diffraction (EBSD) is used for texture analysis. 3D-FIB micro-tomography allows for finding structure related parameters as pore concentration, tortuosity or contact area. On the other side, the evaluation of the electrochemical performance of a given cell design can be carried out using spectroscopy techniques such as Electrochemical Impedance Spectroscopy (EIS). The combination of site-specific sampling, high resolution microscopy techniques, and this kind of spectroscopic studies are thus an excellent group of tools for characterization. It allows establishing clear correlations between the microstructure/composition at the electrode/electrolyte interphase and the measured cell performance. Thus, this information can be used to improve the cell design and to increase the cell performance and its stability after long term operation conditions.

Author details

Analía Leticia Soldati[1,2], Laura Cecilia Baqué[1,3],
Horacio Esteban Troiani[2,4] and Adriana Cristina Serquis[1,2]
[1]*Materials Characterization Department, Bariloche Atomic Center, Atomic Energy Nacional Commission, San Carlos de Bariloche, Argentina*
[2]*CONICET, Buenos Aires, Argentina*
[3]*IDEPA-CONICET, CCT-Comahue, Buenos Aires, Neuquén, Argentina*
[4]*Metals Physics Department, Bariloche Atomic Center, Atomic Energy Nacional Commission, San Carlos de Bariloche, Argentina*

Acknowledgement

The authors gratefully thank Mrs. Anja Schreiber and Dr. Richard Wirth from the GeoFoschungsZentrum (GFZ Potsdam, Germany) for preparing the FIB foils. Mr. Carlos

Cotaro (CNEA, Argentine) is acknowledged for making the FEG-SEM images. Fundación Balseiro, University of Cuyo and CONICET are acknowledged for the financial support and CNEA for supplying the equipment, technical staff and laboratories. Dr Alberto Caneiro from the Materials Characterization Department (CNEA, Argentine) is thankfully recognized for allowing us to use the experimental facilities and for helpful discussions.

Abbreviations

FC = Fuel Cell
EDS = energy dispersive spectroscopy
FIB = focused ion beam
MC = mixed conductor
SEM = scanning electron microscopy
SOFC = solid oxide fuel cell
TEM = transmission electron microscopy
TPB = triple phase boundary
WDS = wavelength dispersive spectroscopy
XRD = x-ray diffraction

Cathodes

LSCF= $La_xSr_{1-x}Co_yFe_{1-y}O_{3-\delta}$
LSM= $La_xSr_{1-x}MnO_{3-\delta}$

Electrolytes

CGO= $Ce_{0.9}Gd_{0.1}O_{2-\delta}$
SDC= $Ce_{0.9}Sm_{0.1}O_{2-\delta}$
YSZ= $Zr_{0.9}Y_{0.1}O_{2-\delta}$

8. References

[1] Ormerod RM (2003) Solid oxide fuel cells. Chem. soc. rev. 32: 17-28.
[2] Steele BCH, Heinzel A (2001) Materials for fuel-cell technologies. Nature 414: 345-352.
[3] Hibino T, Hashimoto A, Inoue T, Tokuno J, Yoshida S, Sano M (2000) A low-operating-temperature solid oxide fuel cell in hydrocarbon-air mixtures. Science 288: 2031-2033.
[4] Larminie J, Dicks A (2003) Fuel cell systems explained John Wiley & Sons.
[5] Adler SB (2004) Factors governing oxygen reduction in solid oxide fuel cell cathodes. Chem. rev. 104: 4791-4843.
[6] Jorgensen MJ, Primdahl S, Bagger C, Mogensen M (2001) Effect of sintering temperature on microstructure and performance of LSM-YSZ composite cathodes. Solid state ionics 139: 1-11.
[7] Macdonald JR (1987) Impedance Spectroscopy. Emphasizing solid materials and systems John Wiley & Sons.
[8] Baqué LC (2011) Preparation and characterization of high performance cathodes for intermediate temperature solid oxide fuel cells.available online at

http://ricabib.cab.cnea.gov.ar, in *Engineering Department*. 2011, Universidad Nacional de Cuyo, Instituto Balseiro: San Carlos de Bariloche.

[9] Armstrong T, Prado F, Manthiram A (2001) Synthesis, crystal chemistry, and oxygen permeation properties of LaSr3Fe3-xCoxO10 (0 <= x <= 1.5). Solid state ionics 140: 89-96.

[10] Backhaus-Ricoult M (2008) SOFC - A playground for solid state chemistry. Solid state sci. 10: 670-688.

[11] Sase M, Suzuki J, Yashiro K, Otake T, Kaimai A, Kawada T, Mizusaki J, Yugami H (2006) Electrode reaction and microstructure of La0.6Sr0.4CoO3-delta thin films. Solid state ionics 177: 1961-1964.

[12] Sasaki K, Susuki K, Iyoshi A, Uchimura M, Imamura N, Kusaba H, Teraoka Y, Fuchino H, Tsujimoto K, Uchida Y, Jingo N (2006) H[sub 2]S Poisoning of Solid Oxide Fuel Cells. J. electrochem. soc. 153: A2023-A2029.

[13] Beckel D, Muecke UP, Gyger T, Florey G, Infortuna A, Gauckler LJ (2007) Electrochemical performance of LSCF based thin film cathodes prepared by spray pyrolysis. Solid state ionics 178: 407-415.

[14] Nielsen J, Jacobsen T (2008) SOFC cathode/YSZ - Non-stationary TPB effects. Proceedings 1314-1319. Elsevier Science Bv.

[15] Soldati AL, Baqué L, Troiani H, Cotaro C, Schreiber A, Caneiro A, Serquis A (2011) High resolution FIB-TEM and FIB-SEM characterization of electrode/electrolyte interfaces in solid oxide fuel cells materials. Int. j. hydrogen energ. 36: 9180-9188

[16] Hildenbrand N, Boukamp BA, Nammensma P, Blank DHA (2011) Improved cathode/electrolyte interface of SOFC. Solid state ionics 192: 12-15.

[17] Nam JH, Jeon DH (2006) A comprehensive micro-scale model for transport and reaction in intermediate temperature solid oxide fuel cells. Electrochim. acta 51: 3446-3460.

[18] Baqué L, Serquis A (2007) Microstructural characterization of La0.4Sr0.6Co0.8Fe0.2O3-delta films deposited by dip coating. Appl. surf. sci. 254: 213-218.

[19] Murray EP, Sever MJ, Barnett SA (2002) Electrochemical performance of (La,Sr)(Co,Fe)O-3-(Ce,Gd)O-3 composite cathodes. Solid state ionics 148: 27-34.

[20] Hansch R, Chowdhury MRR, Menzler NH (2009) Screen printing of sol/gel-derived electrolytes for solid oxide fuel cell (SOFC) application. Ceram. int. 35: 803-811.

[21] Besra L, Liu M (2007) A review on fundamentals and applications of electrophoretic deposition (EPD). Prog. mat. sci. 52: 1-61.

[22] Corni I, Ryan MP, Boccaccini AR (2008) Electrophoretic deposition: From traditional ceramics to nanotechnology. J. eur. ceram. soc. 28: 1353-1367.

[23] Perednis D, Gauckler LJ (2004) Solid oxide fuel cells with electrolytes prepared via spray pyrolysis. Solid state ionics 166: 229-239.

[24] Yoon J, Araujo R, Grunbaum N, Baque L, Serquis A, Caneiro A, Zhang XH, Wang HY (2007) Nanostructured cathode thin films with vertically-aligned nanopores for thin film SOFC and their characteristics. Appl. surf. sci. 254: 266-269.

[25] Napolitano F, Baqué L, Cho SM, Qing S, Wang H, Casanova JR, Lamas D, Soldati AL, Serquis A (2011) Characterization of SOFC cathodes prepared by pulse laser deposition. ECS trans. 35: 2379-2386.

[26] Chen KF, Lu Z, Chen XJ, Ai N, Huang XQ, Wei B, Hu JY, Su WH (2008) Characteristics of NiO-YSZ anode based on NiO particles synthesized by the precipitation method. J. alloy. compd. 454: 447-453.

[27] Li C-x, Xing Y-z, Yang G-j, Li C-j (2005) Effect of anode microstructure on performance of SOFC fabricated by thermal spraying. Chin. j. nonferr. met. 15: 1411-1415.
[28] Barbucci A, Viviani M, Panizza M, Delucchi M, Cerisola G (2005) Analysis of the oxygen reduction process on SOFC composite electrodes. In: editors. Springer. pp. 399-403.
[29] Chen KF, Lu Z, Chen XJ, Ai N, Huang XQ, Du XB, Su WH (2007) Development of LSM-based cathodes for solid oxide fuel cells based on YSZ films. J. power sources 172: 742-748.
[30] Dusastre V, Kilner JA (1999) Optimisation of composite cathodes for intermediate temperature SOFC applications. Solid state ionics 126: 163-174.
[31] Baqué L, Caneiro A, Moreno MS, Serquis A (2008) High performance nanostructured IT-SOFC cathodes prepared by novel chemical method. Electrochem. commun. 10: 1905-1908.
[32] Backhaus-Ricoult M (2006) Interface chemistry in LSM-YSZ composite SOFC cathodes. Solid state ionics 177: 2195-2200.
[33] Bevilacqua M, Montini T, Tavagnacco C, Fonda E, Fornasiero P, Graziani M (2007) Preparation, characterization, and electrochemical properties of pure and composite LaNi0.6Fe0.4O3-Based cathodes for IT-SOFC. Chem. mat. 19: 5926-5936.
[34] Grosjean A, Sanséau O, Radmilovic V, Thorel A (2006) Reactivity and diffusion between La0.8Sr0.2MnO3 and ZrO2 at interfaces in SOFC cores by TEM analyses on FIB samples. Solid state ionics 177: 1977-1980.
[35] Izuki M, Brito ME, Yamaji K, Kishimoto H, Cho D-H, Shimonosono T, Horita T, Yokokawa H Interfacial stability and cation diffusion across the LSCF/GDC interface. J. power sources In Press, Corrected Proof:
[36] Montenegro Hernandez A, Soldati A, Troiani H, Schreiber A, Flavio, Caneiro A (2012) Interface Reaction between Ln2NiO4 cathodes and two comercial electrolytes. In prep.
[37] Williams DB, Carter CB (1996) Transmission Electron Microscopy New York: Plenum Press.
[38] Phaneuf MW (2005) FIB For Materials Science Applications - A Review. In: Giannuzzi LA, Stevie FA editors. Introduction to Focused Ion Beams. Instrumentation, Theory, Techniques and Practice. New York: Springer. pp. 143-172.
[39] Anderson R, Klepeis SJ (2005) Practical Aspects on FIB TEM Specimen Preparation. With Emphasis on Semiconductor Applications. In: Giannuzzi LA, Stevie FA editors. Introduction to Focused Ion Beams. Instrumentation, Theory, Techniques and Practice. New York: Springer. pp. 173-200.
[40] Giannuzzi LA, Kempshall BW, Schwarz SM, Lomness JK, Prenitzer BI, Stevie FA (2005) FIB Lift-out Specimen Preparation Techniques. In: Giannuzzi LA, Stevie FA editors. Introduction to Focused Ion Beams. Instrumentation, Theory, Techniques and Practice. New York: Springer. pp. 201-228.
[41] Wirth R (2009) Focused Ion Beam (FIB) combined with SEM and TEM: Advanced analytical tools for studies of chemical composition, microstructure and crystal structure in geomaterials on a nanometre scale. Chem. geol. 261: 217-229.
[42] Liu YL, Jiao C (2005) Microstructure degradation of an anode/electrolyte interface in SOFC studied by transmission electron microscopy. Solid state ionics 176: 435-442.
[43] Wilson JR, Kobsiriphat W, Mendoza R, Chen H-Y, Hiller JM, Miller DJ, Thornton K, Voorhees PW, Adler SB, Barnett SA (2006) Three-dimensional reconstruction of a solid-oxide fuel-cell anode. Nat. mater. 5: 541-544.

[44] Wilson JR, Cronin JS, Duong AT, Rukes S, Chen H-Y, Thornton K, Mumm DR, Barnett S (2010) Effect of composition of (La0.8Sr0.2MnO3-Y2O3-stabilized ZrO2) cathodes: Correlating three-dimensional microstructure and polarization resistance. J. power sources 195: 1829-1840.

[45] Wilson JR, Duong AT, Gameiro M, Chen H-Y, Thornton K, Mumm DR, Barnett SA (2009) Quantitative three-dimensional microstructure of a solid oxide fuel cell cathode. Electrochem. commun. 11: 1052-1056.

[46] Smith JR, Chen A, Gostovic D, Hickey D, Kundinger D, Duncan KL, DeHoff RT, Jones KS, Wachsman ED (2009) Evaluation of the relationship between cathode microstructure and electrochemical behavior for SOFCs. Solid state ionics 180: 90-98.

[47] Holzer L, Muench B, Wegmann M, Gasser P, Flatt RJ (2006) FIB-Nanotomography of Particulate Systems—Part I: Particle Shape and Topology of Interfaces. J. am. ceram. soc. 89: 2577-2585.

[48] Chae NS, Park KS, Yoon YS, Yoo IS, Kim JS, Yoon HH (2008) Sr- and Mg-doped LaGaO3 powder synthesis by carbonate coprecipitation. Colloid surface A 313-314: 154-157.

[49] Shearing PR, Gelb J, Brandon NP (2010) X-ray nano computerised tomography of SOFC electrodes using a focused ion beam sample-preparation technique. J. eur. ceram. soc. 30: 1809-1814.

[50] Shearing PR, Gelb J, Yi J, Lee WK, Drakopolous M, Brandon NP (2010) Analysis of triple phase contact in Ni-YSZ microstructures using non-destructive X-ray tomography with synchrotron radiation. Electrochem. commun. 12: 1021-1024

[51] Shearing PR, Golbert J, Chater RJ, Brandon NP (2009) 3D reconstruction of SOFC anodes using a focused ion beam lift-out technique. Chem. eng. sci. 64: 3928-3933.

[52] Soldati A, Troiani H, Montenegro-Hernández A, Schreiber A, Soldera F, Caneiro A, Serquis A (2012) Evaluación de efectos secundarios originados por la preparación con fib de interfases entre materiales cerámicos porosos y sólidos. Proceedings of *SAMIC 2012*, p: Buenos Aires:

[53] Siebert E, Hammouche A, Kleitz M (1995) Impedance spectroscopy analysis of $La_{1-x}Sr_xMnO_3$-yttria-stabilized zirconia electrode kinetics. Electrochim. acta 40: 1741-1753.

[54] Grunbaum N, Dessemond L, Fouletier J, Prado F, Caneiro A (2006) Electrode reaction of $Sr_{1-x}La_xCo_{0.8}Fe_{0.2}O_{3-\delta}$ with x=0.1 and 0.6 on $Ce_{0.9}Gd_{0.1}O_{1.95}$ at 600 <= T <= 800ºC Solid state ionics 177: 907-913.

[55] Grunbaum N, Dessemond L, Fouletier J, Prado F, Mogni L, Caneiro A (2009) Rate limiting steps of the porous $La_{0.6}Sr_{0.4}Co_{0.8}Fe_{0.2}O_{3-\delta}$. Solid state ionics 180: 1448-1452.

[56] Teraoka Y, Zhang HM, Okamoto K, Yamazoe N (1988) Mixed ionic-electronic conductivity of $La_{1-x}Sr_xCo_{1-y}Fe_yO_{3-\delta}$ perovskite-type oxides. Mater. res. bull. 23: 51-58.

[57] Soldati AL, Baqué L, Troiani H, Cotaro C, Schreiber A, Caneiro A, Serquis A (2011) $La_{0.4}Sr_{0.6}Co_{0.8}Fe_{0.2}O_{3-\delta}$ / $Ce_{0.9}Gd_{0.1}O_{2-\delta}$ Interface: Characterization by High Resolution SEM and TEM. ECS trans. 35: 657-664.

[58] Wiedenmann D, Hauch A, Grobéty B, Mogensen M, Vogt UF (2010) Complementary techniques for solid oxide electrolysis cell characterisation at the micro- and nano-scale. Int. j. hydrogen energ. 35: 5053-5060.

Chapter 11

High Performance Membrane Electrode Assemblies by Optimization of Processes and Supported Catalysts

Chanho Pak, Dae Jong You, Kyoung Hwan Choi and Hyuk Chang

Additional information is available at the end of the chapter

1. Introduction

Recently, the mitigation of the greenhouse-gases (GHG) is the important issue to solve climate changes caused by the global warming. According to the international energy agency (IEA) report at 2010 (IEA, 2010), 65% of all GHG emissions can be attributed to energy supply and use. In addition, according to the blue scenario of IEA, all areas will need to reduce the CO_2 emission drastically until 2050, when level of CO_2 emission should be halved. In a view of energy supply and use, fossil fuels are used mainly in transport and power sectors which generate electricity through multiple steps. Thus, the highly efficient and clean technologies for these sectors are necessary for saving energy and reducing CO_2 emission (Pak et al., 2010). Among the alternative means, fuel cell technologies have been attracted because they can transform directly the chemical energy of fuel into electricity and emit clean exhaust gases.

Fuel cells have been developed for a long time since the principle of fuel cell has demonstrated by Sir Groove at 1939, who suggested the "gas battery" (Andujar & Segura, 2009). Initially, fuel cells were seen as an attractive technology for the generation of power due to high theoretical efficiency. However, as the efficiency of other alternative technologies was rapidly being increased, the development of fuel cell became almost negligible during the early of 20th century (Perry & Tuller, 2002). Also, since the interest in fuel cell reoccurred by the "space race" between USA and Russia in the late of 1950s and the first actual power generation system of fuel cell was launched in the Gemini at 1962, many types of fuel cells were developed for many applications and categorized by the electrolyte for use.

Among the various kinds of fuel cells, polymer electrolyte fuel cells (PEFCs) have extensively been developed for transport and distributed-power generation applications due to low emission of GHG and high efficiency compared to the internal combustion engine (ICE) and generators. Thus, PEFC technology is considered as a green technology for energy savings and reduction of GHG emission. As a one part of PEFCs, direct alcohol fuel cell (DAFC), which can use the high energy density of alcohol such as methanol and ethanol that could be produced by using the biomass and corn for carbon neutral cycle or directly using solar energy for artificial photosynthesis, is the most promising fuel cell for mobile and portable applications.

Although the PEFC systems have been developed to its current status through several technical breakthroughs over the years, it is now on initial market stage with the help of government for new and renewable energy policy. To expand the market size or thrive in fuel cell market without external supports, further innovations in the areas of cost and durability are demanded. For this innovation, the understanding and improvement of materials and components for membrane electrode assembly (MEA), which considered as the core of fuel cell system, are very important besides development of fabrication process maximizing the performance with the improved materials for PEFC.

A typical MEA as shown in Fig. 1 consists of a polymer electrolyte membrane (PEM), interposed by two electrodes, cathode and anode, which are composed of catalyst layer (CL) and the gas diffusion layers (GDL), respectively. Usually, a microporous layer (MPL) made from porous carbon materials is located between the CL and the GDL. The total thickness of multilayer of MEA is less than 500 μm (Ramasamy, 2009).

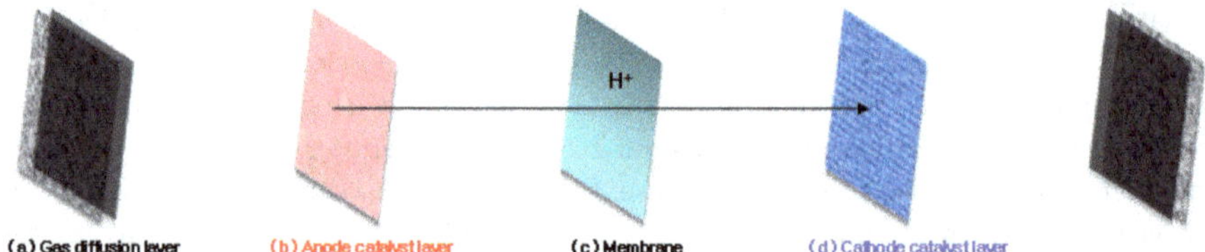

Figure 1. Schematic components of a membrane electrode assembly

The performance of MEA is displayed by the power density (W/cm^2), which is the product of current density (A/cm^2) and voltage (V), which is totally dependent on the choice of components and materials, especially membrane and catalyst. Thus, the research for increasing the performance of MEA is usually focused on the new materials for membranes and catalysts with enhanced properties. However, to reveal the improved performance of materials in the MEA level, the fabrication process for MEA should be optimized.

In this chapter, the processes for MEA preparation are reviewed in the section 2 and the optimized performance of MEA using new supported catalysts will be discussed in the sections 3 and 4. Finally, the chapter is closed with conclusions.

2. Processes for membrane electrode assembly (MEA)

Among the components of a MEA, the performance of the MEA is usually dependent on the CL properties and contact interfacial resistance between the CL and membrane according to its preparation technique. The CL has its own set of criteria to fulfil the chemical reaction with complex functionalities and has a three-dimensional porous structure composed of a network of catalyst particles made of porous carbon supports and catalytic metal nanoparticles, usually and ionomer fragments.

Some common requirements of an idea regarding to CL must have a high electrocatalytic activity for PEFC reactions, a good ionic transport and a high porosity for efficient transport reactant and product (Ramasamy, 2009). Over the years, various slurry formations and coating procedures have been developed for the preparation of CLs in order to realize the better performance of MEA for the commercialization.

One key factor in the preparation of CLs is the selection of solvents to form a homogeneous mixture in the catalyst ink, which was generally made by dispersing the catalyst (supported Pt based catalyst or Pt based black catalyst) with a mixture of Nafion ionomer solution, the solvents and deionized (DI) water. Many researchers have mainly studied to focus on the dispersion of catalyst ink formed by ionomer and the catalyst particles. For example, Uchida et al. demonstrated that the further improvement in cell performance could be obtained by using an intermediate dielectric constant solvents with a range of 3 -10 to form a colloidal suspension of Nafion particles in a water-alcohol mixture. Their experimental results were attributed to the higher number of electrochemical interactions between the Nafion ionomers and catalysts in the extended reaction interface than those using solvents with a high-dielectric-constant, such as water (Uchida et al., 1998). It is suggested that the selection of highly viscous glycol for the catalyst ink as a solvent resulted in the higher performance MEAs in which showed low mass-transfer and electrode ionic resistance due to the formation of homogeneous catalyst particles in the catalyst ink (Wilson et al., 1995). Although the homogeneous ionomer and the catalyst particles are important for the formation of the slurry ink, the appropriate amount ratio of ionomer to catalyst and fine distribution of the ionomer in the CLs are the most critical factor, which leads to the minimized electrode resistance and maximized contact of ionomer with catalytic metal nanoparticles. This ratio should be normally optimized for the best formulation of the catalyst ink.

Another key factor in preparation of the CLs is the selection of coating procedures to minimize the roughness factor of the CLs and the contact resistance between the CL and the membrane. The types of coating procedure for preparation of the CLs can be broadly classified into three categories as followings: (1) catalyst coated on electrode (CCE), (2) decal transfer catalyst coated on membrane (DTM) and (3) direct catalyst coated on membrane (DCM).

2.1. Catalyst coated on electrode method (CCE)

The CCE method is to form the CL on the GDL as shown in Fig. 2. The catalyst ink was coated onto the MPL in the GDL, and then the electrode was dried in the vacuum oven at a specific temperature. Finally, the MEAs were assembled by hot pressing the catalyst coated electrodes with a membrane. The CCE method has widely been used in the formation of the large scale mass production of MEAs due to the simple coating process (Frey & Linardi, 2004). The final hot-pressing is indispensable and important process for the higher performance MEA to make a good interfacial contact between the CLs and membrane in the CCE Method. The main parameters of hot-pressing process are the temperature, the pressure and the time. Zhang et al. investigated that the effect of hot-pressing conditions (temperature, pressure and time) on the performance of MEA using the CCE method for the DMFC. The optimized parameters for temperature, pressures and time are 135℃, 80kg$_f$/cm^2 and 90s, respectively. The highest power density of MEA is attributed to the lowest contact resistance between the membrane and CL (Zhang et al., 2007). In addition, Therdthianwong et al. tried to find systematically the most significant hot-pressing parameter by designing a full factorial analysis of the three main hot-pressing parameters related to the cell performance (Therdthianwong et al., 2007). In this study, MEA prepared with hot-pressing condition of 100 °C, 70.3 kg$_f$/cm^2 and 120s resulted in the highest power density.

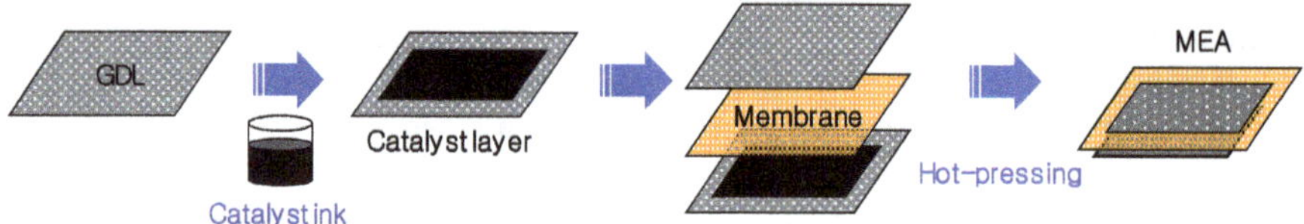

Figure 2. Schematic process of a catalyst coated on electrode (CCE).

Although various conditions for hot-pressing have been employed by researchers in the CCE method, the controlling of hot-pressing temperature to slightly above the glass transition temperature of the membrane might be a critical point to promote good contact within the triple-phase regions of the CL. (Zhang et al., 2007, Tang et al., 2007, Lindermeir et al., 2004). However, the CL prepared in CCE method cannot be effectively transferred to the membrane during the course of hot-pressing the MEA due to the change of the structure in the CL (a distribution of ionomer and porosity) and the dehydration of the membrane, which may lead to an irreversible performance loss of the MEA (Kuver et al., 1994).

2.2. Decal transfer catalyst coated on membrane method (DTM)

The DTM method is to form the CL on the decal substrates as a shown in Fig. 3. The catalyst inks were coated uniformly onto decal blank substrates. The CLs of both electrodes were then transferred from substrates to the membrane by hot pressing under high pressure and temperature for a specific time. The decal substrates can be peeled away from the CCM leaving the CLs fused to membrane, yielding a three-layer CCM. The GDLs can then be added to the CCM by hot-pressing as mentioned in the previous section.

Tang et al. reported that the DTM method could show a better utilization of catalysts and a superior formation of the ionomer network compared to the CCE method, which are all beneficial for improving the performance and long-term durability of the MEA for the DMFC due to a low interfacial resistance between the CL and polymer membrane, a thinner catalyst layer with a lower mass transfer resistance, and a better contact among the electrode components (Tang et al., 2007). However, the process of the DTM method seems to be more complex than CCE method and impossible to control the porosity and the thickness of the CLs due to the dehydration of the membrane during the decal transfer and, it has a possibility of sintering of the catalytic nanoparticles (Song et al., 2005). Furthermore, the ionomer segregation is likely to occur onto the outside of the CL during the transfer step of the CLs from the decal substrates to Na^{+}-Nafion membrane with high hot-pressing temperature in order to increase the transfer ratio of the CL into the membrane (Xie et al., 2004). Recently, a breaking layer composed of carbon powder and Nafion ionomer on the CLs was suggested by two groups to overcome those problems, that is, the CL was sandwiched between the inner thin carbon and the outer ionomer layers (Park et al., 2008, Cho et al., 2010). However, the additional layer could generate a further resistance to proton and mass transports, which may lead to an irreversible performance loss of the MEA. The DTM method must be improved further for the commercialization of MEA.

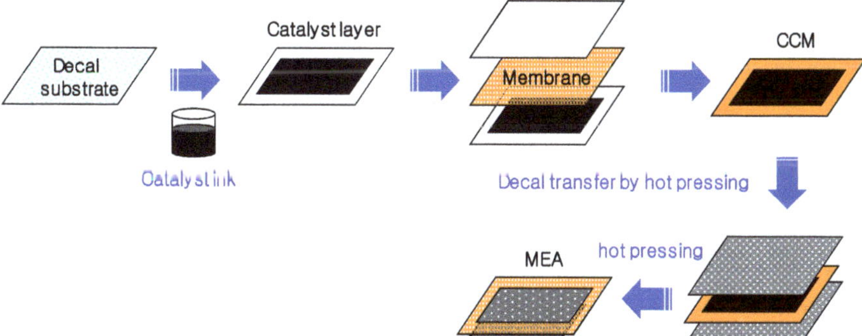

Figure 3. Schematic process of a decal transfer catalyst coated on membrane (DTM)

2.3. Direct catalyst coated on membrane method (DCM)

The DCM method is to form the CL directly onto the membrane as shown in Fig. 4. The DCM method is more simple and efficient than indirect coating process, DTM method and has no risk of uneven and incomplete transfer of catalyst in the CL. Furthermore, it also produces a higher MEA performance than the DTM method due to an easier controllability of the CL thickness as well as a better ionic connection between the CLs and the membrane resulted from a strong attachment of the solvent on the membrane. However, the direct coating of catalyst slurry onto the membrane has a critical problem that the membrane has a

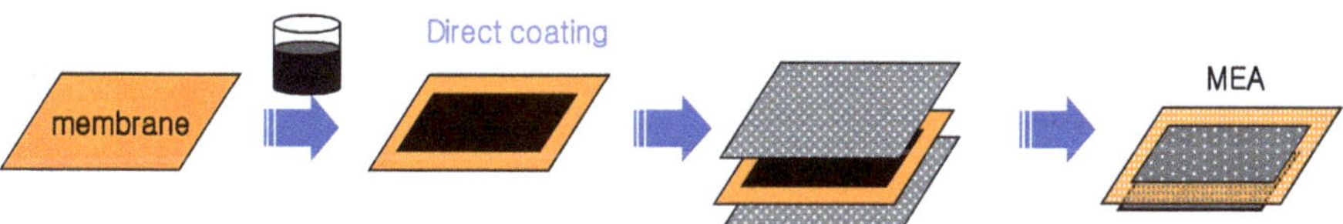

Figure 4. Schematic process of a direct catalyst coated on membrane (DCM)

high tendency to swell or wrinkle with a contact of many solvents in the catalyst slurries, which could give rise to the deformation of the CL by fast volume changes of the membrane.

It could cause the membrane and the CL to be deformed by fast volume changes. Therefore, swelling control of the membrane is very important in the DCM method for the high quality MEA fabrication. To minimize such dimensional changes during the catalyst coating process, many researchers have tried to prevent the membrane from swelling during the coating process. For example, Park et al. suggested a process by employing a pre-swollen Nafion membrane. They soaked the Nafion membrane in EG and sprayed the catalyst slurry onto the pre-swollen membrane. Thus, the prepared MEA showed improvement over a commercially available MEA due to the reduction of a stress problem of membranes by the pre-swelling process (Park et al., 2010). Shao et al. prepared MEA using direct spray deposition of the catalyst ink into Nafion 212 membrane with the aid of a hot-plate at 150 °C, whose condition could decrease the swelling and wrinkling of the Nafion membrane due to the solvent gasification before being absorbed into the Nafion membrane (Shao et al. 2001). Also, in case of our lab, an innovative process preventing the swelling from the solvents by holding a membrane on a porous vacuum plate is developed, which is an efficient way for realizing a high precision in catalyst loading with high reproducibility (You et al. 2010). Considering the CL design, the selection of good solvents and DCM method as a coating process could offer a more efficient and attractive way for high quality and high performance of MEAs.

3. Optimization of DCM processes for MEA

One parameter in MEA design to improve the performance of DMFCs is to increase the catalyst utilization and electrochemical surface area (ESA) of the electrodes by increasing the level of gas access, proton access and electron access to the reaction sites. Hence, the structures of CLs where the electrochemical reactions occur should be optimized for maximizing the triple-phase boundaries. In addition, minimizing a resistance between the catalyst and ionomer in the CL, as well as the interfacial resistance between the electrolyte and the CL are essential in MEA structure. Furthermore, the resistance of the electrolyte membrane itself should be minimized.

Another parameter in MEA design is the control of porosity to maximize their active surface area of catalyst in the electrodes. During DMFC operation, complex flow of reactants and reaction products exists in the porous space of CL. The pores in cathode should allow oxygen to reach the catalyst surface and support efficient transport of water to prevent flooding of the layer. One method achieving a good balance of fuel transport capability with effective product removal is the addition of pore-forming materials into the CL. They help tailor the CL morphology and pore structure to meet the above-mentioned requirements, thereby decreasing the transport resistance.

Considering the above-mentioned issues in the CL design, the optimized design of CLs must have the lowest resistance between the membrane and CL, the proper distance of proton conductor from the catalyst and the optimum porosity of the catalyst layer. The

DCM method with the catalyst slurry composed of EG and pore forming agent is investigated to optimize the structure of CLs by parameters for hot-pressing such as temperature and pressure.

3.1. Preparation of electrode and MEA

All investigated MEAs in this section were prepared with the hydrocarbon membrane (the conductivity and thickness of the membrane was 0.06 S/cm and 32 μm, respectively) and the SGL 25 BC for GDLs of both electrodes. Pt-Ru black (HiSpec 6000, Johnson Matthey) and Pt black (HiSpec 1000, Johnson Matthey) were used as the anode and cathode catalyst, respectively. Catalyst inks, consisting of black catalysts, Nafion solution, DI water, $MgSO_4$, and EG with weight ratios of 0.288: 0.18: 0.155: 0.058: 0.36 for the anode and 0.241: 0.151: 0.12: 0.036: 0.452 for the cathode, respectively, were well dispersed using high speed rotating equipment (conditioning mixer, AR-500) for 10 min.

For the preparation of CLs using the DCM, the anode inks were coated uniformly onto one side of the electrolyte membrane directly, which was held on a vacuum plate with 32 mm × 32 mm mask films to prevent dimensional change of the membrane. The coated membrane was then dried for 24 h in a vacuum oven at 120ºC after removing the mask. The cathode catalyst ink was applied to the opposite side of the anode-catalyst coated membrane in the same manner. The catalyst loading for both electrodes was 5 mg/cm² and the active area of the MEAs was 10 cm².

Hot-pressing after direct coating of CL was performed at three different temperatures (140, 150, and 160ºC) and pressures (0.1, 0.2, and 0.3 ton_f/cm^2) for 10 min to control the porosity and contact resistance of the CLs. And then, The CCM prepared was pre-treated in a solution containing 1M methanol and 1M sulfuric acid at 95ºC for 4 h. Finally, the MEAs were fabricated by placing GDLs onto the corresponding sides of the CCM by hot-pressing at the 125ºC and 0.1 ton_f/cm^2 for 3 min.

The performance of the MEA under the DMFC condition was measured by fuel cell testing system (Won-A Tech) using single cell hardware with an active area of 10 cm². A 1M aqueous methanol solution was fed to the anode side at a flow rate of 0.25 ml/min·A. Dry air was supplied to the cathode side at a flow rate of 45 ml/min·A under ambient pressure. The cell performance was measured at 60ºC and operated in potentiostatic mode at a voltage of 0.45V for 4 h each day. The polarization curves of the MEAs were recorded at the end of the procedure at a constant voltage.

Electrochemical impedance spectroscopy (EIS) of the MEAs were measured at a current of 220 mA/cm² using an electrochemical analysis instrument (VMP2) in the frequency range from 100 kHz to 0.1 Hz with 10 points per decade at 60 ºC. The amplitude of the sinusoidal current signal was 10 mA. To separate the anode and cathode impedance, the cathode side was supplied with a continuous supply of hydrogen, which would function as a dynamic hydrogen reference (DHE) and counter electrode.

3.2. Effects of the temperature for the hot-pressing

The effect of the hot-pressing conditions (temperature and pressure) on the MEA performance for DMFC was investigated to decrease the reaction transfer resistance through the extended catalyst and ionomer interface in the electrode and to increase the interfacial bonding through the strong formation of a proton conducting ionomer network between the CL and membrane. Fig. 5 (a) shows the power densities at 0.45V of the MEAs from the CCMs prepared by the DCM at various hot-pressing temperatures under the same pressing pressure (0.2ton$_f$/cm^2). The performance of the MEA produced at a hot-pressing temperature of 150 ºC was higher than that of the MEA produced at 140 and 160 ºC, respectively.

Electrochemical impedance spectroscopy (EIS) and differential scanning calorimetry (DSC) were performed to elucidate the effect of the hot-pressing temperature on the DMFC MEA performance. Firstly, in EIS analysis, the reaction transfer resistance of the anode, cathode, and the total showed similar values regardless of the hot-pressing temperature, as shown in Fig. 5 (b). This indicates that the pressing temperature for CLs is not related to the reaction transfer resistance between catalyst and ionomer, and pore structure but is associated with changes in the interfacial properties by the strong bonding between the CL and membrane. Secondly, Fig. 5 (c) shows the DSC analysis of the hydrocarbon membrane. The exothermic process was observed at 150 °C, which corresponds to the glass transition temperature (Tg) of the hydrocarbon membrane. At an appropriate temperature, such as 150 °C, side chain movement brings the $-SO_3H$ group out of the bulk to the surface to decrease the surface energy (Liang et al., 2006, Guan et al., 2006, Robertson et al., 2003). This might give rise to intimate bonding between the hydrocarbon membrane and CL resulting in the enhanced proton conductivity. Therefore, the optimum hot-pressing temperature contributes to the significant increase in the MEA performance.

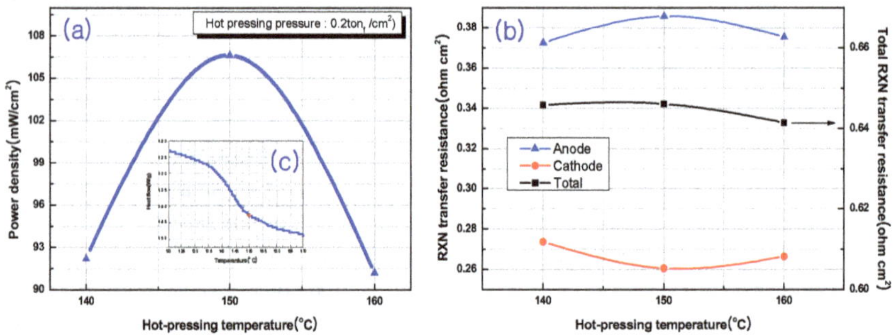

Figure 5. Effect of hot pressing temperature for the CCM prepared by direct coating on (a) power density of MEA and (b) reaction (rxn) transfer resistance from electrochemical impedance spectroscopy (EIS) and (c) DSC analysis of hydrocarbon membrane.

3.3. Effects of pressure for the hot-pressing

Fig. 6 (a) shows the power densities at 0.45V and reaction transfer resistances of the MEA using CCM produced by the DCM under various hot-pressing pressures under the same temperature of 150 °C. The increase of hot-pressing pressure for the CCM to approximately 0.225ton$_f$/cm^2 resulted in significantly improved MEA performance with power density up to 107mW/cm^2 at 0.45V and 60 °C. This might be attributed to the decreased reaction transfer resistance by the improved proton conduction and oxygen transport through the well-connected network of the CLs in the MEA. However, further increases in the hot-pressing pressure led to a decrease in the performance of MEA owing to the destroyed microstructures of the CL by excessive pressing.

This suppose was confirmed by EIS analysis of the MEAs. Fig. 6 (b) shows the effect of the hot-pressing pressure for the CCM prepared by DCM on the reaction transfer resistance. The cathode reaction transfer resistance increased as both CLs were further compressed by increasing the hot-pressing pressure, whereas the anode reaction transfer resistance decreased, even though the thickness of both CLs have decreased. Generally, the thickness of both porous CLs decreases with increasing of density (decreased porosity) as the pressing pressure increases to fabricate the CCM. The dense anode CL may increase the methanol utilization efficiency with the decreasing of the methanol crossover, resulting in the decreased the anode reaction transfer resistance. This phenomenon might be that because the dense anode CL serves as an additional resistance against methanol crossover (Mao et al., 2007, Liu et al., 2006, Park et al., 2008). In contrast, the thick microstructure (high porosity) for the cathode CL is essential for transporting reactant gas effectively from the GDL to CL and for eliminating the water produced by the electrochemical reaction from the CL to GDL (Liu & Wang, 2006, Wei et al., 2002, Song et al., 2005). Therefore, the hot-pressing pressure for the CCM showed the lowest total reaction transfer resistance at 0.2 ton$_f$/cm^2 and resulted in the highest power density of the MEA produced by the DCM. It was suggested that the hot-pressing condition has a significant effect on the electrochemical performance of MEAs, particularly in the reaction transfer resistance.

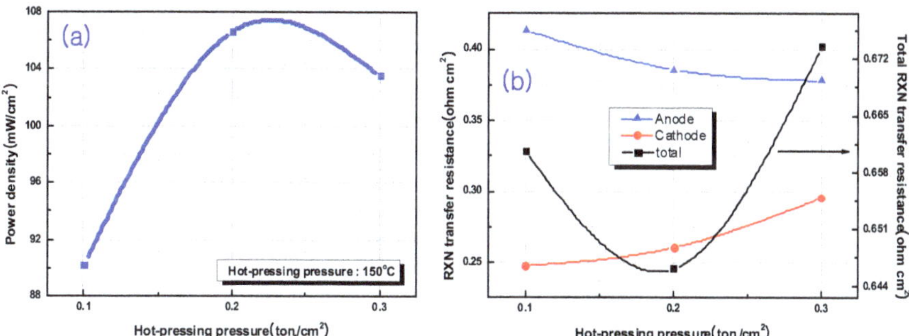

Figure 6. Effect of hot-pressing pressure for the CCM prepared (a) power density of MEA and (b) reaction (rxn) transfer resistance from EIS.

3.4. Effect of pore forming agent in the cathode

The magnesium sulphate ($MgSO_4$) was chosen as a pore forming agent for the preparation of the cathode CL. The $MgSO_4$ is widely used as a drying agent due to its hygroscopic properties (readily absorbs water from the air). Hence, it can be easily removed in the CL by boiling the CCM at DI water after DCM process. The vacant sites in the CL by resulted from removed $MgSO_4$ may play a role as pores. In addition, compared to that of insoluble pore forming agents (e.g. Li_2CO_3) (Tucker et al., 2005), the addition of soluble $MgSO_4$ could form more uniform pore distributions with smaller pore size (approximately 3 nm as a shown in Fig. 7 (b)) in the CL by the homogeneous catalyst inks, since the solubility of $MgSO_4$ was superior in catalyst ink mixtures composed of the water, EG, ionomers and catalysts.

Fig. 7 (a) shows the effect of the pore forming agent ($MgSO_4$) loading in the CL on the power density of the MEA by the DCM. As it can be seen, the addition of $MgSO_4$ in the catalyst slurry from 0 wt% to 30 wt% led to an increase in power density of MEAs at 0.45V. This might be due to the higher degree of catalyst utilization and increase in active ESA because more pores that had formed by the $MgSO_4$ contributed ~~to~~ the supply of air to the catalytic active sites effectively to produce the required amount of power and eliminate the water produced by the electrochemical reaction. Moreover, the effect of the cathode reaction transfer resistance by the addition of $MgSO_4$ showed an opposite trend to the result of the cell performance as shown in Fig. 7 (a). However, further increasing of $MgSO_4$ showed an increase of cathode reaction transfer resistance due to the destroyed microstructures of the weaken CL mechanically by excessive porous structure. Furthermore, the resistance for the proton transport to and from the active sites increased with increasing distance between catalysts and ionomers at the CL. Therefore, the pores generated by the $MgSO_4$ might be an effective channel for air transport inside the CL. In addition, increased pore volumes are expected to enhance rapid mass-transfer near the catalyst surface providing open diffusion paths for the water produced from the CL. Furthermore, the enhanced oxygen supply increased the rate of oxygen reduction because the charge transfer reaction is a function of the reactant concentration.

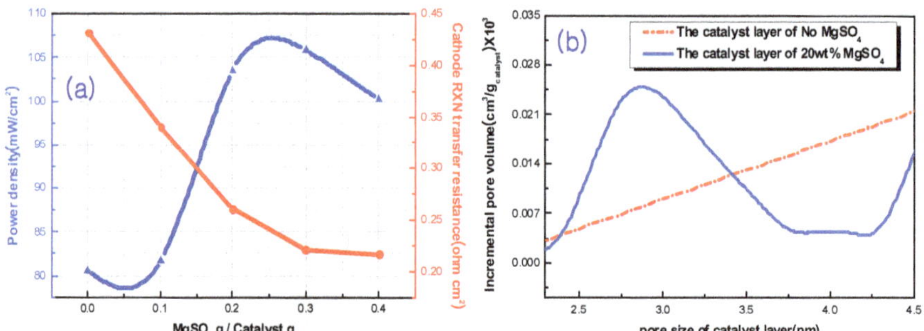

Figure 7. Effect of $MgSO_4$ amount in the catalyst layer of CCM prepared on (a) power density and cathode reaction transfer resistance and (b) pore size distribution of catalyst layer.

4. High performance MEA using new supported catalyst

In this section, the adoption of new supported catalyst (Pt/OMC) consisted of novel ordered mesoporous carbon (OMC) support and highly dispersed Pt to the cathode catalyst layer of the

MEA for DMFC is presented as an example of optimization of properties such as catalyst loading, ionomer concentration and porosity in the electrode (Kim et al., 2008). The Pt/OMC catalysts have been developed in our lab over several years for the DMFC, which is based on the novel OMC supports having very high surface area and ordered array of mesopores inside the particles (Chang et al., 2007, He et al., 2010, Joo et al., 2009., Lee et al., 2009, Pak et al., 2009).

A balance between proton conduction path and mass transport via pore structures of catalyst layer was investigated by changing an amount of ionomer and compressing the MEA with hot press. The performance of MEA as a function of voltage was measured to determine the optimized the conditions for catalyst layer, which governs the power density. Furthermore, the performance of MEA with optimized Pt/OMC catalyst layer was compared to that of unsupported Pt black catalyst layer to prove the possibility for decreasing the Pt amount in the cathode without loss of power density for DMFC.

4.1. Characteristics of OMC support and Pt/OMC catalyst

The SEM and TEM images of ordered mesoporous silica (OMS), which is the hard-template for OMC, and OMC were displayed in the Fig 8. As observed in Fig. 8, the OMS is composed of 200 – 300 nm particles and OMC has similar particle morphology and size, which indicates that the nano-replication (Joo et al., 2001) of OMS into OMC is successfully occurred and the removal of OMS to generate the pore inside of the OMC did not alter the apparent morphology of OMC. The TEM image from OMS (Fig. 8 (b)) shows that the uniform mesopores are hexagonally well-arranged in the particle. For the OMC, the TEM image display that the pores and the walls of OMS are inverted to the carbon-nanorod and mesopore of OMC, respectively. The low angle X-ray diffraction (XRD) patterns (refer to Joo et al., 2006) of OMS and OMC showed three, well resolved peaks corresponding to (1 0 0), (1 1 0) and (2 0 0) diffractions of hexagonal p6mm symmetry. The unit cell dimension of OMS and OMC, estimated from the (1 0 0) diffraction was 12.0 and 11.0 nm, respectively. The OMC has a slightly compressed unit cell because of the structural shrinkage of carbon frameworks during the high temperature carbonization (Jun et al., 2000).

Figure 8. SEM and TEM images of (a, b) OMS and (c, d) OMC, respectively.

The pore structures of OMS and OMC were further characterized by using the nitrogen adsorption and desorption isotherms (Joo et al., 2006). The corresponding pore size distribution estimated from the adsorption branch by BJH method for OMS and OMC samples. The OMS template showed typical Type IV isotherm with H1 hysteresis. The sharp increase of nitrogen uptake in the adsorption branch in the partial pressure of 0.8–0.9 indicates that the mesopore of OMS has uniform distribution through the particles. The BET surface area of OMS template is 451 m^2/g and pore volume is 1.27 cm^3/g, while the pore diameter calculated from the adsorption branch of isotherms is 12.2 nm. The nitrogen isotherms of OMC sample exhibited similar shapes, where capillary condensation occurred in the partial pressure range of 0.4–0.6. The BET surface area of OMC sample is 884 m^2/g and pore volume is 0.86 cm^3/g, while the pore diameter is 4.0 nm.

The unique structural characteristics of OMCs make them suitable as catalyst supports for DMFC application as mentioned earlier. For example, the high surface area of OMC, compared with the conventional carbon blacks such as Vulcan XC-72R and Ketjen Black, can provide sufficient surface functional groups or anchoring sites for the nucleation and growth of metal nanoparticles, thus metal catalysts can be prepared on OMC with high dispersion. Further, uniform mesopore structure of OMC would facilitate the diffusion of reactive molecules for electrochemical reactions.

Pt nanoparticles were supported on the OMC by incipient wetness impregnation of the Pt precursor ($H_2PtCl_6{\cdot}xH_2O$) in acetone solution into the pores of OMC support and subsequent reduction under H_2 flow. The total loading of Pt was controlled as high as 60 wt%, because an electrocatalyst for DMFC application requires very high metal loading (Chang et al., 2007). The TEM image of Pt/OMC (Fig. 9 (a)) indicates that Pt nanoparticles are uniformly scattered on the carbon nanorod of OMC. The average particles size determined from the TEM image is 2.85 nm. The XRD patterns for 60 wt% Pt/OMC presented in Fig. 9 (b) showed distinct peaks at around 39.8°, 46.3° and 67.5°, corresponding to the (1 1 1), (2 0 0) and (2 2 0) planes of a face-centered cubic structure, respectively. The crystalline size of the Pt nanoparticle estimated by the Scherrer equation is 2.86 nm, which is matched well with the value obtained from TEM analysis.

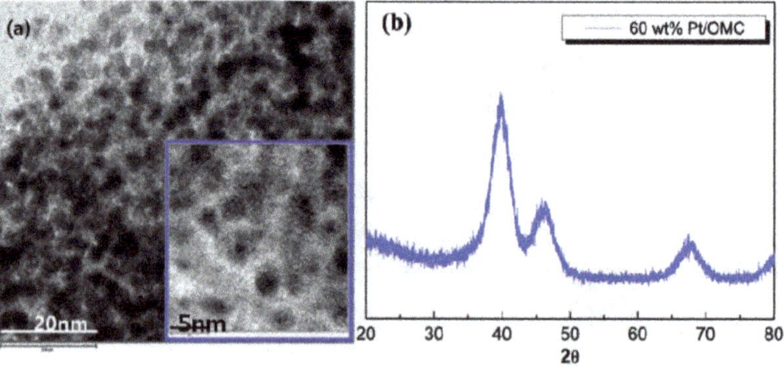

Figure 9. (a) TEM image and (b) XRD pattern for Pt/OMC catalyst.

4.2. Optimization of catalyst layer with Pt/OMC catalyst

To realize the adoption of Pt/OMC in the cathode catalyst layer for DMFC, the effect of the ionomer contents (18, 30 and 45 % compared to the Pt/OMC) and process parameter (compressed vs. uncompressed) were investigated on the morphology of electrode and performance of MEA at 70 °C as summarized in the Table 1. The catalyst ink was sprayed directly on to a Nafion 115 membrane to form the so-called CCM. The membrane was held on a vacuum plate to prevent dimensional change of the membrane during the direct coating of the ink. The cathode catalyst ink was coated on one side of the membrane followed by drying at 60 °C under vacuum for 2 h. On the uncoated side of the cathode-coated membrane, the anode catalyst ink was applied in the same manner. As a reference, the catalyst layer based on unsupported Pt black catalyst (Johnson Matthey, HiSpec® 1000) at a loading level of 6mg/cm^2 was also prepared. The geometric area of the catalyst layers was 25 cm^2. In order to produce a catalyst layer with lower porosity, the cathode-coated membranes were compressed at 30 MPa and at 135 °C for 5 min before the subsequent anode coating was performed. As a diffusion layer, 35 BC (SGL, Germany) was used for both the cathode and the anode. The MEAs were prepared by hot pressing the CCM and two diffusion layers at 125 °C and 51 MPa. The morphology of the catalyst layers was observed by SEM.

Catalyst Layer	Amount of Pt (mg/cm^2)	Ionomer content (%)	Compression
CL18-U	2.39	18	X
CL18-C	2.39	18	O
CL30-U	2.64	30	X
CL30-C	2.64	30	O
CL45-U	2.39	45	X
CL45-C	2.39	45	O

Table 1. Physical parameters of catalyst layers based on Pt/OMC catalyst

The thickness of the uncompressed catalyst layers (CL18-U, CL30-U and CL45-U) is 70, 128 and 132μm, respectively, for corresponding ionomer contents of 18, 30 and 45 % and the thickness of the compressed catalyst layers (CL18-C, CL30-C and CL45-C) is found to be 64, 53 and 48% of the pristine thickness for ionomer amount of 18, 30 and 45%, respectively. In the case of ionomer amount of 12%, the strength of catalyst layer is not enough to adhere on the GDL, which could be attributed that the ionomer content is not enough to bind Pt/OMC catalysts effectively. The ionomer contents become more than 12% and the catalyst layer showed acceptable mechanical strength.

The apparent shape of the Pt/OMC-based catalyst layers was observed by SEM, as displayed in Fig. 10 for representative examples (CL18-U and CL18-C). These are featured by the formation of agglomerates of the Pt/OMC and ionomer and of the pores between these agglomerates. The agglomerate size is in the range of 200–1000 nm. Considering the size of the primary OMC (200–300 nm) particle as shown in Fig. 8, several Pt/OMC particles are included in the agglomerate. The change of ionomer amount did not cause the appreciable changes in the size of the agglomerates in the catalyst layers. For the uncompressed catalyst layers, the porosity appears to be larger at higher ionomer contents.

After compression, densification of the catalyst layer is observed. On the other hand, the size of the agglomerates is little affected by the compression, as shown in Fig. 10. This indicates that macropores between the agglomerates are reduced during the compression process. The compressed catalyst layers do not differ in their pore structures.

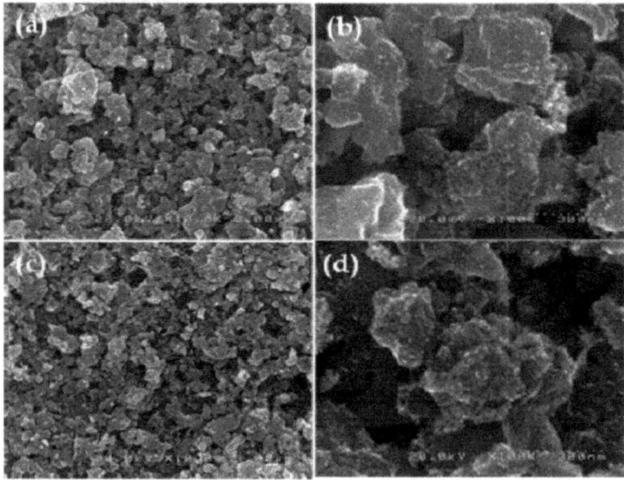

Figure 10. Represnetative SEM images with different magnification of (a, b) CL18-U and (c, d) CL18-C.

Fig. 11 showed polarization curves obtained after five day activation at 70 °C. Among the MEA, CL18U-C case showed the highest power density of 104.2 mW/cm^2 at 0.45 V. The operating voltage of DMFC MEA was chosen based on the balance between power density and energy efficiency. Operation at lower voltage generates high power density, and thus size reduction of the stack is possible. On the other hand, energy efficiency decreases on lowering the voltage, which requires a larger fuel tank for a given energy consumption. To maximize the system efficiency, an operating voltage of 0.45V is chosen for DMFC usually. Dry air and 1M aqueous methanol solution were used as feed stocks for the cathode and the anode, respectively. As shown in Fig. 11, uncompressed MEAs show higher performance at 0.45 V than that of compressed MEAs with same amount of ionomer in the catalyst layer. The variation of the ionomer amount results in a more pronounced effect on the power density than compression of catalyst layer, which is consistent with a result reported earlier in the literature (Frey and Linardi, 2004). As the catalyst layer compressed, the layer becomes more compact, which reduced the mass transport in the catalyst layer. However, the proton conductivity in the catalyst layer should be increased with compression, which was confirmed by the analysis of impedance (Kim et al., 2008).

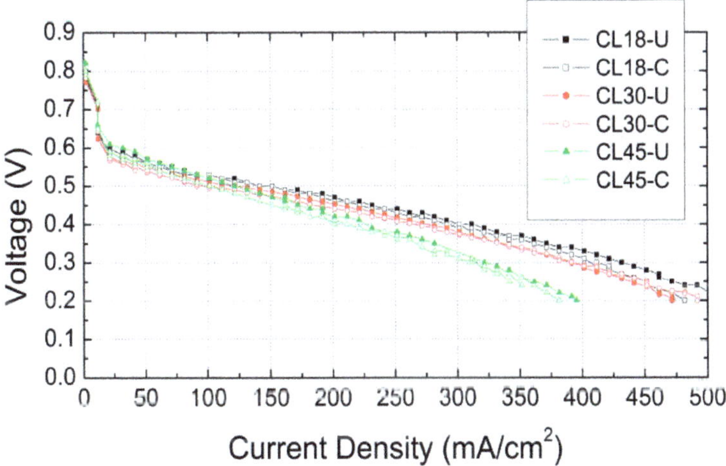

Figure 11. Polarization curves for MEAs based on Pt/OMC at 70 °C (cathode feed: air, anode feed: 1M CH_3OH).

Thus, the decrease in power density with the compression indicates that mass transport is more important than the proton conductivity (ionic transport) for the Pt/OMC based cathode. A similar finding was reported for a PtRu/C supported catalyst in the previous paper, which suggested the compaction of the anode catalyst layer led to a decrease in performance by 23% due to increased mass transport (Zhang et al., 2006).

For both the uncompressed and compressed MEAs, the gap of power density increases with reducing in the ionomer content above 150 mA/cm^2. Below 150 mA/cm^2, however, the difference is less pronounced, where the power density of CL45-U becomes comparable to that of CL18-U. The large difference in power densities at high current densities indicates a considerable mass-transport limitation at higher ionomer amount. When conventional Pt-supported carbon is used, there is an optimum value of ionomer content in the catalyst layer to obtain the best performance. Even though the existence of an optimum at lower ionomer content is expected for a Pt/OMC-based catalyst layer, it is not possible to confirm this because the mechanical integrity of the catalyst layer is not sufficient as mentioned earlier in this section. An improvement in the formation of the three-dimensional network of the ionomer phase which provides mechanical integrity is needed to confirm the existence of an optimum at lower content of ionomer.

The polarization behaviour of a CL18-U catalyst layer at a loading level of 2 mg/cm^2 and a catalyst layer based on Pt black catalyst at a loading of 6mg/cm^2 is compared at Fig. 12. The thickness of the catalyst layer for the CL18-U and Pt-black catalysts is 70 and 45 μm, respectively. The MEAs have identical components, except the cathode catalyst layer. The CL18-U catalyst delivers higher power at high voltages (>0.4 V) and lower power density at low voltages (<0.4 V) than the Pt black-based cathode. Since catalytic activity governs the electrochemical reaction rate at the high voltages (activation region), the higher power density for Pt/OMC indicates that 2 mg/cm^2 of Pt/OMC gives higher catalytic activity than 6 mg/cm^2 of Pt black catalyst, which is of practical importance. With the introduction of an OMC support, the Pt loading in the cathode can be reduced to one-third of Pt black-based catalyst layer, without any negative effect on power performance, and this would significantly contribute to cost reduction of MEAs. The lower power density for the Pt/OMC catalyst layer at high-current density indicates that the mass-transport limitation is greater than that for the Pt black-based catalyst layer.

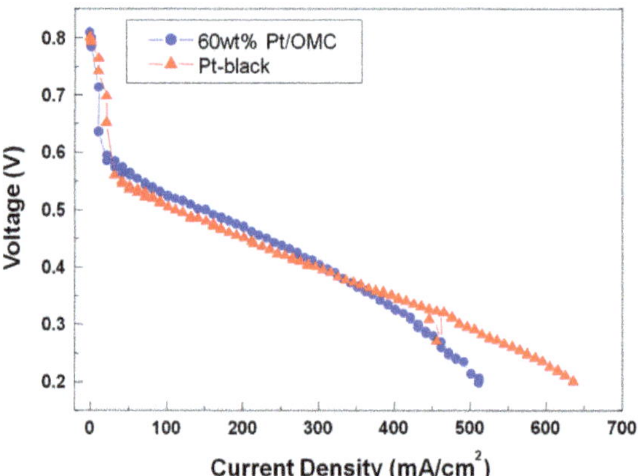

Figure 12. Comparison of polarization curves at 70 ◦C of MEA with the cathode from Pt/OMC catalyst (2 mg/cm^2) and unsupported Pt black catalyst (6mg/cm^2) (cathode feed: air, anode feed: 1M CH_3OH).

5. Conclusion

In this chapter, the fabrication processes for the electrode for MEA was briefly reviewed in a view of optimization of process parameter and application of new supported catalyst. The processes were catalyst coated on the electrode (CCE), decal transfer catalyst coated on membrane (DTM) and direct catalyst coated on membrane (DCM) methods. Among the three processes, the optimization of DCM method for DMFC MEA was presented as an example. The temperature and pressure were the main parameter which should be adjusted for maximizing the performance of MEA using hydrocarbon membrane. The effect of pore generation by pore forming agent on the performance was discussed. In addition, the application of new Pt/OMC catalyst for DMFC MEA was demonstrated by controlling the amount of ionomer and compression of catalyst layer. The application of Pt/OMC catalyst resulted in the decrease of the amount of Pt in the cathode from 6 mg/cm^2 for Pt black catalyst to 2 mg/cm^2 using Pt/OMC without lost the performance.

Author details

Chanho Pak, Dae Jong You, Kyoung Hwan Choi and Hyuk Chang
Energy Lab, Samsung Advanced Institute of Technology, Samsung Electronics, Co. Ltd., Republic of Korea

6. References

Andujar, JM. (2009). Fuel Cells : History and updating. A walk along two centuries. *Renewable and Sustainable Energy Reviews*, Vol.13, No.9, (December 2009), pp. 2309-2322, ISSN 1364-0321

Chang, H. (2007). Synthesis and characterization of mesoporous carbon for fuel cell applications. *Journal of Materials Chemistry*, Vol.17, No.30 (May 2007), pp.3078-3088, ISSN 1364-5501

Cho, J. (2009). Fabrication and evaluation of membrane electrode assemblies by low-temperature decal methods for direct methanol fuel cell. *Journal of Power Sources*, Vol.187, No.2, (February 2009), pp. 378–386, ISSN 0378-7753

Frey, Th. (2004). Effects of membrane electrode assembly preparation on the polymer electrolyte membrane fuel cell performance. *Electrochimica Acta*, Vol.50, No.11, (November 2004), pp. 99-105, ISSN 0013-4686

Guan, R. (2006). Effect of casting solvent on the morphlogy and performance of sulfonated polyethersulfone membranes. *Journal of Membrane Science*, Vol.277, No.29, (June 2006), pp.148-156, ISSN 0376-7388

He, W. (2010). Oxygen reduction on Pd_3Pt_1 bimetallic nanoparticles highly loaded on different carbon supports. *Applied Catalysis B: Environmental*, Vol.97, No.18, (April 2010), pp.347-353, ISSN 0926-3373

IEA (Interational Energy Agency) (2010). Energy Technology Perspectives from http://www.iea.org/papers/2009/ETP_2010_flyer.pdf

Joo, S. (2001). Ordered nanoporous array of carbon supporting high dispersions of platinum nanoparticles. *Nature,* Vol.412, No. 6843 (July 2001), pp.169-172, ISSN 0028-0836

Joo, S. (2006). Ordered mesoporous carbon (OMC) as supports of electrocatalysts for direct methanol fuel cells (DMFC) : Effect of carbon precursors of OMC on DMFC performances. *Electrochimica Acta,* Vol.52, No.15, (May 2006), pp.1618-1626, ISSN 0013-46868

Joo, S. (2009). Preparation of high loading Pt nanoparticles on ordered mesoporous carbon with a controlled Pt size and its effect on oxygen reduction and methanol oxidation reactions. *Electrochimica Acta,* Vol.54, No.14, (May 2009), pp.5746-5753, ISSN 0013-46868

Jun, S. (2000). Synthesis of New, Nanoporous Carbon with Hexagonally Ordered Meso structure. *Journal of American Chemical Society,* Vol.122. No.43 (November 2000), pp. 10712-10713, ISSN 1520-5126

Kim., H. (2008). Cathode catalyst layer using supported Pt catalyst on ordered mesoporous carbon for direct methanol fuel cell. *Journal of Power Sources,* Vol.180, No.2 (March 2008), pp. 724-732, ISSN 0378-7753

Kuver, A. (1994). Distinct performance evaluation of a direct methanol SPE fuel cell. A new method using a dynamic hydrogen reference electrode. *Journal of Power Sources,* Vol.52, No. 1 (November 1994), pp.77-80, ISSN 0378-7753

Lee., H. (2009). Ultrastable Pt nanoparticles supported on sulphur-containing ordered mesoporous carbon via strong metal-support interaction. *Journal of Materials Chemistry,* Vol.19, No. 33 (July 2009), pp.5934-5939, ISSN 1364-5501

Liang, ZX. (2004). FT-IR study of the microstructure of Nafion® membrane. *Journal of Membrane Science,* Vol.233, No.1-2 (April 2004), pp.39-44, ISSN 0376-7388

Lindermeir, A. (2004). On the question of MEA preparation for DMFCs. *Journal of Power Sources,* Vol.129, No. 2 (April 2004), pp. 180–187, ISSN 0378-7753

Liu, F. (2006). Optimization of cathode catalyst layer for direct methanol fuel cells: Part I. Experimental investigation. *Electrochimica Acta,* Vol.52, No.10, (September 2006), pp.1417-1425, ISSN 0013-46868

Liu, FQ. (2006). Low crossover of methanol and water through thin membranes in direct methanol fuel cells. *Journal of Electrochemical Society,* Vol.153, No.3, (January 2006), pp.A543–A553, ISSN 0013-4651

Mao, Q. (2007). Comparative studies of configuration and preparation methods for direct methanol fuel cell electrodes. *Electrochimica Acta,* Vol.52, No.10, (August 2007), pp.6763-6770, ISSN 0013-4686

Pak, C. (2009). Mesoporous Carbon-Supported Catalysts for Direct Methnaol Fuel Cells. *Electrocatalysis of Direct Methanol Fuel Cells,* (October 2009) pp.355-378. ISBN 978-3-527-32377-7

Pak, C. (2010). Nanomaterials and structures for the fourth innovation of polymer electrolyte fuel cell. *Journal of Materials Research,* Vol.25, No.11 (November 2010) pp.2063–2071, ISSN 0884-2914

Park, H-S. (2008). Modified Decal Method and Its Related Study of Microporous Layer in PEM Fuel Cells. *Journal of Electrochemical Society,* Vol.155, No. 5 (May 2008), pp.B455–B460, ISSN0013-4651

Park, I. (2010). Fabrication of catalyst-coated membrane-electrode assemblies by doctor blade method and their performance in fuel cells. *Journal of Power Sources,* Vol.195, No.7, (May 2010), pp. 7078–7082, ISSN0378-7753

Park, J. (2008). Mass balance research for high electrochemical performance direct methanol fuel cells with reduced methanol crossover at various operating conditions. *Journal of Power Sources,* Vol.178, No.1, (March 2008), pp.181-187, ISSN0378-7753

Perry, ML. (2002). A Historical perspective of fuel cell technology in the 20th Century. *Journal of Electrochemical Society,* Vol.149, No. 7, (July 2002), pp.S59-S67, ISSN0013-4651

Ramasamy, RP. (2009). Fuel cell – proton-exchange membrane fuel cells : Membrane-ElectrodeAssemblies, *Encyclopedia of Electrochemical Power source,* (November 2009), pp. 787-805, ISBN 978-0-444-52745-5

Robertson, GP. (2003). Casting solvent interactions with sulfonated poly (ether ether ketone) during proton exchange membrane fabrication. *Journal of Membrane Science,* Vol.219, No.1-2, (July 2003), pp.113-121, ISSN 0376-7388

Song, SQ. (2005). Direct methanol fuel cells: The effect of electrode fabrication procedure on MEAs structural properties and cell performance. *Journal of Power Sources,* Vol.145, No.2, (August 2005), pp.495-501, ISSN 0378-7753

Song, Y. (2005). Improvement in high temperature proton exchange membrane fuel cells cathode performance with ammonium carbonate. *Journal of Power Sources,* Vol.141, No.2 (March 2005), pp. 250-257, ISSN 0378-7753

Sun, L. (2010). Fabrication and evolution of catalyst-coated membranes by direct spray deposition of catalyst ink onto Nafion membrane at high temperature. *International Journal of Hydrogen Energy,* Vol.35, No.10, (June 2010), pp.2921-2925, ISSN 0360-3199

Tang, H. (2007). Performance of direct methanol fuel cells prepared by hot-pressed MEA and catalyst-coated membrane (CCM). *Electrochimica Acta,* Vol.52, No.11, (March 2007), pp.3714-3718, ISSN 0013-4686

Therdthianwong, A. (2007). Investigation of membrane electrode assembly (MEA) hot-pressing parameters for proton exchange membrane fuel cell. *Energy,* Vol.32, No.12, (December 2007), pp. 2401–2411, ISSN 0360-5442

Tucker, MC. (2005). The pore structure of direct methanol fuel cell electrodes. *Journal of Electrochemical Society,* Vol.152, No.9 (September 2005) pp.A1844–A1850, ISSN 0013-4651

Uchida, M. (1998). A Improved preparation process of very low platinum loading electrodes for polymer electrolyte fuel cells. *Journal of Electrochemical Society,* Vol.145, No.11, (November 1998), pp. 3708-3713, ISSN 0013-4651

Wei, Z. (2002). Influence of electrode structure on the performance of a direct methanol fuel cell. *Journal of Power Sources,* Vol.106, No.1-2 (April 2002), pp. 364-369, ISSN 0378-7753

Wilson, M. (1995). Low platinum loading electrodes for polymer electrolyte fuel cells fabricated using thermoplastic ionomers. *Electrochimica Acta,* Vol.40, No.3, (February 1995), pp. 355-363, ISSN 0013-4686

Xie, J. (2004). Ionomer Segregation in Composite MEAs and Its Effect on Polymer Electrolyte Fuel Cell Performance. *Journal of Electrochemical Society*, Vol.151, No.7, (June 2004) pp.A1084–A1093, ISSN 0013-4651

You, D. (2010). High performance membrane electrode assemblies by optimization of coating process and catalyst layer structure in direct methanol fuel cells. *International Journal of Hydrogen Energy*, Vol.36, No.12, (April 2011), pp.5096-5103, ISSN 0360-3199

Zhang, J. (2006). Effects of MEA preparation on the performance of a direct methanol fuel cell. *Journal of Power Sources*, Vol.160, No.2, (April 2006), pp.1035-1040, ISSN 0378-7753

Zhang, J. (2007). Effects of hot pressing conditions on the performances of MEAs for direct methanol fuel cells. *Journal of Power Sources*, Vol.165, No.1, (February 2007), pp73–81, ISSN 0378-7753

Chapter 12

Computational Study of Thermal, Water and Gas Management in PEM Fuel Cell Stacks

Agus P. Sasmito, Erik Birgersson and Arun S. Mujumdar

Additional information is available at the end of the chapter

1. Introduction

The overall high degree of research and development of PEM fuel cells over the last few decades has led to a series of improvements. Noteworthy is the 'jump' in cell performance that was achieved from 50-100 W m^{-2} in 1959-1961 with a phenol sulfonic membrane to 6000-8000 W m^{-2} in 1971-1980 by the introduction of the Nafion membrane (Costamagna, 2001). Around thirteen years later, the first fuel cell was brought from laboratory to power a car in the New Generation of Vehicle program (PNGV) in the US in 1993 (Costamagna, 2001b); moreover, in 2007, the vehicle manufacturer Honda launched a fuel cell car FCX clarity and commercially available in 2008. The PEM fuel cells have recently passed the demonstration phase and have partially reached the commercialization stage because of the rapid development and impressive research effort worldwide. However, despite the promising achievements and plausible prospects for PEM fuel cell, in view of the remaining challenges, it is commonly agreed that there is still a long way to go before they can successfully and economically replace the various traditional energy systems (Hellman et al., 2007).

Currently, there are three major challenges in fuel cell research and development which need to be improved (Martin et al., 2010 and Steiner et al., 2009):

- *Cost*: The current cost of a PEFC for automotive application is approximately 600-1000 $/kW (Tsuchiya and Kobayashi, 2004, Mock and Schmid, 2009) which means that the cost of a vehicle powered by fuel cell is about 10 times than that of the traditional car with an internal combustion engine; the total cost of a PEM fuel cell for stationary applications and combined heat power system is around 23000 Euro per kW (Staffell and Green, 2009). In order to be able to compete with traditional power sources, the US Department of Energy (DOE) has set targets for hydrogen and fuel cell technologies. The targets for automotive fuel cells cost is around 30 $/kW by 2015; whereas, for stationary application, the targets is at fuel cell cost of 750 $/kW (Martin et al., 2010).

- *Durability*: For a fuel cell operating under constant load conditions, e.g., in stationary applications, the degradation rate can be as low as 1-10 μV/h (Wolfgang et al., 2008), this would result in 40-80 mV performance loss over 40000 h, i.e. an efficiency and power loss of 10% over 40000 h lifetime (Brujin et al., 2007). However, the degradation rates can increase by orders of magnitude when conditions include load cycling, start-stop cycles, low humidification or humidification cycling, temperatures of 90 °C or higher and fuel starvation. Such conditions are expected for automotive applications. Currently, the typical durability for fuel cell in automotive application is less than 5000 h (equivalent to 150000 miles); for example, Mercedes-Benz claims a lifetime of above 2000 h without performance degradation for their current fuel cell stacks operated in test-vehicles. Hence, US DOE has also set targets for fuel cell durability to be improved up to 5000 and 40000 h for automotive and stationary applications, respectively, by 2015 (Martin et al., 2010).
- *Performance*: Fuel cell performance has improved rapidly in recent years; the performance improved from 300 W m^{-2} in the NASA Gemini project in 1960 to power densities of about 6000 Wm^{-2} for more recent fuel cell stack, e.g., Ballard MK 900 (Mock and Schmid, 2009). Although power densities of 10000 Wm^{-2} and even higher have already been met in the laboratory, there is still an interdependencies between fuel cell performance and catalyst loading. Hence, US DOE has set a target for power densities of 10000 W m^{-2} with fuel cell efficiency 60 % by 2015 (Martin et al., 2010 and Mock and Schmid, 2009).

Out of all these challenges, this chapter will mainly focus on the performance improvement as fuel cell can be considered as immature and complex technology. Immaturity is reflected by the rapid technological progress in performance, e.g. power density, capacity, sustainability and cost (Hellman et al., 2007). Whereas, complexity refers to the fact that fuel cells require an overall system that comprises of numerous components between which there is a high degree of interdependence. Furthermore, these two issues are correlated with PEM fuel cell performance improvement and cost reduction (Sopian et al., 2006). To achieve these goals, a significant effort on research and development in the worldwide fuel cell area focuses on high-purity hydrogen, cost reduction of the system, including fuel cell material, where the cost of the bipolar plate and the electrode including platinum make up approximately 80% of the total cost of PEM fuel cell (Mock and Schmid, 2009, Tsuchiya and Kobayashi, 2004, Hermann et al., 2005), and various technological problems (Wee et al., 2007).

In order to improve fuel cell system performance, one needs to address three issues that are most vital to the PEM fuel cell performance (Sasmito et al., 2011):

- *Thermal management*: An effective heat removal system is necessary to keep the fuel cell/stack at a uniform operating temperature, as a significant amount of heat is generated in a fuel cell. As it is known that fuel cells are usually about 30-60% electrically efficient at typical operating power density; hence, energy which is not converted into electrical power is dissipated as heat. For a stack, to achieve high stack performance, each and every cell should be running at identical operating conditions. Hence, either an active or a passive cooling system should be applied (Sasmito, et al., 2011a). Active cooling usually either involves coolant channels in the bipolar plate or as separate layers, where coolant liquid is pumped through to remove heat, or external fans to improve the air flow through the fuel cell/stack. For passive cooling, natural convection can be employed to remove heat. However, this mode of cooling is only sufficient for small portable fuel cells with power rating of less than 100 W (Sasmito et al., 2010 and Sasmito 2011b). Another option for passive cooling is edge-cooling, which means that if the fuel cell is made narrow enough, the heat generated may be removed on the sides of the cell by attaching fins and/or selecting high thermal conductivity material (Sasmito et al., 2011).

- *Water management*: Since the PEM fuel cell operates at temperatures below 100 °C, water vapor can condense and form a liquid phase inside the fuel cell (Sasmito et al., 2011c). Hence, to prevent flooding of the various cell components while keeping the membrane sufficiently hydrated for ionic conduction, a proper water management should be applied together with thermal management. Several techniques have been employed in order to mitigate water flooding in cathode PEFC such as proper flow field design, anode water removal, operating condition control and so forth. More advanced techniques can be employed, e.g. hydrophobic bipolar plates together with electro osmotic pumps (Buie et al., 2006). This setup can be used to pump water into the cell if it is too dry or to pump out water if the cell is flooding. Furthermore, a frequent automatic purge can also be implemented to avoid liquid flooding in a dead-end anode PEM fuel cell (Sasmito et al., 2011d).
- *Gas management*: To ensure that enough hydrogen and oxygen reach the active layer to generate current, especially so at high current densities. In addition, uniformly distributed fuel and reactant is a requirement to generate uniform current distribution and thus utilize the catalyst evenly (Sasmito et al., 2011e).

The main objective of this chapter is to address fuel cell issues related to various technological problems, which could improve the fuel cell system performance by optimal thermal, gas, and water management. This research is concerned with the fundamental issues as well as application of PEM fuel cells:

- *Fundamental research*: The fundamental aspects concern the transport phenomena that can be found in a PEM fuel cell: phenomena that are inherently multiscale (time and space) in nature, including heat and mass transfer, electrochemistry, two-phase flow, charge transfer, agglomerate catalyst layers, and transport in the membrane. Here, a one-domain, two-phase-flow model comprising conservation of equations of mass, momentum, species, energy, charge, an agglomerate catalyst layer and a phenomenological membrane model is developed, calibrated and further validated with experimental polarization curves, iR-corrected curves, and local current density distributions. After validation, to ensure the fidelity of model predictions, the model is extended to account for a fuel cell stack together with its surrounding environment and auxiliary equipment, such as fans; thus, allowing for studies that not only capture the essential physics of the fuel cell stack but also the interaction with auxiliary equipment. Furthermore, the model is also extended to account for transient characteristics of PEM fuel cell operating with a dead-end anode

- *Applied research*: The applied aspects address key issues related to thermal, water, and gas management, as well as design and optimization to ensure optimal or near-to optimal fuel cell performance. Four different thermal management strategies – liquid-cooling, forced air-convection cooling, edge-air cooling, and natural-convection cooling – as well as various designs and conditions commonly employed in PEM fuel cell stacks are investigated. Furthermore, a new concept to enhance thermal management in PEFC stack is proposed. In addition, we also address gas and water management issues in a PEM fuel cell with a dead-end anode.

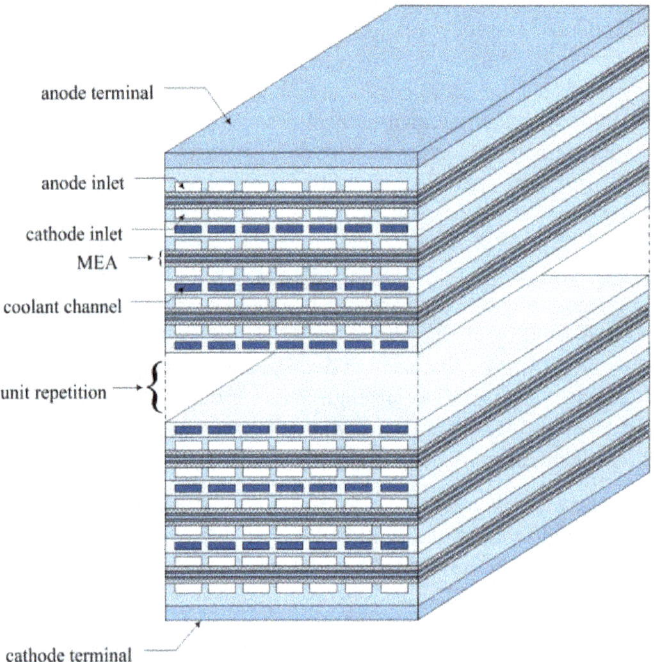

Figure 1. Schematic of PEM fuel cell stack

2. Model formulation

Several coupled transient transport phenomena take place within the PEFC which comprises of two sets of flow channels, anode and cathode gas diffusion layers, anode and cathode catalyst layers and a polymer electrolyte membrane, see Fig. 1 for details. They include:

- Convective heat and mass transfer in the flow channels;

- Diffusion in the gas diffusion layers;
- Two-phase flow/transport in the gas diffusion layers and flow fields (water in vapor and liquid phases);
- Phase change in the gas diffusion layers and flow channels;
- Electrochemical reactions in the catalyst layers;
- Heat generation in the catalyst layers;
- Water production in the cathode catalyst layer;
- Oxygen consumption in the cathode catalyst layer;
- Hydrogen consumption in the anode catalyst layer;
- Electron transport in the gas diffusion layers and bipolar plates;
- Membrane water transport;
- Proton transport in the membrane;
- Electro-osmotic drag in the membrane;
- Diffusion of $H_2/O_2/N_2$ through the membrane;
- Ohmic heating in the membrane, catalyst layers, gas diffusion layers, and bipolar plates;
- Conjugate heat transfer between bipolar plates/cooling channels and bipolar plates/flow channels.

Starting from the basic physics equation and governing equation, electrochemical reactions, two-phase flow and its constitutive relations in a numerical model are summarized. The model accounts for the following transport phenomena in each cell for the stack:

- *Mass, momentum and species transfer*: Conservation of two-phase mass, momentum and species is considered in the whole cell, with certain simplifications for the membrane. The gas phase consists of hydrogen, oxygen, water vapour and nitrogen, whereas the liquid phase is assumed to comprise only of liquid water due to the low solubility of the other gases.
- *Heat transfer*: We consider convection, conduction, evaporation/condensation, ohmic heating, entropy generation and heat generation due to the activation overpotential.
- *Charge transfer*: Conservation of charge and Ohm's law is solved.

The main model assumptions/approximations are:

- *Thermal equilibrium*: We assume local thermal equilibrium between all phases.
- *Membrane*: The membrane model takes into account the flux of water due to electroosmotic drag and diffusion. In this work, a GORE membrane is employed for which we modify the standard phenomenological equations derived for a Nafion® with a correction factor.
- *Two-phase flow*: We assume that the dominating driving force for liquid transport inside the gas diffusion layers and catalyst layers is capillarity. In the flow fields, we consider a mist flow approximation of the liquid flow, for which the liquid velocity is assumed to be the same as the gas velocity, as well as the capillary contribution due the porous nature of the net-type flow fields we solve for here.
- *Catalyst layers*: An agglomerate model is implemented to account for mass transfer inside the cathode catalyst layer. Here, we assume the agglomerate nucleus to be

spherical in shape, which in turn is covered by a thin film of ionomer and water (Sasmito et al., 2011). For the anode, a conventional expression based on the Butler-Volmer equation is employed as the overpotential is significantly lower than at the cathode.

- *End-plates*: We assume that the stack comprises a large number of cells, so that the effects of the two end plates are negligible for the overall stack performance.

The governing equations together with the constitutive relations and appropriate boundary conditions are then solved numerically.

In this chapter, the superscripts (g), (l) and (m) denote properties associated with the gas, liquid, solid and membrane, respectively, and (c) denotes any quantity associated with capillary pressure

2.1. Fuel cell stack

We consider conservation of mass, momentum, species, energy, charge, phenomenological membrane model and two-phase in the separator plates, flow fields, gas diffusion layers, catalyst layers, and membrane in the PEMFC, expressed as

$$\nabla \cdot \left(\rho^{(g)} \mathbf{u}^{(g)} \right) = S_{\text{mass}} - \dot{m}_{\text{H}_2\text{O}} \tag{1}$$

$$\nabla \cdot \left(\rho^{(l)} \mathbf{u}^{(l)} \right) = \dot{m}_{\text{H}_2\text{O}} \tag{2}$$

$$\nabla \cdot \left(\rho^{(g)} \mathbf{u}^{(g)} \mathbf{u}^{(g)} \right) = \nabla \cdot \boldsymbol{\sigma} - \frac{\mu^{(g)}}{\kappa} \mathbf{u}^{(g)} \tag{3}$$

$$\nabla \cdot \left(\rho^{(g)} C_{\text{p}}^{(g)} \mathbf{u}^{(g)} T \right) = \nabla \cdot \left(k_{\text{eff}} \nabla T \right) + S_{\text{temp}} \tag{4}$$

$$\nabla \cdot \mathbf{n}_{\text{i}}^{(g)} = S_{\text{i}} \tag{5}$$

$$\nabla \cdot \mathbf{n}_{\text{H}_2\text{O}}^{(m)} = 0 \tag{6}$$

$$\nabla \cdot \mathbf{i}^{(m)} = S_{\text{pot}} \tag{7}$$

$$\nabla \cdot \mathbf{i}^{(s)} = -S_{\text{pot}} \tag{8}$$

In the above equations, $\rho^{(g,l)}$ denote phase densities, $\mathbf{u}^{(g,l)} = \left(u^{(g,l)}, v^{(g,l)} \right)$ are the phase velocities (in x and y directions), $\dot{m}_{\text{H}_2\text{O}}$ is the interphase mass transfer of water between the gas and liquid phase, $\boldsymbol{\sigma}$ is the total stress tensor, $\mu^{(g)}$ is the gas dynamic viscosity, κ is the permeability, $c_{\text{p}}^{(g)}$ is the specific heat capacity, T is the temperature and k_{eff} is the effective thermal conductivity. Furthermore, $\mathbf{i}^{(m)}$ and $\mathbf{i}^{(s)}$ are the current densities carried by protons and electrons respectively. The governing equations are similar to those implemented by Li *et al.*(2004) and Schwarz and and Djilali (2007).

The total stress tensor, mass fluxes of species, current densities and liquid water velocity are defined as

$$\boldsymbol{\sigma} = -p^{(g)}\mathbf{I} + \mu^{(g)}\left(\nabla\mathbf{u}^{(g)} + \left(\nabla\mathbf{u}^{(g)}\right)^{\mathrm{T}}\right) - \frac{2}{3}\mu^{(g)}\left(\nabla\cdot\mathbf{u}^{(g)}\right)\mathbf{I} \tag{9}$$

$$\mathbf{n}_{\mathrm{i}}^{(g)} = \rho^{(g)}\mathbf{u}^{(g)}\omega_{\mathrm{i}}^{(g)} - \rho^{(g)}D_{\mathrm{i,eff}}^{(g)}\nabla\omega_{\mathrm{i}}^{(g)} \quad (\mathrm{i} = \mathrm{H_2, O_2, H_2O, N_2}) \tag{10}$$

$$\mathbf{n}_{\mathrm{H_2O}}^{(m)} = \frac{n_{\mathrm{d}}M_{\mathrm{H_2O}}}{F}\mathbf{i}^{(m)} - \frac{\rho^{(m)}}{M^{(m)}}M_{\mathrm{H_2O}}D_{\mathrm{H_2O,eff}}^{(m)}\nabla\lambda \tag{11}$$

$$\mathbf{i}^{(m)} = -\sigma_{\mathrm{eff}}^{(m)}\nabla\phi^{(m)} \tag{12}$$

$$\mathbf{i}^{(s)} = -\sigma_{\mathrm{eff}}^{(s)}\nabla\phi^{(s)} \tag{13}$$

$$\mathbf{u}^{(l)} = \begin{cases} \mathbf{u}^{(g)}s \quad D^{(c)}\nabla s & (\mathrm{ff}) \\ -D^{(c)}\nabla s & (\mathrm{gdl,\ cl}) \end{cases} \tag{14}$$

For the total stress tensor in Eq. 9, $p^{(g)}$ is the gas pressure and **I** is the identity matrix. Here, we solve for a species mixture of hydrogen (H_2), water (H_2O), oxygen (O_2) and nitrogen (N_2); $\omega_i^{(g)}$ denotes the mass fraction of species i in the gas phase, and $D_{i,eff}^{(g)}$ represents the effective diffusivity in the gas phase. The flux of water in the membrane, Eq. 11, is expressed with a phenomenological model in terms of the membrane water content, λ, which account for the electroosmotic drag (first term on the right hand side [RHS]) and diffusion (second term on the RHS). Here, n_d is the electroosmotic drag coefficient, F is Faraday's constant, $D_{H_2O,eff}^{(m)}$ is the effective diffusivity of water in the membrane, $\rho^{(m)}$ and $M^{(m)}$ are the density and equivalent weight of the dry membrane, respectively. In Eqs. 12 and 13, $\phi^{(m)}$ and $\phi^{(s)}$ represent the potential of the ionic phase and the solid phase, while $\sigma_{eff}^{(m)}$ and $\sigma_{eff}^{(s)}$ are the electrical conductivities of proton and electron transport, respectively. In Eq. 14, s is the liquid saturation and $D^{(c)}$ is the capillary diffusion.

The source terms in Eqs. 1-8 are given by

$$S_{\mathrm{mass}} = \begin{cases} -\frac{M_{\mathrm{O_2}}}{4F}J_{\mathrm{c}} + \frac{M_{\mathrm{H_2O}}}{2F}J_{\mathrm{c}} - \nabla\cdot\mathbf{n}_{\mathrm{H_2O}}^{(m)} & (\text{cathode cl}) \\ -\frac{M_{\mathrm{H_2}}}{2F}J_{\mathrm{a}} - \nabla\cdot\mathbf{n}_{\mathrm{H_2O}}^{(m)} & (\text{anode cl}) \\ 0 & (\text{elsewhere}) \end{cases} \tag{15}$$

$$S_i \begin{cases} -\dfrac{M_{O_2}}{4F} J_c & (O_2\text{, cathode cl}) \\ +\dfrac{M_{H_2O}}{2F} J_c - \nabla \cdot \mathbf{n}^{(m)}_{H_2O,\text{eff}} - \dot{m}_{H_2O} & (H_2O\text{, cathode cl}) \\ -\nabla \cdot \mathbf{n}^{(m)}_{H_2O,\text{eff}} - \dot{m}_{H_2O} & (H_2O\text{, anode cl}) \\ -\dot{m}_{H_2O} & (H_2O\text{, gdl}) \\ -\dfrac{M_{H_2}}{2F} J_a & (H_2\text{, anode cl}) \\ 0 & (\text{elsewhere}) \end{cases} \tag{16}$$

$$S_{\text{pot}} \begin{cases} -J_c & (\text{cathode cl}) \\ J_a & (\text{anode cl}) \\ 0 & (\text{elsewhere}) \end{cases} \tag{17}$$

$$S_{\text{temp}} = \begin{cases} J_c\left(-T\dfrac{\partial E_{\text{rev}}}{\partial T} + |\eta_c|\right) + \sigma^{(m)}_{\text{eff}}\left(\nabla\phi^{(m)}\right)^2 + \\ \quad +\sigma^{(s)}_{\text{eff}}\left(\nabla\phi^{(s)}\right)^2 + \dot{m}_{H_2O}H_{\text{vap}} & (\text{cathode cl}) \\ J_a\eta_a + \sigma^{(m)}_{\text{eff}}\left(\nabla\phi^{(m)}\right)^2 + \\ \quad +\sigma^{(s)}_{\text{eff}}\left(\nabla\phi^{(s)}\right)^2 + \dot{m}_{H_2O}H_{\text{vap}} & (\text{anode cl}) \\ \sigma^{(m)}_{\text{eff}}\left(\nabla\phi^{(m)}\right)^2 & (\text{m}) \\ \sigma^{(s)}_{\text{eff}}\left(\nabla\phi^{(s)}\right)^2 + \dot{m}_{H_2O}H_{\text{vap}} & (\text{gdl}) \\ \sigma^{(s)}_{\text{eff}}\left(\nabla\phi^{(s)}\right)^2 & (\text{ff, sp}) \end{cases} \tag{18}$$

The source term in Eq. 15 comprises mass consumption and production due to electrochemical reactions and the transport of water through the membrane, whilst the source terms for species conservation, Eq. 16, considers species consumption and production due to electrochemical reactions as well as interphase mass transfer for water and the transport of water through membrane. In catalyst layer, λ is defined by solving Eq. 75 and 76. In Eq. 17, $J_{a,c}$ ($J_{a,c}>0$) denote the volumetric current densities, η_a $(\eta_a > 0)$ and η_c $(\eta_c < 0)$ are the overpotential at the anode and cathode, and E_{rev} is the reversible potential. Heat source from the reversible and irreversible entropic generated by the electrochemical reactions can be found in the first two terms of Eq. 18, the third and forth describe ohmic heating, while the energy transfer due to interphase mass transfer is described in the last term, where H_{vap} is the heat of vaporization of water.

2.2. Coolant plates

We solve for conservation of mass (incompressible liquid), momentum, energy, and charge in the coolant plates, given by

$$\nabla \cdot \left(\rho^{(l)} \mathbf{u}^{(l)}\right) = 0 \tag{19}$$

$$\nabla \cdot \left(\rho^{(l)} \mathbf{u}^{(l)} \mathbf{u}^{(l)}\right) = \nabla \cdot \boldsymbol{\sigma} - \frac{\mu^{(l)}}{\kappa} \mathbf{u}^{(l)} \tag{20}$$

$$\nabla \cdot \left(\rho^{(l)} C_{\mathrm{p}}^{(l)} \mathbf{u}^{(l)} T\right) = \nabla \cdot \left(k_{\mathrm{eff}} \nabla T\right) \tag{21}$$

$$\nabla \cdot \mathbf{i}^{(s)} = 0 \tag{22}$$

where the total stress tensor and current density are defined as

$$\boldsymbol{\sigma} = -p^{(l)} \mathbf{I} + \mu^{(l)} \left(\nabla \mathbf{u}^{(l)} + \left(\nabla \mathbf{u}^{(l)}\right)^{\mathrm{T}} \right) \tag{23}$$

$$\mathbf{i}^{(s)} = -\sigma_{\mathrm{eff}}^{(s)} \nabla \phi^{(s)} \tag{24}$$

3. Constitutive relations

The gas density (assuming ideal gas) is defined as

$$\rho^{(g)} = \frac{p^{(g)} M^{(g)}}{RT} \tag{25}$$

where

$$M^{(g)} = \left(\omega_{\mathrm{H_2}}^{(g)} / M_{\mathrm{H_2}} + \omega_{\mathrm{O_2}}^{(g)} / M_{\mathrm{O_2}} + \omega_{\mathrm{H_2O}}^{(g)} / M_{\mathrm{H_2O}} + \omega_{\mathrm{N_2}}^{(g)} / M_{\mathrm{N_2}}\right)^{-1} \tag{26}$$

The mass fraction of nitrogen is given by

$$\omega_{\mathrm{N_2}}^{(g)} = 1 - \omega_{\mathrm{H_2}}^{(g)} - \omega_{\mathrm{O_2}}^{(g)} - \omega_{\mathrm{H_2O}}^{(g)} \tag{27}$$

The molar fractions are related to the mass fractions as

$$x_{\mathrm{i}}^{(g)} = \frac{\omega_{\mathrm{i}}^{(g)} M^{(g)}}{M_{\mathrm{i}}} \tag{28}$$

The molar concentrations are given by

$$c_{\mathrm{i}}^{(g)} = \frac{\omega_{\mathrm{i}}^{(g)} \rho^{(g)}}{M_{\mathrm{i}}} \tag{29}$$

The gas mixture viscosity, $\mu^{(g)}$, is defined as

$$\mu^{(g)} = \sum_{\alpha} \frac{x_{\alpha}^{(g)} \mu_{\alpha}^{(g)}}{\sum_{\beta} x_{\alpha}^{(g)} \Phi_{\alpha,\beta}} \qquad \text{with } \alpha, \beta = H_2, O_2, H_2O, N_2 \tag{30}$$

where $x_{\alpha,\beta}^{(g)}$ are the mole fraction of species α and β, and

$$\Phi_{\alpha,\beta} = \frac{1}{\sqrt{8}} \left(1 + \frac{M_{\alpha}}{M_{\beta}}\right)^{-1/2} \left[1 + \left(\frac{\mu_{\alpha}^{(g)}}{\mu_{\beta}^{(g)}}\right)^{\frac{1}{2}} \left(\frac{M_{\beta}}{M_{\alpha}}\right)^{\frac{1}{4}}\right]^2 \tag{31}$$

The effective thermal conductivity is defined as

$$k_{\text{eff}} = \varepsilon(1-s)(\sum k_i^{(g)} \omega_i^{(g)}) + \varepsilon s k_{H_2O}^{(l)} + (1-\varepsilon) k^{(s)} \tag{32}$$

where $k^{(l)}$ is the thermal conductivity of liquid water, ε is the porosity, and $k^{(s)} = \left(k_{co}^{(s)}, k_{sp}^{(s)}, k_{ff}^{(s)}, k_{gdl}^{(s)}, k_{cl}^{(s)}, k_{m}^{(s)}\right)$ are the thermal conductivities of the solid phases in various functional layers. The gas mixture specific heat capacity, $c_p^{(g)}$, is written as

$$c_p^{(g)} = \sum_i \omega_i^{(g)} c_{p,i}^{(g)} \tag{33}$$

where $c_{p,i}^{(g)} = \left(c_{p,H_2}^{(g)}, c_{p,O_2}^{(g)}, c_{p,H_2O}^{(g)} . c_{p,N_2}^{(g)}\right)$ are the specific heat capacities of hydrogen, oxygen, water and nitrogen. The mass diffusion coefficients for each species i depends on the local temperature and pressure, defined as

$$D_i^{(g)}\left(T, p^{(g)}\right) = \left(\frac{T}{T_0}\right)^{3/2} \left(\frac{p_0^{(g)}}{p^{(g)}}\right) D_{i,0}^{(g)}\left(T_0, p_0^{(g)}\right) \tag{34}$$

where $D_{i,0}^{(g)}$ is the diffusion coefficient for each species i at a given temperature T_0 and gas pressure $p_0^{(g)}$. Furthermore, in porous media, we apply the Bruggeman correction and consider pore blockage due to the presence of liquid water (Wang and Wang 2006)

$$D_{i,\text{eff}}^{(g)} = (1-s)\varepsilon^{3/2} D_i^{(g)} \tag{35}$$

The relative humidity (%) which determines the water content at the anode and cathode inlet is defined as

$$H = \frac{p_{H_2O}^{(g)}}{p_{H_2O}^{\text{sat}}} \times 100 \tag{36}$$

where $p_{H_2O}^{(g)}$ is the partial pressure of water vapour, defined as

$$p_{\mathrm{H_2O}}^{(\mathrm{g})} = x_{\mathrm{H_2O}}^{(\mathrm{g})} p^{(\mathrm{g})} \tag{37}$$

and $p_{\mathrm{H_2O}}^{\mathrm{sat}}$ is the saturation pressure, given by

$$p_{\mathrm{H_2O}}^{\mathrm{sat}} = p^{\mathrm{ref}} \times 10^{c_1 + c_2(T-T_0) + c_3(T-T_{00})^2 + c_4(T-T_0)^3} \tag{38}$$

The mass fraction of water vapour at the anode and cathode inlet can be determined from

$$\omega_{\mathrm{H2O,a}}^{\mathrm{in}} = \omega_{\mathrm{H2O,c}}^{\mathrm{in}} = \frac{p_{\mathrm{H_2O}}^{\mathrm{sat}} M_{\mathrm{H_2O}} \left(\mathrm{H}_{\mathrm{a,c}}^{\mathrm{in}} / 100\right)}{p^{(\mathrm{g})} M^{(\mathrm{g})}} \tag{39}$$

By retaining the ratio $x_{\mathrm{O_2}}^{(\mathrm{g})} / x_{\mathrm{N_2}}^{(\mathrm{g})} = 21/79$, the mass fraction of oxygen at the cathode inlet can be calculated from

$$\omega_{\mathrm{O_2,c}}^{\mathrm{in}} = \frac{M_{\mathrm{O_2}}}{1 + 79/21} \left[\frac{1}{M^{(\mathrm{g})}} - \frac{\omega_{\mathrm{H_2O,c}}^{\mathrm{in}}}{M_{\mathrm{H_2O}}} \right] \tag{40}$$

while the mass fraction of hydrogen at the anode inlet is defined as

$$\omega_{\mathrm{H_2,a}}^{\mathrm{in}} = 1 - \omega_{\mathrm{H_2O,a}}^{\mathrm{in}} \tag{41}$$

The mass flow inlet in the anode and cathode are defined as

$$\dot{m}_{\mathrm{a}}^{\mathrm{in}} = \xi_{\mathrm{a}}^{\mathrm{in}} \frac{M_{\mathrm{H_2}}}{2F\omega_{\mathrm{H_2,a}}^{\mathrm{in}}} i_{\mathrm{ave}} A_{\mathrm{cl}} \tag{42}$$

$$\dot{m}_{\mathrm{c}}^{\mathrm{in}} = \xi_{\mathrm{c}}^{\mathrm{in}} \frac{M_{\mathrm{O_2}}}{4F\omega_{\mathrm{O_2,c}}^{\mathrm{in}}} i_{\mathrm{ave}} A_{\mathrm{cl}} \tag{43}$$

where $\xi_{\mathrm{a,c}}^{\mathrm{in}}$ is the anode and cathode inlet stoichiometry, A_{cl} is the catalyst surface area and i_{ave} is the average current density is given by

$$i_{\mathrm{ave}} = \frac{1}{L} \int_0^L \mathbf{i}^{(\mathrm{s})} \cdot \mathbf{e}_{\mathrm{y}} dx \tag{44}$$

where L is the fuel cell length.

The interphase mass transfer for condensation/evaporation of water is defined as (Li 2004)

$$\dot{m}_{\mathrm{H_2O}} = c_{\mathrm{r}} \max\left((1-s) \frac{p_{\mathrm{H_2O}}^{(\mathrm{g})} - p_{\mathrm{H_2O}}^{\mathrm{sat}}}{RT} M_{\mathrm{H_2O}}, -s\rho^{(\mathrm{l})} \right) \tag{45}$$

where c_r is the condensation/evaporation rate constant. The capillary diffusion for two-phase flow is given by

$$D^{(c)} = \frac{\kappa s^3}{\mu^{(l)}} \frac{dp^{(c)}}{ds} \tag{46}$$

where the capillary pressure is defined as

$$p^{(c)} = \tau \cos\theta \left(\frac{\varepsilon}{\kappa}\right)^{\frac{1}{2}} \mathrm{J} \tag{47}$$

where τ is the surface tension, θ is the wetting angle, and the Leverett function, J, is defined as

$$\mathrm{J} = 1.417(1-s) - 2.12(1-s)^2 + 1.263(1-s)^3 \tag{48}$$

4. Electrochemistry and agglomerate model

We consider an agglomerate model for the electrochemistry at the cathode side and retain a simple Butler-Volmer-type expression for the anode catalyst layer (Sasmito and Mujumdar, 2011)

$$J_a = j_a^{ref} \left(\frac{c_{H_2}^{(g)}}{c_{H_2,ref}^{(g)}} \right)^{1/2} \left[\exp\left(\frac{\alpha_a^{ox} F}{RT} \eta_a \right) - \exp\left(\frac{-\alpha_c^{ox} F}{RT} \eta_a \right) \right] \tag{49}$$

$$J_c = j_c^{ref} \left(\frac{c_{O_2}^{(g)}}{c_{O_2,ref}^{(g)}} \right) \left[-\exp\left(\frac{\alpha_a^{rd} F}{RT} \eta_c \right) + \exp\left(-\frac{\alpha_c^{rd} F}{RT} \eta_c \right) \right] (1-\gamma_{cl}) \left(1 - \frac{\gamma^{(p)}}{\gamma^{(agg)}} \right) \frac{RT}{H_{O_2}^{(p)}} \xi_1 \frac{1}{1+\xi_2+\xi_3} \tag{50}$$

The agglomerate model introduces additional mass transfer resistances in the cathode catalyst layer via mass transport inside the spherical agglomerate and the polymer and liquid water films. In Eqs. 47 and 48, $j_{a,c}^{ref}$ and $\alpha_{a,c}^{ox,rd}$ are the volumetric exchange current density and transfer coefficient for anode/cathode oxidation/reduction reaction, respectively. In Eq. 48, $H_{O_2}^{(p)}$ is Henry's constant for the air-ionomer interface, $c_{H_2,ref}^{(g)}$ and $c_{O_2,ref}^{(g)}$ are the reference concentration for hydrogen and oxygen, and ξ_1, ξ_2 and ξ_3 are the correction factors due to resistances of the agglomerate, the polymer and liquid water films, respectively. Furthermore, γ_{cl} represents the volume fraction of pores in the catalyst layer, while $\gamma^{(p)}$ and $\gamma^{(agg)}$ are the volume fraction of polymer and agglomerate, respectively. The cathode volumetric reference exchange current density, j_c^{ref}, is corrected for temperature via Arrhenius-type relation

$$j_c^{ref} = j_{c,0}^{ref} \exp\left[-\frac{E_a}{R}\left(\frac{1}{T} - \frac{1}{T_1} \right) \right] \tag{51}$$

The overpotentials are defined as

$$\eta_a = \phi^{(s)} - \phi^{(m)} \tag{52}$$

$$\eta_c = \phi^{(s)} - \phi^{(m)} - E_{rev} \tag{53}$$

Where the reversible potential, E_{rev}, is written as

$$E_{rev} = E_{rev,0} - e_1 (T - T_2) + \frac{RT}{4F} \ln x_{O_2}^{(g)} \tag{54}$$

The reference concentration, $c_{i,ref}^{(g)}$, is given by

$$c_{i,ref}^{(g)} = \frac{p^{ref}}{H_i^{(p)}} \tag{55}$$

The correction factor due to the agglomerate is defined as the effectiveness of the mass transfer of oxygen through the spherical agglomerate nucleus, and is given as:

$$\xi_1 = \frac{1}{\Phi} \left(\frac{1}{\tanh(3\Phi)} - \frac{1}{3\Phi} \right) \tag{56}$$

$$\Phi = \frac{r^{(agg)}}{3} \sqrt{\frac{k_c}{D_{O_2,eff}^{(agg)}}} \tag{57}$$

where $r^{(agg)}$ is the agglomerate radius, and k_c is the reaction rate constant, defined as

$$k_c = \frac{j_c^{ref} \left(1 - \frac{\gamma^{(p)}}{\gamma^{(agg)}} \right) \left(-\exp\left(\frac{\alpha_a^{rd} F}{RT} \eta_c \right) + \exp\left(-\frac{\alpha_c^{rd} F}{RT} \eta_c \right) \right)}{4F c_{O_2}^{ref}} \tag{58}$$

The effective diffusion coefficient of oxygen in polymer inside agglomerate, $D_{O_2,eff}^{(agg)}$, is given by the diffusion coefficient of oxygen in polymer film, $D_{O_2}^{(p)}$, with the Bruggeman correlation

$$D_{O_2,eff}^{(agg)} = D_{O_2}^{(p)} \left(\frac{\gamma^{(p)}}{\gamma^{(agg)}} \right)^{1.5} \tag{59}$$

The correction factor due to polymer film is calculated as

$$\xi_2 = \frac{\delta^{(p)}}{D_{O_2}^{(p)}} \frac{\xi_1}{a^{(p)}} k_c \tag{60}$$

where the polymer film thickness, $\delta^{(p)}$, is defined as

$$\delta^{(p)} = \sqrt[3]{\left(r^{(agg)}\right)^3 (1 + \frac{\gamma^{(p)}}{\gamma^{(PtC)}})} - r^{(agg)} \quad (61)$$

and the agglomerate surface area per unit volume of catalyst layer, $a^{(p)}$, is given by

$$a^{(p)} = 4\pi n^{(agg)} (r^{(agg)} + \delta^{(p)})^2 \quad (62)$$

where $n^{(agg)}$ is the number of agglomerates per unit volume, defined as

$$n^{(agg)} = \frac{3\gamma^{(agg)}}{4\pi\left(r^{(agg)} + \delta^{(p)}\right)^3} \quad (63)$$

The correction factor due to liquid water film

$$\xi_3 = \frac{\delta^{(l)}}{D_{O_2}^{(l)}} \frac{\xi_1}{a^{(l)}} k_c \frac{H_{O_2}^{(l)}}{H_{O_2}^{(p)}} \quad (64)$$

where $H_{O_2}^{(l)}$ is Henry's constant for the air-water interface, $D_{O_2}^{(l)}$ is the diffusion coefficient of oxygen in the liquid water , $\delta^{(l)}$ is the thickness of liquid film and $a^{(l)}$ is the surface area of the agglomerate including liquid water per unit volume, defined as

$$\delta^{(l)} = \sqrt[3]{\left(r^{(agg)} + \delta^{(p)}\right)^3 \left(1 + \frac{\gamma^{(l)}}{\gamma^{(agg)}}\right)} - \left(r^{(agg)} + \delta^{(p)}\right) \quad (65)$$

$$a^{(l)} = 4\pi n^{(agg)} (r^{(agg)} + \delta^{(p)} + \delta^{(l)})^2 \quad (66)$$

where the volume fraction of liquid water, $\gamma^{(l)}$, is expressed as function of the liquid saturation, s, by

$$\gamma^{(l)} = \frac{V^{(l)}}{V_{tot}} = s\gamma_{cl} \quad (67)$$

The carbon loading, $L^{(C)}$, and mass fraction of polymer loading in the catalyst layer, $\omega^{(p)}$ are

$$L^{(C)} = \frac{L^{(Pt)}}{\omega^{(Pt)}} - L^{(Pt)} \quad (68)$$

$$\omega^{(p)} = \frac{L^{(p)}}{L^{(Pt)} + L^{(C)} + L^{(p)}} \quad (69)$$

providing the following relationship for the volume fraction of platinum and carbon

$$\gamma^{(\mathrm{PtC})}=\frac{V^{(\mathrm{PtC})}}{V_{\mathrm{tot}}}=\left[\frac{1}{\rho^{(\mathrm{Pt})}}+\frac{1-\omega^{(\mathrm{Pt})}}{\rho^{(\mathrm{C})}\omega^{(\mathrm{Pt})}}\right]\frac{L^{(\mathrm{Pt})}}{h_{\mathrm{cl}}} \tag{70}$$

Here, volume fraction of polymer and agglomerate given by

$$\gamma^{(\mathrm{p})}=\frac{V^{(\mathrm{p})}}{V_{\mathrm{tot}}}=\frac{\omega^{(\mathrm{p})}}{1-\omega^{(\mathrm{p})}}\frac{1}{\rho^{(\mathrm{m})}}\frac{L^{(\mathrm{Pt})}}{\omega^{(\mathrm{Pt})}h_{\mathrm{cl}}} \tag{71}$$

$$\gamma^{(\mathrm{agg})}=\frac{V^{(\mathrm{agg})}}{V_{\mathrm{tot}}}=\gamma^{(\mathrm{p})}+\gamma^{(\mathrm{PtC})} \tag{72}$$

The porosity of the catalyst layer is defined as

$$\gamma_{\mathrm{cl}}=\frac{V_{\mathrm{void}}}{V_{\mathrm{tot}}}=1-\gamma^{(\mathrm{agg})} \tag{73}$$

with

$$\begin{aligned}&V^{(\mathrm{agg})}=V^{(\mathrm{PtC})}+V^{(\mathrm{p})},\ V_{\mathrm{void}}=V^{(\mathrm{g})}+V^{(\mathrm{l})}\\&V_{\mathrm{tot}}=V^{(\mathrm{agg})}+V_{\mathrm{void}}-V^{(\mathrm{PtC})}+V^{(\mathrm{p})}+V^{(\mathrm{l})}\end{aligned} \tag{74}$$

5. Membrane model

For the membrane the activity and water contents per sulfonic group are given by

$$a=\frac{p_{\mathrm{H_2O}}^{(\mathrm{g})}}{p_{\mathrm{H_2O}}^{\mathrm{sat}}}+2s \tag{75}$$

$$\lambda=\begin{cases}0.043+17.81a-39.85a^2+36a^2 & \mathrm{a}\le 1\\ 14+1.4(a-1) & 1<\mathrm{a}\le 3\end{cases} \tag{76}$$

The ionic conductivity is defined as

$$\sigma_{\mathrm{eff}}^{(\mathrm{m})}=\beta^{(\mathrm{m})}\sigma^{(\mathrm{m})} \tag{77}$$

where $\sigma^{(\mathrm{m})}$ is defined as

$$\sigma^{(\mathrm{m})}=\left(0.514\lambda-0.326\right)\exp\left[1268\left(\frac{1}{303.15}-\frac{1}{T}\right)\right] \tag{78}$$

The diffusivity of membrane water content per sulfonic group

$$D_{\mathrm{H_2O,eff}}^{(\mathrm{m})}=\beta^{(\mathrm{m})}D_{\mathrm{H_2O}}^{(\mathrm{m})} \tag{79}$$

$$D_{H_2O}^{(m)} = \begin{cases} 3.1\, x\, 10^{-7} \times \lambda\left(\exp(0.28\lambda) - 1\right)\exp\left[-2436 / T\right] & \text{for } \lambda \leq 3 \\ 4.17\, x\, 10^{-8} \times \lambda\left(1 + 161\exp(-\lambda)\right)\exp\left[-2436 / T\right] & \text{for } \lambda > 3 \end{cases} \quad (80)$$

The electro-osmotic drag coefficient is defined as

$$n_d = 2.5\frac{\lambda}{22} \quad (81)$$

6. Numerical procedure

The model geometry was created using the Gambit pre-processor software. This software can be used to draw the geometry from 1D to 3D and from simple to complex geometry (CAD). Furthermore, it can be used to create mesh either structured or unstructured mesh, with triangular, tetrahedral, hexahedral and polyhedral meshes. Advanced meshing is achieved by applying boundary layer meshing and size function meshing. Labeling boundary condition and defining zone type, either fluid or solid, for further numerical solution are also available.

The mathematical model of the PEM fuel cell was solved with the commercial multi-purpose CFD software FLUENT (Fluent, 2009). The software is based on the finite volume method where the computational domains are divided into a finite number of control volumes (cells) and all variables are stored at the centroid of each cell. Essentially, the software solves for the standard Navier-Stokes equation together with the scalar transport equations. The latter is heavily modified with User Define Scalar (UDS) and User Define Functions (UDFs). The second order upwind is chosen for the discretization. Furthermore, an iterative solver based on the SIMPLE algorithm is employed to solve the pressure velocity coupling. Note that the open-source CFD code can also be utilized to solve the mathematical framework of PEM fuel cell derived here.

The accuracy of numerical solutions is strongly related to mesh density. A grid independence study was undertaken to ensure mesh independent results. In general, a finer mesh can be expected to produce results that are more accurate; however, it comes at the cost of higher memory requirements and longer computing time. In view of this, a grid independence study was performed to obtain an optimum mesh density that gives sufficiently accurate results at acceptable computational cost.

7. Results and discussion

7.1. Validation

When developing and implementing mathematical models to predict the behaviour of a running PEM fuel cell one needs to pay special attention to validation due to the inherent complexity of the coupled physical and chemical phenomena inside the various functional layers of each cell in the stack. As seen in Fig. 2, the iR-corrected and full polarization curves

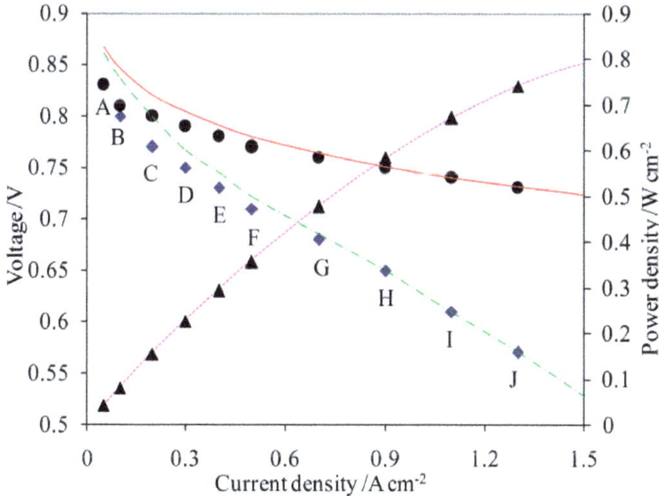

Figure 2. Polarization curves: [♦] is the experimentally measured potential; [●] is the iR-corrected experimentally measured potential; [▲] is the experimentally obtained power density; [ ] is the predicted potential of the model; [—] is the predicted iR-corrected potential of the model (both case a and b); [- - -] is the predicted power density of the model.

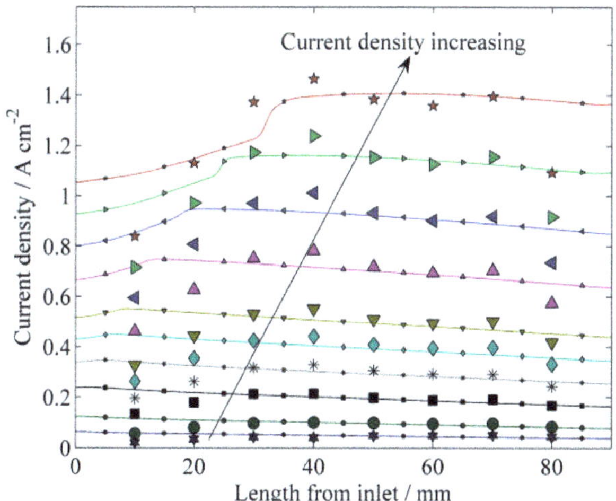

Figure 3. Local current density distribution at the anode terminal in the streamwise direction; symbols correspond to the experimental measurements and lines are model predictions; 1.3 Acm^{-2} (★), 1.1 Acm^{-2} (▶), 0.9 Acm^{-2} (◀), 0.7 Acm^{-2} (▲), 0.5 Acm^{-2} (▼), 0.4 Acm^{-2} (♦), 0.3 Acm^{-2} (✱), 0.2 Acm^{-2} (■), 0.1 Acm^{-2} (●), 0.05 Acm^{-2} (✱).

are in good agreement with experimental data measured by Noponen et al., 2004. Furthermore, the predicted local current density distribution along top of the anode terminal is also compared with its experimental counterpart. For each measured average current density in Fig. 2, denoted by A to J, the local current density distribution was also determined, as depicted in Fig. 3. We see that the model predictions agree well with both iR-corrected and full polarization curves in Fig. 2. We note that at low current densities, the model prediction slightly deviates with maximum relative error of ~ 4% which is good enough for the purposes of this article. Furthermore, the predicted local current density distributions are also in good agreement at lower current densities and start to deviate at the close to the inlet and outlet region at higher current densities. The deviation can most likely be attributed to the placement of the inlet/outlet holes in the experimental setup (see Noponen et al. 2004 for details), which would require a three-dimensional computational domain to resolve properly. This implies that the model correctly account for the fundamental physics associated with PEM fuel cell.

7.2. Thermal management

Thermal management is one of the important keys to efficient operation of PEM fuel cell stacks. In an ideal situation, each fuel cell in the stack would operate under identical operating condition; in reality, however, this is not feasible, as variations inevitably arise due to the design of stack manifold, position of the cell in the stack and choice of thermal management strategy. Efficient and cost-effective cooling of each cell in the stack is necessary to ensure high overall performance. If one removes too much heat, the reaction kinetics are adversely affected and, if water vapor partial pressure exceeds the saturation pressure, water vapor tends to condense which, in turn, lowers stack performance. Conversely, if cooling is not adequate, the stack temperature rises beyond its allowable operating temperature causing the membrane water content and its protonic conductivity to drop, which deteriorates stack performance. It is therefore of interest to understand how various thermal management affect overall stack performance and which strategy to select for a given application.

Broadly speaking, depending on the type of the cathode manifold, the oxygen and/or air flow to the cathode can be supplied directly from ambient air (open cathode) or through a gas manifold. For the former, ambient air is directly supplied to the cathode to provide both the oxidant as well as cooling air. The air flow can be provided by forced convection using a fan, or natural convection due to temperature gradient between the PEM fuel cell stack and the ambient. The former usually can be applied to stack with power rating up to sub kW, while the latter can only be used for small stacks of few cells (depends on the cell geometry) with power rating less than 100 W. For a closed manifold stack, on the other hand, pure oxygen or air is usually supplied by forced convection through a stack manifold and is pre-conditioned, e.g., filtered, heated, and humidified. Therefore, an additional thermal management strategy is usually required. The most common cooling designs include forced convection in specially designed cooling plates/channels with either liquid or air as the coolant, edge cooling with or without fins, and cooling coupled with phase-change materials used for thermal storage.

In practice, the choice of the thermal management strategy depends strongly on details of the specific application and its constraints; for example, in combined heat and fuel cell stationary power systems, liquid cooling (as illustrated in Fig. 1) is preferred since it has larger heat removal rate. Moreover, in this application there is no limitation on the size, weight and complexity of the system as it is used in stationary applications. In small automotive and power generator applications where the weight, size and complexity of the system become the limiting factors, air cooling is preferred due to its simplicity as no coolant loop and heat exchanger are required by the system. An open-cathode PEM fuel cell stack with natural convection cooling has been considered for portable electronic power such as handphone power due to its simplicity.

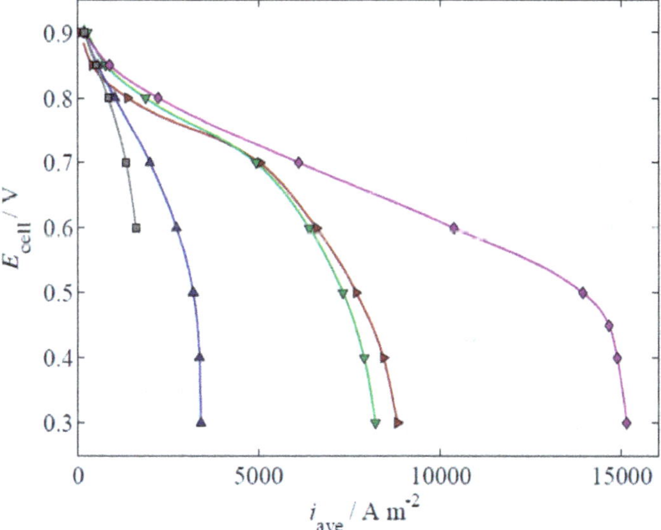

Figure 4. Polarization curves for PEFC stack with liquid cooling (◊); edge cooling (▶); air cooling (▼); open-cathode with forced (▲) and natural-air convection (■).

Now, let us have a look at Fig. 4 which displays and compares the predicted stack performance for different thermal management strategies. Here, several features are apparent; foremost is that stack performance is greatly affected by the cooling strategy chosen. It is seen that liquid cooling yields the best results among various alternatives. The natural convection open-cathode stack gives the lowest performance. The computed limiting current density for each case is found to be similar to that found in published experimental investigation: liquid cooling up to 15000 A m^{-2} (Noponen et al., 2004), air-cooling up to 8000 A m^{-2} (Shimpalee et al., 2009), edge-cooling up to 8500 A m^{-2} (Fluckiger et al., 2007), and forced and natural convection with open-cathode up to 3500 A m^{-2} (Wu et al., 2009) and 1500 A m^{-2} (Urbani et al., 2007), respectively. Conversely, when the complexity, cost, size, weight, and parasitic load are of interest, the order is reversed; for example, stack with liquid cooling requires a coolant loop, a radiator, a pump, more space and weight as well as additional costs to build the supporting equipment. Natural convection cooling, on the other hand, does not require any additional auxiliary equipment and hence is the least expensive.

7.3. Water management

A PEM fuel cell with a sealed off anode flow field (dead-end anode) has several advantages and also disadvantages. The advantages of this design are that one does not need to consider any excess hydrogen that will exit from the anode outlet if the flow channel is open; presumably all hydrogen is consumed in the fuel cell. However, proper water management is essential in order to keep the membrane sufficiently humidified whilst ensuring that the anode does not flood from water accumulation. In addition, nitrogen can cross over from the cathode side and accumulate at the anode. This can result in a deterioration of cell performance with time as has been reported frequently in the literature.

To avoid operating the anode under adverse conditions, one can purge the accumulated gas-vapor mixture to clear the anode flow field of liquid water and nitrogen. However, this can defeat the main purpose of the dead-end mode as some hydrogen will be lost during the purge and reduce hydrogen utilization. Moreover, it adds complexity to the design as it requires valves and control schemes which result in a parasitic load on the fuel cell. Hence, an optimal or near-optimal purge frequency and duration need to be developed to minimize hydrogen loss, reduce parasitic load, and maintain good fuel cell performance with time.

Starting from purge frequency, two different purge frequencies -- purge every five minutes and every ten minutes -- which is commonly used in a typical PEM fuel cell with a dead-end mode were modelled for 30-second purge duration in each case. With respect to cell performance, it can be expected that purge will recover the cell to its highest performance. This is indeed the case, as seen in Fig. 5, which shows the cell performance drop during dead-end operation and then recovers during purge. It is noted that more frequent purge will keep the cell at a higher performance level compared to that with less frequent ones. In this particular case, a 3 % drop in cell voltage (0.6 V) is observed for the case of more frequent purge (every five minutes), whereas, an 11 % drop (0.55 V) is found for the case of less frequent purge (every 10 minutes) as compared to the open-end case (0.62 V). Further, the performance drop during dead-end operation and performance recovery during purge is mirrored by depletion of the average hydrogen concentration at the anode and its increase during the purge as well as liquid water and nitrogen accumulation and their release (flush-out) during purge (see Fig. 6). More frequent purge results in lower hydrogen depletion, reduced water accumulation, and decreased nitrogen crossover.

Though more frequent purges can maintain higher cell performance, the hydrogen loss is also higher which diminishes the advantage of utilizing a dead-end mode; more parasitic power is also consumed to control the valves, which reduces level of the net power generated. Conversely, if the purge is less frequent, hydrogen utilization is higher which the purpose of dead-end anode mode is. However, the performance can be lower due to greater water and nitrogen accumulation. In addition, purge frequency is also a function of the volume of anode flow field and gas diffusion layer, flow field type, and operating conditions. Clearly, purge frequency requires careful consideration to ensure a good cell performance whilst maintaining high hydrogen utilization.

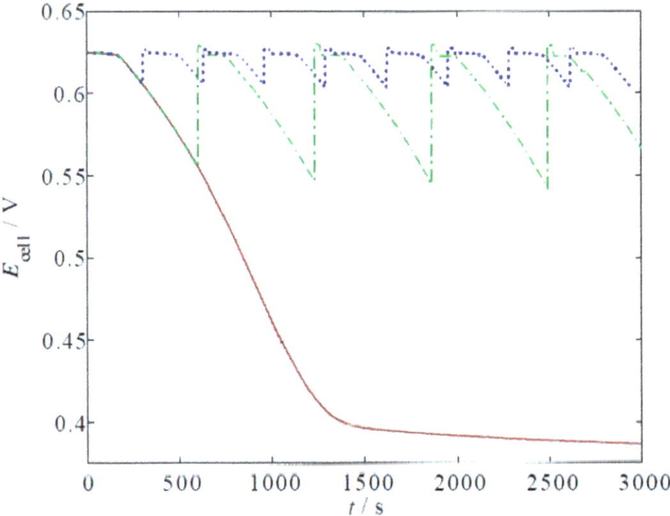

Figure 5. Effect of purge frequency to the cell performance for purging frequency of (···) 5 minutes; (-·-) 10 minutes; and (—) no purging (base case).

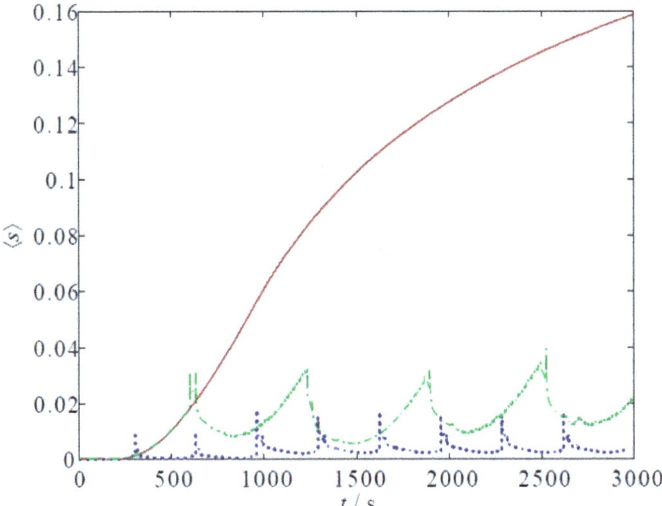

Figure 6. Effect of purge frequency to the liquid water accumulation for purging frequency of (···) 5 minutes; (-·-) 10 minutes; and (—) no purging (base case).

7.4. Gas management

Another point of interest in this study is gas management. A good channel design should be able to provide sufficient gas (fuel and oxidant) supply throughout catalyst layer for electrochemical reaction. Here, the oxygen distribution in the cathode catalyst layer is examined. As can be seen in Fig. 7, oxygen concentration is depleted from inlet to outlet as it is consumed for electrochemical reaction. On closer inspection, it reveals that severe oxygen depletion exists at the middle of parallel channel (Fig. 7a). This is attributed to the low velocity and non-uniform oxidant distribution in each passage of the channel, especially at the middle area. Less severe oxygen depletion is observed in the oblique-fin channel, especially at the right-side region (see Fig. 7c for details); this can be due to the presence of secondary flow which leads to flow separation toward outlet region. Serpentine channel, on the other hand, yields the most uniform oxygen distribution and no oxygen depletion is found throughout the cell. This, in combination, indicates that serpentine channel yields the best management among others.

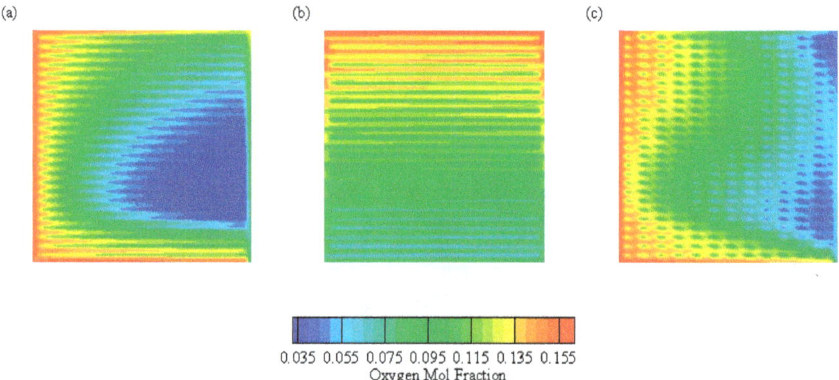

Figure 7. Oxygen mol fraction distribution at cathode catalyst layer for (a) parallel; (b) serpentine; (c) new oblique-fin.

Thus far, the three different channel designs have been evaluated in terms of thermal water and gas management. Now we look further to the fuel cell performance which is represented by current generation in the catalyst layer shown in Fig. 8. Clearly, a good thermal, water and gas management in serpentine channel is mirrored by high stack performance with uniformly distributed current generation (Fig. 8b). In contrast, the oxygen depletion, liquid accumulation and "hot-spot" temperature in parallel channel results in non-uniform and low current generation (~26% lower than serpentine channel). Closer inspection reveals that almost no current is generated in the region where oxygen depletion and liquid accumulation exist. The performance of oblique-fin channel lies in between serpentine and parallel channel: the current density is around 10% lower than serpentine channel and around 16% higher with the degree of uniformity improves up to around 34 % than parallel channel.

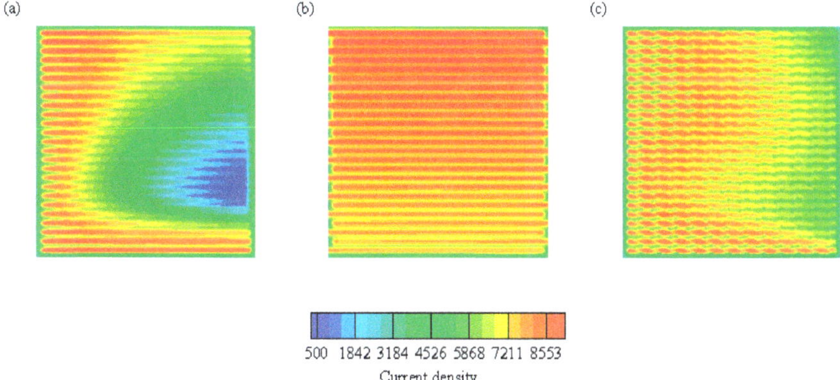

Figure 8. Current density distribution at cathode catalyst layer for (a) parallel, i_{ave} = 6200 A/m^2; (b) serpentine, i_{ave} = 8400 A/m^2; (c) new oblique-fin channel, i_{ave} = 7400 A/m^2.

8. Concluding remarks

A computational study was carried out of the transport phenomena in PEM fuel cell for a single cell as well as a stack of fuel cells. The model has good agreement with experimental data for global and iR-corrected curves as well as local current density distribution. Several thermal management strategies, e.g., liquid-cooled, air-cooled, edge-cooled and open-cathode fuel cell with forced and natural convection cooling, were evaluated. It is found that liquid-cooled fuel cell yields the best performance among others; however, it requires the most complex auxiliary system, size and weight.

A transient performance of PEM fuel cell with a dead-end anode was simulated. It is noted purge can maintain fuel cell at high performance as it flush-out accumulated liquid water and other gases which, in turn, recovers hydrogen concentration in the anode side.

Furthermore, three different gas and gas channel designs, e.g., parallel, serpentine and oblique-fin channels were investigated via mathematical model. The serpentine design gives the most uniform oxygen distribution throughout the catalyst layer which is mirrored by high cell performance.

Finally, we note that mathematical modeling plays significant contribution to speed-up the commercialization of the fuel cell as it can assist experimental on addressing main issues in fuel cell research and development, e.g., thermal, gas and water management.

Author details

Agus P. Sasmito, Erik Birgersson and Arun S. Mujumdar
Minerals Metals Materials Technology Centre (M3TC), National University of Singapore, Singapore

Agus P. Sasmito
Mechanical Engineering, Masdar Institute of Science and Technology, Abu Dhabi, United Arab Emirates

Erik Birgersson
Department of Chemical and Biomolecular Engineering, National University of Singapore, Singapore

Arun S. Mujumdar
Mechanical Engineering Department, National University of Singapore, Singapore

Nomenclature

a Water activity
$a^{(l)}$ Surface are of the agglomerates including water per unit volume, m^{-1}
$a^{(p)}$ Surface area of the agglomerates per unit volume of catalyst layer, m^{-1}
A_{cl} Catalyst area, m^2
$c_i^{(g)}$ Molar concentration of species i, mol m^{-3}
$c_{i,ref}^{(g)}$ Reference molar concentration of species i, mol m^{-3}
$c_p^{(g)}$ Specific heat capacity of gas mixture, J kg^{-1} K^{-1}
$c_{p,i}^{(g)}$ Specific heat capacity of species i, J kg^{-1} K^{-1}
c_r Condensation/evaporation rate constant, s^{-1}
c_1, c_2, c_3, c_4 Constants for the saturation pressure of water; -, K^{-1}, K^{-2}, K^{-3}
$D^{(c)}$ Capillary diffusion, m^2 s^{-1}
$D_i^{(g)}, D_{i,eff}^{(g)}$ Diffusivity and effective diffusivity of species i, m^2 s^{-1}
$D_{H_2O}^{(m)}, D_{H_2O,eff}^{(m)}$ Diffusivity and effective diffusivity of water in the membrane, m^2 s^{-1}
$D_{O_2,eff}^{(agg)}$ Effective diffusion coefficient of oxygen in the agglomerate, m^2 s^{-1}
$D_{O_2}^{(l)}, D_{O_2}^{(p)}$ Diffusion coefficient of oxygen in liquid water and in polymer film, m^2 s^{-1}
$\mathbf{e}_x, \mathbf{e}_y, \mathbf{e}_z$ Coordinate vectors
E_{cell}, E_{stack} Cell and stack voltage, V
E_a Activation energy, J mol^{-1}
E_{rev} Reversible cell potential, V
F Faraday constant, C mol^{-1}
h_j Height of layer j, m
$H_{O_2}^{(l)}, H_{O_2}^{(p)}$ Henry's constant for air-water and air-polymer interfaces, Pa m^3 mol^{-1}
H_{vap} Heat of vaporization, J kg^{-1}
H Relative humidity, %
i, $\mathbf{i}$ Current density, A m^{-2}
$j_{a,c}^{ref}$ Anode and cathode volumetric reference exchange current density, A m^{-3}
J Volumetric current density, A m^{-3}
J Leverett function
k Thermal conductivity, W m^{-1} K^{-1}
k_c Reaction rate constant, s^{-1}
L Length of channel, m
$L^{(C)}, L^{(p)}, L^{(Pt)}$ Carbon, polymer and platinum loading, kg m^{-3}

$\dot{m}_{H_2O}$ Interphase mass transfer due to condensation/evaporation of water, kg m^{-3} s^{-1}
$M^{(g)}$ Mean molecular mass of the gas phase, kg mol^{-1}
M_i Molecular mass of species i, kg mol^{-1}
$M^{(m)}$ Equivalent weight of the dry membrane, kg mol^{-1}
$n^{(agg)}$ Number of agglomerates per unit volume, m^{-3}
n_d Electroosmotic drag coefficient
$\mathbf{n}_i^{(g)}$, $\mathbf{n}_{H_2O}^{(m)}$ Mass flux of species i and water in the membrane, kg m^{-2} s^{-1}
$P^{(c)}$, $p^{(g)}$ Capillary and gas pressure, Pa
$p_{H_2O}^{sat}$ Saturation pressure of water, Pa
R Gas constant, J mol^{-1} K^{-1}
$r^{(agg)}$ Radius of agglomerate, m
Re Reynolds number
s Liquid saturation
S Source term
T_0, T_1, T_2 Constant, K
T Temperature, K
$\mathbf{u}$, u, v, U Velocities, m s^{-1}
V Volume, m^3
$x_i^{(g)}$ Molar fraction of species i
x, y, z Coordinates, m
$\omega_i^{(g)}$ Mass fraction of species i
$\omega^{(p)}$ Mass fraction of polymer loading
$\omega^{(Pt)}$ Mass fraction of platinum loading on carbon

Greek

α Transfer coefficient
$\beta^{(m)}$ Membrane modification coefficient
γ Volume fraction
δ Thickness of the film, m
ε Porosity
η Overpotential, V
θ Wetting angle
κ Permeability, m^2
λ Membrane water content
μ Dynamic viscosity, kg m^{-1} s^{-1}
ξ Stoichiometry
ξ_1, ξ_2, ξ_3 Correction factors for agglomerate model
ρ Density, kg m^{-3}
τ Surface tension, Pa
$\boldsymbol{\sigma}$ Total stress tensor, Pa
σ Conductivity S m^{-1}
ϕ Potential, V
$\Phi_{\alpha,\beta}$ Dimensionless quantities

Φ	Thiele modulus

Superscripts

(agg)	Agglomerate
(c)	Capillary
(C)	Carbon
cool	Coolant
(g)	Gas phase
in	Inlet
(l)	Liquid phase
(m)	Membrane
ox	Oxidation
(p)	Polymer phase
(Pt)	Platinum
(PtC)	Platinum and carbon
rd	Reduction
ref	Reference
(s)	Solid
sat	Saturation
set	Setting

Subscripts

α, β	Index for species
a	Anode
ave	Average
c	Cathode
cc	Current collector
cl	Catalyst layer
co	Coolant channel
eff	Effective
cool	Coolant
eff	Effective
ff	Flowfield
gdl	Gas diffusion layer
H_2	Hydrogen
H_2O	Water
i	Species i
iR	iR-corrected
j	Functional layer j
m	Membrane
mass	Mass
N_2	Nitrogen
O_2	Oxygen

pot Potential
ref Reference
sp Separator plate
temp Temperature
tot Total
void Void
0 Standard conditions

9. References

Brujin, F. A. de; Dam, V. A. T.; Janssen, G. J. M. Review: Durability and degradation issues of PEM fuel cell components, Fuel Cells 8 (2007) 3-22.

Costamagna, P.; Srinivasan, S. Quantum jump in the proton exchange membrane fuel cell science and technology from 1960 to the year 2000, part 1: Fundamental scientific aspect, J. Power Sources 102 (2001a) 242-252.

Costamagna, P.; Srinivasan, S. Quantum jump in the proton exchange membrane fuel cell science and technology from 1960 to the year 2000, part 2: Engineering, technology development and applications aspect, J. Power Sources 102 (2001b) 253-269.

Fluckiger, R.; Tiefenauer, A.; Ruge, M.; Aebi, C.; Wokaun, A.; Buchi, F. N. Thermal analysis and optimization of a portable edge-air-cooled PEFC stack, J. Power Sources 172 (2007) 324-333.

FLUENT PEMFC module, http://www.fluent.com, accessed 2009.

Hellman, H. L.; Hoed, R. V. D. Characterizing fuel cell technology challenge of the commercialization process, Int. J. Hydrogen Energy 32 (2007) 305-315.

Hermann, A.; Chaudhuri, T.; Spagnol, P. Bipolar plates for PEM fuel cells: A review, Int. J. Hydrogen Energy 30 (2005) 1297-1302.

Li, S. and Becker, U., A Three Dimensional Model for PEMFC, ASME Proc. 2nd Fuel Cell Science Engineering and Technology, Rochester, New York 2 (2004) 157-164.

Noponen, M., Birgersson, E., Ihonen, J., Vynnycky, M., Lundblad, A., Lindbergh, G., A Two-phase Non-isothermal PEFC, Model: theory and validation, Fuel Cells, 4 (2004) 365-377.

Martin, K. E.; Kopasz, J. P.; McMurphy, K. W. Status of fuel cells and the challenges facing fuel cell technology today, American Chemical Society, 2010.

Mock, P.; Schmid, S. A. Fuel cells for automotive power trains: A techno-economic assessment, J. Power Sources 190 (2009) 133-140.

Sasmito, A.P.; Mujumdar, A.S. Transport phenomena models for polymer electrolyte fuel cell stacks: thermal, water and gas management – From fundamental to application, *Lambert Academic Publishing,* Germany 2011. ISBN: 978-3-8443-9063-6.

Sasmito, A.P.; Birgersson, E.; Mujumdar, A.S. Numerical Evaluation of Various Thermal Management Strategies for Polymer Electrolyte Fuel Cell Stacks, International Journal of Hydrogen Energy 36 (2011a) 12991-13007.

Sasmito, A.P.; Lum, K.W.; Birgersson, E.; Mujumdar, A.S. Computational study of forced air-convection in open-cathode polymer electrolyte fuel cells stacks, Journal of Power Sources 195 (2010) 5550-5563.

Sasmito, A.P.; Birgersson, E.; Lum, K.W.; Mujumdar, A.S. Fan selection and stack design for open-cathode polymer electrolyte fuel cell stacks, Renewable Energy 37 (2012b) 325-332.

Sasmito, A.P.; Birgersson, E.; Mujumdar, A.S. Numerical investigation of liquid water cooling for a proton exchange membrane fuel cell stack, Heat Transfer Engineering 32 (2011c) 151-167.

Sasmito, A.P.; Mujumdar, A.S.; Performance Evaluation of a Polymer Electrolyte Fuel Cell with a Dead-End Anode: A Computational Fluid Dynamic Study, International Journal of Hydrogen Energy 36 (2011d) 10917-10933.

Sasmito, A.P.; Kurnia, J.C.; Birgersson, E.; Mujumdar, A.S. Numerical Evaluation of Performance of Oblique-Fin Channel for PEM Fuel Cell Stacks Relatives to Conventional Channels, 3rd International Conference on Fuel Cell and Hydrogen Technology 2011, Kuala Lumpur, Malaysia, 2011e.

Schwarz, D. H.; Djilali, N., 3D Modeling of Catalyst Layer in PEM Fuel Cells Effect of Transport Limitation, J. Electrochem. Soc., 154 (2007) B1167 - B1178.

Shimpalee, S.; Ohashi, M.; Zee, J. W. V.; Ziegler, C.; Stoeckmann, C.; Sadeler, C.; Hebling, C. Experimental and numerical studies of portable PEMFC stack, Electrochim. Acta 54 (2009) 2899-2911.

Staffell, I.; Green, R. J. Estimating future prices for stationary fuel cells with empirically derived experienced curves, Int. J. Hydrogen Energy 34 (2009) 5617-5628.

Steiner, N. Y.; Mocoteguy, P.; Candusso, D.; Hissel, D. A review on polymer electrolyte membrane fuel cell catalyst degradation and starvation issues: causes, consequences and diagnostic for mitigation, J. Power Sources 194 (2009) 130-145.

Sopian, K.; Daud, W. R. W. Challenges and future developments in proton exchange membrane fuel cells, Renewable Energy 31 (2006) 719-727.

Tsuchiya, H.; Kobayashi, O. Mass production cost of PEM fuel cells by learning curve, Int. J. Hydrogen energy 29 (2004) 985-990.

Urbani, F.; Squadrito, G.; Barbera, O.; Giacoppo, G.; Passalacqua, E.; Zerbinati, O. Polymer electrolyte fuel cell mini power unit for portable application, Journal of Power Sources 169 (2) (2007) 334-337.

Wang, Y. and Wang, C.Y., A Non-isothermal, Two-phase Model for Polymer Electrolyte Fuel Cells, J. Electrochem. Soc., 153 (2006) A1193-A12000.

Wee, J. H. Applications of proton exchange membrane fuel cell systems, Renewable Sustainable Energy Rev. 11 (2007) 1720-1738.

Wu, J.; Galli, S.; Lagana, I.; Pozio, A.; Monetelone, G.; Yuan, X. Z. An air-cooled proton exchange membrane fuel cell with combined oxidant and coolant flow, J. Power Sources 188 (2009) 199-204.

Chapter 13

Ammonia as a Hydrogen Source for Fuel Cells: A Review

Denver Cheddie

Additional information is available at the end of the chapter

1. Introduction

The concept of a hydrogen economy was revived in the 1990s as interest in fuel cell technology surged. There was an explosion of research into fuel cells since then mainly because of its status as a hydrogen technology, and as such both concepts shared a symbiotic relationship. However there are a number of problems with the direct use of hydrogen in fuel cells. Firstly hydrogen does not exist naturally. Secondly it is not easy to store or transport because of its low volumetric energy density and its small molecular size.

Recently the concept of an ammonia economy has gained eminence [1]. Like hydrogen, ammonia is carbon free and can be produced from any energy resource. However there are also some significant advantages in terms of storage and transport. Ammonia can be liquefied at room temperature at pressures of 8-10 bar and stored in a similar manner to propane, whereas hydrogen requires expensive cryogenic storage. In addition, ammonia allows for safer handling and distribution than hydrogen. Although it is toxic, its smell can be detected even at safe concentration levels (< 1 ppm). Ammonia has a narrower flammable range than hydrogen and is actually considered nonflammable when being transported, whereas hydrogen burns with an invisible flame. Ammonia is the second most widely produced commodity chemical in the world (second to sulfuric acid), with over 100 million tons per year being transported [2], and as such its worldwide distribution system is well established. Such is not the case for hydrogen. In fact, one major drawback with hydrogen technologies is the fact that the necessary hydrogen infrastructure does not presently exist. Essentially the ammonia economy can achieve the same benefits of a hydrogen economy, but using infrastructure that already exists.

Ammonia provides a source of hydrogen for fuel cells. It contains 17% hydrogen by weight, which can be extracted via thermal catalytic decomposition or electro-oxidation. Alternatively ammonia may be oxidized directly in fuel cells without the need for a separate reactor. Table 1 compares the storage capabilities of various fuels based on their higher heating value (HHV) [3]. Hydrogen has a very low energy density (per volume) because of its low density. Ammonia's energy density is comparable to that of compressed natural gas (CNG) and methanol, but lower than gasoline and liquefied propane gas (LPG). Per unit volume, the cost of hydrogen energy is lower than that of ammonia energy, but hydrogen has less energy stored per volume than ammonia. Per unit energy, ammonia is the cheapest energy source listed in Table 1 – estimated at US$13.3/GJ. Note that these values are based on the HHV of the fuel and do not account for conversion of this energy to useful forms. The life cycle production cost of energy from ammonia is estimated at US$1.2/kWh compared to US$3.8/kWh for methanol and US$25.4/kWh for hydrogen [4]. Thus ammonia presents a very viable and cost effective fuel for fuel cells.

This chapter reviews the progress of ammonia fuel cells – those that use ammonia directly or indirectly. Ammonia fuel cells have been previously reviewed [5-7]. Ref [5] was published in 2004 and provides a mini-review focusing only on decomposition catalysts. Ref [6] was published in 2008 and provides a good review up to that point, however it only addresses decomposition catalysts, ammonia fed SOFCs and SOFC modeling. Ref [7] was published more recently (2011) but gave a more general overview rather than integrate research findings in the different areas of research in ammonia fuel cells. The present work seeks to integrate the research findings and provide a wide picture of the research conducted in ammonia fuel cells, and to show the development of the field. It also highlights areas that warrant further investigation to fully develop the field. Section 2 discusses developments in hydrogen generation for fuel cells via thermal decomposition of ammonia, electro-oxidation of ammonia, and from ammonia products. Section 3 outlines the development of direct ammonia fuel cells, citing experimental studies and their results. Section 4 reviews the various works done on mathematical modeling and simulation of ammonia fuel cells.

Fuel / Storage System	P (bar)	Energy Density (GJ/m³)	Specific Volumetric cost (US$/m³)	Specific Energy Cost (US$/GJ)
Ammonia gas / pressurized tank	10	13.6	181	13.3
Hydrogen / metal hydride	14	3.6	125	35.2
Gasoline (C_8H_{18}) / liquid tank	1	34.4	1000	29.1
LPG (C_3H_8) / pressurized tank	14	19.0	542	28.5
CNG (CH_4) / integrated storage system	250	10.4	400	38.3
Methanol (CH_3OH) / liquid tank	1	11.4	693	60.9

Table 1. Energy Storage Capabilities of Various Fuels [3]

2. Hydrogen generation from ammonia

Hydrogen can be produced from ammonia for use in fuel cells in various ways. Most of the literature is devoted to thermal decomposition or catalytic cracking of ammonia into nitrogen and hydrogen, with fewer articles addressing electrolysis or electro-oxidation. Some papers also address hydrolysis of ammonia products such as ammonia borane.

2.1. Catalytic decomposition of ammonia

$$2NH_3 + 92.4kJ \rightarrow N_2 + 3H_2 \quad (1)$$

Ammonia is unstable at high temperatures and begins to decompose at 200 °C [7]. The slightly endothermic decomposition reaction is shown in equation 1. Thermodynamically, 98-99% conversion of ammonia to hydrogen is possible at temperatures as low as 425 °C. However in practice, the rate of conversion depends on temperature as well as catalysts.

Thermal decomposition or catalytic cracking is the most common means of hydrogen generation from ammonia. Lipman and Shah [8] report that for large scale hydrogen generation (> 1000 m^3/hour), reformation of natural gas remains the most cost effective process, however for small scale generation, (< 10 m^3/hour), ammonia cracking becomes slightly more economical than natural gas reformation (see Table 2). This study is based on lifecycle cost analysis, taking into account investment and operation costs.

Scale of H_2 production (m^3/hour)	Cost of H_2 production, US$ / (m^3/hour)			
	Water Electrolysis	Natural Gas Reformation	Methanol Reformation	Ammonia Cracking
10	0.943	0.390	0.380	0.343
100	0.814	0.261	0.285	0.279
1000	0.739	0.186	0.226	0.241

Table 2. Life Cycle Cost of Hydrogen Production via Various Processes [8]

The early studies done on ammonia decomposition focused more on ammonia synthesis, and as such considered iron based catalysts. Since then various metals, alloys, and compounds of noble metal characters have been tested for ammonia decomposition. These include Fe, Ni, Pt, Ru, Ir, Pd, Rh; alloys such as Ni/Pt, Ni/Ru, Pd/Pt/Ru/La; and alloys of Fe with other metal oxides including Ce, Al, Si, Sr, and Zr [5]. Various catalysts have been investigated for decomposing ammonia to produce hydrogen for alkaline fuel cells. These include WC, Ni/Al_2O_3, NiCeO_2/Al_2O_3, Cr_2O_3, Ru/ZrO_2, and Ru on carbon nano-fibres. Caesium-promoted ruthenium supported on graphite was also found to be very promising [7]. For these catalysts, a minimum temperature of 300 °C is required for efficient release of ammonia for hydrogen production.

The performance of the catalysts can be quantified using the rate of hydrogen production, conversion fraction of ammonia (fraction of ammonia that is converted to hydrogen), and activation energy. The rate of formation of hydrogen from ammonia decomposition has been measured experimentally, typically in units of millimoles of hydrogen produced per minute per gram of catalyst loaded (mmol/min/g). The performance of various catalysts for ammonia decomposition, reviewed in this section, is summarized in Table 3.

Catalyst / Support	Temp. (°C)	Rate of H_2 Gen. (mmol/min/g)	Conv. Eff. (%)	Ref.
Nano-sized Ni/Santa Barbara Amorphous (SBA)-15 support	450	8.4	25.0	[11]
	500	17.4	52.1	
	550	26.8	80.1	
	600	31.9	95.2	
	650	33.2	99.2	
Ni/SBA-15	550	12.7	37.8	[12]
Ni/SiO_2	400	0.4	1.4	[10]
	500	3.3	10.5	
	550	6.8	21.6	
	600	11.4	36.4	
	650	21.1	70.0	
Ni/SiO_2	550	11.6	34.6	[12]
Ni/Al_2O_3	550	12.7	37.8	[13]
Ni/Al_2O_3	500	24.1	71.9	[14]
Ni/Al_2O_3 coated cordierite monolith	550	16.5	50.0	[15]
Ni/Al_2O_3 (unsupported particles < 200 μm)		13.2	40.0	
Ir/SiO_2	400	1.2	3.9	[10]
	500	5.7	18.2	
	600	17.6	56.0	
	700	30.6	98.0	
Ru/SiO_2	400	4.5	14.3	[10]
	500	20.0	64	
	600	30.3	97	
	650	30.9	99	
Ru/ZrO_2	550	25.8	77.0	[16]
Ru/Al_2O_3		23.5	73.7	
Ru/CNT	400	6.2	3.7	[16][a]
Ru/K-CNT		12.2	7.3	
$Ru/K\text{-}ZrO_2$-BD		8.5	5.3	
Ru/ZrO_2		3.7	2.2	
Ru/Al_2O_3		3.8	2.3	
Ru/MgO		5.4	3.2	
Ru/TiO_2		4.3	2.6	
Ru/CNT	400	6.0	9.0	[17][b]
Ru/MgO–CNT		8.7	13.0	
Ru/CNT treated with KNO_3	400	33.3	49.7	[18][b]
Ru/CNT treated with KOH		31.6	47.2	
Ru/CNT treated with K_2CO_3		31.3	46.7	

All studies are based on an ammonia gas hourly space velocity (GHSV) of 30,000 ml/h/g of catalyst except (a) GHSV = 150,000 ml/h/g and (b) GHSV = 60,000 ml/h/g.

Table 3. Summary of Ammonia Decomposition Catalysts Performance Reported

Papapolymerou and Bontozoglou [9] studied the rate of decomposition at 225 – 925 °C and 133 kPa ammonia partial pressure. They used the catalyst in the form of polycrystalline wires of foils, and ranked them in decreasing order of reaction rate: Ir > Rh > Pt > Pd. Choudhary et al [10] performed similar studies at 400 – 700 °C with pure ammonia and ranked them: Ru > Ir > Ni. Comparing Ni, Ir and Ru supported in silica, Ru based catalysts have been reported to produce the highest decomposition rates as well as the highest conversion rate of ammonia. Yin et al [5] studied the effect of Ru loading within the silica support (in the range 0-35 wt.%) and found that the conversion rate of ammonia reached a peak at 15% weight loading of Ru. It increased with Ru loading from 0-15%, but above this, the sublayers of Ru were inaccessible thus rendering them redundant.

Different supports have also been investigated. The purpose of the support is to enhance the dispersion and increase the effective area of the active catalyst. The support should be stable under reaction conditions and have a high specific surface area. For Ru catalyst, the various supports include silica, alumina, graphitized carbon, carbon nanotubes, and nitrogen doped carbon nanotubes [10,19-25]. Yin et al [16] ranked the supports for Ru in order of decreasing activity measured by ammonia conversion rate: Carbon nanotube (CNT) > MgO > TiO_2 > Al_2O_3 > ZrO_2 > AC > ZrO_2/BD. It was proposed that CNTs performed the best because they allowed the best dispersion of Ru and also because of their high purity. CNTs have the added advantage of high conductivity which aids in electron transfer thus facilitating the recombinative nitrogen desorption step (see section 2.2). They further showed that using a MgO-CNT support resulted in better performance of the Ru catalyst than using a MgO base or CNT base alone[17]. Temperature programmed hydrogenation results showed that MgO resulted in even greater stability for the CNT.

Studies have shown that acidic conditions are not suitable for ammonia decomposition. Yin et al [16] prepared CNT with KOH and found that they resulted in better catalytic performance measured by reaction rate and conversion efficiency. N_2-temperature programmed desorption (TPD) results showed that the stronger the basicity, the better the catalyst performance [16]. They later studied the effects of promoter cations and the amount of potassium on the morphological structure and catalysis of Ru/CNT [18]. Essentially they found that ammonia conversion increased as the electro-negativity of the promoter decreased. When Ru/CNT is treated with potassium nitrate, potassium hydroxide or potassium carbonate, the conversion rate of ammonia and the rate of hydrogen evolution are significantly improved (see Table 3). Yin et al [5] concluded that the best catalyst for ammonia decomposition is Ru supported on alkaline promoted CNT.

The problem with Ru is that it is a noble metal which will significantly increase the cost of the fuel cell system. For mass production, it is preferable to use less expensive materials. Ru is widely accepted as the most active catalyst for ammonia decomposition, however the performance of Ni is very close [26]. It is possible to substitute the noble catalyst by other more economic active phases. Plana et al [15] studied the effect of having the catalyst in the form of a structured reactor with the hope of extracting greater activity from less expensive metals. The small scales of microstructured devices have inherent advantages, including high heat and mass transfer coefficients and high surface area to volume ratios. They considered cordierite monoliths, which are structured reactors with multiple channels of several hundreds of microns in diameter. Monoliths have uniform flow distribution and low pressure drop which is crucial for the energy-efficiency of the process. Furthermore, they are commercially available, and they can withstand high temperatures and their coating with catalyst layer is a mature technology [27]. They used coated cordierite monoliths with mesoporous calumina, on which they dispersed Ni by electrostatic adsorption. This catalytic structured reactor was thoroughly characterized by transmission electron microscopy (TEM), X-ray diffraction (XRD), N_2 phisisorption and temperature programmed reduction (TPR), and it was tested in NH_3 decomposition for in situ H_2 generation under realistic conditions such as pure NH_3 feed and high space velocity. The structured catalyst reactor consisted of Ni supported on alumina-coated monoliths. After prolonged reaction, Ni remained well dispersed with particle sizes of 6 nm and mesopores between 4-5 nm. Ni remained anchored within the alumina matrix and did not plug the pores. 100% conversion of ammonia was observed at 600 °C. They found that at temperatures exceeding 500 °C, the monolith reactors showed better performance than a packed bed catalyst – higher conversion of ammonia and more robustness.

Table 3 shows that the best ammonia conversion and hydrogen generation rates via thermal decomposition are obtained using Ru/CNT catalysts treated with potassium based alkalis. Ni produces very good results as well but requires higher temperatures (500 – 600 °C) to produce equivalent performance of Ru at 400 °C. The advantage of Ni is that it is less expensive than Ru and can be loaded at high concentrations to achieve the desired results. An anode supported SOFC (with an anode thickness of 500 μm, 40% porosity and 50% Ni by volume) requires a Ni loading of 0.134 g/cm^2. If it is operated at a current density of 5000 mA/cm^2, it consumes hydrogen at the rate of 11.6 mmol/min/g of catalyst. If the cell operates at 600 °C, then Ni can safely decompose ammonia at the required rate.

2.2. Reaction mechanism of ammonia decomposition

Various studies have investigated the reaction mechanism of ammonia decomposition. The reaction steps include 1) adsorption of ammonia onto catalyst sites, 2) cleavage of N-H bond on adsorbed ammonia, 3) recombinative desorption of N_2 atoms [28]. These three steps are respectively illustrated in equations 2-4, where * refers to an active site and X^* refers to species X adsorbed onto an active site.

$$NH_3 + {}^* \rightarrow NH_3^* \tag{2}$$

$$NH_3^* + {}^* \rightarrow NH_2^* + H^* \tag{3}$$

$$2N^* \rightarrow N_2 + 2^* \tag{4}$$

Early studies observed that the rate of ammonia decomposition over Pt and Fe, shifted from zero order with respect to ammonia partial pressure at low temperatures (< 500 °C) to first order at high temperatures [29]. Tsai and Weinberg [27] proposed that on Ru crystal catalysts, below approximately 400 °C the recombinative desorption of nitrogen atoms (step 3) is rate limiting, whereas above 400 °C the cleavage of the N–H bond of adsorbed NH_3 (step 2) is rate limiting. This was based on the observation that the apparent activation energy decreased from 180 kJ/mol at the low temperatures to 21 kJ/mol at high temperatures. It should be noted however, that these early studies did not consider the effects of hydrogen inhibition.

Later studies observed that at low temperatures and low ammonia partial pressures, the released hydrogen acted as an inhibitor to the decomposition reaction. Bradford et al [19] sought to gain information on H_2 inhibition on NH_3 decomposition over Ru/C catalyst. NH_3 partial pressure was varied from 1.3-12.0 kPa with temperatures between 370-390 °C, and a first order dependence of the reaction rate on NH_3 was observed. They proposed the following equation where α varies from 0.69 to 0.75, and β varies from -1.5 to -2, while the activation energy was 96.6 kJ/mol.

$$r_{H_2} = kp_{NH_3}^{\alpha} p_{H_2}^{\beta} \tag{5}$$

Egawa et al [30] used deuterated NH_3 on Ru single crystal surfaces and determined that the inhibition by H_2 was a consequence of an equilibrium established among adsorbed nitrogen atoms, gas-phase NH_3, and gas-phase H_2; and that recombinative desorption of adsorbed nitrogen atoms was the rate determining step. Vitvitskii et al [31] came to a similar conclusion based on experimental results acquired with diluted NH_3. Boudart et al [28] proposed that over W and Mo catalysts, N-H bond cleavage and recombinative desorption of surface nitrogen atoms are slow irreversible steps in NH_3 decomposition, NH_3 is activated via a direct dissociative adsorption step, and the adsorbed N atoms are the most abundant reactive intermediate.

Skodra et al [32] found that at higher temperatures (350 – 650 °C) and low ammonia partial pressures (0.5 – 2.0 kPa) over a Ru catalyst, hydrogen inhibition was no longer significant. They also observed a second order dependence of the rate of decomposition on ammonia partial pressure. This was explained by assuming step 3 above was the rate determining step. Shustorovich and Bell [33] suggested, based on the BOC Morse potential method, that the rate-determining step of ammonia decomposition is recombinative desorption of N_2. Chellappa et al [34] investigated pure ammonia (high concentration) over Ni-Pt/Al_2O_3 catalyst at 520 – 690 °C, and H_2 inhibition was not observed. The reaction was first order with respect to NH_3 pressure, and the activation energy was 196.2 kJ/mol.

Thus it appears that H_2 inhibition is only significant at low NH_3 concentrations and low temperatures [5]. Earlier studies reported a shift in reaction order from 0 to 1 with respect to ammonia partial pressure as temperature increases, however more recent studies report a shift in reaction order from 1 to 2 with temperature. β varies between -1.5 and -2 at low temperatures and low ammonia concentrations, but shifts to 0 as temperature and ammonia concentration increase. There also appears to be a consensus among researchers that the recombinative desorption of nitrogen atoms is the rate determining step in the decomposition reaction.

2.3. Electrolysis of Ammonia

Electrolysis or electro-oxidation is another method of extracting hydrogen from ammonia. It has the advantage of scalability and versatility to interface with renewable energy sources including those whose electricity production varies with time [35]. Hydrogen can also be produced at moderate temperatures. It was first discussed by Vitse et al [35], who proposed the coupling of ammonia oxidation in an alkaline medium at the anode with the reduction of water at the cathode.

$$2NH_3 + 6OH^- \rightarrow N_2 + 6H_2O + 6e^-, E^0 = -0.77V / SHE \tag{6}$$

$$6H_2O + 6e^- \rightarrow 3H_2 + 6OH^-, E^0 = -0.82V / SHE \tag{7}$$

The thermodynamic potential for ammonia electrolysis in alkaline media is -0.77 V compared with -1.223 V for the electrolysis of water. The theoretical thermodynamic energy consumption is 1.55 Wh/g of H_2 from electrolysis of NH_3 compared to 33 Wh/g of H_2 from H_2O [36]. This means that theoretically, ammonia electrolysis consumes 95% less energy to produce a quantity of hydrogen than water electrolysis. This however, does not account for kinetics of the reaction.

The most widely accepted mechanism of ammonia oxidation is 1) the adsorption of ammonia on to Pt surfaces, 2) dehydrogenation of ammonia into various adsorbed intermediates (N, NH, NH_2), 3) reaction of the intermediates to form $N_2H_{2,ad}$, $N_2H_{3,ad}$ and $N_2H_{4,ad}$ which then react with OH^- to produce nitrogen [37]. The reaction of $N_2H_{2,ad}$ is considered the rate determining step. Vidal-Iglesias et al [38-39] conducted differential Electrochemical Mass Spectrometry (DEMS) studies on ammonia oxidation and suggested also the presence of an azide intermediate species (N_3^-) at certain potentials. Of the various adsorbed intermediates, NH and NH_2 are active, however, N remains adsorbed (N_{ad}) and acts as a poison.

Although ammonia electrolysis is thermodynamically favorable, kinetics are slow. In practice, high overpotentials are required to drive the ammonia oxidation reaction, and deactivation of the Pt catalyst is observed at high current densities [40-41]. With Pt, N_{ad} is only formed at very high potentials, thus making Pt the best choice of catalyst for electro-oxidation of ammonia. Alloys of Pt have been found effective as catalysts for ammonia oxidation, with the other metals in the alloy chosen for their ability to dehydrogenate ammonia. Endo et al [42] studied combined catalysts of Pt with other metals including Ir, Cu, Ni, and Ru. They concluded that only Ru and Ir can improve the catalytic properties of Pt. In another study, an alloy of bulk Pt with bulk Ir was tested, and the performance was found to be better than Pt alone, however oxidation current densities were still less than 1 mA/cm^2 [43]. De Vooys et al [40] studied ammonia oxidation and intermediates on various polycrystalline catalyst surfaces – Pt, Pd, Rh, Ru, Ir, Cu, Ag, and Au. They concluded that only Pt and Ir combine a good capability to dehydrogenate ammonia with a low affinity to produce N_{ad}. In another study, a Pt-Ir powder mixture (50 wt.%) impregnated in Teflon and painted on a platinum screen was found to provide much lower overpotentials for the oxidation of ammonia than platinum black [44]. However, in these studies, a very high loading of precious metal catalysts was used (up to 51 mg/cm^2) rendering them uneconomical for fuel cell use.

Botte et al [35] studied the use of Pt-Ru alloys for ammonia oxidation catalysts. Individually, Pt and Ru resulted in fast dehydrogenation of ammonia at low potentials which resulted in fast deactivation of the catalyst. However, when combined, the Pt allowed for a significant rate of recombination of adsorbed nitrogen. A low loading of Ru prevented the fast ammonia dehydrogenation from prevailing over the nitrogen recombination step. They reported that catalyst preparation using co-electrodeposition allows for a low loading of noble metals (~2.5 mg/cm^2).

In another study, they evaluated the electrolysis of ammonia on a high surface area Raney Nickel substrate plated with Pt and Rh [45]. The electrodes were characterized by scanning electron microscopy, energy dispersive X-ray spectroscopy, and X-ray photoelectron spectroscopy. All tested electrodes demonstrated that Rh produced a synergistic effect when paired with Pt as a catalyst for ammonia electro-oxidation. Hydrogen was successfully produced from a 1M NH_3/5M KOH solution at 14.54 Wh/g H_2 at a current density of 2.5 mA/cm^2 by an anode containing 1 mg/cm^2 Rh and 10 mg/cm^2 Pt at ambient temperature and pressure. When the Pt loading was reduced to 5 mg/cm^2, the required energy for electrolysis was 16.83 Wh/g of H_2. They did not report results when Pt alone is used as catalyst. Their results indicated that rhodium can increase the kinetics of the electrolysis reaction while allowing a reduced loading of precious catalysts. Their XPS results indicated that 1 mg/cm^2 is the optimum loading of Rh, since it maximized the proportion of the noble metal coverage to exposed substrate metal. Nevertheless, the energy required to produce hydrogen is nearly 10 times higher than the theoretical thermodynamic value.

In a follow up study, they considered the use of carbon fiber substrate electrodes instead of Raney Nickel [46]. An observed decrease in current density with Raney Nickel at polarization potentials indicated that blockage of active sites by OH^- occurred. This blockage was possibly due to non-uniform coverage of the substrate. It resulted in a reduced surface area of the active catalyst. It was proposed that OH^- competes with NH_3 for adsorption on to the Pt surface, thereby decreasing the number of available sites for electrolysis. Rh added to Pt has been shown to solve this problem by reducing the number of unused catalyst sites compared with Pt alone. It was also observed that the reactivity of the catalyst decreased over time, indicating that the Ni substrate was not stable. Better results were obtained using carbon fiber electrodes [46-47], which allowed for uniform surface coverage of the noble metal, prevented blockage of active sites, and were light weight compared with Ni. They also obtained promising results using Pt-Ir-Rh and Pt-Ir catalysts on the carbon fiber substrate, which resulted in 91-92% conversion of ammonia to hydrogen at room temperature and low ammonia concentrations, and with electrolysis occurring at current densities up to 25 mA/cm^2 and a precious metal loading of 5.5 mg/cm^2. This corresponds to an energy consumption of 18.15 Wh/g of H_2 which is higher than those reported by Cooper and Botte [45], and a hydrogen generation rate of 1.4 mmol/min/g of catalyst.

The previous studies were based on bulk catalysts. Other studies have considered the effects of nano-sized Pt particles, but they found that the oxidation of ammonia was more sensitive to the structure of Pt particles rather than their size. Vidal-Iglesias et al [48-50] studied ammonia oxidation on stepped electrodes consisting of Pt (1 0 0) terraces and Pt (1 1 1) steps. They used voltammetry, chronoamperometry, and in situ infrared spectrometry to characterize the electrodes and concluded that electrocatalytic activity is increased by a factor of up to 7 when Pt (1 0 0) is used rather than Pt (1 1 1) or Pt (1 1 0) as the preferential orientation of nano-particles. They found that the oxidation was highly structure sensitive and that it took place exclusively on the Pt (1 0 0) sites.

Vidal-Iglesias et al [51] further considered further the effect of adding nano-sized alloys to Pt. Ir, Pd, Rh and Ru were tested. Ru and Pd were found to decrease the oxidation current. In fact, as Ru content increased, the oxidation current decreased. They explained this result by proposing that Pd and Ru decreased the density as well as the dimensions of the Pt (1 0 0) sites. However, Ir and Rh were found to enhance the oxidation current at low potentials. They also studied the effect of particle size and found that 9 nm Pt particles produced better oxidation results than 4 nm particles. This is because the 9 nm particles had a larger number of Pt (1 0 0) sites. They thus concluded that oxidation of ammonia on nano-particles is highly structure sensitive.

Much work has been done in developing catalysts to electrolyze ammonia. Calculations show that the rate of hydrogen generated via electrolysis is in the order of 0.1 to 1 mmol/min/g of catalyst, which is several orders of magnitude lower than what is reported for ammonia decomposition. Also the energy consumption required to produce hydrogen ranges from 14-18 Wh/g. This energy consumption needs to be reduced to 5.4 Wh/g in order to produce H_2 at a realistic cost of US$2/kg [45]. This means that oxidation overpotentials must be reduced to below 200 mV at much higher current densities than those reported in the literature.

2.4. Hydrogen production from ammonia borane

Products of ammonia have also received some attention in the literature as sources of hydrogen, with most of the studies focusing on ammonia borane (NH_3BH_3) or AB. Ammonia-borane complex has a high material hydrogen content (about 19.6 wt%) with a system-level H_2 energy storage density of about 2.74 kWh/L (versus 2.36 kWh/L for a liquid hydrogen). Hydrogen can be evolved via hydrolysis of AB.

$$NH_3BH_3 + 2H_2O \rightarrow NH_4^+ + BO_2^- + 3H_2 \tag{8}$$

To employ H_2 as a direct fuel supply for PEMFCs a suitable catalyst is needed to accelerate the hydrolysis of AB. Various catalysts with excellent catalytic performance have been developed [52-63]. These include noble metal based catalysts such as Pt, Ru, Rh, Pd, Pt and Au supported on alumina; combinations of Pt with Ir, Ru, Co, Cu, Sn, Au and Ni supported on carbon, Rh(0) nano-clusters [52-57]; and also non-noble metal based catalysts such as Ni and Co on alumina, and Ni and Co nano-particles, Cu/Cu_2O, Poly(N-vinyl-2-pyrrolidone) (PVP) stabilized Ni, Ni-SiO_2 and Fe–Ni alloys [58-63]. Unfortunately, most of the aforementioned catalysts, except for the magnetic Fe–Ni alloy catalyst, are difficult to use repeatedly in solution because they are in a powdery form or are supported weakly on a substrate. The development of catalysts with high durability is thus important for practical use.

Mohajeri et al [64] studied the room temperature hydrolysis of ammonia borane using K_2PtCl_6 and found the reaction rate to be third order (second order with respect to catalyst concentration and first order with respect to AB concentration) with an activation energy of 86.6 kJ/mol. Their average hydrogen generation rate was 590.3 mmol/min/g of catalyst, although this rate varied throughout the test. Good results were also obtained using non-precious metal catalysts. Eom et al [65] considered the effect of an electroless-deposited Co–P/Ni foam catalyst on H_2 generation kinetics in AB solution and investigated the cyclic behavior (durability) of the catalyst. The activation energy for the hydrolysis of AB using the Co–P/Ni foam catalyst was calculated to be 48 kJ/mol. Their hydrogen generation rates were an order of magnitude lower than ref [64] at room temperature, but increased with the temperature of the AB solution. After six cycles, the H_2 generation rate dropped to about 70% of the initial values.

Xu et al [62-63,66-67] obtained excellent results for hydrolysis of an ammonia borane / sodium borohydride ($NaBH_4$) mixture in a 5:1 mass ratio. Their various works utilized different non-precious nano-catalysts including Ni/silica, Co/silica nano-spheres, unsupported Co nano-particles and Fe-Ni nano-particles. For unsupported Co and Fe-Ni, they reported extremely high hydrogen generation rates at room temperature. Some hydrogen generation rates they obtained are calculated using data provided in their references. 10 nm unsupported Co nano-particles evolved hydrogen at the rate of 775 mmol/min/g, while $Fe_{0.5}Ni_{0.5}$ generated 178 mmol/min/g, both at room temperature.

Although AB has 19.1 wt% hydrogen content, the gravimetric (mass) hydrogen storage capacity (GHSC) is relatively low. The GHSC of the AB – H_2O system is only 9% when hydrolysis is intended in stoichiometric conditions and even lower with excess H_2O. Practical studies showed that the effective GHSC is typically only 1% [58]. Demirci and Miele [68] considered the effect that storing AB in a solid form and regulating the supply of H_2O would have on the effective GHSC. They used $CoCl_2$ as catalyst, and found that when water was supplied in stoichiometric quantities, an effective GHSC of 7.8% was achieved at 25 °C. They reported the hydrogen generation rate to be 85.9 mmol/min/g of catalyst under these conditions.

Results in this section show that ammonia borane is extremely promising as a hydrogen source. Hydrolysis of AB has yielded order of magnitude higher rates of hydrogen generation than ammonia decomposition, and it can be done at room temperatures using non-precious metal catalysts. These results are shown in Table 4, which shows significant variations in the performance of the hydrolysis catalysts. For example the hydrogen generation rate using unsupported Co nano-particles is 3 orders of magnitude higher than for Co nano-particles supported on silica. It should also be pointed out that in Ref [62], when the Ni loading increased, the rate of hydrogen generation increased but by an amount less than proportional to the increase in catalyst loading. In other words, as the Ni loading increased, the rate of hydrogen generation per gram of catalyst decreased. Thus there is a lot that is not yet fully understood. The effects of particle size, loading, support as well as other unknown factors need to be investigated in greater detail.

Catalyst	Conditions	Hydrogen Generation Rate (mmol/min/g)	REF
Co-P/ Cu sheet	30 °C	38.7	[65]
Co-P/Ni foam	30 °C 40 °C 50 °C 60 °C	35.8 69.3 130.0 220.1	[65]
K_2PtCl_6 salt	25 °C	590.3	[64]
20-30 nm Ni/Si_2O_3	25 °C	7.0	[62]
15-30 nm Co/ Si_2O_3 nano-spheres	25 °C, NH_3BH_3 / $NaBH_4$ mixture	0.7	[66]
10 nm unsupported Co particles	25 °C, NH_3BH_3 / $NaBH_4$ mixture	775.0	[67]
$Fe_{0.5}Ni_{0.5}$ nano-particles	25 °C, NH_3BH_3 / $NaBH_4$ mixture	178.1	[63]
$CoCl_2$	25 °C, NH_3BH_3 / $NaBH_4$ mixture	85.9	[68]

Table 4. Performance of Various Catalysts for Ammonia Borane Hydrolysis

2.5. Other means of hydrogen generation from ammonia and ammonia products

Urea is an ammonia product that can be used to generate hydrogen. Urea rich waste water is widely abundant and releases ammonia into the atmosphere when purged into rivers and lakes. Preliminary work has been conducted in urea electrolysis in alkaline media (equations 9,10). Typically a KOH electrolyte is used with a Ni based anode catalyst. One of the problems encountered in the literature is the instability and deactivation of nickel oxide sites during the oxidation of organic compounds [69-70]. Recent research suggested that Rh added to Ni enhances stability, reduce overpotentials and increase current densities by a factor of 200 [71]. Typically KOH is used as the electrolyte, although King and Botte have obtained good results using a polymer gel electrolyte, poly(acrylic acid) (PAA) cross-linked polymer [72]. They found that an electrolyte containing 8M KOH and 15% PAA (by weight) showed a good combination of good conductivity, mechanical strength, and ease of preparation.

$$CO(NH_2)_2 + 6OH^- \rightarrow N_2 + 5H_2O + CO_2 + 6e^- \tag{9}$$

$$6H_2O + 6e^- \rightarrow 3H_2 + 6OH^- \tag{10}$$

Another ammonia product, ammonia hydride was studied by Sifer and Gardner [73] and was proposed for use in military applications requiring low power for long durations. Their study showed that an ammonia hydride system could provide 483 Wh/kg and operate for over 50 hours. This compares well with other batteries used for military applications in terms of energy density.

It is also possible to extract hydrogen by reacting ammonia with metal hydrides such as MgH_2 [74], $LiAlH_4$ [75] or $NaAlH_4$ [76]. This can be done at moderate temperatures (75-150

°C) and has been reported to have 2-3 times higher specific energy and energy density than ammonia cracking at 650 °C. However additives are required such as $PdCl_2$ and $PtCl_4$ to enhance the rate of hydrogen formation [74]. Paik et al [77] discussed the mechanochemical production of hydrogen from ammonia via milling at room temperature on $BaTiO_3$ and $SrTiO_3$ catalysts, however H_2 production rates are relatively low.

3. Direct ammonia fuel cells

When hydrogen is produced from ammonia, there are usually traces of unconverted ammonia and nitrogen oxides in the feed. This is sometimes not amenable for use in fuel cells. The ammonia acts as a poison to Nafion membrane used in PEM fuel cells. For higher temperature fuel cells such as SOFCs, ammonia can be decomposed directly in the fuel cell thus negating the need for an external reactor. This section discusses some research conducted on the direct use of ammonia in fuel cells. It considers PEM fuel cells, SOFCs, alkaline fuel cells, and fuel cells directly utilizing ammonia borane.

3.1. Polymer Electrolyte Membrane (PEM) Fuel Cells Using Ammonia

Ammonia has been proposed for use in PEM fuel cells, however, because of the low temperature of PEM fuel cell operation, internal ammonia decomposition is not thermodynamically favorable. Ammonia must be decomposed externally at higher temperatures, then the hydrogen supplied to the fuel cell. If there is not 100% conversion of ammonia to hydrogen, then there will be trace amounts of ammonia in the hydrogen feed. Unfortunately ammonia has been shown to act as a poison to the Nafion membrane typically used in PEM fuel cells.

Uribe et al [78] studied the effect of ammonia on PEM fuel cells with 0.15-0.2 mg/cm^2 Pt loading using high frequency resistance (HFR). They exposed the fuel cell intermittently to hydrogen and ammonia at the anode and found that when exposed to 30 ppm ammonia for 1 hour, the performance of the cell degraded, but when it was switched back to neat hydrogen, full recovery of performance was observed after 18 hours. When it was exposed to ammonia for 17 hours, full recovery was not observed within 4 days after switching back to hydrogen. Cyclic voltammetry (CV) did not indicate the presence of any adsorbed species at the anode or the cathode, thus the poisoning mechanism of ammonia on PEM fuel cells was not determined. HFR showed that the cell resistance doubled in 15 hours of ammonia exposure. Soto et al [79] performed a similar study but using higher catalyst loading – 0.45 mg/cm^2 Pt/Ru at the anode and 0.6 mg/cm^2 Pt at the cathode. Using the current interrupt technique, they showed that after 10 hours of exposure to 200 ppm ammonia, the cell resistance increased by 35%. As in Ref [78] CV showed no evidence of any adsorbed species at the anode or cathode. They proposed that ammonia affects the anode catalyst layer rather than the cathode catalyst layer, primarily because it is supplied at the anode. In these studies, calculations imply that ohmic losses were not sufficient to explain the loss in fuel cell performance observed.

In another test using 0.45 mg/cm^2 Pt/Ru at the anode and 0.4 mg/cm^2 Pt at the cathode with a Nafion membrane, the cell was exposed to 10 ppm ammonia [80]. The cell resistance gradually increased reaching a steady state value of twice the original value after 24 hours of exposure. During this time, the voltage dropped by 160 mV, of which ohmic losses could only account for 8 mV. The cell recovered to normal operation after 2-4 hours of neat hydrogen. Exposure to 1 ppm of ammonia produced a degradation in performance that was slower than for 30 ppm, but the cell resistance tended to the same steady state value of double the original resistance after more than 1 week of exposure. Using Pt vs Pt/Ru at the anode did not significantly alter the poisoning profile, thus they concluded that it was unlikely to be the anode catalyst layer that was affected. Szymanski et al [81] studied the effect of ammonia on phosphoric acid fuel cells. They found that for a PAFC operating at 191 °C, it was the oxygen reduction reaction at the cathode that was most affected by the presence of ammonia. When 1% of H_3PO_4 was converted to $(NH_4)H_2PO_4$, the cathode activity decreased by 84%.

Halseid et al [80] performed a unique test using a symmetric H_2/H_2 PEM cell – hydrogen supplied at both electrodes. Then one of them was switched to 10 ppm ammonia. They observed that the cell resistance did not decrease when the ammonia supply was stopped, and that a limiting current existed not due to mass transport, but due to a reaction limiting current possibly attributed to the Tafel step in the Tafel-Volmer hydrogen oxidation reaction (HOR) mechanism. They interpreted this to mean that ammonium remains in the membrane phase, confirming the low volatility of ammonium in PFSA ionomers. Performance degradation followed a first order response therefore could not be the result of ammonia adsorption on the carbon of the GDL. Oxidation of ammonia on Pt in acidic solutions to form N_2 or NO_x was also unlikely to be significant at the anode. At high cell potentials, they proposed that a platinum oxide is formed at the catalyst sites, thus reducing the effectiveness of the catalyst. They proposed that ammonium transfers across the PEM membrane within minutes, affecting not only the anode side but the cathode as well. They postulated that ammonium does not adsorb on to the anode catalyst, but shifts the potential of the H_2 adsorption process in Pt solutions, but not H_2 desorption [82]. Ammonium in sulfuric acid (low concentrations 10% NH_4) has been shown to increase the cathode overpotential by up to 100 mV at any given current density. They suggested the possibility of an adsorbed species from the electrochemical oxidation of ammonium that blocks active sites, or the result of a mixed potential at the cathode due to simultaneous oxygen reduction and ammonium oxidation [80].

Hongsirikarn et al [83-84] performed further studies on the effects of ammonia on Nafion, measuring the membrane conductivity in the liquid and gas phases. They prepared their membranes using HCl and NH_4Cl to simulate the desired concentrations of H^+ and NH_4^+ in the membrane. They measured conductivity via a two probe technique using a frequency response analyzer. The room temperature conductivity of Nafion 117 in distilled water decreased almost linearly from 115 mS/cm to 24 mS/cm when NH_3 content increased from 0 to 100%. Ammonia tolerance was also seen to improve with temperature. The effect of ammonia on conductivity was more severe in the gas phase than the aqueous phase because in the gas phase, only small amounts of water vapor were present, and the strong anion sulfonic sites stabilized the ammonium ion in the structure [80]. They also performed tests exposing Nafion membranes to 5-30 ppm NH_3 gas, similar conditions to operating fuel cells. Conductivity decreased from 30 mS/cm to a steady state value of 2.5 mS/cm over time, with the rate of degradation increasing with ammonia concentration. 30 ppm required roughly 6 hours to reach a steady state poisoning whereas 5 ppm took 36 hours [83].

NH_3 poisoning of PEM fuel cells is a slow process unlike CO poisoning. Recovery is also slow. This is due to the relatively slow diffusion of ammonium in the membrane, and the slow process of ammonium oxidation which results in ammonium sinks (for diffusion). For use in PEM, ammonia must be removed from the fuel stream. Saika et al [85] suggested an ammonia recirculation system that can reduce the ammonia content in the fuel stream from 300 ppm to 0 ppm. This involves dissolving ammonia in water (H_2 and N_2 do not dissolve) and recirculating the ammonia. However, much more work is needed to refine this particular area of research. The alternative is enhancing the catalysts (perhaps the cathode rather than the anode) with other noble metals (perhaps Ir) to enhance ammonium oxidation [80].

To date, no studies have been reported showing the effects of ammonia on polybenzimidazole (PBI) membranes, which are an alternative intermediate temperature polymer electrolyte membrane to Nafion. PBI is normally doped in phosphoric acid and operates up to 200 °C. These conditions are not suitable for internal ammonia decomposition. However it is not known how traces of ammonia in the hydrogen feed would affect the performance of PBI membranes. This remains a subject for future work.

3.2. Direct ammonia solid oxide fuel cell

Ammonia has proven to be problematic for PEM fuel cells involving Nafion since both the conductivity of the membrane and the activity of the catalysts are adversely affected by trace amounts of ammonia in the fuel feed. This requires either 100% conversion of ammonia to hydrogen which is not always guaranteed, or total clean up of ammonia which is not always practical. Various researchers have investigated solid oxide fuel cells for direct ammonia oxidation. In fact ammonia can actually allow for smaller scale operation of SOFCs [86]. Farhad and Hamdullahpur reported a 100 W SOFC system where 1 liter of ammonia can provide nearly 10 hours of sustained power [87].

Ammonia decomposes readily at the high temperatures of SOFCs and has not been shown to act as a poison to the ceramic electrolytes utilized in SOFCs. Research into ammonia SOFCs can be divided into two categories – SOFC-O (oxygen ion conducting electrolytes) and SOFC-H (proton conducting electrolytes). SOFC-O entails an oxygen ion conducting electrolyte, such as yittria stabilized zirconia (YSZ) or samarium doped ceria (SDC), and water formation at the anode. SOFC-H entails a proton conducting ceramic electrolyte such as barium cerate and water formation at the cathode.

3.2.1. *Ammonia Fed SOFC-O*

$$H_2 + O^{2-} \rightarrow H_2O + 2e^- \quad (11)$$

$$0.5O_2 + 2e^- \rightarrow O^{2-} \quad (12)$$

The anode and cathode half cell equations for the SOFC-O fuel cell are shown respectively in equations 11 and 12. Pioneering research in this field was conducted by Wojcik et al [88]. They used an SOFC with a YSZ electrolyte at 800 °C with various catalysts – Fe, Ag, and Pt. Their objective was to test the various electrodes (catalysts). Their work raised the possibility of NO formation at the anode according to equation 13.

$$4NH_3 + 5O_2 \rightarrow 4NO + 6H_2O \quad (13)$$

The objective of the catalyst is to decompose NH_3 to H_2 faster than NO can be produced. Their research has ranked the catalysts in the following order of decreasing performance: Pt > Fe > Ag. They found that with Pt as the catalyst, fuel cells operating with NH_3 showed very little difference in performance to those operating with H_2, indicating that complete decomposition of NH_3 occurs over Pt. They discussed the possibility of using Ni although they did not test it in their fuel cells.

Detailed experimental investigation by Sammes and Boersma [86] have shown that Ni outperformed Ag and Pt in the temperature range 500 – 800 °C, with 90% conversion of ammonia occurring at 800 °C. Choudhary and Goodman [89] suggested that Ni electrodes can produce 5-10 times higher power densities than Ag or Pt. As the temperature increases, the performance of the ammonia fed cell approaches that of the hydrogen fed cell since conversion of ammonia approaches 100%, confirming the findings of Wojcik et al [88]. With the Ni-YSZ/YSZ/Ag (Ni-YSZ anode, YSZ electrolyte and Ag cathode) system at 700 °C, the ammonia cell actually performed better that the hydrogen cell. At 800 °C, a planar SOFC showed a peak power density of 75 mW/cm^2 (using a 0.4 mm thick electrolyte) while a tubular SOFC showed a peak power density of 10 mW/cm^2 (using 1 mm thick electrolytes). Fournier et al found that at 800 °C, a peak power density of 60 mW/cm^2 was achieved [90]. Dekker and Rietveld [91] used an anode supported system – NiO-YSZ/YSZ/LSM at 700 °C and obtained a peak power density of 55 mW/cm^2. They suggested that the temperature be at least 700 °C for sufficient ammonia conversion.

Most of the early work used electrolyte supported systems, which contained thick electrolytes with high ohmic overpotentials. As a result their peak power densities were in the order of 10 mW/cm^2. Zhang et al [92] prepared thin electrolytes (15 μm thick) by coating onto the anode substrate by a vacuum assisted dip-coating method using a YSZ slurry. Their anode supported NiO-YSZ tube was prepared by extrusion method. Their results showed a peak power density of 200 mW/cm^2 at 800 °C with NH_3 as fuel with no NO_x emissions detected. The equivalent peak power density with H_2 as fuel was 202 mW/cm^2. Ma et al [93] tested anode supported SOFCs based on a YSZ thin film electrolyte (30 μm thick), prepared by dry pressing, which they reported to be a reproducible way of producing thin film electrolytes by controlling the amount of powders used. With liquid ammonia as the fuel, they obtained power densities of 299 and 526 mW/cm^2 at 750 and 850 °C respectively. These results reveal a high temperature sensitivity. Open circuit voltage (OCV) analysis showed that ammonia oxidation occurs via two stages – cracking of ammonia and oxidation of hydrogen. The ammonia and hydrogen fueled cells had the same electrolyte resistances, however the ammonia cell had a higher interfacial polarization resistance at temperatures less than 750 °C.

SOFC-O System (anode/electrolyte/cathode)	Electrolyte Thickness (μm)	Temperature	Power Density (mW/cm^2)	REF
Pt-YSZ/YSZ/Ag	200	1000 900 800	125 90 50	[88]
Ni-YSZ/YSZ/Ag planar Ni-YSZ/YSZ/Ag tubular	400 1000	800	75 10	[89]
NiO-YSZ/YSZ/Ag	400	800	60	[90]
NiO-YSZ/YSZ/LSM	150	700	55	[91]
Ni-YSZ/YSZ/YSZ-LSM	30	750 850	299 526	[93]
Ni-YSZ/YSZ/YSZ-LSM	15	800	200	[92]
Ni-SDC/SDC/SSC-SDC	50	500 600 700	65 168 253	[96]
NiO-SDC/SDC/SSC-SDC	24	650	467	[94]
Ni-SDC/SDC/BSCF	10	700	1190	[95]

Table 5. Summary of SOFC-O Peak Power Densities

Another oxygen ion conducting electrolyte tested in the literature is SDC. Liu et al [94] compared liquid methanol, ammonia and hydrogen as fuels in a SOFC with a NiO-SDC anode, a 24 μm SDC electrolyte and a $Sm_{0.5}Sr_{0.5}CoO_3$ (SSC)-SDC cathode. At 650 °C, they obtained peak power densities of 698 mW/cm^2 (methanol), 870 mW/cm^2 (H_2) and 467 mW/cm^2 (NH_3) mW/cm^2. Meng et al [95] fabricated a 10 μm thick SDC electrolyte using a glycine-nitrate process, with Ni-SDC as the anode and $Ba_{0.5}Sr_{0.5}Co_{0.8}Fe_{0.2}O_{3-\delta}$ BSCF as the cathode. A peak power density of 1190 mW/cm^2 was obtained at 700 °C. At this temperature, the performance of a hydrogen fueled cell was 1872 mW/cm^2. Unlike previous works, they observed marked differences between the ammonia and hydrogen fueled cells at the respective temperatures. Their explanation was that the thermocouple reading was higher than the actual cell temperature because of the endothermic nature of the ammonia decomposition reaction. Nevertheless, their observed power densities were among the highest reported in the literature, primarily because of their thin electrolytes. Also noteworthy is that with the thinner SDC electrolytes, high power densities were observed at lower temperatures. No endurance tests were reported for such thin electrolytes.

With thin electrolytes, the power density of ammonia fed SOFC-O systems has been tremendously improved. At higher temperatures, the conversion of ammonia to hydrogen increases, and as a result, the performance of the ammonia fed system approaches that of the hydrogen fed system. Ni has been shown to be a very active catalyst for the ammonia decomposition and hydrogen oxidation reactions. However, due to the slow diffusion of oxygen ions through the electrolyte, the anode reaction is the rate limiting step [7]. This allows for the production of NO_x at the anode, although the use of Fe based catalysts can aid in reducing NO_x production. Table 5 summarizes the results of SOFC-O research presented.

3.2.2. *Ammonia Fed SOFC-H*

SOFC-O requires relatively high temperatures of operation (800-1000 °C), otherwise the conductivity of the YSZ electrolyte suffers, although better peak power densities have been observed with SDC electrolytes than YSZ. At intermediate temperatures (400-600 °C), the conductivity of SOFC-O electrolytes are significantly diminished. This problem can be resolved by utilizing thinner electrolytes which have been achieved, or by using proton conducting electrolytes. At these intermediate temperatures, proton conducting electrolytes such as $BaCeO_3$ and $SrCeO_3$ have better ionic conductivity than YSZ, thus making SOFC-H an attractive alternative at these temperatures. Another advantage of using SOFC-H is that since oxygen ions are not conducted through the electrolyte, the chances of producing NO_x are significantly reduced. Some of the common electrolytes proposed include $BaCeO_3$ and $BaZrO_3$, with the former (barium cerate) receiving a lot of attention in the literature. Commonly reported doping materials for barium cerate include gadolinium, praseodymium, and europium. The anode and cathode half cell reactions for the SOFC-H system are shown respectively in equations 14 and 15.

$$H_2 \rightarrow 2H^+ + 2e^- \tag{14}$$

$$0.5O_2 + 2H^+ + 2e^- \rightarrow H_2O \tag{15}$$

Maffei et al [96-97] published some early work on ammonia fed SOFC-H systems. They used barium cerate doped with gadolinium and praseodymium (BCGP) as the electrolyte, and obtained a peak power density of 35 mW/cm² at 700 °C. This was essentially the same peak power density observed when hydrogen was used as the fuel. It should be noted that their electrolyte was 1.3 mm thick. Gas chromatography analysis showed no NO_x emission. Stability was observed for 100 hours of operation. With a barium cerate doped with gadolinium (BCG) electrolyte, the peak power density was 25 mW/cm². In a subsequent publication [98], they reported barium cerate doped with europium (BCE) as the electrolyte. The BCE powders prepared by conventional solid state synthesis and the final electrolytes pressed to 1 mm thickness. Pt was used as the electrodes. The IV curves showed very little electrode polarization implying that most of the voltage drops were due to ohmic losses in electrolyte. The OCV was less than 0.7 V due to significant electronic conduction in the electrolyte. At 700 °C, the peak power densities observed were 32 mW/cm² for the NH_3 cell and 38 mW/cm² for the H_2 cell. They attributed this difference to the reduced partial pressure of hydrogen due to nitrogen formation in the NH_3 fed cell. Sustained operation was observed for over 200 hours. In yet another publication [99], they used a novel cermet anode consisting of Ni, europium doped barium cerate, a mixed ionic and electronic solid anode (BCE-Ni), and a BCGP electrolyte (1 mm thick). The BCGP and BCE components were fabricated using conventional solid state synthesis techniques. They found the BCE-Ni anode to be superior to the Pt anode in that a mere 1% weight composition of Ni was observed to completely decompose ammonia at 650 °C. Peak power densities at 600 °C for the NH_3 fed system were 28 mW/cm^2 using the NiO anode compared to 23 mW/cm^2 using Pt. Sustained operation was observed for over 500 hours at 450 °C. Thus they obtained sustained performance in their works, but low power densities, primarily because of the extremely thick electrolytes used in their tests.

Other researchers were able to produce thinner proton conducting electrolytes to obtain much higher current densities. Ma et al [100] used BCG electrolytes and obtained a peak power density of 355 mW/cm^2 at 700 °C. they observed OCV values of 1.102 V at 600 °C and 0.985 V at 700 °C. These values were consistent with theoretical predictions indicating that complete decomposition of ammonia took place. Zhang et al [101] also used BCG electrolytes dry pressed over a $Ce_{0.8}Gd_{0.2}O_{1.9}$ (CGO)-Ni anode substrate, and obtained peak power densities of 147 mW/cm^2 at 600 °C. Their OCV was 1.12 V at 600 °C and 1.10 V at 650 °C. These values are slightly lower than those for hydrogen fed cell because of the reduced partial pressure of the hydrogen due to nitrogen.

Ma et al [93,102] used 50 μm thin film BCG electrolytes with Ni-BCG anode and LSC-BCG cathode, and obtained peak power density of 355 mW/cm^2 and an OCV of 0.975 V at 700 °C. Gas chromatography showed that the partial pressure of NO was in the order of 10^{-12} atm. Lin et al [103] had similar observations with a 35 μm $BaZr_{0.1}Ce_{0.7}Y_{0.2}O_{3-\delta}$ (BZCY) electrolyte with BZCY/Ni anode and $Ba_{0.5}Sr_{0.5}Co_{0.8}Fe_{0.2}O_{3-\delta}$ (BSCF) cathode. Both BZCY and BSCF oxides were synthesized by a combined ethylenediaminetetraacetic acid (EDTA)–citrate complexing sol-gel process. Their peak power densities were 420 and 135 mW/cm^2 at 700 and 450 °C respectively. At 450 °C, the OCV was 0.98 V with ammonia compared to 1.1 V with hydrogen, the difference being attributed to reduced partial pressure. Based on impedance tests, they concluded that the actual operating temperature of the reaction may be lower (by about 35-60 °C) because of the endothermic nature of the reaction.

SOFC-H System (anode/electrolyte/cathode)	Electrolyte Thickness (μm)	Temperature	Power Density (mW/cm^2)	REF
Pt/BCGP/Pt	1300	700	35	[96]
Pt/BCG/Pt			25	[97]
Pt/BCE/Pt	1000	700	32	[98]
Ni-BCE/BCGP/Pt	1000	600	23	[99]
Ni-BCG/BCG/LSCO	50	700	355	[100]
Ni-BZCY/BZCY/BSCF	35	450	135	[103]
		700	420	
Ni-CGO/BCG/BSCF	30	600	147	[101]
NiO-BCNO/BCNO/LSCO	20	700	315	[104]

Table 6. Summary of SOFC-H Peak Power Densities

Zhang and Yang [101] were able to further reduce the electrolyte thickness to 30 μm based on a CGO-Ni/BCG/BSCF system and obtained a peak power density of 147 mW/cm² at 600 °C. The thinnest proton conducting electrolyte reported was a 20 μm $BaCe_{0.9}Nd_{0.1}O_{3-\delta}$ (BCNO) electrolyte fabricated with a suspension spray method, with a NiO-BCNO anode and a $La_{0.5}Sr_{0.5}CoO_{3-\delta}$ (LSCO)-BCNO cathode [104]. Their peak power density was only 315 mW/cm² however. They attributed this lower than expected value to the dense anode layer which resulted in incomplete reduction of NiO into Ni, thereby reducing the activity of the anode catalyst. Tests with a less dense anode were not reported. Table 6 summarizes the results of SOFC-H systems reported.

A significant advantage of direct ammonia SOFCs is that non-precious metal catalysts can be used. Nickel has been shown to be the most effective anode catalyst for both SOFC-O and SOFC-H – for both ammonia decomposition and hydrogen oxidation. This is critical for cost reduction. Perhaps SOFCs hold the best prospects for direct ammonia use in fuel cells. Thin electrolytes have been achieved with both SOFC-O and SOFC-H and both have reported high power densities. Most importantly, the performance of the SOFC has not been reported to be significantly diminished using ammonia instead of hydrogen as the fuel. There is no clear consensus on which of the two SOFC types is better. Although, some studies have reported concerns about NO_x formation at the anode of SOFC-O, in practice this has not been observed. NO formation is more likely in SOFC-O at high temperatures and high current densities when both the ammonia decomposition rate and the oxygen ion flow rate are high. However, in one study, NO was not detected up to 800 °C [102]. With anode supported SOFC-O systems (thick anodes and thin electrolytes), gas chromatography studies have shown that NH_3 completely decomposes by the Ni catalyst long before it reaches the triple phase boundary (TPB) [92].

3.3. Other Types of Ammonia Fuel Cells

Anhydrous ammonia has been proposed for use in alkaline fuel cells [105]. Aqueous alkaline electrolytes are tolerant to ammonia, unlike Nafion used in PEM fuel cells [10,20,34,106]. In a test cell, an AFC was run on ammonia for over 100 hours without any signs of poisoning [107]. However, at the low temperatures in which AFCs operate, ammonia will not readily release hydrogen atoms, therefore a separate external reformer is required [105]. Ganely [108] developed a test cell using a eutectic mix of NaOH and KOH fueled with pure ammonia and compressed air. The peak power density observed was only 40 mW/cm² at 450 °C. The electrodes were Ni tubes dipped into the molten electrolyte. One problem observed during the operation of the cell is that at the cathode, Ni oxidized to NiO and suffered reduced conductivity. They solved the problem by pretreating the Ni with 3M LiOH solution at 100 °C for 24 h while maintaining an anodic current of 1 mA/cm². The Nickel was thermally and electrochemically converted into a hydrated Nickel oxide, which was then electrochemically oxidized and lithiated by cationic exchange to produce $LiNiO_2$. This lithiated nickel was more stable and did not polarize during a 2-4 hour test. Nevertheless, the liquid electrolyte system suffers from high ohmic losses due to large electrode separation compared with solid state technology. Also the active catalyst sites are reduced due to the bulk nature of the electrodes. This problem is inherent to AFCs. Another potential problem with AFCs is that they require air that is free of CO_2, because of an unfavorable reaction between KOH and CO_2 [109].

Ammonia borane has already been discussed as a hydrogen source. However, there has been preliminary work suggesting that it can be utilized directly in fuel cells. This way, a greater reversible potential (1.616 V at 25 °C) can be achieved and thus greater power if it is used directly in fuel cells [110]. The anode and cathode half cell reactions are shown respectively in equations 16 and 17.

$$NH_3BH_3 + 6OH^- \rightarrow BO_2^- + NH_4^+ + 4H_2O + 6e^- \tag{16}$$

$$1.5O_2 + 3H_2O + 6e^- \rightarrow 6OH^- \tag{17}$$

In a test cell using a Nafion 117 membrane with Vulcan XC-72 30% Pt catalyst (0.15 mg/cm² loading) and carbon cloth electrodes, a peak power density of 14 mW/cm² was achieved at room temperature [110]. However, in the test cell, it was noticed that hydrogen gas was evolved at the anode indicating that hydrolysis of AB took place. The authors suggested that this could be inhibited by using thorium as an additive. Alternatively increasing the pH could also be used to avoid hydrolysis. In a follow up work, the Nafion membrane was replaced by a 28 μm thick anion exchange membrane from Tokuyama Corp. This time, peak power densities of 110 mW/cm² and 40 mW/cm² were achieved at 45 °C and 25 °C respectively. They also were able to run the fuel cell at 50 mA/cm² and 120 mA/cm² for over 20 hours without noticeable degradation in performance. However, energy dispersive X-ray spectroscopy (EDX) measurements showed that AB migrated across the anion exchange membrane and was directly oxidized at the cathode, thus greatly reducing the cell performance at higher current densities [111]. Kiran et al [112] tested a AB fuel cell using a TiC anode, Pt/C cathode with a Nafion 117 membrane, and obtained a peak power density of 110 mW/cm², 85 mW/cm² and 45 mW/cm² at 80 °C, 60 °C and 25 °C. respectively. This work allowed for the use of a non-precious catalyst, however the performance of the cell at 80 °C was not comparable to Ref [111].

Based on the half cell reactions, a significant amount of hydrogen is not extracted from ammonia borane. Zhang et al [111] suggests that BO_2^- can be reverted to BH_4^- through a reaction with a saline hydride (MgH). Then it can be converted back to NH_3BH_3 using a diethyl ether at room temperature and recycled. However, this just adds to the overall cost of the system. Another major drawback to ammonia borane fuel cells is that ammonium is produced in the anode reaction and this has already proven to be problematic for Nafion membranes. Based on present research, ammonia borane is more effective as a hydrogen source via hydrolysis than as a direct fuel in fuel cells. Much more work is needed to develop this technology.

4. Modeling and simulation of ammonia fed fuel cells

Research into ammonia fuel cells is relatively new and most of the work reported in the literature has been experimental. There have been a few works devoted to modeling of ammonia fuel cells, however this remains a relatively underdeveloped field. Mathematical or computational modeling is useful in that it can help attain information on the fuel cell that cannot be measure in situ. Most of the models reported thus far have been thermo-electrical models, with very few reporting comprehensive CFD type modeling. All of these models use parameters reported in the experimental literature.

Ni et al [113] published a thermodynamic model comparing the theoretical performance of SOFC-O and SOFC-H. Their computations showed that SOFC-H had a higher equilibrium potential (1000 vs 850 mV) and higher efficiency (72 vs 61 %) than SOFC-O at 800 °C. This assumed 80% fuel utilization and 20% oxygen utilization, 100% conversion of ammonia, and no formation of NO_x. Efficiency was defined as the ratio of electrical work to the lower heating value (LHV) of ammonia. The efficiency of both fuel cells decreased with temperature because of decreasing reversible potential, however the SOFC-H had an 11% higher efficiency. The reason for the better performance of SOFC-H is that water is not produced at the anode and therefore does not dilute the hydrogen. This model dealt primarily with thermodynamic equilibrium or reversible conditions.

This was followed by a 1D electrical model of a SOFC-H comparing the performance with NH_3 and H_2 as fuel [114]. This model incorporated the Nernst equation, Butler-Volmer equation, ohm's law, and the dusty gas model to determine partial pressures at the electrode/electrolyte interface, with the finite difference method used to solve the system of differential equations. Their results showed that the NH_3 fed cell always maintained a lower potential than the H_2 fed cell. These differences were insignificant for electrolyte supported systems (thick electrolytes), however with anode supported systems (thin electrolytes), the differences became pronounced. N_2 dilution of H_2 was cited as the reason for this observation. At 800 °C, their peak power densities for the SOFC-H / NH3 system were 68, 440 and 820 mW/cm^2 respectively for 500, 50 and 10 μm electrolyte thicknesses. In another 1D electrical model, the same authors compared the effect of electrolyte type on NH_3 fed SOFC systems, i.e. SOFC-O vs SOFC-H [115]. Surprisingly they found that under operating conditions (non-equilibrium), the potential of the SOFC-H was lower than that of the SOFC-O especially at higher current densities (respective peak power densities at 800 °C for a 50 μm electrolyte thicknesses were 360 and 420 mW/cm^2). This result is not consistent with Ma et al [102], who experimentally compared both systems at 700 °C and found their SOFC-H to have higher power densities than their SOFC-O (355 vs 252 mW/cm^2, both with 50 μm electrolytes). It is possible that as temperature increases, the SOFC-O performance improves compared to the SOFC-H. It is also possible that Ref [115] used a higher electrolyte resistance for the SOFC-H than those reported in the literature since the results of a model depend heavily on the parameters inputted to the model. They also noted that the SOFC-H had a higher anode concentration overpotential than the SOFC-O but a lower cathode concentration overpotential. It appears that the formation of water increases the concentration overpotential at the electrode where it is formed (anode for SOFC-O and cathode for SOFC-H). Dincer et al [116] performed a similar study and confirmed this finding. However, unlike Ref [115], they found that the SOFC-H outperformed the SOFC-O in both open and closed circuit conditions. Respective peak power densities at 800 °C for 30 μm electrolytes were 700 and 590 mW/cm^2. It should be noted that Ref [116] used a lower electrolyte resistance for the SOFC-H than Ref [115], which may account for their contradicting conclusions.

Ni et al [117] published an improved electrical model. The previous models assumed complete decomposition of ammonia, which is not realistic at lower temperatures utilized in SOFC-H. This work accounted for reactive transport of ammonia through the porous anode utilizing correlations reported by Chellappa et al [34], consistent with Ni/Al_2O_3 anode catalysts. Simulations showed that at 800 °C, complete conversion of ammonia to hydrogen occurred in the outer 10% of the anode. At 600 °C, ammonia permeated through 60% of the anode, but decomposed fully. At 400-500 °C however, only 10-40% conversion of ammonia was predicted. They thus recommended that SOFC-H be operated above 600 °C to avoid unconverted ammonia. One weakness of this work is that it did not compare the fuel cell performance with the previous models which assumed complete conversion of ammonia at the intermediate temperatures. This would have allowed a quantification of the effect of unconverted ammonia. The model also did not account for the inhibition effect of hydrogen on ammonia decomposition at the lower temperatures which were reported in the experimental literature (Section 2.2).

Ni et al [118] reported the first CFD based model of ammonia fuel cells. They developed a 2D thermo-electric model that combines a thermodynamic model, an electrochemical model, a chemical model for ammonia decomposition, and a CFD model for heat and mass transfer. This work focused on a planar anode supported SOFC-O based on a Ni-YSZ anode supplied with gases at 600 °C. Results demonstrated that high temperature gradients existed inside the cell – decreasing by up to 100 °C over 5 mm of channel length in some cases. High temperatures occurred near the anode and cathode inlets, but decreased because of the endothermic nature of the ammonia decomposition reaction. The fact that the rate of decomposition increases with temperature exacerbates the thermal gradients. Increasing the current density (decreasing cell potential from 0.8 to 0.3 V) decreased the temperature variations slightly, but not significantly. Increasing the flow velocity also was not shown to be effective in reducing the temperature variation. It was observed that increasing flow velocity by a factor of 10 resulted in a decrease in current density and Nernst potential, thus decreased the heat generation by the electrochemical reactions. This effect was more significant than the increased convection associated with the higher flow rates. The temperature variation actually increased when the flow velocity increased. Decreasing the flow rate, however, decreased the temperature variation to 30-40 °C (between inlet and outlet). They also observed a "cold spot" upstream in the electrolyte, which was approximately 67 °C colder than the inlet temperatures. This observation is consistent with what has been reported in some experimental works [95,103]. This simulation was based on parallel flow operation. No simulation was shown for counter flow operation or for tubular SOFCs with injection tubes, which may help to better distribute the temperature within the fuel cell.

Dincer et al [119] performed an exergy analysis on a combined heat and power system (CHP) with energy storage and heat recovery for vehicular application. The CHP entails an ammonia fed SOFC-H, a microturbine, a compressor, three compact heat exchangers, and a heat recovery unit. Their proposed system used ammonia (stored at 10 bar) and exiting the compressor at 600 °C before entering the fuel cell. They found that the system efficiency ranged from 60-90 % while energy efficiency ranged from 40-60 %. The rate of entropy generation decreased by 25% for every 100 °C increase in operating temperature. Exergy efficiency increased with current density for both the SOFC and the CHP system. As temperature increased, they observed that energy efficiency decreased while exergy efficiency increased. The authors recommended that the SOFC be operated with an exhaust temperature between 1200 and 1300 °C. It is questionable however whether such high temperatures are appropriate for vehicular applications.

Other works have focused on numerical simulation of the decomposition reactor [120-122]. Chein et al [123] theoretically and numerically predict H_2 generation over packed beds of Ni-Pt/Al_2O_3 particles using a chemical reaction model to account for varying rates of ammonia decomposition. Their results indicated that the porosity and permeability of the packed bed did not significantly affect the performance of the reactor, at least for the range of values investigated, and that their results gave the same predictions as a 1D plug flow model. They recommended low flow rates of ammonia to increase the residence time in the reactor, thus allowing for higher conversion efficiencies to be attained at lower temperatures. However, one major assumption in their model is that of an isothermal reactor. Ni [118] has shown that there are significant temperature variations as a result of the endothermic nature of the ammonia decomposition reaction. It is also possible that the model reported in Ref [123] was too simplistic, and did not account for the effects of intermediate species, which may depend on porosity and permeability.

Alagharu et al [124] modeled an ammonia decomposition reactor for a 100 W PEM fuel cell. It was shown that operating the reactor adiabatically results in a sharp decrease in temperature due to endothermic reaction, which resulted in low conversion rates of ammonia. For this reason, the reactor was heated electrically to provide heat for the endothermic reactions. It was observed that when the reactor was operated non-adiabatically, it was possible to get over 99.5% conversion of ammonia. In this simulation, fuel entered the reactor at 520 °C, decreasing to 480 °C within the first 5% of the reactor, then exiting at nearly 600 °C. They proposed operating conditions where ammonia is supplied to the reactor at 0.9 mmol/s, supplying a power of 142 W to a stack of 20 PEM fuel cells, of which 42 W is used to heat the reactor.

Thus far, modeling of ammonia fuel cells is a new field of research. The present models account for direct internal ammonia decomposition in the fuel cell anode, however, all the models have reported SOFC performance at elevated temperatures (600-800 °C). It is also possible to operate SOFC-H at lower temperatures (400-500 °C), however no model has as yet taken into account the effect of hydrogen inhibition on the rate of ammonia decomposition that has been reported at these temperatures (section 2.2). Fuel cell models should also propose strategies for dealing with temperature drops inside the fuel cell due to the endothermic nature of ammonia decomposition. Reduced temperature adversely affects the performance of the cell, while high thermal gradients adversely affect its longevity. Future models must also consider the effect of the nano and micro structure on fuel cell performance – the effect of catalyst particle size, particle loading with the supports, and support conductivity. Also no work has as yet simulated the ammonia poisoning effect in PEM fuel cells or the behavior of direct ammonia borane fuel cells. These are all subjects for future work.

5. Conclusions

Ammonia has been shown to be a cost effective means of hydrogen storage for fuel cells. Its energy cost based on high heating value is less than hydrogen, natural gas, propane and methanol. Also the life cycle cost of power is estimated to be lower for ammonia than for methanol or hydrogen. Studies have shown that for small scale systems, hydrogen generation from ammonia cracking is lower in cost than natural gas reformation and water electrolysis. Ammonia has the added advantage of being widely produced globally and possessing a transport and storage infrastructure that already exists – a significant advantage over hydrogen. Thus ammonia has proven to be a cost effective alternative to hydrogen for fuel cells.

Hydrogen generation from ammonia and ammonia products has been widely studied. Much research has been devoted to developing suitable catalysts for ammonia decomposition. It is widely agreed that Ru is the best catalyst for decomposition at 400 °C, while Ni based catalysts perform comparably at 600 °C. Studies have shown that carbon nanotubes function very effectively as a support for Ru. Treating the support with potassium based alkalis has resulted in a 5-fold increase in catalyst activity. Ni has the advantage of being non-precious. Although the decomposition rate (measured per gram of catalyst) is lower for Ni than Ru at a given temperature, higher loading of Ni is permissible because of its lower cost. Ni also allows for the use of nano-sized particles, which have shown to produce decomposition rates comparable to Ru. There is much scope for improving the low temperature performance of Ni through the use of nano-technology. The mechanism for ammonia decomposition has been widely studied over Ru based catalysts, however there is a general lack of research studying the corresponding mechanism over Ni. Such studies have resulted in the development of supports and promoters that optimize the performance of Ru as a decomposition catalyst. Various supports for Ni have been investigated (silica, alumina, SBA) without any clear consensus on which is optimal. The performance of Ni can be optimized by studying the reaction mechanism and investigating the effect of various promoters.

The hydrolysis of ammonia borane has also been shown to be a very effective means of hydrogen generation, with hydrogen generation rates (per gram of catalyst) being reported to be an order of magnitude higher than for ammonia decomposition. Further, hydrolysis of ammonia borane can occur at room temperature. Excellent results have been obtained with Pt as well as Co, Ni and Fe based catalysts. However, the results reported in the literature vary greatly. For example, two studies by the same research group of ammonia borane hydrolysis using Co based nano-catalysts have produced room temperature hydrogen generation rates that differ by three orders of magnitude. It is not yet fully understood what are the optimal particle size, loading and support required to produce effective hydrolysis catalysts. Much more work is needed to refine this technology, although it holds tremendous potential for low temperature fuel cells.

Ammonia has also been considered for direct use in fuel cells. PEM fuel cells are adversely affected by ammonia because of reduced conductivity of Nafion and decreased activity of the catalyst layers (especially the cathode). Alkaline fuel cells are unaffected by ammonia but are unlikely to be commercially pursued because of the inherently poor performance associated with non-solid state technologies. High power densities have been reported for SOFCs (100-1000 mW/cm^2 for both proton conducting and oxygen ion conducting electrolytes). Ammonia is internally decomposed at the high operating temperatures of SOFCs. Early works have focused on Pt based catalysts, however later works have achieved comparable results using Ni based catalysts. The general consensus among researchers is that at higher temperatures, the performance of ammonia fed SOFCs approach that of hydrogen fed SOFCs, differing only due to nitrogen dilution at the anode. However, the works reported suggest that temperatures over 700 °C are required for SOFCs to produce sufficiently high power densities. At lower temperatures, the ammonia to hydrogen conversion efficiency decreases with Ni based catalysts. It is possible and desirable to operate an SOFC-H at temperatures as low as 400 °C, however good performance at such low temperatures has not been reported. The use of nano-sized Ni particles in the SOFC anode has not as yet been reported, but may hold tremendous potential for improving the low temperature performance of SOFC-H. Future work needs to study the preparation and dispersion methods of such catalysts.

There is also much scope for mathematical modeling of ammonia fed SOFCs. So far, few models have been developed, with most of them focusing on thermodynamic considerations. Only one CFD based model has been reported, which revealed that a cold spot exists within the SOFC which reduces its performance. CFD based models can provide strategies for eliminating or reducing the effects of this cold spot. All of the models reported so far have simulated internal ammonia decomposition at relatively high SOFC temperatures (~800 °C). It is desirable to operate the SOFC at much lower temperatures, where the effect of hydrogen inhibition on the decomposition rate has been reported in the experimental literature. No mathematical models have thus far considered this effect. Future models must simulate the behavior of lower temperature SOFCs and the effect of hydrogen inhibition on the ammonia decomposition reaction. They must also consider the effect of catalyst particle size and loading within the support, as well as conductivity of the support on internal ammonia decomposition. Further all of the models focus on SOFCs. Models of the ammonia poisoning effect on PEM electrolytes and catalysts may reveal strategies for reducing this poisoning effect and optimizing the performance of ammonia fed PEM fuel cells. This also remains a subject for future research.

Ammonia can achieve all the benefits of a hydrogen economy with existing infrastructure. Combined with the development of proton conducting SOFC electrolyte technology and Ni based nano-catalyst technology, ammonia allows for smaller scale operation of SOFCs at reduced temperatures (~ 400 °C), which makes it a very attractive power source for small scale stationary and portable applications. Ammonia may very well be the fuel that makes fuel cell technology commercially viable.

Author details

Denver Cheddie
Center for Energy Studies, University of Trinidad and Tobago, Brechin Castle, Couva, Trinidad and Tobago

6. References

[1] Christensen CH, Johannessen T, Sorensen RZ, Norskov JK (2006). Catal. Today 111:140–144.
[2] Jensen JO, Vestbo AP, Li Q, Bjerrum NJ (2007). J. Alloys Compd. 446-447:723–728.
[3] Zamfirescu C, Dincer I (2008). J. Power Sources 185:459–465.
[4] Kaye IW, Bloomfield DP (1998). Conference of Power Source, Cherry Hill:408–409.
[5] Yin SF, Xu BQ, Zhou XP, Au CT (2004). Applied Catalysis A: General 277:1–9.
[6] Ni M, Leung MKH, Leung DYC (2009). Int. J. Energy Res 33:943–959.
[7] Lan R, Irvine JTS, Tao S (2012). Int. J. Hydrogen Energy 37:1482 -1494.
[8] Lipman T, Shah N (2007). UC Berkeley: UC Berkeley Transportation Sustainability Research Center Retrieved from: http://escholarshiporg/uc/item/7z69v4wp.
[9] Papapolymerou G, Bontozoglou V, Mol J. (1997). Catal. A: Chem. 120:165.
[10] Choudhary TV, Svadinaragana C, Goodman DW (2001). Catal. Lett. 72:197.
[11] Liu HC, Wang H, Shen JH, Sun Y, Liu ZM (2008). Applied Catalysis A: General 337:138–147.
[12] Li XK, Ji WJ, Zhao J, Wang SJ, Au CT (2005). J. Catal. 236:181–189.
[13] Zhang J, Xu HY, Jin XL, Ge QJ, Li WZ (2005). Appl. Catal. A 290:87–96.
[14] ZhEng. WQ, Zhang J, Ge QJ, Xu HY, Li WZ (2008). Applied Catalysis B: Env 80:98.
[15] Plana C, Armenise S, Monzón A, García-Bordejé E (2010). Journal of Catalysis 275:228–235.
[16] Yin SF, Zhang QH, Xu BQ, Zhu WX, Ng CF, Au CT (2004). J. Catal. 224:384.
[17] Yin SF, Xu BQ, Wang SJ, Ng CF, Au CT (2004). Catal. Lett. 96:113.
[18] Wang SJ, Yin SF, Li L, Xu BQ, Ng CF, Au CT (2004). Appl. Catal. B: Environ 52:287.
[19] Bradford MCJ, Fanning PE, Vannice MA (1997). J. Catalysis 172:479.
[20] Ganley JC, Seebauer EG, Masel RI (2004). J. Power Sources 137:53.
[21] Pyrz W, Vijay R, Binz J, Lauterbach J, Buttrey D (2008). Topics in Catalysis 50:180.
[22] Sorensen RZ, Klerke A, Quaade U, Jensen S, Hansen O, Christensen CH (2006). Catalysis Letters 112:77.
[23] Rarog-Pilecka W, Szmigiel D, Kowalczyk Z, Jodzis S, Zielinski J. (2003). J. Catalysis 218:465.
[24] Chen J, Zhu ZH, Wang S, Ma Q, Rudolph V, Lu GQ (2010). Chem. Eng. J. 156:404.
[25] García-García FR, Álvarez-Rodríguez J, Rodríguez-Ramos I, Guerrero-Ruiz A (2010). Carbon 48:267.

[26] Ganley JC, Thomas FS, Seebauer EG, Masel RI (2004). Catalysis Letters 96:117.
[27] Tsai W, Weinberg WH (1987). J. Phys. Chem. 91:5307.
[28] Boudart M, Djega-Mariadassou G (1984). Princeton, NJ:Princeton University Press p 98.
[29] Löffler DG, Schmidt LD (1976). J. Catal. 44:244.
[30] Egawa C, Nishida T, Naito S, Tamaru K (1984). J. Soc. Chem. Faraday Trans. 180:1595.
[31] Vitvitskii AI, Gaidei TP, Toporkova ME, Kiseleva EM, Melikhov EN (1990). J. Appl. Chem. USSR 63:1883.
[32] Skodra A, Ouzounidou M, Stoukides M (2006). Solid State Ionics 177:2217–2220.
[33] Shustorovich E, Bell AT (1991). Surf. Sci. Lett. 259:791.
[34] Chellappa AS, Fisher CM, Thomson WJ (2002). Applied Catalysis A: General 227: 231–240.
[35] Vitse F, Cooper M, Botte GG (2005). J. Power Sources 142:18–26.
[36] Simons EL, Cairns EJ, Surd DJ (1969). J. ElectroChem. Soc. 115:556–561.
[37] Gerischer H, Mauerer A (1970). J. Electroanal. Chem. 25:421.
[38] Vidal-Iglesias FJ, Solla-Gullon J, Feliu JM, Baltruschat H, Aldaz A (2006). J. Electroanal. Chem. 588:331–338.
[39] Vidal-Iglesias FJ, Solla-Gullon J, Perez JM, Aldaz A (2006). Electrochem. Comm. 8:102–106.
[40] de Vooys ACA, Koper MTM, van Santen RA, van Veen JAR (2001). J. Electroanal. Chem. 506:127–137.
[41] Endo K, Nakamura K, Miura T (2004). Electrochim. Acta 49:2503–2509.
[42] Endo K, Nakamura K, Miura T (2004). Electrochim. Acta 49:1635.
[43] Ge J, Johnson DC (1995). J. Electrochem. Soc. 142:3420–3423.
[44] McKee DW, Scarpellino AJ Jr, Danzig IF, Pak MS (1969). J. Electrochem. Soc. 116:562–568.
[45] Cooper M, Botte GG (2006). J. Electrochem. Soc. 153:A1894-A1901.
[46] Bonnin EP, Biddinger EJ, Botte GG (2008). J. Power Sources 182:284–290.
[47] Boggs BK, Botte GG (2010). Electrochimica Acta 55:5287–5293.
[48] Vidal-Iglesias FJ, Garcıa-Araez N, Montiel V, Feliu JM, Aldaz A (2003). Electrochemistry Communications 5:22–26.
[49] Vidal-Iglesias FJ, Solla-Gullon J, Rodrıguez P, Herrero E, Montiel V, Feliu JM, Aldaz A (2004). Electrochemistry Communications 6:1080–1084.
[50] Rodrıguez P Herrero E, Solla-Gullon J, Vidal-Iglesias FJ, Aldaz A, JM Feliu JM (2005). Electrochimica Acta 50:4308–4317.
[51] Vidal-Iglesias FJ, Solla-Gullon J, Montiel V, Feliu JM, Aldaz A (2007). J. Power Sources 171:448–456.
[52] Chandra M, Xu Q (2006). J. Power Sources 156:190.
[53] Chandra M, Xu Q (2007). J. Power Sources 168:2135.
[54] Chandra M, Xu Q (2007). J. Alloys Compd. 446:729.
[55] Yao CF, Zhuang L, Cao YL, Ai XP, Yang HX (2008). Int. J. Hydrogen Energy 33:2462.
[56] Basu S, Brockman A, Gagare P, ZhEng. Y, Ramachandran PV, Delgass WN, Gore JP (2009). J. Power Sources 188:238.
[57] Zahmakıran M, Ozkar S (2009). Appl. Catal. B: Environ. 89:104.

[58] Chandra M, Xu Q (2006). J. Power Sources 163:364.
[59] Kalidindi SB, Indirani M, Jagirdar BR (2008). Inorg. Chem. 47:7424.
[60] Kalidindi SB, Sanyal U, Jagirdar BR (2008). Phys. Chem. Chem. Phys. 10:5870.
[61] Umegaki T, Yan JM, Zhang XB, Shioyama H, Kuriyama N, Xu Q (2009). Int. J. Hydrogen Energy 34:3816.
[62] Umegaki T, Yan JM, Zhang XB, Shioyama H, Kuriyama N, Xu Q (2009). J. Power Sources 191:209–216.
[63] Yan JM, Zhang XB, Han S, Shioyama H, Xu Q (2009). J. Power Sources 194:478–481.
[64] Mohajeri N, Raissi AT, Adebiyi O (2007). J. Power Sources 167:482–485.
[65] Eom K, Kim M, Kim R, Nam D, Kwon H (2010). J. Power Sources 195:2830–2834.
[66] Umegaki T, Yan JM, Zhang XB, Shioyama H, Kuriyama N, Xu Q (2010). J. Power Sources 195:8209–8214.
[67] Yan JM, Zhang XB, Han S, Shioyama H, Xu Q (2010). J. Power Sources 195:1091–1094.
[68] Demirci UB, Miele P J. (2010). Power Sources 195:4030–4035.
[69] Schafer HJ (1987). Top. Curr. Chem. 142:101–129.
[70] Vaze A, Sawant S, Pangarkar V (1997). J. Appl. ElectroChem. 27:584–588.
[71] King RL, Botte GG (2011). J. Power Sources 196:9579– 9584.
[72] King RL, Botte GG (2011). J. Power Sources 196:2773–2778.
[73] Sifer N, Gardner K (2004). J. Power Sources 132:135–138.
[74] Li L, Hurley JA (2007). Int J. Hydrogen Energy 32:6 – 10.
[75] Resan M, Hampton MD, Lomness JK, Slattery DK (2005). Int. J. Hydrogen Energy 30:1413–6.
[76] Kircher O, Fichtner M (2005). J. Alloys Comp. 404–406:339–42.
[77] Paik B, Tsubota M, Ichikawa T, Kojima Y (2010). Chem. Commun. 46:3982-4.
[78] Uribe FA, Zawodzinski TA Jr, Gottesfeld S (2002). J. ElectroChem. Soc. 149:A293.
[79] Soto HJ, Lee WK, Van Zee JW, Murthy M (2003). ElectroChem. Solid-State Lett. 6:A133.
[80] Halseid R, Vie PJS, Tunold R (2006). J. Power Sources 154:343–350.
[81] Szymanski ST, Gruver GA, Katz M, Kunz HR (1980). J. ElectroChem. Soc. 127:1440.
[82] Halseid R, Wainright JS, Savinell RF, Tunold R (2007). J. Electrochemical Soc. 154:B263.
[83] Hongsirikarn K, Goodwin JG Jr, Greenway S, Creager S (2010). J. Power Sources 195:30–38.
[84] Hongsirikarn K, Mo X, Goodwin JG Jr (2010). J. Power Sources 195:3416–3424.
[85] Saika T, Nakamura M, Nohara T, Ishimatsu S (2006). JSME Int. J. 49:78.
[86] Sammes NM, Boersma R (2000). J. Power Sources 86:98–110.
[87] Farhad S, Hamdullahpur F, (2010). J. Power Sources 195:3084–3090.
[88] Wojcik A, Middleton H, Damopoulos I, Van herle J. (2003). J. Power Sources 118:342–348.
[89] Choudhary TV, Goodman DW (2002). Catal. Today 77:65–78.
[90] Fournier GGM, Cumming IW, Hellgardt K (2006). J. Power Sources 162:198–206.
[91] Dekker N, Rietveld B (2004). Proc. 6th European Solid Oxide fuel cell forum: Lucerne, Switzerland.
[92] Zhang LM, Cong Y, Yang WS, Lin LW (2007). Chinese Journal of Catalysis 28:749–751.
[93] Ma Q, Ma J, Zhou S, Yan R, Gao J, MEng. G (2007). J. Power Sources 164:86–89.

[94] Liu M, PEng. R, Dong D, Gao J, Liu X, MEng. G (2008). J. Power Sources 185:188–192.
[95] Meng G, Jiang C, Ma J, Ma Q, Liu X (2007). J. Power Sources 173:189–193.
[96] Pelletier L, McFarlan A, Maffei N (2005). J. Power Sources 145:262–265.
[97] Maffei N, Pelletier L, Charland JP, McFarlan A (2005). J. Power Sources 140:264–267.
[98] Maffei N, Pelletier L, Charland JP, McFarlan A (2006). J. Power Sources 162:165–167.
[99] Maffei N, Pelletier L, McFarlan A (2008). J. Power Sources 175:221–225.
[100] Ma Q, PEng. R, Lin Y, Gao J, MEng. G (2006). J. Power Sources 161:95–98.
[101] Zhang LM, Yang WS (2008). J. Power Sources 179:92–95.
[102] Ma QL, PEng. RR, Tian LZ, MEng. GY (2006). Electrochemistry Communications 8:1791–1795.
[103] Lin Y, Ran R, Guo Y, Zhou W, Cai R, Wang J, Shao Z (2010). Int. J. Hydrogen Energy 35:2637.
[104] Xie K, Ma QL, Lin B, Jiang YZ, Gao JF, Liu XQ (2007). J. Power Sources 170:38-41.
[105] Kordesch K, Hacker V, Gsellmann J, Cifrain M, Faleschini G, Enzinger P, Fankhauser R, Ortner M, Muhr M, Aronson RR (2000). J. Power Sources 86: 162–165.
[106] Metkemeijer R, Achard P (1994). J. Power Sources 49:271–282.
[107] Hejze T, Besenhard JO, Kordesch K, Cifrain M, Aronsson RR (2008). J. Power Sources 176:490–493.
[108] Ganley JC (2008). J. Power Sources 178:44–47.
[109] Lan R, Tao SW (2010). ElectroChem. Solid State Lett.13:B83e6.
[110] Zhang XB, Han S, Yan JM, Chandra M, Shioyama H, Yasuda K, Kuriyama N, Kobayashi T, Xu Q (2007). J. Power Sources 168:167–171.
[111] Zhang XB, Yan JM, Han S, Shioyama H, Yasuda K, Kuriyama N Xu Q (2008). J. Power Sources 182:515–519.
[112] Kiran V, Kalidindi SB, Jagirdar BR, Sampath S (2011). Electrochimica Acta 56:10493–10499.
[113] Ni M, Leung DYC, Leung MKH (2008). J. Power Sources 183:682–686.
[114] Ni M, Leung DYC, Leung MKH (2008). J. Power Sources 183:687–692.
[115] Ni M, Leung DYC, Leung MKH (2008). Int. J. Hydrogen Energy 33:5765–5772.
[116] Ishak F, Dincer I, Zamfirescu C (2012). J. Power Sources 202:157– 165.
[117] Ni M, Leung DYC, Leung MKH (2008). J. Power Sources 185:233–240.
[118] Ni M (2011). Int. J. Hydrogen Energy 36:3153-3166.
[119] Baniasadi E, Dincer I (2011). Int. J. Hydrogen Energy 36:11128-11136.
[120] Deshmukh SR, Mhadeshwar AB, Vlachos DG (2004). Ind. Eng. Chem. Res 43:2986–99.
[121] Abashar MEE, Al-Sughair YS, Al-Mutaz IS (2002). Appl. Catal. A; 236:35–53.
[122] Waghode AN, Hanspal NS, Shigidi IMTA, Nassehi V, Hellgardt K (2005). Chem. Eng. Sci. 60:5862–77.
[123] Chein RY, Chen YC, Chang CS, Chung JN (2010). Int. J. Hydrogen Energy 35589–597.
[124] Alagharu V, Palanki S, West N (2010). J. Power Sources 195:829–833.

Section 5

Hydrogen Sensing and Safety

Chapter 14

Hydrogen Sensing Characteristics of a Quartz Oscillator

Atsushi Suzuki

Additional information is available at the end of the chapter

1. Introduction

A variety of hydrogen sensing technologies have been developed, which mainly involve detection of hydrogen leaked into air and represent key technologies for realization of the "hydrogen society" (Hubert T. et al., 2011). These sensors can be classified into those using chemical sensing and those using physical sensing. Here, chemical sensing is defined as technologies in which chemical reactions with hydrogen are essential. On the other hand, physical sensing is defined as methods other than chemical sensing.

Most conventional hydrogen sensing methods belong to the chemical sensing category. Heat obtained by combustion of hydrogen, which results from the reaction with oxygen, is used for catalytic combustion type hydrogen sensors (Gentry S. J. and Jones T. A., 1986; Han C.-H. et al., 2007; Jones M. G. and Nevell T. G., 1989; Katti V. R. et al., 2002). This heat produced by combustion is transformed into electrical energy using thermoelectric materials, such as Si, SiGe, SnO_2, Pt/Al_2O_3, Pt/activated carbon fiber cloth, and $BaCe_{0.95}Y_{0.05}O_{3-\delta}$(BCY), in thermoelectric sensors (Mcaleer J. F. et al., 1985; Nishibori M. et al., 2007; Nishibori M. et al., 2010; Qiu F. B. et al., 2003; Röder-Roith U. et al., 2011; Shin W. et al., 2001; Zhang J. S. et al., 2007). Changes in the resistivity and surface conductivity on the oxide semiconductor induced by the reaction between surface oxygen and hydrogen detects hydrogen in oxide semiconductor sensors (Lundstrom K. I. et al., 1975). Generally, these conventional chemical sensors have extremely low detection limits for hydrogen on the order of several parts per million. However, these chemical sensing methods frequently require thermal energy to enhance the chemical reactions for detection, and therefore, are not safe because of heat generation. In addition, their response times are normally on the order of seconds because rates of chemical reactions for sensing are slow and accompanied with a slow rate of heat translation. Moreover, they cannot identify concentrations of leaked hydrogen over 10 vol%. Other noble chemical sensing methods are those using field effect transistors, which measure threshold voltages affected by reactions between hydrogen and oxygen on the surface (Lundstrom I. et al., 1975)

Physical hydrogen sensing methods defined in this chapter do not include chemical reactions of hydrogen. One representative technique is thermal conductive sensing, making use of the high thermal conductivity of hydrogen (Hale J. et al., 1992; Simon I. and Arndt M., 2002). Optical sensing of hydrogen optically detects changes associated with hydrogen exposure, such as volume expansion, alterations in refractive index, and optical absorption (Butler M. A., 1984). Surface acoustic wave sensing detects hydrogen by measuring the acoustic properties of surface waves induced by hydrogen adsorption (Jakubik W. P. et al., 2002). These physical sensing methods have difficulty in detecting low hydrogen concentrations on the order of parts per million. On the other hand, physical sensing tends to have a quick response, probably because the sensing methods are independent of chemical reactions and thermal energy transfer.

Hydrogen sensing using quartz oscillators is a physical sensing method. Energy is transferred to the quartz oscillator from collision with gas molecules when exposed to a gas. As the energy depends on pressure, viscosity, and molecular weight of the gas, output can be obtained related to these properties of the investigated gas (Kokubun K. et al., 1984). Hydrogen leakage and hydrogen concentration can then be measured using this output. The principles underlying hydrogen sensing are described in detail in the following section.

We have applied this sensing method to measure the partial pressure of binary gas mixtures because viscosity and molecular weight, which can be measured by the quartz oscillator, are only dependent on the partial pressures of each gas in the mixture. For example, the partial pressures of ozone and oxygen in their gas mixture at 100 kPa can be measured with sensitivity of 0.2 kPa (Kurokawa A. et al., 2004). For hydrogen and silane gas mixtures, which are the source gases of thin silicon solar cell materials, the partial pressures of each gas can be measured with the sensitivity of several % for 133-1,333 Pa (Suzuki A. and Nonaka H., 2008). Hydrogen concentration in air can also be derived by an identical procedure, as discussed later.

As hydrogen sensing with the quartz oscillator is classified as a physical sensing method, it has merits of fast response and recovery times and non-consumption of the measured gas. In addition, hydrogen sensing with a quartz oscillator has a number of other advantages. This sensing method works effectively at room temperature and does not require any external energy to enhance the detection efficiency, except that required to induce resonation of the quartz oscillator, which is below several microwatts. The quartz oscillator does not require heating for detection, and therefore flammable hydrogen can be measured safely without temperature elevation. Another advantage is the small size of the quartz oscillator (1 × 4 mm), which allows its use in various locations, such as hydrogen storage, hydrogen stations, hydrogen transfer facilities, and fuel cell vehicles.

There are many demands on hydrogen sensing such as low detection limit (0.01 vol%), wide concentration range (1-100 vol%), safe, reliable and stable measurement on accuracy and sensitivity (<5%), fast response and recovery time (<1 s), low interference by other gases, low sensitivity to environment of pressure (80-110 kPa), temperature (-30-80 ºC) and humidity (10-98RH%), long lifetime (> 5 years), low power consumption (<100 mW) and cost (<100 euro per system), small size, simple operation and system integration and interface, and maintenance free (Hübert T. et al., 2011). Despite the variety of the hydrogen sensing methods, practical sensing methods that fulfill all of the requirements mentioned above have not been developed. From this viewpoint, hydrogen sensing using a quartz oscillator is one candidate as a practical hydrogen sensing method in the future hydrogen society because of its unique characteristics. Therefore, this chapter presents a review of novel hydrogen sensing methods using the quartz oscillator together with the test apparatus and procedures.

2. Principle for measurement

As mentioned in the Introduction, output from the quartz oscillator depends not only on the pressure but also on the viscosity and molecular weight of the measured gas. In this section, the correlation between the output from the quartz oscillator and the properties of the gas, such as viscosity and molecular weight, will be explained from a theoretical perspective. The output from the actual device used in this study will also be presented based on this explanation.

2.1. Theory for the quartz oscillator

The source for energy to the quartz oscillator is given by force from gas molecules hitting its surface. The electrical impedance of the quartz oscillator in the resonant state for gases can be described theoretically from the dissipation energy and the drag force (Kokubun K. et al., 1984). First, the electrical impedance Z can be expressed as follows.

$$Z = \frac{cV^2 \cos\theta}{A^2} \bullet f \tag{1}$$

Here, c, V, θ, A, and f are the constant, the constant supplied voltage, the phase difference between the voltage and current, the amplitude of the forced vibration, and the coefficient of the drag force between the surface of the quartz oscillator and the investigated gas, respectively. Equation (1) indicates that Z is proportional to f.

The drag force can be determined based on a string-of-beads model of the tuning-fork-shaped quartz oscillator. That is, the quartz oscillator is treated as a group of spherical objects. Using this "string of beads" model, the drag forces can be given by different formula depending on the nature of gas flow.

In the molecular flow region, where the mean free path of the gas molecules is sufficiently larger than the size of the spherical beads, the drag force is obtained from the kinetic theory as follows:

$$f = R^2 \sqrt{\frac{8\pi M}{R_0 T}} \bullet pu \tag{2}$$

Where R, M, p, u, R_0, and T are the radius of the sphere, molecular weight, gas pressure, velocity, gas constant, and absolute temperature, respectively. As seen in equation (2), f is proportional to square root of M.

In the higher pressure region where the gas reaches viscous flow, the mean free path of the gas molecules is smaller than the size of the sphere and the gas medium can be regarded as a continuous fluid. Thus, the drag force can be calculated by fluid mechanics as follows:

$$f = 6\pi\eta R + \sqrt{2\eta\rho\omega} \tag{3}$$

Here, η, ϱ, and ω are the coefficient of gas viscosity, density, and the angular frequency of the sphere, respectively. The formula of equation (3) means that the drag force in the viscous gas flow region is proportional to the square root of the product of the gas viscosity and the density.

Finally, from equations (1), (2), and (3), the expression on the impedance of the quartz oscillator can be summarized for the molecular and the viscous flow region as follows (Kokubun K. et al., 1987):

$$Z = CR^2\sqrt{\frac{8\pi M}{R_0 T}} \bullet p \tag{4}$$

$$Z = C(6\pi\eta R + \sqrt{2\eta\rho\omega}) \tag{5}$$

where C is a constant. It should be noted that the density (ϱ) can be rewritten as:

$$\rho = M \bullet p \tag{6}$$

By substituting equation (6) into equation (5), equation (5) can be converted as:

$$Z = 6\pi\eta R + \sqrt{2\eta M p\omega} \tag{7}$$

As above, in equations (4) and (7) it is clear that the electrical impedance of the quartz oscillator depends on the molecular weight in the molecular flow region and on the viscosity and the molecular weight of the measured gas in the viscous flow region, respectively. As the impedance of the quartz oscillator also depends on pressure, it must be pressure-calibrated to exclude the influence of pressure to obtain information on the properties of the measured gas. Finally, we can derive information on the viscosity and the molecular weight of the measured gas through this pressure-calibrated electrical impedance of the quartz oscillator.

2.2. Principle for hydrogen sensing

Normally, hydrogen sensing involves the detection of hydrogen that has leaked into air. Therefore, if a sensor can identify differences in gas properties between pure air and hydrogen-leaked air, it can be used for hydrogen sensing. As described in the previous subsection, the electrical impedance of the quartz oscillator depends on the viscosity and molecular weight of the measured gas. The viscosity and molecular weight of hydrogen-leaked air will decrease compared to those of pure air because the values for hydrogen (8.35 μ Pa s, 2.02) are smaller than those for air (17.08 μ Pa s, 28.97), respectively (Golubev I. F., 1970). Therefore, measurement using the quartz oscillator can be used as a hydrogen sensing method because viscosity and molecular weight of the measured gas would change when hydrogen is leaked into air. In particular, the viscosity and molecular weight of hydrogen are markedly smaller than those of other gases, and therefore hydrogen can be detected with high sensitivity.

2.3. Principle for measurement of hydrogen concentration in air

One important advantage of hydrogen sensing with the quartz oscillator is the measurement of hydrogen concentration in air up to 100 vol% hydrogen. Hydrogen concentration can be measured by application of partial pressure measurement to binary gas mixtures (Kurokawa A. et al., 2004; Suzuki A. and Nonaka H., 2008). This partial pressure measurement can measure the partial pressures of each gas because viscosity and molecular weight of binary gas mixtures are only dependent on the composition ratio of each gas. Average molecular weight of binary gases depends on and changes monotonically with the partial pressure ratio. Therefore, if viscosity of binary gases changes monotonically with partial pressure ratio, the viscosity of binary gas mixtures determines the partial pressure ratio. In the case of binary gas mixtures, the partial pressure ratio determines the partial pressures of each gas, and thus the partial pressure can be measured for binary gases.

Hydrogen concentration in air can be measured if hydrogen-leaked air is assumed to be a binary gas mixture of hydrogen and air. Of course, air itself is also composed of multiple gases, but it can be treated as a single kind of gas with constant viscosity and molecular weight because the kinds and densities of gases in air are normally constant.

As mentioned above, the quartz oscillator can be used to measure hydrogen concentration up to 100 vol% if viscosity of the binary gas mixtures changes with the partial pressure ratio without any saturation of output from the hydrogen sensor as seen in chemical hydrogen sensors.

3. Experimental setup

Figure 1 shows the experimental setup used in this study, consisting of a vacuum chamber, gas supply, and equipment for data processing (Suzuki A. et al., 2006). The main vacuum chamber was a 90-cm^3 stainless steel tube, where the quartz friction pressure gauge (VP Co. model GC-210) including the quartz oscillator and a diaphragm gauge (MKS Instruments, Baratron model 222BA) were fixed. Outputs from the gauges, temperature, and the signals from gas valves were sent to a digital multimeter and analyzed on a personal computer. Gas inlet and position of the quartz oscillator depended on the type of test. Therefore, they will be presented in the following section as well as details of the quartz friction pressure gauge and the quartz oscillator. Figure 2 shows the quartz oscillator and quartz friction pressure gauge. The part of the left hand side where the quartz oscillator is fixed is included in the right hand side of the quartz friction pressure gauge. It should be noted that the size of the connection plug at the quartz friction pressure gauge is 3/8". The size of the quartz oscillator is less than 1 × 4 mm.

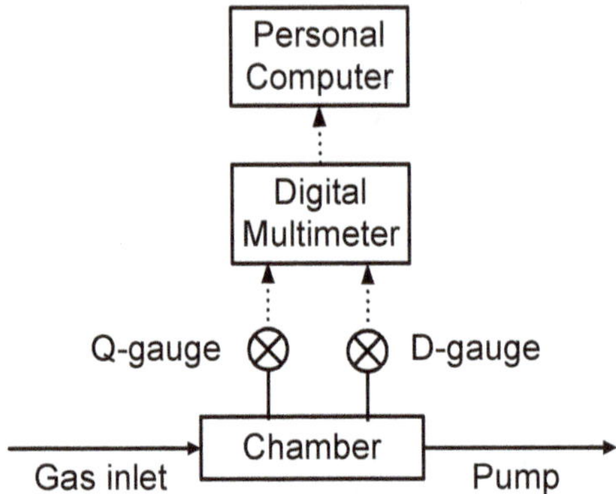

Figure 1. Schematic diagram for experimental setup: Q-gauge: quartz friction pressure gauge; D-gauge: diaphragm pressure gauge

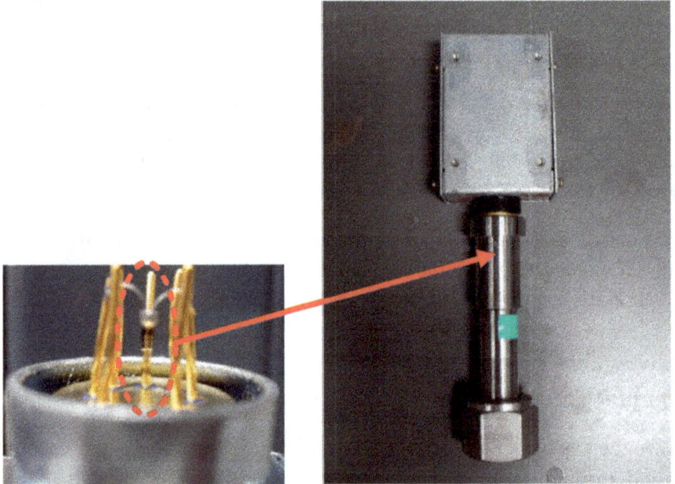

Figure 2. (Outlook) image of the quartz oscillaor (left) and the quartz friction pressure gauge (right)

3.1. Output from the quartz oscillator based on the electrical impedance

The electrical impedance of the quartz oscillator mentioned in Section 2.1 is difficult to measure experimentally. Instead, the voltage was obtained by measuring the current of the quartz oscillator for simplicity (Ono M. et al., 1985).

Pressure dependence of the impedance-converted-voltage in the pressure range up to 100 kPa was investigated for air in Figure 3. The inset in Figure 3 shows the results for the range up to 0.3 kPa. The impedance-converted-voltage decreased with pressure when measured for one kind of gas, but depended on the kinds of measured gas because the electrical impedance of the quartz oscillator depends on viscosity and molecular weight of the measured gas as discussed in Section 2.1.

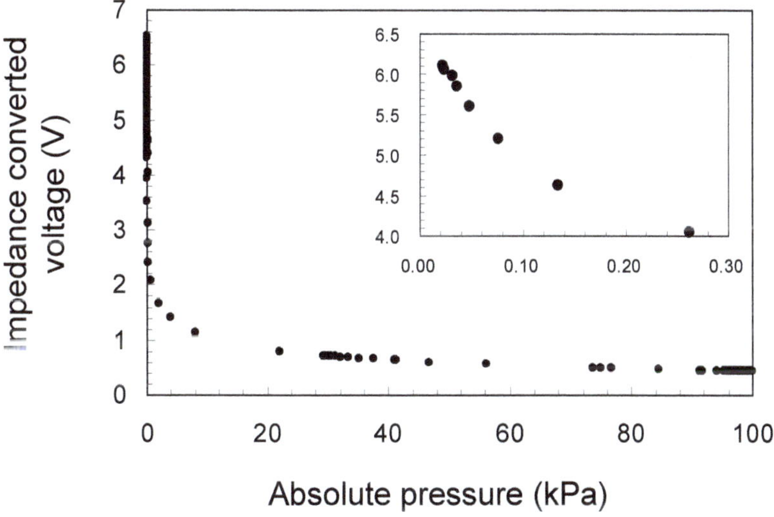

Figure 3. Pressure dependence of impedance-converted-voltage measured for air

One advantage of the quartz oscillator for pressure measurement shown in Figure 3 is wide range pressure measurement of 0.01-100 kPa with one gauge. This almost five orders of magnitude range of pressure measurement is useful because such pressures can be measured with one pressure gauge. In fact, such a pressure gauge applicable to the pressure range of 0.01-100 kPa has been developed (Kobayashi T. et al., 1996; Ono M. et al., 1985). Details of this gauge are presented in the following subsection.

3.2. Pressure readings from quartz friction pressure gauge

The pressure gauge with the quartz oscillator is named the "quartz friction pressure gauge" (Ono M. et al., 1985). The pressure reading from this quartz friction pressure gauge is derived from the impedance-converted-voltage through pressure calculation (Kokubun K. et al., 1987). The calculation is based on equations (4) and (7) depending on the nature of gas flow. The pressure around the border between the molecular and viscous flow region was obtained by fitting of equations (4) and (7) to the experimental results (Kokubun K. et al., 1987). It should be noted that the pressure reading from the quartz friction pressure gauge depends on the kinds of gas measured because the electrical impedance of the quartz oscillator, which is the source for the pressure reading, depends on the viscosity and molecular weight of the measured gas. The pressure reading from the quartz friction pressure gauge is originally not absolute pressure, but a nominal pressure. By introducing the gas-dependent-constant, the pressure reading can be the absolute pressure for the measured gas.

This nominal pressure mentioned above measured for air is plotted against the absolute pressure in Figure 4. The absolute pressure was measured with a diaphragm gauge presented in Figure 1. Here, the kind of gas is set to air for the quartz friction pressure gauge, and therefore the pressure reading from the quartz friction pressure gauge mostly agrees with absolute pressure.

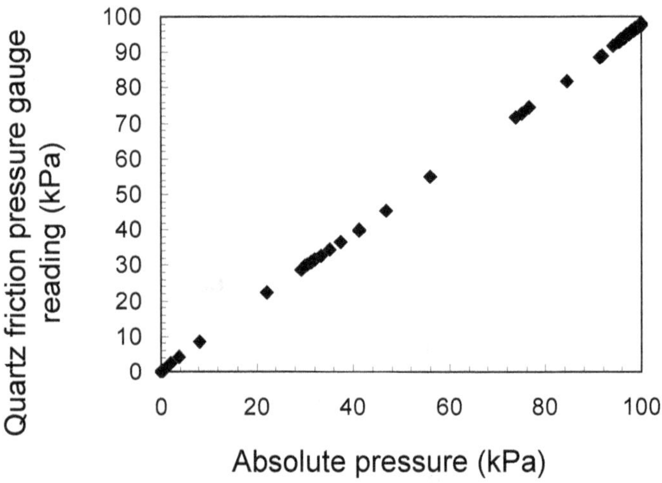

Figure 4. Absolute pressure dependence of pressure reading from quartz friction pressure gauge measured for air

4. Hydrogen sensing with the quartz friction pressure gauge

The characteristics of hydrogen sensing with the quartz oscillator were investigated using the quartz friction pressure gauge shown in Figure 2. The quartz friction pressure gauge generates output in the form of nominal pressure, but it is sufficient to determine the capability for hydrogen sensing.

4.1. Hydrogen sensing in hydrogen-leaked air

For preliminary tests, atmospheric air and hydrogen were introduced into the chamber with the quartz friction pressure gauge (Suzuki A. et al., 2006). Figure 5 shows a schematic diagram of the chamber and gas introduction. Air was introduced into the chamber with the quartz friction pressure gauge and a diaphragm gauge from atmospheric air without any purification. On the other hand, hydrogen was introduced into the chamber through a mass flow controller to control hydrogen concentration. This method of gas introduction is similar to hydrogen leakage into atmospheric air, and therefore this test should be practical.

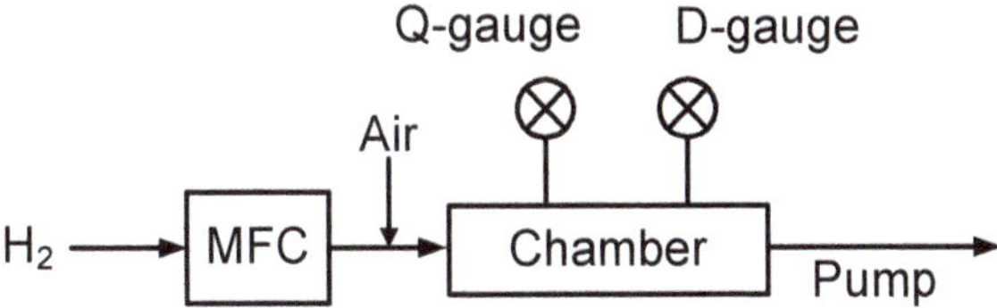

Figure 5. Experimental setup for test on hydrogen-leakage detection with the quartz friction pressure gauge; MFC: mass flow controller

The pressure readings of the quartz friction and the diaphragm pressure gauges when hydrogen was introduced into the vacuum chamber and terminated are shown in Figure 6. At the beginning, when only atmospheric air was introduced into the vacuum chamber, both the quartz friction gauge and the diaphragm pressure gauge readings were equal. This is reasonable because the quartz friction pressure gauge reading is calibrated for air. About 10 minutes later, 70 sccm of hydrogen gas was introduced into the vacuum chamber, and the quartz friction pressure gauge readings decreased markedly, while the diaphragm pressure gauge reading remained constant[1]. After about 35 minutes, the quartz friction pressure gauge reading reached a constant value. After saturation of the quartz friction pressure reading, introduction of hydrogen into the vacuum chamber was interrupted. The quartz friction pressure gauge reading began to return to the value before the introduction of hydrogen.

As there were almost no changes in the diaphragm pressure gauge reading, the change in quartz friction pressure gauge reading at the commencement and termination of hydrogen introduction do not result from changes in absolute pressure. Therefore, these changes in quartz friction pressure gauge reading must be attributable to the differences in viscosity and molecular weight between hydrogen-leaked and pure air. This quartz friction pressure gauge reading is proportional to the viscosity of the measured gas and calibrated as air. Introduction of low viscosity and molecular weight hydrogen into air reduced the viscosity and molecular weight of hydrogen-leaked air and resulted in a decrease in the quartz friction pressure gauge reading. After termination of hydrogen introduction, the hydrogen-leaked air was replaced by pure air, and therefore the viscosity and molecular weight of the measured gas and the quartz friction pressure gauge reading returned to the earlier values.

[1] Flow rate in m^{-6} cariblated at 1.103×10^5 Pa and 273 K

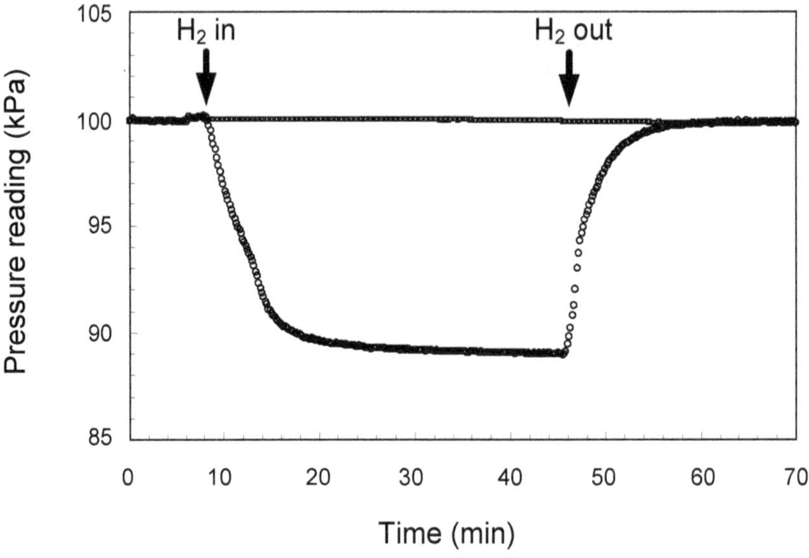

Figure 6. Temporal change of pressure readings of the quartz friction (circle) and daiphragm (square) presssure gauges

4.2. Long-term detection of hydrogen leakage into air

Hydrogen sensing normally must be used and work 24 h per day. To investigate the repeatability of hydrogen sensing using the quartz friction pressure gauge and hydrogen gas flow dependence, the long-term hydrogen sensing was tested. The results of long-term range hydrogen sensing are shown in Figure 7. Arrows and numbers at the top of Figure 7 a) indicate the time at which hydrogen was introduced and the flow rates in sccm, respectively.

Figure 7 a) shows the time-evolution of quartz friction and the diaphragm pressure gauge readings when hydrogen was introduced and interrupted several times with different flow rates. For almost all times in a day, the quartz friction pressure gauge reading decreased with introduction of hydrogen and returned with its termination, indicating that the repeatability of hydrogen sensing with the quartz friction pressure gauge is reliable.

Figure 7 b) shows the pressure-calibrated quartz friction pressure gauge reading ratio with hydrogen introduction and termination. The pressure-calibration of the quartz friction pressure gauge readings was carried out by dividing the quartz friction pressure gauge readings by absolute pressure measured with the diaphragm pressure gauge. The diaphragm pressure gauge reading was also normalized with the initial pressure with t=0 to determine the temporal changes in atmospheric pressure. It is clear that the pressure-calibrated quartz friction pressure gauge reading ratio responded to hydrogen introduction and termination in Figure 7 b) as well as in Figure 7 a), even if the atmospheric pressure

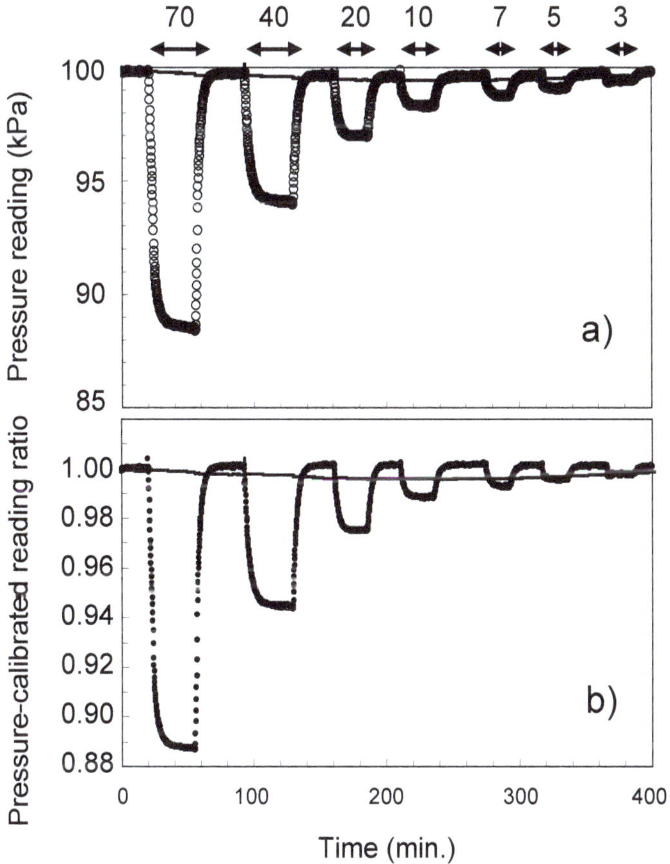

Figure 7. Temporal change of pressure readings of the quartz friction (circle) and daiphragm (line) presssure gauges a) and the pressure-normalized quartz friction (dot) and the initial pressure-normalized pressure readings for hydrogen intoroduction and termination b).

fluctuated. Therefore, this method was shown to be practical for long-time hydrogen sensing using the pressure-calibrated quartz friction pressure gauge reading ratio although the atmospheric pressure fluctuated.

Both raw and pressure-calibrated quartz friction pressure gauge readings during hydrogen introduction depended on the introduced hydrogen flow rates. This pressure-calibrated quartz friction pressure gauge reading ratio should only depend on viscosity and molecular weight of the measured gas because the influence of pressure was excluded. Therefore, the results shown in Figures 7 a) and b) indicate the possibility of measurement of hydrogen concentration in hydrogen-leaked air using this pressure-calibrated quartz friction pressure gauge reading ratio.

4.3. Hydrogen concentration measurement with quartz friction pressure gauge

The hydrogen concentration dependence of the pressure-calibrated quartz friction pressure gauge reading at a total pressure of 100 kPa was investigated to measure hydrogen concentration in hydrogen-air gas mixtures based on the partial pressure measurement for binary gas mixtures described in the Introduction. For this purpose, gas mixtures containing known hydrogen concentrations in air must be prepared using the experimental setup in Figure 8 and the following procedure.

Known concentrations of hydrogen in hydrogen-air gas mixtures were prepared based on the diaphragm pressure reading. One example of this diaphragm pressure gauge reading during preparation of hydrogen-air gas mixtures with known hydrogen concentration is plotted in Figure 9 to explain the procedure for preparing hydrogen-air gas mixtures with 10 vol% hydrogen and 100 kPa total pressure. First, the pressure in the vacuum chamber was zero. Then, at t=9 min for this case, hydrogen was introduced into the chamber until the diaphragm pressure gauge reading reached 10 kPa with shutting the vacuum pump by closing a valve on the vacuum pump. This means that the chamber contained 10 kPa hydrogen. Next, air was introduced into the chamber at t=22 min in this case until the diaphragm pressure gauge reading reached 100 kPa, meaning that the chamber contained (100-10) kPa air. Finally, 10 vol% = 10 kPa/(10+90) kPa hydrogen and total pressure of 100 kPa hydrogen-air gas mixture was prepared in the chamber.

The quartz friction pressure gauge reading at t=40 min in Figure 9 for 100 kPa and 10 vol% hydrogen-air gas mixture was about 95 kPa. Other hydrogen-air gas mixtures with known hydrogen concentrations were also prepared in the same way as described above. Finally, the hydrogen concentration dependence of the pressure-calibrated quartz friction pressure gauge reading ratio is summarized in Figure 10. The results indicated that the pressure-calibrated quartz friction pressure gauge reading ratio depends on hydrogen concentration in hydrogen-air gas mixtures at 100 kPa. Here, the pressure-calibrated quartz friction pressure gauge reading ratio was used instead of raw quartz friction pressure gauge readings to eliminate the effect of pressure difference for each gas mixture. Using Figure 10, the hydrogen concentration in hydrogen-air gas mixtures can be determined by measurements using a quartz friction gauge and a diaphragm pressure gauge.

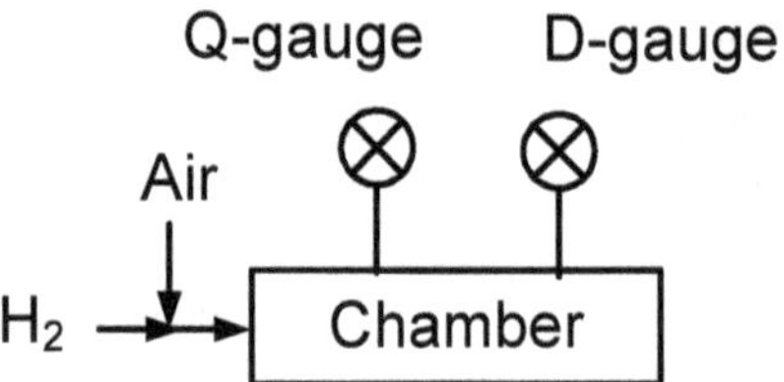

Figure 8. Experimental setup for preparetion for the hydrogen-air gas mixtures with known concentration hydrogen

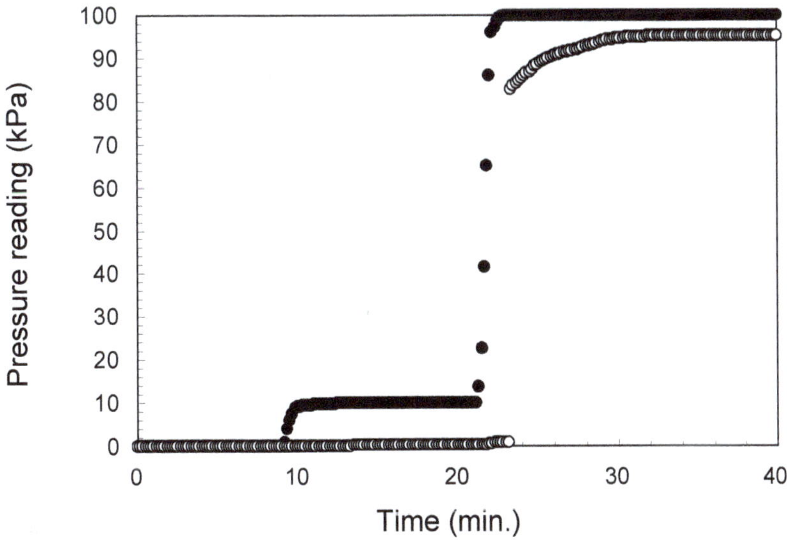

Figure 9. Diaphragm (closed) and quartz friction (open) pressure gauge reading during preparation for the 100 kPa hydrogen-air gas mixtures with 10vol.% hydrogen

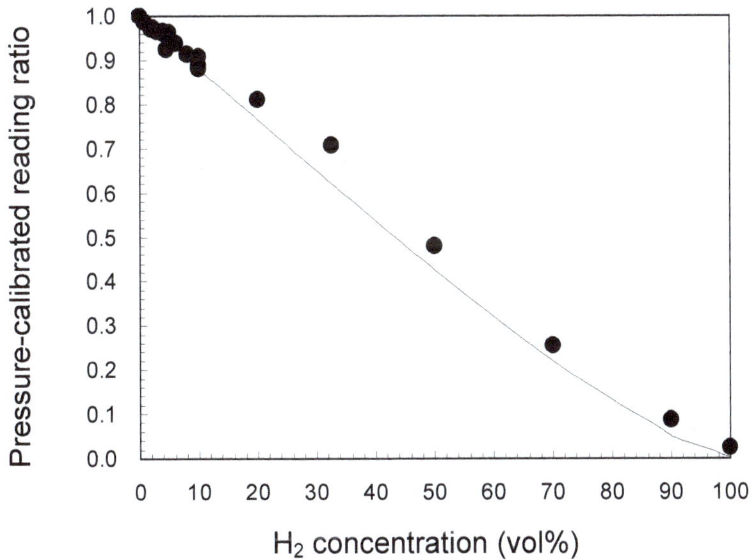

Figure 10. Hydrogen concentration dependence of the pressure-calibrated quartz friction pressure gauge reading ratios

To understand this hydrogen concentration dependence of the pressure-calibrated quartz friction pressure gauge reading ratio, we discuss the viscosity of gas mixtures based on a previous report by Wilke because the reading depends on the viscosity of the measured gas, as seen in equation (7) presented in Section 2.1 (Wilke C. R., 1950). Based on Wilke's work, the viscosity of binary gas mixtures (η_m) can be written as follows:

$$\eta_m = \sum_{i=1}^{2} \frac{\eta_i}{1 + \frac{1}{x_i} \sum_{\substack{j=1 \\ j \neq i}}^{j=2} x_j \phi_{ij}} \tag{8}$$

Here, η and x are viscosity and fraction of each gas in the binary gas mixtures. In addition, ϕ is given by

$$\phi_{ij} = \frac{[1 + (\eta_i / \eta_j)^{1/2} (M_j / M_i)^{1/4}]^2}{(4/\sqrt{2})[1 + (M_i / M_j)]^{1/2}} \cdots \tag{9}$$

where M is the molecular weight of each gas. By assuming viscosity at 0 and 100 vol% hydrogen concentration as those of pure air and hydrogen, respectively, equation (8) is expressed by the line in Figure 10, which agrees well with the pressure-calibrated quartz friction pressure gauge reading ratio and indicates the viscosity of the hydrogen-air gas mixture.

From differences in the pressure-calibrated quartz friction pressure gauge readings of 0.024 and 1.0 between 0 and 100 vol% hydrogen concentration and its fluctuation of 0.001, the minimum detectable hydrogen concentration is derived as (1.0-0.024)/100/0.001=0.1vol%, which is sufficiently low for practical hydrogen sensing (DOE, 2007)

4.4. Response and recovery time measurement for hydrogen sensing

As shown in Figure 6, the response time for hydrogen introduction, which is defined as the time to reach 90% of saturated readings, seems to be on the order of several minutes. However, this is not because of intrinsic characteristics of the quartz friction pressure gauge but due to the slow introduction of hydrogen. Therefore, the response and recovery time for hydrogen introduction and termination were measured using combined valves with which the gas flow can be quickly switched.

Temporal changes in the pressure calibrated quartz friction and the diaphragm pressure gauge reading ratios and signals for 1 vol% hydrogen introduction and termination using quick gas switching are shown in Figure 11 (Suzuki A. et al., 2006). From Figure 11, it is clear that the response and recovery times for hydrogen introduction and termination were below 1 s (about 600 ms). This response and recovery times also fulfilled the requirements for hydrogen sensing.

It should be noted that these response and recovery times were independent of hydrogen concentration, as indicated in Figure 12, which shows the temporal changes in the quartz

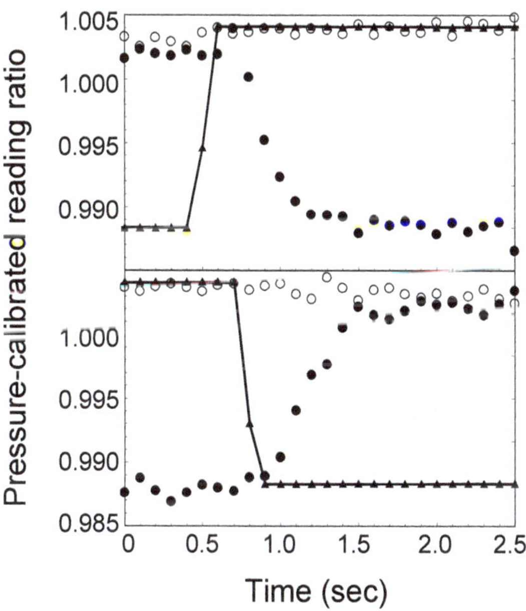

Figure 11. Temporal changes in the pressure calibrated quartz friction (closed) and the diaphragm (open) pressure gauge reading ratios and signal for hydrogen introduction and termination (triangle)

friction pressure gauge reading for various hydrogen concentrations for hydrogen introduction and termination. Figure 13 summarizes the hydrogen concentration dependence of response and recovery times, indicating values below 1 s even for higher hydrogen concentrations. Thus, the quartz friction pressure gauge can be used to measure hydrogen concentration as quickly as hydrogen-leakage detection. This simultaneous hydrogen concentration measurement will be useful with regard to safety.

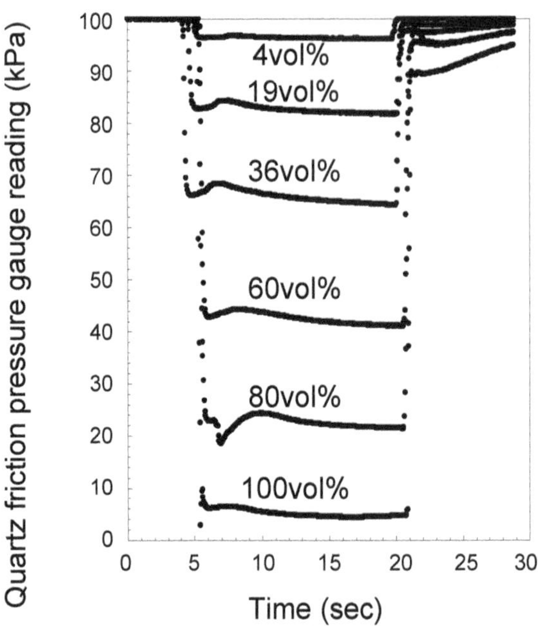

Figure 12. Temporal changes in the pressure-calibrated quartz friction pressure gauge readings for various hydrogen concentration (in vol%)

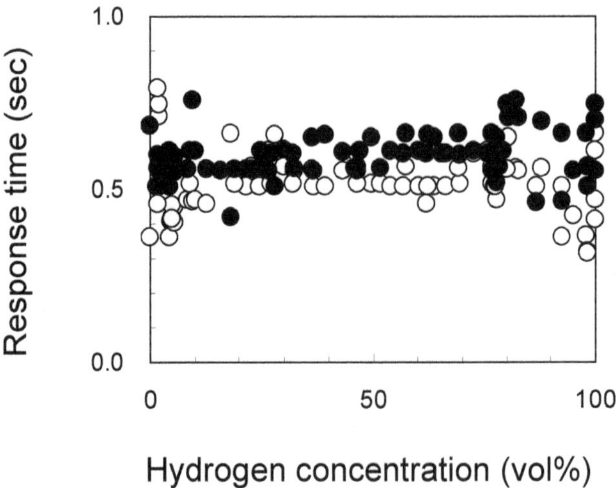

Figure 13. Response time to the hydrogen introduction (open) and recovery time (closed) for 0.1-100vol% hydrogen concentration

4.5. Influences of temperature and humidity

As hydrogen sensors are used outside, the influences of temperature and humidity on hydrogen sensing with the quartz friction pressure gauge were investigated. In this study, the influences on background of the quartz friction pressure readings without hydrogen were investigated. The resonant frequency of the quartz oscillator, which affects the electrical impedance of the quartz oscillator and the quartz friction pressure gauge readings, is dependent on temperature (Spassov L. et al., 1997). Humidity should also affect the quartz friction pressure gauge reading because viscosity and molecular weight of water (12.55 μ Pa s, 18.02), which is the source of humidity, are smaller than those of air.

Figures 14 and 15 show the temperature and humidity dependences of the pressure-calibrated quartz friction pressure gauge reading ratio, respectively. At constant humidity and temperature, the pressure-calibrated quartz friction pressure gauge reading ratios depends on temperature and humidity. The changes in pressure-calibrated reading ratio between 22°C and 34°C in Figure 14 and between 37% and 100% relative humidity (RH) in Figure 15 correspond to 0.36 and 1.74 vol% hydrogen in atmospheric air, respectively.

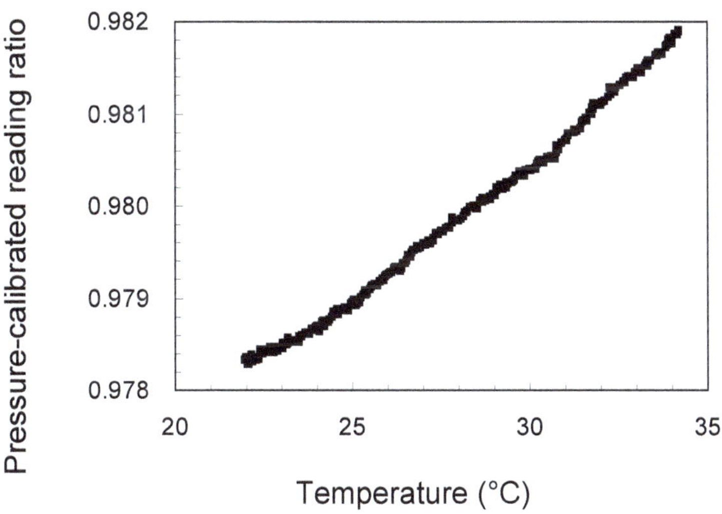

Figure 14. Temperature dependence of pressure-calibrated quartz friction pressure gauge reading ratio at 30% relative humidity

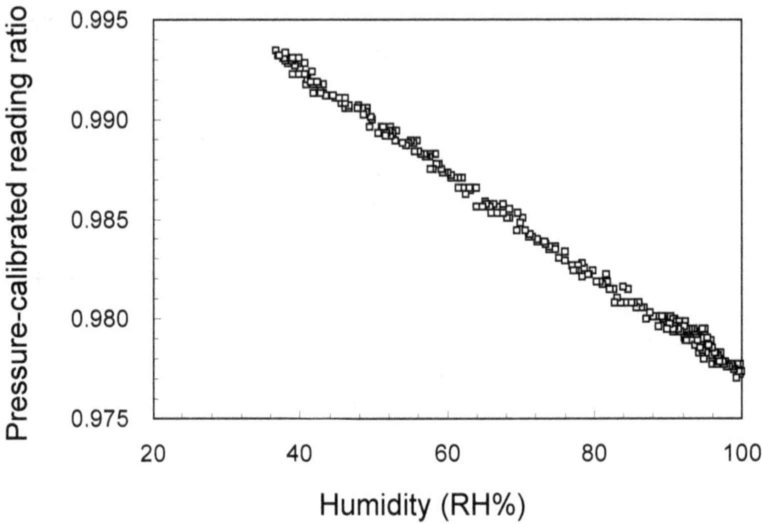

Figure 15. Relative humidity dependence of pressure-calibrated quartz friction pressure gauge reading ratio at 32.5 ºC

A hydrogen concentration of 1.74 vol% is greater than the lower explosion limit, and therefore this change induced by humidity may cause errors in hydrogen sensing, and must be avoided.

These influences can be excluded by measuring temperature and humidity at the same time. By calibrating the pressure-calibrated reading ratio using temperature and humidity dependence, these influences can be reduced down to 0.005, corresponding to 0.5 vol% hydrogen and is below lower explosion limit. However, further improvements are required because it is necessary to better suppress the influences of temperature and humidity and the present calibration procedure and measurement apparatus are complicated.

4.6. Interference by other gases

There are two possible types of gas that can interfere with hydrogen sensing using the quartz friction pressure gauge. The first is gas that can directly cause deterioration of the quartz itself, such as fluorinated gases. The other type of interfering gas significantly affects viscosity and molecular weight of the measured gas. One such gas was already mentioned as humidity, i.e., water. From this viewpoint, helium would have the greatest effect because its viscosity and molecular weight (18.60 μ Pa s, 4.00) are as small and are similar to those of hydrogen. Leakage of helium into air would reduce the quartz friction pressure gauge reading to almost the same extent as hydrogen leakage (A. Suzuki and H. Nonaka, 2008).

5. Conclusions

The present specifications of hydrogen sensing using the quartz friction pressure gauge are summarized in Table I. Almost all specifications meet the requirements for hydrogen sensing. Therefore, it can be concluded that hydrogen sensing using the methods described here is sufficiently practical for the hydrogen society.

Measurement range	0.1-100 vol%
Operating temperature	15°C-40°C
Response time	< 1.0 s
Accuracy	0.5%
Gas environment	Ambient air, 0%-90% relative humidity range
Interference resistive	Helium, water (humidity)
Sensor head size	1 × 4 mm
Power	Microwatts

Table 1. Present specifications of hydrogen sensing with the quartz friction pressure gauge

6. Future research

Several problems remain for the practical use of hydrogen sensing using the quartz friction pressure gauge as follows.

6.1. Reduction of influences of temperature and humidity

The most serious problem for hydrogen sensing with the quartz friction pressure gauge is the requirement of suppressing the influences of temperature and humidity when used outside. Solutions to these problems may be as follows.

6.1.1. Reduction of the influence of temperature

The resonant frequency of the quartz oscillator depends on temperature. Therefore, the quartz friction pressure gauge reading can be temperature-calibrated using the resonant frequency. This can be applied to hydrogen sensing with the quartz friction pressure gauge.

6.1.2. Reduction of the influence of humidity

To reduce the influence of humidity, some filters may be effective to prevent water coming into contact with the head of the quartz friction pressure gauge. Such effects must be studied with varying humidity and hydrogen concentration simultaneously.

6.2. Improvement of sensitivity for hydrogen sensing

The quartz friction pressure gauge reading is calculated with the digitized electrical impedance of the quartz oscillator. Analog-digital conversion of the electrical impedance of the quartz oscillator reduces the resolution from 24 to 8 bit. In addition, pressure calculation also includes about 10% deviation because of the approximation for the pressure derivation. Therefore, use of the electrical impedance of the quartz oscillator may reduce the minimum detection concentration of hydrogen by skipping these inaccurate processes.

Author details

Atsushi Suzuki
Research Institute of Instrumentation Frontier,
National Institute of Advanced Industrial Science and Technology AIST, Japan

Acknowledgement

This work was financially supported by the Ministry of Economy, Trade, and Industry (METI), Japan.

7. References

Butler, M. A. (1984). Optical fiber hydrogen sensor. *Applied Physics Letters* 45(10): 1007-1009.

DOE (2007). "(http://www1.eere.energy.gov/hydrogenandfuelcells/mypp/pdfs/safety.pdf)."

Gentry, S. J. and Jones, T. A. (1986). The role of catalysis in solid-state gas sensors. *Sensors and Actuators* 10(1–2): 141-163.

Golubev, I. F. (1970). Viscosity of Gases and Gas Mixtures, a Handbook. Jerusalem: Israel Program for Scientific Translations.

Hübert, T., Boon-Brett, L., Black, G. and Banach, U. (2011). Hydrogen sensors – A review. *Sensors and Actuators B: Chemical* 157(2): 329-352.

Hale, J., Stehle, G. and Bals, I. (1992). Gas analysis using a thermal conductivity method. *Sensors and Actuators B: Chemical* 7(1-3): 665-671.

Han, C.-H., Hong, D.-W., Han, S.-D., Gwak, J. and Singh, K. C. (2007). Catalytic combustion type hydrogen gas sensor using TiO2 and UV-LED. *Sensors and Actuators B: Chemical* 125(1): 224-228.

Jakubik, W. P., Urbańczyk, M. W., Kochowski, S. and Bodzenta, J. (2002). Bilayer structure for hydrogen detection in a surface acoustic wave sensor system. *Sensors and Actuators B: Chemical* 82(2–3): 265-271.

Jones, M. G. and Nevell, T. G. (1989). The detection of hydrogen using catalytic flammable gas sensors. *Sensors and Actuators* 16(3): 215-224.

Katti, V. R., Debnath, A. K., Gadkari, S. C., Gupta, S. K. and Sahni, V. C. (2002). Passivated thick film catalytic type H2 sensor operating at low temperature. *Sensors and Actuators B: Chemical* 84(2–3): 219-225.

Kobayashi, T., Hojo, H. and Ono, M. (1996). Gas concentration analysis with a quartz friction vacuum gauge. *Vacuum* 47(6-8): 479-483.

Kokubun, K., Hirata, M., Murakami, H., Toda, Y. and Ono, M. (1984). A bending and stretching mode crystal oscillator as a friction vacuum gauge. *Vacuum* 34(8-9): 731-735.

Kokubun, K., Hirata, M., Ono, M., Murakami, H. and Toda, Y. (1987). Unified Formula Describing the Impedance Dependence of a Quartz Oscillator on Gas-Pressure. *Journal of Vacuum Science & Technology a-Vacuum Surfaces and Films* 5(4): 2450-2453.

Kurokawa, A., Odaka, K. and Ichimura, S. (2004). Partial-pressure measurement of atmospheric-pressure binary gas using two pressure gauges. *Vacuum* 73(2): 301-304.

Lundstrom, I., Shivaraman, S., Svensson, C. and Lundkvist, L. (1975). A hydrogen - sensitive MOS field - effect transistor. *Applied Physics Letters* 26(2): 55-57.

Lundstrom, K. I., Shivaraman, M. S. and Svensson, C. M. (1975). A hydrogen-sensitive Pd-gate MOS transistor. *Journal of Applied Physics* 46(9): 3876-3881.

McAleer, J. F., Moseley, P. t., Bourke, P., Norris, J. O. W. and Stephan, R. (1985). Tin dioxide gas sensors: use of the seebeck effect. *Sensors and Actuators* 8(3): 251-257.

Nishibori, M., Shin, W., Houlet, L. F., Tajima, K., Itoh, T., Izu, N. and Matsubara, I. (2007). Preparation of micro-thermoelectric hydrogen sensor loading two kinds of catalysts to enhance gas selectivity. *Journal of the Ceramic Society of Japan* 115(1347): 748-750.

Nishibori, M., Shin, W., Izu, N., Itoh, T., Matsubara, I., Watanabe, N. and Kasuga, T. (2010). Thermoelectric hydrogen sensors using Si and SiGe thin films with a catalytic combustor. *Journal of the Ceramic Society of Japan* 118(1375): 188-192.

Ono, M., Hirata, M., Kokubun, K., Murakami, H., Tamura, F., Hojo, H., Kawashima, H. and Kyogoku, H. (1985). Design and Performance of a Quartz Oscillator Vacuum Gauge with a Controller. *Journal of Vacuum Science & Technology a-Vacuum Surfaces and Films* 3(3): 1746-1749.

Qiu, F. B., Shin, W., Matsumiya, M., Izu, N. and Murayama, N. (2003). Hydrogen sensor based on RF-sputtered thermoelectric SiGe film. *Japanese Journal of Applied Physics Part 1-Regular Papers Short Notes & Review Papers* 42(4A): 1563-1567.

Röder-Roith, U., Rettig, F., Sahner, K., Röder, T., Janek, J. and Moos, R. (2011). Perovskite-type proton conductor for novel direct ionic thermoelectric hydrogen sensor. *Solid State Ionics* 192(1): 101-104.

Shin, W., Imai, K., Izu, N. and Murayama, N. (2001). Thermoelectric thick-film hydrogen gas sensor operating at room temperature. *Japanese Journal of Applied Physics Part 2-Letters* 40(11B): L1232-L1234.

Simon, I. and Arndt, M. (2002). Thermal and gas-sensing properties of a micromachined thermal conductivity sensor for the detection of hydrogen in automotive applications. *Sensors and Actuators A: Physical* 97–98(0): 104-108.

Spassov, L., Yossiffov, E., Georgiev, V. and Vergov, L. (1997). A rotated Y-cut quartz resonator with a linear temperature-frequency characteristic. *Sensors and Actuators a-Physical* 58(3): 185-189.

Suzuki, A. and Nonaka, H. (2008). Measured gas concentrations and flow properties in SiH4-H2 mixtures. *Japanese Journal of Applied Physics Part 1-Regular Papers Brief Communications & Review Papers* 47(5): 3661-3665.

Suzuki, A., Kurokawa, A., Nonaka, H. and Ichimura, S. (2006). A possible hydrogen sensing method with dual pressure gauges. *Sensors and Actuators A: Physical* 127(1): 37-40.

Zhang, J. S., Luan, W. L., Huang, H., Qi, Y. S. and Tu, S. T. (2007). Preparation and characteristics of Pt/ACC catalyst for thermoelectric thin film hydrogen sensor. *Sensors and Actuators B-Chemical* 128(1): 266-272.

Permissions

The contributors of this book come from diverse backgrounds, making this book a truly international effort. This book will bring forth new frontiers with its revolutionizing research information and detailed analysis of the nascent developments around the world.

We would like to thank Professor Dragica Minić, for lending her expertise to make the book truly unique. She has played a crucial role in the development of this book. Without her invaluable contribution this book wouldn't have been possible. She has made vital efforts to compile up to date information on the varied aspects of this subject to make this book a valuable addition to the collection of many professionals and students.

This book was conceptualized with the vision of imparting up-to-date information and advanced data in this field. To ensure the same, a matchless editorial board was set up. Every individual on the board went through rigorous rounds of assessment to prove their worth. After which they invested a large part of their time researching and compiling the most relevant data for our readers. Conferences and sessions were held from time to time between the editorial board and the contributing authors to present the data in the most comprehensible form. The editorial team has worked tirelessly to provide valuable and valid information to help people across the globe.

Every chapter published in this book has been scrutinized by our experts. Their significance has been extensively debated. The topics covered herein carry significant findings which will fuel the growth of the discipline. They may even be implemented as practical applications or may be referred to as a beginning point for another development. Chapters in this book were first published by InTech; hereby published with permission under the Creative Commons Attribution License or equivalent.

The editorial board has been involved in producing this book since its inception. They have spent rigorous hours researching and exploring the diverse topics which have resulted in the successful publishing of this book. They have passed on their knowledge of decades through this book. To expedite this challenging task, the publisher supported the team at every step. A small team of assistant editors was also appointed to further simplify the editing procedure and attain best results for the readers.

Our editorial team has been hand-picked from every corner of the world. Their multi-ethnicity adds dynamic inputs to the discussions which result in innovative outcomes. These outcomes are then further discussed with the researchers and contributors who give their valuable feedback and opinion regarding the same. The feedback is then collaborated with the researches and they are edited in a comprehensive manner to aid the understanding of the subject.

Apart from the editorial board, the designing team has also invested a significant amount of their time in understanding the subject and creating the most relevant covers. They scrutinized every image to scout for the most suitable representation of the subject and create an appropriate cover for the book.

The publishing team has been involved in this book since its early stages. They were actively engaged in every process, be it collecting the data, connecting with the contributors or procuring relevant information. The team has been an ardent support to the editorial, designing and production team. Their endless efforts to recruit the best for this project, has resulted in the accomplishment of this book. They are a veteran in the field of academics and their pool of knowledge is as vast as their experience in printing. Their expertise and guidance has proved useful at every step. Their uncompromising quality standards have made this book an exceptional effort. Their encouragement from time to time has been an inspiration for everyone.

The publisher and the editorial board hope that this book will prove to be a valuable piece of knowledge for researchers, students, practitioners and scholars across the globe.

List of Contributors

Vladimir A. Blagojević and Dragica M. Minić
University of Belgrade, Faculty for Physical Chemistry, Serbia

Dejan G. Minić
Kontrola LLC, Austin, TX, USA

Jasmina Grbović Novaković
Laboratory for Material Sciences, Institute for Nuclear Science Vinča, University of Belgrade, Belgrade, Serbia

J.M. Olivares-Ramírez and Á. Marroquín de Jesús
Universidad Tecnológica de San Juan del Río, San Juan del Río, Querétaro, México

O. Jiménez-Sandoval
Centro de Investigación y de Estudios Avanzados del Instituto Politécnico Nacional, Unidad Querétaro, México

R.C. Pless
Centro de Investigación en Ciencia Aplicada y Tecnología Avanzada del IPN, Unidad Querétaro, México

Dominic Deo Androga
Department of Biotechnology, Middle East Technical University, Ankara, Turkey

Ebru Özgür
Micro-Electro-Mechanical Systems Research and Application Center, Middle East Technical University, Ankara, Turkey

Inci Eroglu
Department of Chemical Engineering, Middle East Technical University, Ankara, Turkey

Ufuk Gündüz and Meral Yücel
Department of Biological Sciences, Middle East Technical University, Ankara, Turkey

Doki Yamaguchi, Liangguang Tang, Nick Burke and Ken Chiang
CSIRO Earth Science and Resource Engineering, Australia

Lucas Rye
CSIRO Marine and Atmospheric Research, Australia

Trevor Hadley and Seng Lim
CSIRO Process Science and Engineering, Australia

Iosif Mariakakis, Carsten Meyer and Heidrun Steinmetz
Institute for Sanitary Engineering, Water Quality and Solid Waste Management (ISWA), University of Stuttgart, Germany

Raúl Pérez-Hernández, Demetrio Mendoza-Anaya and Albina Gutiérrez Martínez
Instituto Nacional de Investigaciones Nucleares, Carr. México-Toluca S/N La Marquesa, Ocoyoacac, México

Antonio Gómez-Cortés
Instituto de Física-Universidad Nacional Autónoma de México, D.F., México

Mario Toledo Torres and Carlos Rosales Huerta
Department of Mechanical Engineering, Technical University Federico Santa Maria, Valparaiso, Chile

Bilge Albayrak Çeper
Erciyes University Faculty of Eng., Dept. of Mech Eng., Kayseri, Turkey

Małgorzata Norek
Department of Advanced Materials and Technology, Military University of Technology, Warsaw, Poland

Horacio Esteban Troiani
CONICET, Buenos Aires, Argentina
Metals Physics Department, Bariloche Atomic Center, Atomic Energy Nacional Commission, San Carlos de Bariloche, Argentina

Analía Leticia Soldati and Adriana Cristina Serquis
Materials Characterization Department, Bariloche Atomic Center, Atomic Energy National
Commission, San Carlos de Bariloche, Argentina
CONICET, Buenos Aires, Argentina

Laura Cecilia Baqué
Materials Characterization Department, Bariloche Atomic Center, Atomic Energy National Commission, San Carlos de Bariloche, Argentina
IDEPA-CONICET, CCT-Comahue, Buenos Aires, Neuquén, Argentina

Chanho Pak, Dae Jong You, Kyoung Hwan Choi and Hyuk Chang
Energy Lab, Samsung Advanced Institute of Technology, Samsung Electronics Co. Ltd., Republic of Korea

Erik Birgersson
Minerals Metals Materials Technology Centre (M3TC), National University of Singapore, Singapore

Agus P. Sasmito
Mechanical Engineering, Masdar Institute of Science and Technology, Abu Dhabi, United Arab Emirates

Erik Birgersson
Department of Chemical and Biomolecular Engineering, National University of Singapore, Singapore

Arun S. Mujumdar
Mechanical Engineering Department, National University of Singapore, Singapore

Denver Cheddie
Center for Energy Studies, University of Trinidad and Tobago, Brechin Castle, Couva, Trinidad and Tobago

Atsushi Suzuki
Research Institute of Instrumentation Frontier, National Institute of Advanced Industrial Science and Technology AIST, Japan

www.ingramcontent.com/pod-product-compliance
Lightning Source LLC
Chambersburg PA
CBHW070712190326
41458CB00004B/955

* 9 7 8 1 6 3 2 3 9 2 0 5 3 *